THE CAMBRIDGE HANDBOOK OF HEALTH RESEARCH REGULATION

The first ever interdisciplinary handbook in the field, this vital resource offers wide-ranging analysis of health research regulation. The chapters confront gaps between documented law and research in practice, and draw on legal, ethical and social theories about what counts as robust research regulation to make recommendations for future directions. The handbook provides an account and analysis of current regulatory tools – such as consent to participation in research and the anonymisation of data to protection participants' privacy – as well as commentary on the roles of the actors and stakeholders who are involved in human health research and its regulation. Drawing on a range of international examples of research using patient data, tissue and other human materials, the collective contribution of the volume is to explore current challenges in delivering good medical research for the public good and to provide insights on how to design better regulatory approaches. This title is also available as Open Access on Cambridge Core.

GRAEME LAURIE is Professorial Fellow and Founding Director of the J. Kenyon Mason Institute, Edinburgh Law School, University of Edinburgh. He was the Principal Investigator on the Liminal Spaces Project (2014–2021).

EDWARD DOVE is Lecturer in Health Law and Regulation and a Deputy Director of the Mason Institute.

AGOMONI GANGULI-MITRA is Lecturer in Bioethics and Global Health Ethics at the Mason Institute.

CATRIONA MCMILLAN is a Senior Research Fellow in Medical Law and Ethics and a Deputy Director of the Mason Institute.

EMILY POSTAN is Senior Research and Teaching Fellow in Bioethics and a Deputy Director of the Mason Institute.

NAYHA SETHI is Chancellor's Fellow in Data Driven Innovation and a Deputy Director of the Mason Institute.

ANNIE SORBIE is Lecturer in Medical Law and Ethics and a Deputy Director of the Mason Institute.

The Cambridge Handbook of Health Research Regulation

Edited by

GRAEME LAURIE

University of Edinburgh

EDWARD DOVE

University of Edinburgh

AGOMONI GANGULI-MITRA

University of Edinburgh

CATRIONA MCMILLAN

University of Edinburgh

EMILY POSTAN

University of Edinburgh

NAYHA SETHI

University of Edinburgh

ANNIE SORBIE

University of Edinburgh

CAMBRIDGE
UNIVERSITY PRESS

CAMBRIDGE
UNIVERSITY PRESS

University Printing House, Cambridge CB2 8BS, United Kingdom

One Liberty Plaza, 20th Floor, New York, NY 10006, USA

477 Williamstown Road, Port Melbourne, VIC 3207, Australia

314–321, 3rd Floor, Plot 3, Splendor Forum, Jasola District Centre, New Delhi – 110025, India

79 Anson Road, #06–04/06, Singapore 079906

Cambridge University Press is part of the University of Cambridge.

It furthers the University's mission by disseminating knowledge in the pursuit of
education, learning, and research at the highest international levels of excellence.

www.cambridge.org
Information on this title: www.cambridge.org/9781108475976
DOI: 10.1017/9781108620024

© Cambridge University Press 2021

First published 2021

Printed in the United Kingdom by TJ Books Limited. Padstow, Cornwall

A catalogue record for this publication is available from the British Library.

ISBN 978-1-108-47597-6 Hardback

regulators should be up front about that. However, enforcing reasonable rules that are propor-tionate to the burdens they accrue is a means to ensure that the research that does occur is responsible in terms of the benefits to those affected. The term 'facilitation' gives explicit emphasis on the need to ensure that the regulations are as minimally intrusive as necessary to achieve a given protective aim.

This framing has both inward-facing and outward-facing benefits. Looking inwards, regulators are reminded of the need to consider burdens of regulation along with benefits, and the balancing effort between the two in the proportionality assessment. This will help avoid blatantly one-sided approaches to regulation. Also, looking outwards, expressing this attitude in engage-ment with stakeholders can help assure them that their interests are being adequately accounted for. Such engagement is not merely limited to top-down communication of regulatory decisions, but active engagement as will be discussed further below.

A related procedural approach is actually doing the work of a proportionality assessment – that is, providing **rigorous justification** of a rule or policy's proportionality. It may be tempting to give up in the face of the uncertainties and ambiguities discussed. Nevertheless, responsible regulation must proceed. Ignoring proportionality can lead to one-sided policies, which either produce overly protective regimes with unacceptably burden research, or overly permissive regimes that do not adequately provide protections out of fear of inhibiting research.

And it will be work, indeed. When a given rule is under consideration, a non-trivial amount of research and analysis will be needed. Is there evidence on the magnitude of the harms or wrongs being prevented? What about the effectiveness of the proposed rule? And on the flip side, what effects will it have on the research enterprise? What are the quantifiable and non-quantifiable costs? Finally, when all those considerations are taken into account, can the regulation's protective effects truly justify the burdens imposed? And if not, can it be refined so that it does?

The final justificatory step may be the most uncertain and challenging. In some ways, it is an ethical or normative question relating to the values promoted and inhibited by a given policy. Regulators are not typically trained in philosophical analysis that may assist here, but some features of decision-making can be highlighted. These include articulation of the competing values at stake; scrutiny of any empirical evidence adduced; consistency between different judgments; clarity in terms of the reasons a given rule is justified, or not.

There is not space to elaborate here on such analytical tools. Indeed, no single article could adequately do so. Instead, it may be that regulators – or at least, some individuals in the regulatory process – should receive training in these analytical tools. As it stands, many relevant degrees like Masters of Public Policy or Masters of Public Administration do not routinely integrate such analytical training into their curriculums, focusing instead on social sciences. Reform of these curriculums might help boost competence in performing proportionality assessments. Alternative educational systems should also be considered, such as short courses, blended learning modules and ad hoc training workshops that may be more practicable for working professionals.

Especially because of the difficulty of making proportionality assessments, **transparency** in justificatory analyses will be crucial. Transparency here refers to some public promulgation of the reasoning process behind the decision that is reached. This would not only be easily accessed by stakeholders, but promulgated to relevant stakeholder groups so they are aware it exists in the first place.

Almost any rule will involve some trade-offs between protection of individuals and minimising burdens on research. As such, criticism from some affected stakeholders is inevitable. Having the reasoning and evaluation of a proportionality assessment will not eliminate that criticism, but it can go some way towards blunting suspicion that such an assessment was one-sided or ignored their concerns.

Moreover, there is good reason to suppose that stakeholders are owed this sort of transparency. For researchers, regulations have coercive force – failure to abide by them will result in penalties, whether criminal, civil, or – in the case of instructional policies – professional. It is a matter of respect to those individuals who are liable to such punishments that the reasoning process behind the rules is laid out in full. Other individuals like research subjects have a different relationship with regulations; while regulations do not directly bind them, they are carried out in their name. And if a regulator decides against enacting a given protective rule, that regulator is deciding to permit a certain degree of risk of harm to accrue to participants or others. Those affected individuals deserve to know the reasoning process behind this decision, as they may well be harmed by it.[18]

Another benefit of transparency is that it can prompt regulators to ensure their reasoning is truly defensible. Behind closed doors, there may be a temptation to wave away concerns that are too difficult or complex. By making their reasoning public, they are compelled to seriously reckon with all the considerations that stakeholders may find relevant. If not, they will be open to – legitimate – scrutiny and critique for inadequate analysis that will undermine confidence in the rules that are put forth.

Promulgation of reasoning and justification from regulators to stakeholders is important, but limited insofar as it is top-down and one-way. A more thoroughgoing and robust way to ensure adequate consideration of competing interests and earn public trust in proportionality assessments is **to directly engage** with those groups, to allow the co-creation of rules and collaborate assessment of the thorny issue of proportionality.

There are a myriad of ways that stakeholders can be engaged in proportionality assessments. For more details on approaches to and justifications for public engagement, see Aitken and Cunningham-Burley, Chapter 11 in this volume (on public engagement and access), and Burgess, Chapter 25 in this volume, (on public engagement and health research regulation).

These approaches are especially valuable for complex and uncertain issues like proportionality assessments. A small group of regulators may have parochial approaches or biased analyses that can be avoided by the involvement of a larger body of stakeholders. It may also relieve some of the pressure to make such complex judgments on their own, by soliciting assistance from a wider group.

This engagement should not be seen as one-off, or only occurring prior to rulemaking. A truly proportional approach to regulation must recognise the potential fallibility of initial judgments, and the fact that the situation on the ground may change. Protections previously seen as adequate could become threatened. For example, DNA profiles have recently been shown to be re-identifiable, which means previous protections merely stripping names and other extraneous information from such profiles are no longer sufficient to guarantee anonymity.[19] Previously burdensome compliance can be made easier by new technologies, as arguably occurred with the advent of digital compilation of ethics review documents allowing for more rapid collation and assessment.

For this reason, engagement should be a continual process, with the proportionality of a given rule periodically up for review and re-evaluation. Regulators may not be equipped to maintain

[18] N. Daniels, 'Accountability for Reasonableness', (2000) *BMJ*, 321(7272), 1300–1301.
[19] Y. Erlich et al., 'Identity Inference of Genomic Data Using Long-Range Familial Searches', (2018) *Science*, 362(6415), 690–694.

such active review, so instead being open to updates and comments from stakeholders may be optimal. This both relieves regulators of some burden to keep regulations' proportionality up to date, and ensures stakeholders have a continued ability to positively impact the rules that affect them.

To be sure, there are limitations on how much engagement can do. It was noted earlier that regulators may need additional training to adequately undertake proportionality assessments. This would already be practically difficult with regulators; with broader stakeholder groups, it is probably impossible. As such, there may be some limit on the extent to which co-creation is achievable for matters as complex as proportionality assessments. Still, we should not allow the perfect to be the enemy of the good; engagement has substantial value, as explained, that can supplement the deep analysis that regulators are responsible for.

3.7 CONCLUSION

In this chapter, I have explored the notion of proportionality in the context of health research regulation. Proportionality was defined in terms of a justificatory relationship: the benefits afforded by a given rule must serve to justify the burdens imposed by it. Assessing proportionality is no easy task; it is beset by uncertainties and challenges of analysis at a variety of levels, and involves weighing of different values – relating to beneficence, non-maleficence, justice and autonomy – that are non-commensurate and often non-quantifiable. The task of proportionality assessment is not impossible, however. Indeed, it is a necessary part of responsible regulation of health research. I have suggested several procedural approaches that can help improve the reliability and legitimacy of those assessments: a facilitative attitude; rigorous justificatory analysis; transparency in reasoning; and engagement in decision-making. These procedures recognise that we cannot formulaically produce an answer as to whether a given regulation is proportionate, and judgement is required. Hopefully, the contents of this chapter – in conjunction with the other material in this volume – can go some way to assisting those involved in regulation in understanding the nature, importance and practice of proportionality assessments.

4

Social Value

Johannes J. M. van Delden and Rieke van der Graaf

4.1 INTRODUCTION

This chapter starts from the assumption that science is a matter of co-creation. To open up science to democracy means that we have to think about the social value of research, which in itself we cannot leave to science to evaluate. This raises detailed questions around patient and public involvement (PPI) in deciding which research to perform, and about how to handle conflicts between individual and public interests. These are addressed elsewhere in this volume.[1]

In this chapter we focus on social value in health-related research involving humans, including data driven research. We first describe the background to the concept of social value and its meaning. Then we examine the concept itself and define the social value of an intervention as the value that an intervention could eventually have on the well-being of groups of patients and/or society. We also discuss some of the open issues in the scholarly debate about the concept of social value.

We find that to state a requirement for social value is one thing; to actually evaluate the social value of a research project in a Research Ethics Committee (REC) is another. We therefore elaborate on how the requirement of social value can be applied. We argue, first, that it is important to have this requirement as a separate condition. To increase systematisation, we further discuss how social value can be assessed in the steps that together constitute the risk-benefit task of RECs.

Returning to our opening statement, we argue that the addition of the requirement of social value can be seen as a consequence of a change in the sociology of science. It illustrates the move away from a science–internal understanding of scientific validity into an inclusive understanding of social value. Accepting social value as a requirement for research to be evaluated by a REC means that social value has matured from an attractive but illusive idea into something that has to be assessed, evaluated and optimised and can be used to address some of the justice issues in healthcare.

4.2 SOCIAL VALUE IN THE 2016 CIOMS GUIDELINES

Social value is a key principle in the 2016 version of the International Ethical Guidelines for Health-related Research prepared by the Council for International Organizations of Medical

---

[1] See Burgess, Chapter 25, and Aitken and Cunningham-Burley, Chapter 11, in this volume.

Sciences (CIOMS) in collaboration with the World Health Organization (WHO). The account of social value in this chapter has been largely influenced by the wording in the 2016 CIOMS Guidelines. Its very first guideline reads:

> The ethical justification for undertaking health-related research involving humans is its scientific and social value: the prospect of generating the knowledge and the means necessary to protect and promote people's health. Patients, health professionals, researchers, policy-makers, public health officials, pharmaceutical companies and others rely on the results of research for activities and decisions that impact individual and public health, welfare, and the use of limited resources. Therefore, researchers, sponsors, research ethics committees, and health authorities, must ensure that proposed studies are scientifically sound, build on an adequate prior knowledge base, and are likely to generate valuable information.
>
> Although scientific and social value are the fundamental justification for undertaking research, researchers, sponsors, research ethics committees and health authorities have a moral obligation to ensure that all research is carried out in ways that uphold human rights, and respect, protect, and are fair to study participants and the communities in which the research is conducted. Scientific and social value cannot legitimate subjecting study participants or host communities to mistreatment, or injustice.[2]

The entry of the requirement of social value in the 2016 CIOMS International Ethical Guidelines for Health-related Research involving humans was certainly not unprecedented. Many scholars trace its origins back to the Nuremberg Code of 1947, which states that 'The experiment should be such as to yield fruitful results for the good of society'.[3] Also, it is commonly understood that the social value of a research project may be part of the evaluation of risks and benefits of such a project.[4] The concept also plays a key role in the Belmont Report, the World Medical Association's Declaration of Helsinki, and the Common Rule. Furthermore, social value is considered to be of relevance when international collaborators are conducting health research in resource-limited settings. The concept also plays a key role in frameworks for research ethics, such as the '7- principle-framework' of Emanuel and colleagues[5] and the component analysis framework of Weijer and Miller.[6]

4.3 SOCIAL VALUE AS INDICATION FOR A CHANGE IN SOCIOLOGY OF SCIENCE

The addition of social value to the 2016 CIOMS International Ethical Guidelines at this point in history can be understood as part of a broader movement within the sociology of science, which describes how people come to accept certain scientific statements. Elements of this movement can also be seen in other guidelines within the 2016 CIOMS Guidelines, such as those on Community Engagement (7) and Public Accountability for Health-related Research (24). A first example of this broader movement within the sociology of science is the current critique of

---

[2] Council for International Organizations of Medical Sciences, 'International Ethical Guidelines for Health-related Research involving Humans', (CIOMS, 2016), 1.

[3] The Nuremberg Code (1947), (1996) *British Medical Journal*, 313, 1448.

[4] See Coleman, Chapter 13 in this volume.

[5] E. J. Emanuel et al., 'What Makes Clinical Research Ethical?', (2000) *JAMA*, 283(20), 2701–2711.

[6] C. Weijer, 'When Are Research Risks Reasonable in Relation to Anticipated Benefits?', (2004) *Nature Medicine*, 10(6), 570–573; A. Binik and S. P. Hey, 'A Framework for Assessing Scientific Merit in Ethical Review of Clinical Research', (2019) *Ethics & Human Research*, 41(2), 2–13.

science and scientific knowledge.[7] Part of the critique concerns the replicability of research results, which in some areas is disturbingly low. Another part concerns the way in which scientists are evaluated: in many areas of science this is done, at least until recently, by looking at the number of articles produced and/or the number of times an article is cited – e.g. combined into the Hirsch-index – creating an incentive to produce enormous quantities of papers. But the most important critique – also implied in the former point – is that science appears to be concerned more with producing science as such, than with furthering socially valuable goals through research. The term 'research waste' was coined to describe the result of this way of doing research.

In response, we currently see programmes such as the EU programme on Responsible Research and Innovation, movements such as that for Open Science – which is certainly about more than just open access publishing – and Science in Transition.[8] These programmes try to reinvent the sociology of science in order to enable it to perform the tasks society has entrusted to scientists. They also encourage the involvement of all stakeholders in the production of science, including patients and publics, in order to increase the relevance of research results. Present-day problems in society are simply too complex to think we can solve them without cooperating across borders. Science cannot continue to take its own interests as primary, instead of living up to its societal task. Science needs to earn and deserve a so-called social licence for research.[9] PPI in research is an essential means to mitigate concerns on research waste.

There are a number of reasons why we need PPI in research – as addressed in more detail elsewhere in this volume.[10] First, this is because research is about all of us! And nothing should be done 'about us, without us'. We therefore need a model in which patients consider themselves as partners in a trustworthy system, not just passive sources of information. Second, the purpose of patient involvement is ultimately to improve our health. By this we do not mean through individual healthcare. Rather, we suggest that this can come about by ensuring that those who conduct research projects ask the right questions, use the right endpoints, make the right choices and effectively implement their findings. This illustrates the efficiency argument as applied to input from patients – and wider publics – who are similarly motivated to find answers to health and disease-related questions. It is believed that this will help science to become more socially valuable and thus to reduce research waste.

These developments also point to important questions in the area of the philosophy of science. It is common to think that science produces facts that are independent of public preferences. Shouldn't science inform democratic decision-making rather than being influenced by it? What is left of scientific independence if we allow PPI in research? It is generally understood why democracies need science, but why would science need democracy?[11]

To answer these questions we turn to Science and Technology Studies (STS) where several schools of thought can be discerned. The first (1900–1960) was a positivistic one: it was believed that science was a way of knowledge-making and that its knowledge was absolute and universalistic.[12] The correctness of scientific research needed no social explanation, it was simply

---

[7] D. Moher et al., 'Increasing Value and Reducing Waste in Biomedical Research: Who's Listening?', (2016) *Lancet*, 387(10027), 1573–1586.

[8] F. Miedema, *Science 3.0* (Amsterdam University Press, 2010).

[9] P. Carter et al., 'The Social Licence for Research: Why care.data Ran into Trouble', (2015) *Journal of Medical Ethics*, 40(5), 404–409.

[10] See Burgess, Chapter 25, and Aitken and Cunningham-Burley, Chapter 11, in this volume.

[11] H. Collins et al., *Why Democracies Need Science* (Cambridge: Polity, 2017).

[12] Ibid.

true. What needed explanation was how false beliefs were mistakenly taken to be correct, typically by pointing at prejudice, bias and so on. This is what Nowotny calls Mode 1 research.[13] Although this view is no longer supported by social science, it remains the common-sense view of many scientists and the public. One needs only to watch an episode of *CSI* to see how a forensic scientist reveals 'the truth' about the case.

The second school of thought (1960–2000) started when others took the work of Kuhn and other researchers to show that scientific truth is best seen as an outcome of negotiation and agreement located within social groups. Science is a human activity subject to all the strengths and flaws of humans. Nowotny speaks about Mode 2 research in which interaction between science and society is taken as a starting point and science has become a matter of co-creation.[14] Science needed to be democratised. This second school illuminated the constructivist side of science, in order to deconstruct science, but did less to provide an alternative.[15] A risk of this type of thinking is that this may produce the kind of relativism in which scientific claims have become 'just another opinion' and alternative facts are as good as any other account.

To counter this, the third school (after 2000) emphasises that we do not need to end up in relativism, and that there are more arguments in favour of some claims about states of the world than there are for others. Textbook science is not perfect, and remains open to revision, but is more reliable than primary research, because we have more reasons to accept the claims in a textbook than in primary research. In ethics, the Rawlsian understanding of ethical claims as provisional fixed points captures the same idea: claims are always open to revision (hence 'provisional') but we have good reasons to accept them (hence 'fixed'). It is important to note that the last school of thought accepts the rationale established by the former, but tries to make the next, constructive step.

We think that the addition of the requirement of social value into the CIOMS Guidelines can be seen as a consequence of this change in the sociology of science. It clearly illustrates the move away from a science–internal understanding of scientific validity into an inclusive understanding of social value. It sends the message that science needs to be cognisant of its societal role and should explain how it aims to fulfil that role. That message is reinforced by guidelines on community consultation and public accountability. Placing social value as a requirement in a list of conditions to be evaluated by a REC means that social value has matured from an attractive but illusive idea into something that has to be assessed, evaluated and optimised. In other words: social value has gained 'teeth'.

4.4 MEANING OF SOCIAL VALUE

We will now zoom in on the meaning of the concept 'social value' itself. According to Wendler and Rid, the standard view on social value is that 'it is an ethical requirement for the vast majority of clinical studies'.[16] They also argue that there is 'strong support' that social value of research is important 'for protecting participants who cannot consent, preventing inappropriate research that poses high net risks, and promoting appropriate investigator behaviour'[17] (see also below).

---

[13] H. Nowotny et al., *Rethinking Science* (Cambridge: Polity, 2001).
[14] Ibid.
[15] Collins et al., *Why Democracies Need Science*.
[16] D. Wendler and A. Rid, 'In Defense of a Social Value Requirement for Clinical Research', (2017) *Bioethics*, 31(2), 77–86, 77.
[17] Ibid., 86.

Here is the description of the meaning of the term social value according to the 2016 CIOMS Guidelines:

> Social value refers to the importance of the information that a study is likely to produce. Information can be important because of its direct relevance for understanding or intervening on a significant health problem or because of its expected contribution to research likely to promote individual or public health. The importance of such information can vary depending on the significance of the health need, the novelty and expected merits of the approach, the merits of alternative means of addressing the problem, and other considerations.[18]

We next examine separately the concepts of value and social value. We understand value to mean the potential of a study to improve health, broadly construed as biological, psychological or social well-being.[19] Health value can be categorised along two dimensions: immediate versus future health value, and the population that receives this value.[20] It is also important to note that social value is attributed both to information that has direct relevance in promoting health, and to the contribution this information may have for subsequent valuable research.

The concept 'value' has been scrutinised in many different research fields such as sociology and philosophy. However, little agreement exists on how 'value' should be defined. Consensus does exist on the fact that values arise out of human experience. Whereas the term 'benefit' refers to an advantage or profit gained from something, the concept of value refers to the regard that something is held to deserve. The latter is thus a relational concept; both the object to be valued, and an evaluator are necessary preconditions for value to exist.[21]

Turning next to 'social value', this functions in two main ways in our everyday use. First, social value can be seen as values shared by a community of individuals; they are values held by society and are contrasted with individual (non-shared) values. By social value, we refer to socially collective beliefs and systems of beliefs that operate as guiding principles in life. Second, besides values *of* society, the concept can also be used to refer to values *for* society. Here, social value is an assigned predicate or property of an object, and, in our case, of health-related research.[22] This implies that we have to assess the importance of the information in terms of the nature and magnitude of the expected improvement an intervention – as assessed in the study – is expected to have on society. Note that benefit for the individual research participant would be called a direct benefit. Social value is not about rewarding careers for scientists, employment for citizens or a sense of fulfilment for participants.[23]

We conclude that the social value of an intervention encompasses the value that an intervention could eventually have on the well-being of groups of patients and/or society. In case of early phase trials, this value may lie in the distant future; in those cases, RECs may also assess the ability of trials to promote progression to later stages of research in which successful clinical translation becomes more likely.

It is important to note that the CIOMS guideline on social value also explicitly talks about what social value cannot do, as follows:

[18] CIOMS, 'International Ethical Guidelines', 1.
[19] D. J. Casarett and J. D. Moreno, 'A Taxonomy of Value in Clinical Research', (2002) *IRB: Ethics & Human Research*, 24(6), 1–6; C. Grady, 'Thinking Further about Value: Commentary on "A Taxonomy of Value in Clinical Research"', (2002) *IRB: Ethics & Human Research*, 24(6), 7–8.
[20] Casarett and Moreno, 'A Taxonomy of Value'.
[21] M. Habets et al., 'The Social Value of Clinical Research', (2014) *BMC Medical Ethics*, 15, 66.
[22] Ibid.
[23] Wendler and Rid, 'In Defense of a Social Value Requirement'.

Although scientific and social value are the fundamental justification for undertaking research, researchers, sponsors, research ethics committees and health authorities have a moral obligation to ensure that all research is carried out in ways that uphold human rights, and respect, protect, and are fair to study participants and the communities in which the research is conducted. Scientific and social value cannot legitimate subjecting study participants or host communities to mistreatment, or injustice.[24]

This provision is a reformulation in human rights language of the so-called primacy principle. This is the ethical principle stating that the individual shall have priority over science, found, for instance, in guideline 8 of the 2013 Declaration of Helsinki: 'While the primary purpose of medical research is to generate new knowledge, this goal can never take precedence over the rights and interests of individual research subjects'.[25] There is an ongoing debate about the tenability of this primacy principle[26] which deserves a separate discussion.

4.5 SOCIAL VALUE IN SCHOLARLY DEBATE

Whereas the merits of the social value requirement have been largely uncontested, over the past few years the concept of social value has received increasing scholarly attention. Among others, the journal *Bioethics* launched a Special Issue (2017, 31(2)) on social value. Also Danielle Wenner's[27] analysis of social value in the Hastings Center Report led to several responses.[28] The attention has not only led to improved understanding of the meaning and scope of social value but also to more critique. Next, we will consider some of the key points from this ongoing debate.

Traditionally, social value has been located in the context of clinical research, but more recently the concept has also been introduced in health systems research and into the global health ethics debate.[29] Whereas the concept, as discussed above, in clinical research focuses on the knowledge to be gained for society in general, in public and global health ethics the requirement seems to have a different role. For instance, according to Nicola Barsdorf and Joseph Millum, social value should be seen as 'a function of expected benefits of the research and the priorities that beneficiaries deserve'.[30] Social value then also becomes a means to address questions of priority setting,[31] promotion of health equity and addressing health inequality.[32]

---

[24] CIOMS, 'International Ethical Guidelines', 1.

[25] World Medical Association, 'Declaration of Helsinki – Ethical Principles for Medical Research Involving Human Subjects', (WMA, 2013).

[26] G. Helgesson and S. Eriksson, 'The Moral Primacy of the Human Being: A Reply to Parker', (2011) *Journal of Medical Ethics*, 37(1), 56–57.

[27] D. M. Wenner, 'The Social Value Requirement in Research: From the Transactional to the Basic Structure Model of Stakeholder Obligations', (2018) *The Hastings Center Report*, 48(6), 25–32.

[28] D. Wendler, 'Locating the Source(s) of the Social Value Requirement(s)', (2018) *The Hastings Center Report*, 48(6), 33–35; D. B. Resnik, 'Difficulties with Applying a Strong Social Value Requirement to Clinical Research', (2018) *The Hastings Center Report*, 48(6), 35–37; F. S. Holzer, 'Rawls and Social Value in Research', (2019) *The Hastings Center Report*, 49(2), 47.

[29] A. Rid and S. K. Shah, 'Substantiating the Social Value Requirement for Research: An Introduction', (2017) *Bioethics*, 31(2), 72–76; Wenner, 'The Social Value Requirement'.

[30] N. Barsdorf and J. Millum, 'The Social Value of Health Research and the Worst Off', (2017) *Bioethics*, 31(2), 105–115, 105.

[31] Rid and Shah, 'Substantiating the Social Value Requirement'.

[32] D. Wassenaar and A. Rattani, 'What Makes Health Systems Research in Developing Countries Ethical? Application of the Emanuel Framework for Clinical Research to Health Systems Research', (2016) *Developing World Bioethics*, 16(3), 133–139.

At the same time, in the context of health systems research, some argue that its social value can also be justified 'in pragmatic systems rather than linked only to priority setting'.[33]

Further discussion centres on whether the concept of social value should be located in the traditional account of research ethics that has a focus on clinical trials and observational research. According to Wendler and Rid, there are eight reasons that 'taken together provide strong support' that social value must be obtained in the context of clinical research: (1) to protect participants who cannot consent; (2) to ensure the acceptability of high-risk research with competent adults; (3) to maintain researcher integrity; (4) to avoid participant deception; (5) to safeguard against exploitation; (6) to exercise stewardship of public resources; (7) to promote public trust; and (8) support for clinical research.[34] Others, like Wenner,[35] Wertheimer[36] and Resnik,[37] ground the social value requirement in other principles and outside of the traditional scope of research ethics. According to Wenner, the current view on research ethics is primarily about protection. Instead, she believes it should be grounded in justice-based considerations. She argues that certain developments in research, such as the inclusion of pregnant women, cannot be understood only from a protectionist view towards research subjects but has to be explained from underlying issues of justice.[38]

Whereas some, like Wertheimer and Resnik, argue that studies must have 'significant' social value, Wendler and Rid[39] argue that studies should have 'sufficient' social value. The first group of authors expresses concern that without the qualification of significance, the concept becomes too weak, whereas Wendler and Rid argue that their understanding is also able to distinguish between studies with and without social value. Whether a study has sufficient social value should always be determined in relation to the risks of research. In some cases participants may face significant risks. However, if there is no social value to be gained, they argue that the study should not be approved even if participants consent to participation. At the same time, if the social value is limited but the risks are minimal as well, they argue it is not unethical to offer participation.

4.6 APPLICATION

In the preceding analysis we have considered both what the term social value means and the discussions that it has sparked. As such, we can now go on to look at its role in the set of requirements for acceptance of a research protocol. First, we would like to point to the importance of having this as a separate requirement. It could be argued that the social value of a research project is already being taken into account in the classical requirement in research ethics to have a favourable balance of benefits over risks and burdens. The 2013 version of the Declaration of Helsinki for instance reads: 'Medical research involving human subjects may only be conducted if the importance of the objective outweighs the risks and burdens to the research subjects'.[40] One could conclude from this that it would not be necessary to have a separate guideline on social value. However, the problem with including social value in this

---

[33] Wassenaar and Rattani, 'What Makes Health Systems Research in Developing Countries Ethical?', 136.
[34] Wendler and Rid, 'In Defense of a Social Value Requirement'.
[35] Wenner, 'The Social Value Requirement'.
[36] A. Wertheimer, 'The Social Value Requirement Reconsidered. The Social Value Requirement Reconsidered', (2015), *Bioethics*, 29(5), 301–308.
[37] Resnik, 'Difficulties with Applying a Strong Social Value Requirement'.
[38] Wenner, 'The Social Value Requirement'.
[39] Wendler and Rid, 'In Defense of a Social Value Requirement'.
[40] The Declaration of Helsinki (2013).

so-called risk/benefit ratio is that in research projects without risks or burdens, a lack of anticipated benefit would not be sufficient grounds for a REC to deny approval of the project. If one thinks that the main aim of research ethics guidelines is to protect the individual, then one might be satisfied. If one takes a broader view and includes justice among the ethical principles that are relevant to such a deliberation, then allowing a project without benefit is unacceptable from a societal perspective. Projects still use time, money and energy in addition to contributing to more research waste. Therefore we argue that it is necessary to have social value as a separate requirement.

Some might object on the basis that social value cannot be a necessary requirement for research to be ethical since certain medical discoveries have been made by coincidence, and that requiring social value may limit medical advancement. However, accidental findings cannot be planned, nor does requiring social value mean that we will no longer find accidental findings by restricting clinical research to interventions with expected social value.

Having made the preceding claim, we now turn to the role of RECs, which are currently tasked with judging whether a favourable risk-benefit balance is achieved to ultimately decide whether a research project can proceed. This judgement has to be systematic, transparent and grounded in evidence. Evaluating the social value of a particular research project can be seen as part of this task. To increase systematisation we draw upon insights from decision-theory and propose that the risk-benefit tasks are divided into the following steps: (1) analysis; (2) evaluation; (3) treatment; and (4) decision-making.[41]

4.6.1 *Benefit Analysis*

It is the primary responsibility and expertise of investigators to map and characterise benefits, including the social value of research. However, evaluators should be able to judge whether they agree with the reasoning that supports the presented characterisation of benefits.[42] To map benefits, we divide these into direct, collateral and aspirational benefits.[43] Social value can be regarded as one of the aspirational benefits. We further divide social value into: (1) the direct social value of the intervention; (2) the progressive value; and (3) the translational value of a trial.

In characterising the social value of an intervention we draw upon the proposal by Habets and colleagues.[44] They argue that at least three steps should be followed. First, the nature and magnitude of efficacy of the intervention studied in humans has to be critically assessed. Second, the anticipated clinical improvement in actual patients should be assessed, assuming that the intervention is efficacious. This means that it has to be asked whether treatment effects are meaningful, both from a medical and individual perspective, and that they have to be weighed against factors that may hamper beneficial effects, such as adverse effects and ease of use. Third, the nature and magnitude of the anticipated improvement on the well-being of patients, individuals in society and society should be evaluated. This assessment is contextual: the social value of the intervention is the expected improvement relative to other considerations, such as

---

[41] R. Bernabe et al., 'The Risk-Benefit Task of Research Ethics Committees: An Evaluation of Current Approaches and the Need to Incorporate Decision Studies Methods', (2012) *BMC Medical Ethics*, 13(1), 6.

[42] Ibid.

[43] N. King, 'Defining and Describing Benefits Appropriately in Clinical Trials', (2000) *The Journal of Law, Medicine, and Ethics*, 28(4), 332–343.

[44] M. Habets et al., 'The Unique Status of First-in-Human Studies: Strengthening the Social Value Requirement', (2016) *Drug Discovery Today*, 22(2), 471–475.

treatment alternatives, number of patients and costs etc. Ultimately, determining what has social value constitutes a moral judgment.[45]

To characterise progressive value[46] we argue that at least two elements should be evaluated: (1) whether there is a reasonable probability that an intervention could progress to the next stages of research at all; and (2) whether the trial is designed such that the yielded results can contribute to progression to the next stage of research (typically Phase II). The assessment of estimated efficacy can contribute to the assessment of both elements. Evaluators should therefore judge whether they find the estimated efficacy as presented by investigators to be substantive.

For trials to have translational value they should be hypothesis-driven. Preclinical and reference class evidence form the basis for the generation of hypotheses and the context for the subsequent interpretation of both positive and negative findings.[47] For instance, if a positive result in animals is followed by a negative result in humans, this difference can lead to further explorations of this difference and/or which modifications to the intervention have to be made to overcome translational hurdles. Furthermore, the determination and evaluation of reference class evidence helps researchers to put their findings in a broader context and to communicate their findings to other areas of research. Evaluators should thus judge whether investigators base their hypotheses on a solid assessment of preclinical and reference class evidence.[48]

4.6.2 Benefit Evaluation

We contend that investigators and evaluators should be transparent about the weight they ascribe to the different types of benefits (and harms). Progressive and translational value are not necessarily mutually exclusive, however, they may require a different trial design.[49] Therefore, it should be made explicit how a trade-off between different types of benefits and harms are made.

4.6.3 Benefit Treatment

After benefit assessment, RECs need to judge whether measures need to be implemented to modify – and ideally to maximise – benefits. The following measures can be taken to enhance the translational value of a trial. If hypotheses are insufficiently supported by evidence, investigators can be prompted to conduct additional preclinical testing. Alternatively, evaluators can demand more thorough gathering and assessment of existing preclinical and reference class evidence. Methods of PPI can show whether or not patient-relevant outcome measures have been used. Furthermore, open sharing of the assessed preclinical and reference class evidence can enhance the collateral value of a trial. Additionally, amendments to the trial design can spur the translational value.

4.6.4 Decision-Making

Finally, RECs have to decide whether benefits truly outweigh the risks. The three steps of benefit analysis, evaluation and treatment contribute to the transparency of decision-making. It

---

[45] S. Boers, 'Organoid Technology. An Identification and Evaluation of the Ethical Challenges', *PhD thesis* (Utrecht University, 2019).

[46] J. Kimmelman, *Gene Transfer and the Ethics of First-in-Human Research* (Cambridge University Press, 2009).

[47] Kimmelman, *Gene Transfer*.

[48] Boers, 'Organoid Technology'.

[49] Kimmelman, *Gene Transfer*.

has been claimed that it matters whether the research is funded with public money or not. We disagree: even when privately funded, we can see no justification for burdening participants with research that has no social value.

4.7 CONCLUSION

The term 'social value' strikes the necessary balance between scientific advancement, equitably responding to human conditions and realising the human right to health. The requirement of social value bridges the gap between conducting commendable science and making a contribution to the health of the populations where health research is being carried out. The concept of social value is the ethical justification for doing health research involving humans.

5

Solidarity in Health Research Regulation

Katharina Kieslich and Barbara Prainsack

5.1 INTRODUCTION

This chapter explores the analytical and normative roles that solidarity can play when designing health research regulation (HRR) regimes. It provides an introduction to the meanings and practical applications of solidarity, followed by a description of the role solidarity plays in HRR, especially in fostering practices of mutual support between patient organisations and between countries. We illustrate our argument in a case study of HRR, namely the European Union (EU) regulatory regime for research on rare diseases and orphan drugs. The current regime aims to decrease barriers to research on orphan drugs by creating, predominantly financial, incentives for research institutions to take on the perceived increased risks in this area. We show how the concept of solidarity can be used to reframe the purpose of regulation of research on orphan drugs from a market failure problem to a societal challenge in which the nature of barriers is not just financial. This has specific implications for the types of policy instruments chosen to address the problem. Solidarity can be used to highlight the political, social, economic and research value of supporting research on rare diseases and orphan drugs.

5.2 THE MEANING OF SOLIDARITY

The concept of solidarity underpins many social and healthcare systems in Europe.[1] While it could be argued that solidarity – in the form of policies and institutional structures facilitating mutual support, with special emphasis on supporting the vulnerable – has come under pressure with the spread of nativist and other sectarian political ideologies, there are also forceful counter-movements under way. These include people standing up with and for others,[2] may it be newcomers to our society, victims of wars and natural disasters or people who suffer from our economic and political system. As such, it is fair to say that solidarity is seen by many as having a lot to offer to how we frame and address societal challenges.

---

[1] K. Kieslich, 'Social Values and Health Priority Setting in Germany', (2012) *Journal of Health Organization and Management*, 26(3), 374–383; L. D. Brown and D. P. Chinitz, 'Saltman on Solidarity', (2015) *Israel Journal of Health Policy Research*, 4(27), 1–5; R. Saltman, 'Health Sector Solidarity: A Core European Value but with Broadly Varying Content', (2015) *Israel Journal of Health Policy Research*, 4(5), 1–7; R. ter Meulen, *Solidarity and Justice in Health and Social Care in Europe*, (Springer, 2001).

[2] A. Dawson and B. Jennings, 'The Place of Solidarity in Public Health Ethics', (2012) *Public Health Reviews*, 34(1), 65–79.

What is solidarity? At first sight, it might seem an elusive concept. For decades, solidarity has been used to justify a wide variety of policies and practices ranging from vaccination pro- grammes to biobanks to the penalisation of undesirable behaviours. Another reason for the elusiveness of solidarity lies in the practical and embodied nature of solidarity. Solidarity is, first and foremost, a relational practice: its full meaning unfolds only when it is enacted, in concrete practice, by – at least one – giver and a receiver, and its nature cannot be exhaustively captured by language. For the same reason that poetry, art or nature are so much more powerful in conveying the meaning of love or friendship, words alone struggle to convey the full meaning of solidarity.

Acknowledging that part of the meaning of solidarity resides in its embodied- and enactedness does not mean, however, that we cannot spell out what makes solidarity different from other types of prosocial practice. Building upon a long history of scholarship on solidarity we have, in our own work, proposed that solidarity is best understood as a practice that reflects a person's – or persons' – commitments to support others with whom the person(s) recognise(s) similarity in a relevant respect.[3] The similarities with others that people recognise are, however, not 'object- ively' existing properties, but they are characteristics that we have learned to attribute to ourselves and to others. The first step in this process is that we use categories that have been developed to sort people in different groups, such as separating them into women and men, children and adults, Jews, Buddhists and Muslims, or Koreans and Croatians. While these categories clearly have an expression in material reality, such as the correspondence of national labels with specific territories, or – in the case of children and adults, even stages in human biology – these categories are not merely material. To whom the label of 'Korean' or 'Croatian' is applied has not been stable in history but it has depended on changing territorial rule, changing understand- ings of nationality and different perspectives on who can legitimately claim belonging to such a label. Similarly, the notions of children and adults are not clearly delineated in biology in the sense that every person neatly fits into one or the other category. In this way, the categories that we use to describe characteristics that we and others hold are lenses through which we have learned to see reality.

For solidarity this means that when a woman supports another person because she recognises her as a fellow woman, then 'being a woman' is the 'similarity in a relevant respect' that gives rise to solidaristic action – despite the fact that the two people in question are many more things than women. They may be different in almost every other way. In this sense, the recognition of similarities in a relevant respect is a subjective process – I recognise something in you that you may not recognise in yourself because you have not learned to see it. At the same time it concerns shared social meaning – as societies have shared conventions about how they classify people.

Solidarity happens when people are guided in their practices by the similarities they recognise with each other, despite everything that sets them apart. It is the similarities, and not the differences, that give rise to action in the sense that they prompt people to do something to support somebody else. This 'doing something' could consist of something big – such as donating an organ – or something small, such as offering somebody a seat on a bus.

In sum, what makes solidarity different from other pro-social practice is the symmetry between people in the moment of enacting solidarity. This symmetry is not an essentialist ontological statement that glosses over claimed or ascribed differences and structural inequalities. Instead, it

[3] B. Prainsack and A. Buyx, 'Solidarity: Reflections on an Emerging Concept in Bioethics', (Nuffield Council on Bioethics, 2011); B. Prainsack and A. Buyx, *Solidarity in Biomedicine and Beyond* (Cambridge University Press, 2017).

is the description of a relational state in the moment of enacting solidarity. In this way, solidarity is distinct from other pro-social supportive behaviours such as cooperation and charity, for example. The notion of cooperation describes pro-social supportive behaviour without saying anything about how and why people engage in it. The notion of charity describes an asymmetrical interaction between a stronger entity giving something and a weaker entity receiving something. In contrast, solidarity refers to entities that are different in many respects but make the thing they share in common the feature upon which they act: I do something for you because I recognise you as a fellow woman, a co-worker who struggles to make ends meet, as I do, or a fellow human in need of help.

5.3 THE THREE TIERS OF SOLIDARITY: APPLICABILITY AND ADJUSTMENTS IN THE CONTEXT OF HEALTH RESEARCH REGULATION

Having defined solidarity as practices that reflect commitments to support others with whom a person – or persons – recognise(s) similarities in a relevant respect, in previous work one of us identified three main tiers of solidarity, capturing the societal levels where solidaristic practice takes place.[4] Tier 1 is the interpersonal level where solidarity is practised between two or more people without that practice having become more widespread. An example from the field of health research would be a person with diabetes signing up to a biobank researching the disease because she wants to support others with similar health problems.

If this practice were to become more widespread, so that it became common or even normal behaviour within a group, then we speak of solidarity at Tier 2 solidarity, which is solidarity at the group level. The group within which solidarity is practised could be a pre-existing group – such as a self-help group around diabetes where it becomes normal practice, for example, to also volunteer for disease research – or a group that is created through the solidaristic practice itself. An example for the latter would be a patients' rights organisation created in response to the effects of harmful medical practices such as the blood contamination scandal in the 1970s and 1980s in the United Kingdom (UK).

If solidaristic practices become so commonplace that they are reflected in legal, administrative or bureaucratic norms, then we speak of Tier 3 solidarity. This is the 'hardest' form of solidarity because it has coagulated into enforceable norms. Tier 3 solidarity could be seen to contradict the idea held by many scholars in the field that solidarity cannot be demanded, but only appealed to.[5] In this understanding, contractual and legal obligations are incompatible with solidarity. While we agree with these authors that solidarity is typically a more informal, voluntary 'glue' between the bricks of formal institutional arrangements, we also believe solidarity to be a toothless, if not empty, concept if it cannot also denote practices that are so deeply engrained in society that they become legally enforceable in some cases.

Ruud ter Meulen and colleagues very helpfully distinguish between solidarity as a community value and solidarity as a system value:[6] the latter can contain articulations of solidarity in formal, often legal arrangements. The key here is to consider enforceable – and thus not always voluntary – solidarity in conjunction with more informal, voluntary forms of solidarity, and not see them as isolated from one another. An example would be tax or contribution-based financing of universal healthcare where those with higher incomes contribute more than others.

---

[4] Prainsack and Buyx, 'Solidarity: Reflections'; Prainsack and Buyx, *Solidarity in Biomedicine and Beyond*.
[5] J. Dean, *Solidarity with Strangers: Feminism after Identity Politics* (Berkeley: University of California Press, 1996), p. 12;
[6] ter Meulen, *Solidary in Health and Social Care*, p. 11.

A problem arises when legally enforceable solidarity is still in place while the actual practices that used to underpin them are breaking away. This is becoming apparent at the moment in many countries where certain features of welfare states, such as transfer payments in the form of as child allowances or income support for those considered undeserving, have come under attack. The argument is often that the people benefitting from this are 'free riders' as they have not contributed towards the system that they are now using – perhaps because they are new immigrants or people who have never been in paid employment. What is happening here is that the basis for solidaristic practice – namely the 'recognition of similarity in a similar respect' (see above) – is breaking away. The people who are receiving financial support, or benefitting from a solidaristic healthcare system, are no longer seen as belonging to 'us' – because of something that they supposedly did, or failed to do, or because they do not have the same passport as we do.

While it will often be the case that solidarity prescribed at Tier 3, in the form of legal, contractual, bureaucratic and administrative norms, will have evolved out of solidarity practised at group (Tier 2) and interpersonal (Tier 1) levels, the reverse is not necessarily true: interpersonal solidarity can, but does not necessarily, scale upwards. The 'higher' the level of solidarity, the more important reciprocity becomes. Here we refer not to direct reciprocity, where one gives something in return for something else – this would be a business transaction instead of solidaristic practice – but indirect, systemic reciprocity. Institutional arrangements of solidarity work best when people give because they want to support others, but they also know that when they are in need they will be supported as well.

5.4 SOLIDARITY IN HEALTH RESEARCH REGULATION

How do the aforementioned conceptualisations of solidarity apply to HRR regimes? The first aspect we need to acknowledge is that HRR regimes are complex and varied. There is no such thing as one regime that applies to all areas of HRR, but rather there are multiple and sometimes overlapping legal and ethical requirements that need to be fulfilled by those planning, funding, supporting and undertaking research. HRR is a multidisciplinary endeavour that involves different actors such as policymakers, researchers, health professionals, industry and patients. HRR also spans a large variety of 'objects' that are regulated, such as data, tissue, embryos, devices or clinical trials.[7] This means that it occupies regulatory spaces beyond health, such as in data regulation, research financing, in fostering innovation and in the obligation to protect research recruitees.

At the start of this chapter we suggested that solidarity can be thought of as 'enacted commitments to accept costs to assist others with whom a person or persons recognise similarity in one relevant respect'.[8] Thus the question arises: what are the shared practices that reflect a commitment to carry costs – emotional, financial, societal – in HRR, and what are the similarities that give rise to these practices? The two tiers of solidarity most relevant in HRR are Tiers 2 and 3. Tier 2, or group solidarity, is reflected, for example, in the way patients, patient groups and other stakeholders advocate for, inform about, and partake in research endeavours and regulatory steps to make them happen. The question of who partakes in research is not just important for methodological reasons but is also connected to the concept of solidarity. It is considered good scientific practice to carry out research in the populations

---

[7] G. Laurie, 'Liminality and the Limits of Law in Health Research Regulation: What Are We Missing in the Spaces In-Between?', (2016) *Medical Law Review*, 25(1), 47–72.

[8] Prainsack and Buyx, *Solidarity in Biomedicine and Beyond*, p. 43.

for whom an intervention is intended, but there may be instances in which it is justified to conduct research in populations other than the intended beneficiaries. According to the Council for International Organizations of Medical Sciences (CIOMS) and the World Health Organization (WHO) such instances are 'important demonstration[s] of solidarity with burdened populations',[9] for example in 2014 when Ebola vaccines were tested in communities not affected by the Ebola outbreak.

The costs and the similarities that are at the heart of these – predominantly clinical – research practices are comparatively easy to identify. The costs commonly consist of individuals giving up their time to become research participants or to become involved in a patient advocacy group. They accept the burden of cumbersome regulatory steps to partake in research, such as navigating consent forms, risk assessments, data ownership and other issues. The similarity that motivates people to assist others despite the costs they incur is often the experience of suffering from a particular disease or the acknowledgement that we, as members of society or those close to us, all run the potential risk of illness in the future. It is a recognition that temporary sacrifices can result in long-term gains from the generation of new knowledge about health conditions and treatments.

A feature that distinguishes HRR from other areas of policy, regulatory and societal processes is that group solidarity is often not just confined to a small group of patients who are afflicted by the same illness. Rather, other members of the public – so-called healthy recruits – partake in the solidaristic practice of research and are directly affected by the associated regulatory procedures. The underlying 'similarity in a relevant respect' that, in Prainsack and Buyx's definition of solidarity gives rise to solidaristic practice, is then typically a broad sense of human vulnerability that we all have in common. In other words, the nature of Tier 2 solidarity in HRR is not necessarily restricted to suffering from the same illness, but it can arise from the recognition that in a universally funded healthcare system, we all carry a commitment to carry costs because we all carry the risk that we might one day become ill.

To explore how Tier 3 solidarity, or institutional solidarity, is reflected in HRR, we trace the logic that forms the basis for understanding HRR through the lens of solidarity. The logic runs something like this: A solidaristically financed healthcare system is built on the principles of fair access to healthcare, protection against financial risks due to illness and quality. Ensuring access, provision and high-quality healthcare requires efforts to advance knowledge through research. Implicitly entailed in the social contract between governments, citizens and residents is the acceptance that mandatory financial contributions – i.e. costs – in the form of taxes or health insurance contributions will not only be used for the day-to-day provision of services but also for the fostering of research activities. With this implicit acceptance of carrying costs collectively comes a recognition that the health research area needs to be regulated to safeguard against unethical, harmful, and wasteful practices, and to foster innovation. This recognition translates into public policies that regulate the field.

But there are also regulatory burdens arising from such public policies that might negatively affect solidarisic practices in HRR. For example, the cumbersome, and often time-intensive, process of giving consent for a research participant's data to be used for research purposes might deter some people from taking part in a study, especially if the use of the data is not explained or communicated clearly. Moreover, the predominant lens through which data ownership – in a moral and in a legal sense – is currently viewed is that of the rights of individuals, who, in turn,

---

[9] Council for International Organizations of Medical Sciences, and World Health Organization, 'International Ethical Guidelines for Health-related Research Involving Humans', (CIOMS, 2016).

are conceptualised as bounded and independent entities.[10] This view is problematic because it fails to acknowledge the deeply engrained relational characteristics of data. This is so because the meaning of most data only unfolds once the data is interpreted in relation to other data, and that this meaning is often relevant for a wider range of people than only the person from whom they came. Currently, this relational nature of data is not reflected in most data governance frameworks in the health domain; even those frameworks that give people more control over how their data is used typically give this control to individuals. Instruments of collective control and shared ownership of personal data are rare. The 'individualisation' of data governance sits squarely within a system that relies on people's willingness to make data about themselves available for research. It is a missed opportunity for showing how control and use of data can reflect both personal and collective interests and rights.

5.5 SOLIDARITY IN RESEARCH ON RARE DISEASES AND ORPHAN DRUGS

An example of how solidarity can be used to change the way we approach a policy problem in HRR can be found in rare diseases and orphan drugs research. The European Commission (EC) defines a rare disease as 'any disease affecting fewer than 5 people in 10,000 in the EU'.[11] It estimates that there are approximately 5,000–8,000 rare diseases in the world. The challenge around rare diseases is that the comparatively small numbers of people affected by them translate into the neglect or the unavailability of diagnoses and treatment options. It can be explained by drawing on the notion of issue characteristics, famously developed by political scientist Theodore Lowi.[12] Lowi posited that different types of policies – e.g. regulatory, distributive or redistributive policies – give rise to different policymaking or decision-making processes through which distinct patterns of political and societal relationships and behaviours emerge. Just as the categories we use to describe characteristics that we hold – women and men, adults and children, Koreans or Croatians – we can use categories to describe characteristics that policies or policy fields hold. For example, the depiction of European healthcare and welfare systems as solidaristic has arisen from their embeddedness in redistributive policies that allow the state to redistribute taxes and other welfare contributions in the pursuit of policy goals. Different types of policies give rise to different forms of state action, but also to different types of public participation, or even political controversy and contestation. The latter is what we frequently observe when a change in redistributive policies is suggested. Following Lowi's rationale, the key to understanding patterns of behaviours, in this case the lack of attention given to rare diseases, is to identify the characteristics of the issues to which they give rise. The more complex the regulatory or policy area, the more difficult it is to develop policy solutions.

The issue characteristics for rare diseases are complex. We know relatively little about the factors and processes that underlie these diseases. This stems from a lack of basic research into rare diseases[13] which is mostly due to a lack of available funding for research that a relatively small number of people suffer from. From a public policy perspective, the question of how and if to prioritise research for rare diseases is an intrinsically complex issue because of the low

---

[10] B. Prainsack, 'Research for Personalised Medicine: Time for Solidarity', (2017) *Medicine and Law*, 36(1), 87–98.

[11] European Commission, 'Rare Diseases', (European Commission, 2018), www.ec.europa.eu/health/non_communic able_diseases/rare_diseases_en

[12] T. J. Lowi, 'American Business, Public Policy, Case-Studies and Political Theory', (1964) *World Politics*, 16(4), 677–715.

[13] EURORDIS-Rare Diseases Europe, 'EURORDIS' Position on Rare Disease Research', (EURORDIS, 2010), www .eurordis.org/sites/default/files/EURORDIS_Rapport_Research_2012.pdf

numbers of patients and the high costs for research and treatment. It begs the (redistributive) policy question how spending a large proportion of overall research or healthcare budgets on a few patients can be justified if the opportunity costs are such that other patients may lose out as a result. The low patient numbers also result in difficulties in the design of clinical trials that meet the evidentiary hurdles of most regulatory agencies in Europe.[14]

Solidarity offers a lens through which these difficult questions surrounding research on rare diseases can be reframed. Patients suffering from rare diseases are characteristically vulnerable (please see Rogers' Chapter 1 in this volume for more detail on the concept of vulnerability). Their vulnerability results from the severity and the chronicity of their conditions, the inadequate access to appropriate diagnoses and treatment options, societal isolation and a lack of representation of their interests.[15] Coming back to the importance of Tier 3 solidarity in HRR (the institutional and legal level), the solidaristic principles upon which healthcare systems in Europe rest suggest a duty to care for society's most vulnerable members, which patients with rare diseases undoubtedly are. Policies or regulations to support research and service provision for patients with rare diseases can therefore be viewed as solidaristic practices.

However, despite initiatives such as the introduction of Regulation (EC) 141/2000 on orphan medical products, access to adequate services and research for patients is still falling short of expectations. Following Lowi's approach, as outlined above, we can observe that the more complicated the issues to which a regulatory or policy area give rise, the less policymakers are inclined to act because of the perceived lack of policy options. This might also explain why the challenges around fostering research activity on rare diseases are predominantly framed as a regulatory policy problem rather than a distributive or redistributive one. Interestingly, the perceived lack of policy options and responses corresponds with a flourishing of solidaristic practices below the level of public policy that span borders and countries at the EU level. For example, there seems to be an emerging recognition of 'similarity in a relevant respect' among EU countries in the sense that the issue characteristics of rare diseases are such that no country can stem the challenge of protecting vulnerable patients suffering from rare diseases on its own. Here, Tier 2 solidarity does not just apply to the level of interaction and collaboration among patient groups, but also to the level of cooperation between nation states. The similarity is the recognition that all countries face the same challenge in finding adequate research and treatments on rare diseases – the policy problem – and that countries are similar in their failure to find policy solutions. This can lead to the fostering of solidaristic practices such as the EC's advocacy for a European Platform on Rare Diseases Registration that would bring together patient registries and databases to encourage and simplify clinical research in the area.

An unresolved question in the application of a solidarity-based approach to the field of HRR is the role of industry, especially in fostering or hindering solidaristic practices. It is frequently argued that pharmaceutical manufacturers do not invest enough resources into the research and development of rare diseases and orphan drugs because the small patient numbers lead to a low return on investment (RoI).[16] The response of EU member states has been to create incentives through policy instruments such as fee waivers for regulatory procedures or a 10-year market exclusivity for authorised products.[17] The introduction of such measures in the Regulation (EC) 141/2000 on orphan medical products has increased the number of orphan drugs being

---

[14] Ibid.
[15] Ibid.
[16] E.g. ibid.
[17] European Commission, 'Rare Diseases'.

authorised. But is it also a sign that pharmaceutical industries are engaging in solidaristic practices to benefit some of the most vulnerable patients?

We argue that it is not. We must assume that pharmaceutical companies are motivated by the incentives offered through this regulation rather than a recognition of similarity with entities that seek to promote public benefit, or with people suffering from illness. The perception that some people, as taxpayers or patients, are expected to contribute to supporting others who suffer from rare diseases, while some corporate actors do the bare minimum required by law, may have a significant negative effect on the people of other actors to contribute. This may be exacerbated by the payment by corporations of hefty dividends to their shareholders. Institutionalised solidarity requires some level of reciprocity – the understanding that each actor makes a contribution adequate to their nature and ability. As a result, if large multinational companies are seen to get away with 'picking the raisins' this is a serious impediment to solidarity.

In a field that is still very dependent on the investment of pharmaceutical companies into drug research, resolving this challenge of asymmetry is not easy to rectify in the short term. Its solution would require legislation that forces companies to cut their profits and support rare disease patients in more significant ways than they are doing at present. A for-profit company cannot reasonably be expected to be motivated by the desire to help people; it is to be expected, and justified, that they put profits first. This is why it is the role and responsibility of legislators to ensure that companies are contributing their fair share. This is not only a necessity for moral and ethical reasons, but also to avoid the hollowing out of solidaristic practices among people who may, as argued above, be deterred by the expectation to accept costs to help others, while others are making huge profits.

The concept of solidarity can and should be used to reframe the regulation of research on orphan drugs from a market failure problem that requires financial incentives, to a societal problem that requires more than market measures. This will require a reframing of the issue as a redistributive policy problem rather than a purely regulatory one, in the hope that this will instigate political debates, as well as patient and public participation that would help bring the challenges of research on rare diseases and orphan diseases more to the centre of the policy process. Using the concept of solidarity to help reframe the policy issue has the potential to draw it out of the comparatively confined policy spaces it currently occupies. This helps to illuminate its political and public salience. The joined-up working of patient groups for rare diseases and the mutual efforts of EU member states – also as regulators that impose rules of fair play on pharmaceutical companies – are needed to facilitate – and where they already exist, stabilise – solidaristic practices. To make these practices more powerful and meaningful, priority-setting mechanisms for the prioritisation of research funding need to be developed,[18] and more public money should be invested, especially into basic research, in an effort to decrease the dependence on the pharmaceutical industry.

5.6 CONCLUSION

In this chapter, we have used research on rare diseases and orphan drugs to highlight the application of solidarity to HRR. It is an example of a space where solidaristic practices are already taking place, but also illustrates that there is room for improvement. Solidarity is an integral part of health research, and it is enacted every time a person takes part in a clinical trial

---

[18] C. Gericke et al., 'Ethical Issues in Funding Orphan Drug Research and Development', (2005) *Journal of Medical Ethics*, 31(3), 164–168.

or other research because they want to support the creation of public benefits. Regulation is important to ensure that research is carried out in an ethical manner, but, equally, it is important that decision-makers who define the regulatory spaces for HRR recognise the need to support solidaristic practices rather than undermine them through overly cumbersome bureaucratic hurdles to enrol in research.

ACKNOWLEDGEMENTS

We are grateful to Alena Buyx for helpful discussions on an earlier version of this manuscript. The usual disclaimer applies.

6

The Public Interest

Annie Sorbie

6.1 INTRODUCTION

This chapter provides an introduction to the concept of 'the public interest' in health research regulation (HRR). It considers two key ways that the public interest is constructed in HRR: namely as a legal device and through empirical evidence of the views of publics. To appreciate the scope of this concept, the public interest is set in its broader context, i.e. beyond HRR, highlighting that, historically, it has been a contested concept that is difficult to define in the abstract. Next, the public interest is situated within HRR, paying attention first to how it features in the HRR legal landscape and then how this is constructed through the views of publics (with specific reference to the use of identifiable health data for research). Both conceptualisations are analysed with reference to the key challenges and opportunities that they present before a holistic concept of the public interest in HRR is proposed and consideration given to how this may be operationalised in practice.

6.2 THE PUBLIC INTEREST: A CONTESTED CONCEPT

Although the public interest is fully embedded in HRR, it is by no means exclusive to this context. The following brief consideration of wider perspectives on this contested concept point to persistent debates not only on what the public interest 'is', but also to tensions as to how this concept should be understood. Appeals have been made variously to the values it invokes, the process it requires, and/or the views of (some or all) of 'society' at large that it reflects.[1]

Political and social scientists, philosophers and lawyers, among other disciplines, have contemplated this elusive concept without reaching consensus on its meaning or usefulness. During a period of scholarly interest in the public interest in post-World War II America, it was both lauded as 'a central concept of a civilised polity'[2] and dismissed as a concept so vague and ambiguous that it is no more than a rhetorical device.[3] This ambivalence can be seen in Sorauf's work in which, despite his scepticism, he initially concedes a 'modest conception' of the public interest that is rooted in 'our interest in the democratic method and its settlement of conflict by

---

[1] A. Sorbie, 'Sharing Confidential Health Data for Research Purposes in the UK: Where Are 'Publics' in the Public Interest?', (2020) *Evidence & Policy*, 16(2), 249–265

[2] S. Bailey, 'The Public Interest: Some Operational Dilemmas' in C. Friedrich (ed.), *Nomos V: The Public Interest* (New York: Atherton Press, 1962), pp. 96–106.

[3] G. Schubert, *The Public Interest: A Critique of the Theory of a Political Concept* (Glencoe, Illinois: Free Press, 1960).

orderly rules and procedures'.[4] He recognises too the potential function of the public interest as a 'hair shirt' that serves as 'an uncomfortable and persistent reminder of the unorganized and unrepresented (or underrepresented) interests of politics'.[5] Over time, however, his position hardens and becomes more negative. He later posits that the public interest promotes 'oversimplification', as it purports to "solve" the dilemmas of … pluralism'.[6] Turning to the regulatory role of the public interest, Feintuck also points to a continued reluctance to define the public interest beyond what 'will vary according to time, place and the specific values held by a particular society'.[7] He characterises the public interest as an 'empty vessel' and argues for an account that looks 'to the fundamental value laden, democratic imperatives that underlie society: human dignity, parity of esteem, and the ability to participate actively in society'.[8]

Whether the public interest is best understood modestly as a procedural mechanism, ambitiously as protecting fundamental values in society including those that may otherwise be overlooked, or in utilitarian terms as the views of the majority, there is little doubt that this is a contested concept that is 'much used but ill defined'.[9] This chapter proposes that while there is need for further conceptual clarity here, there is also value to be found in such contestation and flexibility.

6.3 APPEALS TO THE PUBLIC INTEREST IN HRR

In HRR, the concept of the public interest is embedded in law and in policy, often as a counterpoint to individual interests. In medical research involving human subjects – including research on identifiable human tissue and data – consideration of the relationship between individual and public interests can be traced back to the original Declaration of Helsinki.[10] More recently, the legal mandate of the Health Research Authority (HRA) in the United Kingdom, as set out in the Care Act 2014, prescribes twin objectives to protect and promote the interests of both individual participants (and potential participants) and the interests of wider publics in safe and ethical health and social care research.[11]

However, reflecting the broader literature on public interest, Taylor notes in his consideration of genetic data and the law, that the public interest remains a 'notoriously uncertain idea'.[12] This chapter proceeds with an account of two key ways in which the concept of the public interest appears in HRR (with a focus on the use of identifiable health data for research), as constructed in law and through publics' views. It considers the key challenges and opportunities presented by

---

[4] F. J. Sorauf, 'The Public Interest Reconsidered', (1957) *The Journal of Politics*, 19(4), 616–639, 633.

[5] Ibid., 639.

[6] F. Sorauf, 'The Conceptual Muddle' Dilemmas' in C. Friedrich (ed.), *Nomos V: The Public Interest* (New York: Atherton Press, 1962), pp. 183–190, p. 189.

[7] M. Feintuck, *'The Public Interest' in Regulation* (Oxford University Press, 2004), p. 34, quoting A. Ogus, *Regulation: Legal Form and Economic Theory* (Oxford: Clarendon, 1989), p. 2.

[8] Feintuck, *'The Public Interest'*, p. 57.

[9] J. Bell, 'Public Interest: Policy or Principle?' in R. Brownsword (ed.), *Law and the Public Interest: Proceedings of the 1992 ALSP Conference* (Stuttgart: Franz Steiner Verlag, 1993) pp. 27–36.

[10] J. R. Williams, 'The Declaration of Helsinki and Public Health', (2008) *Bulletin of the World Health Organization*, 86(8), 650–652.

[11] Care Act 2014, Section 110(2) states: (2) The main objective of the HRA in exercising its functions is – (a) to protect participants and potential participants in health or social care research and the general public by encouraging research that is safe and ethical, and (b) to promote the interests of those participants and potential participants and the general public by facilitating the conduct of research that is safe and ethical (including by promoting transparency in research).

[12] M. Taylor, *Genetic Data and the Law: A Critical Perspective on Privacy Protection* (Cambridge University Press, 2012).

the public interest in each framing. Having identified the benefits and shortcomings of each, a holistic concept of the public interest is proposed, the relationship between the public interest as constructed within and beyond the law is examined, and consideration is given to how, in a more concrete way, public interest might be operationalised in HRR practice.

6.4 THE PUBLIC INTEREST AS LEGAL DEVICE

When health research is conducted on identifiable personal data, the public interest is a striking feature of the legal landscape. For example, in the realm of data protection, the public interest forms one of the routes to the lawful processing of personal data in health and social care research. Thus, the General Data Protection Regulation[13] (GDPR) provides a lawful basis to process personal data where this is a 'task in the public interest'.[14] Health Research Authority (HRA) guidance confirms that, for the purposes of the GDPR, this is the appropriate legal basis that should be used by public authorities, such as NHS bodies or universities, in order to process data for health and social care research.[15] In UK law, the Data Protection Act 2018[16] (DPA 2018) purports to add further detail to the interpretation of 'a task in the public interest', although concerns have been raised that the drafting of this legislation does little to add clarity to how this concept should be understood in practice.[17] A late addition to the Explanatory Note to the Act indicates, by way of an example, that 'a university undertaking processing of personal data necessary for medical research purposes in the public interest should be able to rely on [a task in the public interest]'[18], thus providing some guidance on the *context*, if not the *content*, of the public interest in these circumstances.

Two other prominent features of the health data legal landscape are: (i) the common law duty of confidentiality and (ii) the legislative regime which established the predecessor body to the HRA's Confidentiality Advisory Group (CAG). The common law duty of confidentiality provides that where confidential information is imparted to another person, in circumstances giving rise to an obligation of confidentiality, this must not be disclosed without consent or justification.[19] One such justification is where disclosure is 'in the public interest'. This duty, and its exceptions, apply not only in the context of the traditional doctor/patient relationship, but also where it is proposed that the information in question may be used for purposes beyond direct care, such as for health or social care research. The interpretation of this duty of confidentiality (and, importantly for this chapter, the meaning of the public interest) has

[13] Regulation (EU) 2016/679 of the European Parliament and of the Council of 27 April 2016 on the protection of natural persons with regard to the processing of personal data and on the free movement of such data, and repealing Directive 95/46/EC.

[14] Article 6(1)(e).

[15] Consent retains its ethical significance and legal importance under wider legal frameworks, but it is explicitly stated that: 'For the purposes of the GDPR, the legal basis for processing data for health and social care research should NOT be consent. This means that requirements in the GDPR relating to consent do NOT apply to health and care research'. Health Research Authority, 'Consent in research', (NHS Health Research Authority, 2018), www.hra.nhs.uk/planning-and-improving-research/policies-standards-legislation/data-protection-and-information-governance/gdpr-guidance/what-law-says/consent-research/.

[16] Data Protection Act 2018, Section 8.

[17] Wellcome, 'Data Protection Bill – Second Reading Briefing for the House of Lords by the Wellcome Trust', (Wellcome, 10 October 2017), www.wellcome.ac.uk/sites/default/files/data-protection-bill-second-reading.pdf; 'Data Protection Bill – Lords' Committee Stage Day 1', www.wellcome.ac.uk/sites/default/files/data-protection-bill-lords-committee.pdf

[18] Data Protection Act 2018, Explanatory note to Section 8.

[19] The essential elements were established in *Coco v. A N Clark (Engineers) Ltd* [1969] RPC 41.

emerged as a result of decisions made on the facts of cases that have come before the courts. These judgements indicate, for example, that there is not only a personal interest in an individual's confidentiality being maintained, but also a wider public interest in doing so in order that patients (in general) are not discouraged from consulting with healthcare practitioners.[20] Case law, in relation to whether disclosure of deceased patients' records to a public inquiry was in the public interest,[21] recognises that the public interest (which was distinguished from 'what the public found interesting')[22] is multifaceted and can encompass both individual and collective interests. These include interests in: disclosure, maintaining the patient's confidentiality and maintaining confidence in the institutions under investigation.[23]

As with the legislation, there is no fixed definition of the public interest in case law; where this lies must be decided on the individual facts of each scenario. This perception of a lack of certainty led to concerns from some clinicians that routine activities, such as providing information to registries that collect and analyse data on specific diseases, might be vulnerable to challenge in the absence of specific consent.[24] These worries about the legality of such practices, among other matters, led to the enactment of legislation in England and Wales in 2001 that forms another key feature of the data sharing landscape, namely the establishment of the predecessor to the CAG. In summary, this legislation allows the Secretary of State for Health to make regulations to explicitly 'set aside' the common law duty of confidentiality for defined medical purposes, including medical research, where this is 'in the interests of improving patient care, or in the public interest'. These powers are now found in Section 251 of the NHS Act 2006 (as enabled by the Health Service (Control of Patient Information) Regulations 2002) and referred to colloquially as 's251 support'. In sum: where seeking consent is neither possible nor practical, researchers can obtain s251 support to use confidential patient information for medical research by make an application to the HRA's CAG. The effect of such an application is that, if granted, the researcher need not be concerned whether (in the admittedly unlikely event of litigation) a court would agree that their use of identifiable patient information without consent was indeed in the public interest.

In common with the broader literature on the public interest, the preceding whistle-stop tour of the public interest in law reveals anxieties around how this concept is interpreted in practice. It also speaks to the strengths and limitations of a narrow legal construction of the public interest decided on a case-by-case basis, but for which precedents can be established over time. These are explored further in the passages that follow.

We return first to Taylor's description of the public interest as a 'notoriously uncertain idea'.[25] It is of note that Parliamentary debate on the DPA 2018[26] on this topic resurrected many of the concerns around the public interest that had arisen some fifteen years previously, at the time of the promulgation of the CAG regime. These included the potential for the public interest to be interpreted widely to deliver 'sweeping powers'.[27] Nonetheless the CAG regime, which was first

---

[20] W v. *Egdell* [1989] EWCA Civ 13.
[21] *Lewis v. Secretary of State for Health* [2008] EWHC 2196, Paragraph 58.
[22] Ibid., Paragraph 59.
[23] Ibid., Paragraph 58.
[24] M. Coleman et al., 'Confidentiality and the Public Interest in Medical Research – Will We Ever Get It Right?', (2003) *Clinical Medicine*, 3(3), 219–228.
[25] Taylor, *Genetic Data*, p. 29
[26] For example, see *Hansard*, HL, vol. 785, col. 146, 10 October 2017; *Hansard*, HL, vol. 785, col. 1236, 30 October.
[27] These included the wide scope of the public interest provisions that provided the Secretary of State with 'sweeping powers to collect confidential data on named patients without consent' (*Hansard*, HC, vol. 622, col. 997, 26 February 2001, Earl Howe).

proposed as a temporary solution as the NHS geared up to apply a 'consent or anonymise' binary to its use of health data, has become an example of good governance and established itself as part of the data sharing landscape.[28] This can be attributed, in part, to a growing recognition from stakeholders in HRR – including researchers and publics – that consent is not necessarily the 'magic bullet' to legitimise HRR governance that it might once have been presumed to be. For example, Wellcome's research, as commissioned from Ipsos MORI, on public attitudes to commercial access to health data for research purposes found that, when considering data uses, 'a strong case for public benefit is the most important factor for many people: without it, data use by any organisation is rarely acceptable'.[29] This tends to suggest that while concerns about the uncertainty of the application of the public interest in HRR persist, it is a concept that also, in some ways, benefits from its inherent flexibility and its ability to adapt to changing interests over time.

A further critique that arises from this legal construction of the public interest is that this looks inwards to derive its legitimacy from its institutional origins and is disconnected from actual publics' views. For example, in the case of legislation – such as the DPA 2018 and the legislation underpinning the CAG regime – legitimacy comes from Parliament. Notwithstanding, the public interest in (legal) text tells us little about its context. Even when amplified by its Explanatory Note, the DPA 2018 does not elaborate on the legitimate content of the public interest in HRR.

Turning to case law, the public interest is conceptualised by the courts on the facts of each case, following precedents in previous decisions. This inward-looking legal construction of the public interest is consistent with the long established 'intellectual tradition'[30] within the law of invoking fictional persons to provide a barometer of what 'reasonable' members of the public would expect in any given situation. The paradigm is the fictional 'man on the Clapham Omnibus',[31] who in English law is deployed to represent the reasonable person. Elsewhere in the law, other fictional reference points include the 'right-thinking member of society' (in defamation law) or even the 'officious bystander' (in contract law).[32] It has thus been confirmed by the Supreme Court that: 'The spokesman of the fair and reasonable man, who represents after all no more than the anthropomorphic conception of justice, is and must be the court itself'.[33] This underlines why the law historically has not been centrally concerned with empirical evidence of the views of actual members of the public when it deploys the legal notion of the public interest in civil law cases.

However, this legal self-referential conception of the public interest in HRR is increasingly under pressure, as exemplified by the high-profile failure of care.data. As described more fully in this volume by Burgess (Chapter 25), this was an NHS England initiative that sought to make patient data available for specified purposes, including audit and research, in a format that was stripped of identifiable information. However, following widespread concerns about the scheme – including around its transparency and oversight – the programme closed in 2016.[34]

[28] G. Laurie et al., 'On Moving Targets and Magic Bullets: Can the UK Lead the Way with Responsible Data Linkage for Health Research?', (2015) *International Journal of Medical Informatics*, 84(11), 933–940.

[29] Wellcome, 'Public Attitudes to Commercial Access to Health Data', p. 1, referring to Ipsos MORI, 'The One-Way Mirror: Public Attitudes to Commercial Access to Health Data', (Wellcome Trust, 2016), www.wellcome.ac.uk/sites/default/files/public-attitudes-to-commercial-access-to-health-data-summary-wellcome-mar16.pdf

[30] *Healthcare at Home Limited (Appellant)* v. *The Common Services Agency (Respondent) (Scotland)* [2014], par. 2.

[31] Ibid., para 1.

[32] Ibid., para 1.

[33] Ibid., para 2.

[34] M. Taylor, 'Information Governance as a Force for Good? Lessons to be Learnt from care.data', (2014) *SCRIPTed*, 11(1), 1–8.

Here, a legal framework was in place to facilitate data sharing but, as argued by Carter et al.,[35] the social licence to do so was not. This failure underlines the message that 'legal authority does not necessarily command social legitimacy'.[36] It follows that where the law alone is unable to fully legitimise and animate the public interest, something else must fill this void. The following section suggests that a richer relationship between this legal concept and the views of publics could be a worthy candidate.

6.5 THE PUBLIC INTEREST AS THE VIEWS OF ACTUAL PUBLICS

The potential benefits of responsible access to health data by researchers, as well as the perils of getting this wrong, have led to a renewed focus on the public acceptability of data sharing initiatives and a growing body of literature that explores public attitudes towards sharing health data for research purposes.[37] Aitken et al. note the desire of stakeholders in HRR to optimise the use of existing data in health research and: 'the recognition of the importance of ensuring that data uses align with public interests or preferences'.[38] This commitment to using patient data responsibly is shared by funders, as exemplified by Wellcome's 'Understanding Patient Data' initiative, which works to champion responsible uses of data and improve stakeholder engagement around how and why data is used for care and research.[39]

Consider too the call in HRR for more and better public and patient involvement (PPI). The National Institute for Health Research (NIHR) recently issued 'Standards for Public Involvement in Research', which provide 'a framework for reflecting on and improving the purpose, quality and consistency of public involvement in research'.[40] In particular, Standard 6 on Governance states that '[w]e involve the public in our governance and leadership so that our decisions promote and protect the public interest'. Here, the role of publics is positioned not only as shaping and supporting research, but also as a means of legitimising HRR and grounding the broader public interest.

This approach has the benefit of being anchored to actual publics' views, something that is lacking from the narrow legal account set out above. In this way, it has the potential to provide at least some of the social legitimacy that was lacking in care.data. However, public engagement activities also attract criticisms of exclusivity and tokenism,[41] raising 'questions of representativeness, articulation, impacts and outcomes'.[42] Thus, to simply equate these outputs with 'the public interest' more broadly also runs the risk of reinforcing underlying inequalities in the delivery of a majoritarian account of the concept. Reports of instances of 'personal lobbying by

[35] P. Carter et al., 'The Social Licence for Research: Why care.data Ran into Trouble', (2015) *Journal of Medical Ethics*, 41(5), 404–409.
[36] Ibid., 408
[37] M. Aitken et al., 'Moving from Trust to Trustworthiness: Experiences of Public Engagement in the Scottish Health Informatics Programme', (2016) *Science and Public Policy*, 1–11; M. Aitken et al., 'Public Responses to the Sharing and Linkage of Health Data for Research Purposes: A Systematic Review and Thematic Synthesis of Qualitative Studies', (2016) *BMC Medical Ethics*, 17(73), 1–24; M. Aitken et al., 'Public Preferences Regarding Data Linkage for Health Research: A Discrete Choice Experiment', (2018) *International Journal of Population Data Science*, 3(11), 1–13.
[38] Aitken et al., 'Public Responses', 2
[39] 'About Us', (Understanding Patient Data), www.understandingpatientdata.org.uk/about-us.
[40] NIHR, 'Standards for Public Involvement in Research', (NIHR, 2019), www.invo.org.uk/posttypepublication/national-standards-for-public-involvement/
[41] J. Ocloo, and R. Matthews, 'From Tokenism to Empowerment: Progressive Patient and Public Involvement in Healthcare Improvement', (2016) *BMJ Quality and Safety*, 25(8), 626–632.
[42] J. Stilgoe and S. Lock, 'Why Should We Promote Public Engagement with Science?', (2014) *Public Understanding of Science*, 23(1), 4–15.

volunteers for pet causes'[43] point to the dangers of 'assuming that the perspectives of a small number of involved patients necessarily reflect the perspectives of a larger patient community'.[44] Indeed, McCoy et al.'s analysis of the recent NIHR 'Standards for Public Involvement' suggests that 'it is simplistic to assume that including public representatives on governance and leadership bodies will necessarily promote the public interest'.[45] They highlight the likelihood that the interests of differing 'publics' will, in any event, diverge, and call for more attention to be paid to *who* is being asked to contribute, *at what stage* in a research project, and *for what purpose*.

This is not, of course, to discount the important contributions that can be made to shaping and delivering responsible HRR through the thoughtful involvement of patients and wider publics.[46] However, whereas it is advanced above that the law alone is not enough to legitimise the public interest, this analysis also suggests that an additive approach to publics' views in HRR is also insufficient to provide a lasting and justifiable account of this concept. Something more is required.

6.6 THE PUBLIC INTEREST: A HOLISTIC CONCEPT

Taken together, the preceding examples illustrate the prevalence of the public interest in HRR and how this concept may be constructed both through the law and through the views of publics. On the one hand, the tendency of the law to approach the public interest as a legal test draws the criticism that this narrow notion of what purports to be in the public interest is wholly disconnected from the views of publics and can lack social legitimacy. On the other, to claim that the public interest can simply be extrapolated from the outputs of public involvement work is equally problematic. Nonetheless, despite this disjuncture, common themes emerge and, in this section, two further contributions to the debate on the role of the public interest are offered. The first is a proposal for a holistic concept of the public interest that is able to account for a plurality of interests and views. The second is that, despite the apparent impasse, legal and empirical notions of the public interest are not mutually exclusive. It is argued that these do bear upon one another and that if the public interest is to be effectively deployed in HRR, this relationship should be both acknowledged and made more overt.

The first proposal is to recognise that both the legal and empirical constructions of the public interest call for a conception of the public interest that is able to account for a range of diverse interests. In law, the potential for this approach is evident in an arc of case law that emphasises that the public interest is a multifaceted and flexible concept that is able to account for both individual and collective interests, including wider publics and institutional stakeholders. Similarly, the analysis above suggests that the value of public involvement is optimised when attention is paid to the multiple interests of differing patients and publics, including who is being asked to contribute, when, and for what purpose. This also tracks a move in HRR literature away from a narrow account of the public interest that pits individual interests against collective benefits. For example, Rid describes this 'pluralistic conception of public interest' as an account that is capable of recognising that multiple interests are in play.[47] Taylor's work also proposes

[43] M. McCoy et al., 'National Standards for Public Involvement in Research: Missing the Forest for the Trees', (2018) *Journal of Medical Ethics*, 44(12), 801–804, p. 802, quoting A. Prince et al., 'Patient and Public Involvement in the Design of Clinical Trials: An Overview of Systematic Reviews', (2018) *Journal of Evaluation in Clinical Practice*, 24(1), 240–253.
[44] McCoy et al., 'National Standards', 802.
[45] Ibid., 803
[46] See Burgess, Chapter 25, and Cunningham-Burley and Aitken, Chapter 11, of this volume.
[47] See A. Rid in A. Sorbie, 'Conference Report: Liminal Spaces Symposium at IAB 2016: What Does It Mean to Regulate in the Public Interest?', (2016) *SCRIPTed*, 13(3), 374–381.

that individual and public interests need not be balanced against one another, but rather that the need for legitimacy requires that each should account for each other.[48] Together, this forms the basis for a holistic concept of the public interest in HRR that is able to account for multiple interests and views. This approach does not, in the words of Sorauf, aim to 'solve' pluralism. Quite the opposite: it embraces the messy realities and subjectivities, both of the law, as broadly conceived, and of outputs from public involvement activities.[49]

The second contribution is to suggest that, despite the messiness, these accounts are not mutually exclusive and do, in fact, bear upon one another (though this relationship is far from clear). For example, I suggested earlier that shifting public views on health data sharing (and a move away from a 'consent or anonymise' binary) have contributed to the longevity of the CAG, which was originally proposed only as a temporary measure. Similarly, I have referred to how lobbying from the HRR community during the promulgation of the DPA 2018 led to an amendment of the Explanatory Note to clarify that 'a task in the public interest' is an appropriate route for public authorities such as universities to use when processing health data for research purposes. Lessons from care.data exemplify the importance of 'social licence' to the success of otherwise legal data sharing initiatives. In turn, there is an on-going need for deeper understanding of public acceptability to realise the potential of new and novel uses of health data.[50] Given the impetus to deliver clear and transparent governance of health data, it is proposed that this this relationship ought to be both acknowledged and made more overt, in order that it may be exposed to debate in HRR. Three concrete suggestions are made in this regard. The first is that the public interest, along with other concepts that operate at the intersection of public involvement and governance in HRR, should be examined to identify their potential to bridge the divide between the outputs from public engagement and the implementation of these in practice. The second is that initiatives such as CAG, where there is 'evidence' of the public interest being given effect to facilitate responsible HRR, should be further mobilised. The third is that instances where appeals to the public interest are made in HRR should be captured and articulated publicly, in order to promote transparency and accountability around how and why these have (or indeed have not) been justified.

6.7 CONCLUDING REMARKS

This chapter advocates for a holistic conception of the public interest, where interests are accounted for, rather than polarised. HRR governance has moved on from a 'consent or anonymise' binary and now needs novel and bold mechanisms that do not seek to over-play the role of legal mechanisms, nor suggest that public views alone can deliver good governance solutions. While the concept of the public interest remains contested and highly contextual, there is an increasing drive towards maximising the potential of this embedded concept in order to deliver a step-change in HRR.

---

[48] See M. Taylor in A. Sorbie, 'Conference Report', and Taylor and Whitton, Chapter 24 of this volume.
[49] Although outside the scope if this chapter, this holistic model also calls for scrutiny of the values in which it is grounded. Candidates may include, e.g. citizenship (Feintuck, 'The Public Interest') or solidarity (Kieslich and Prainsack, Chapter 5 of this volume).
[50] For example, health data, such as that held by the NHS, may be of 'immense value' to researchers developing artificial intelligence for use in healthcare settings. However, the question of how this value is realised remains 'a crucial one to get right because of the implications for public confidence' (Select Committee on Artificial Intelligence, 'AI in the UK: ready, willing and able?', (House of Lords, 2018), www.publications.parliament.uk/pa/ld201719/ldselect/ldai/100/100.pdf.

7

Privacy

David Townend

7.1 INTRODUCTION: THE MODERN DIFFICULTY[1]

Privacy is a well-established element of the governance and narrative of modern society. In research, it is a mainstay of good and best practice; major research initiatives all speak of safeguarding participants' rights and ensuring 'privacy protecting' processing of personal data. However, while privacy protection is pervasive in modern society and is at the conceptual heart of human rights, it remains nebulous in character. For researchers who engage with people in their studies, the need to respect privacy is obvious, yet how to do so is less so. This chapter offers first an explanation of why privacy is a difficult concept to express, how the law approaches the concept and how it might be explored as a broader normative concept that can be operationalised by researchers. In that wider scheme, I show how individuals respond to the same privacy situation in different ways – that we have a range of privacy sensitivities. I think about four privacy elements in the law: human rights, privacy in legal theory, personal data protection and consent. Finally, I consider how law participates in the broader normative understanding of property as the private life lived in society.

7.2 PRIVACY AS A NORMATIVE DIFFICULTY

A good starting point is to ask: what do we mean when we talk about 'privacy'? It would be difficult for a modern research project to suggest that it was not 'privacy respecting' or 'privacy preserving'. However, the concept is somewhat ill-defined, and that claim to be privacy respecting or preserving might, in reality, add little to the protection of individuals. In part, this problem stems from the colloquial, cultural aspect of the concept: we each have our own idea of what constitutes our privacy – our private space.

Imagine setting up a new data sharing project. You hypothesise that linking data that different institutions already gather could address a modern health problem – say, the growth of obesity and type 2 diabetes. Such data, current and historical, could be used by machine learning to create and continuously revise algorithms to help identify and 'nudge' those at risk of developing the condition or disease. The already-gathered data could be from general practitioners and hospitals, supermarkets and banks, gym memberships, and health and lifestyle apps on smart phones, watches and other 'wearables'. But how would individuals' privacy be protected within such a project? Many will

[1] I am grateful to Graeme Laurie, Annie Sorbie and all the editors and colleagues who commented on this chapter. Errors are mine.

be uneasy about such data being stored in the first place, let alone retaining it and linking it for this purpose. Many will see that there might be a benefit, but would want to be convinced of technical safeguards before opting into such a project. Many will be happy, having 'nothing to hide' and seeing the benefits for their health through such an app. Some would see this initiative as socially desirable, as part, perhaps, of one's general duty and the basis of personalised medicine, so that such processing would be a compulsory part of registration for healthcare; an in-kind payment to the healthcare system alongside financial payments, necessary for the continued development of modern healthcare that is a general societal and personal good.

Our difficulty is that each one of the people taking these different positions would see their response as a 'privacy preserving' stance.[2] As explored elsewhere in this volume, this observation underlines the diversity of 'publics' and their views (see Aitken and Cunningham Burley, Chapter 11, and Burgess, Chapter 25, in this volume). Under the label 'privacy' there is a wide spectrum of conceptualisations, from the enthusiastic adopter and compulsion for all, through allowing people to opt-out, generally leaving participation to opting in, to wanting nothing to do with such projects. How then can a researcher frame a 'privacy' policy for their research? Are we creating the problem by using the term 'privacy' informally and colloquially? Does the law provide a definition of the term that avoids or militates against the problem?

7.3 PRIVACY AS A HUMAN RIGHT

A logical starting point might be human rights law. Privacy and the right to respect for private life is enshrined in human rights law. Unfortunately, it does not give much assistance in the definition of those rights. Two examples show the common problem clearly.[3] Article 12 of the Universal Declaration of Human Rights states:

> No one shall be subjected to arbitrary interference with his privacy, family, home or correspondence, nor to attacks upon his honour and reputation. Everyone has the right to the protection of the law against such interference or attacks.[4]

Article 8 of the European Convention on Human Rights creates the right in this way:

1. Everyone has the right to respect for his private and family life, his home and his correspondence.
2. There shall be no interference by a public authority with the exercise of this right except such as is in accordance with the law and is necessary in a democratic society in the interests of national security, public safety or the economic well-being of the country, for the prevention of disorder or crime, for the protection of health or morals, or for the protection of the rights and freedoms of others.[5]

[2] See the range of sensitivities expressed in public opinion surveys about privacy. For example, the Eurobarometers on data protection, Eurobarometers numbers 147 and 196 (2003), 225 and 226 (2008), 359 (2011), and 431 (2015), and on biotechnology, Eurobarometers numbers 61 (1991), 80 (1993), 108 (1997), 134 (2000), 177 (2003), 244b (2006), and 341 (2010), all available at 'Public Opinion', (European Union), www.ec.europa.eu/commfrontoffice/publicopinion/index.cfm.
  For a discussion of a broader literature, see D. Townend et al.,'Privacy Interests in Biobanking: A Preliminary View on a European Perspective' in J. Kaye and M. Stranger (eds), *Principles and Practice in Biobanking Governance* (Farnham: Ashgate Publishing Ltd., 2009), pp. 137–159.
[3] See also, Articles 7, 8 and 52 of the European Union, Charter of Fundamental Rights of the European Union, 26 October 2012, 2012/C 326/02.
[4] UN General Assembly, 'Universal Declaration of Human Rights', 10 December 1948, 217 A (III).
[5] Council of Europe, European Convention for the Protection of Human Rights and Fundamental Freedoms, as amended by Protocols Nos 11 and 14, 4 November 1950, ETS 5.

Two observations can be made about these 'privacy' rights: (1) privacy is not an absolute right, i.e. there are always exceptions and (2) 'privacy' and 'respect for private life' require a great deal of further definition to make them operational. As to the first observation, the rights are held in relation to the competing rights of others: a right against 'arbitrary interference' and 'no interference ... except such'. The concepts of privacy in human rights legislation acknowledge that the rights are held in balance between members of society; privacy is not absolute, because on occasion one has to give way to the needs of others.

As to the content of privacy – and reflecting the broad conceptualisations in the research project example above – we see that what is available from the human right to privacy is international recognition of a space where an individual can exist, free from the demands of others; there is a normative standard that recognises that people must be respected as individuals.

The European Court of Human Rights has ruled extensively on the human right to respect for private life, and a line of caselaw has been created. This produces a canon of decisions where particular disputes have been settled where the particular parties have been unable to resolve their conflict between themselves. However, does that line of cases produce a *normative* definition of privacy, i.e. one that sits with and accommodates the range of sensitivities expressed above? I think not. A courtroom determination arguably defines a point on the range of sensitivities as 'privacy', pragmatically for the parties. Our problem comes when we try to use caselaw as indicative of more than how judges resolve conflicts between intractable parties when a privacy right is engaged. Does this mean the law adds little to the broader normative question about how we, as researchers, should respect the privacy of those with whom we engage in our work?

Two North American contributions could help to understand this. The first expression of the legal right to privacy is usually recognised as Warren and Brandeis' 1890 idea that we can agree that individuals have the right to be left alone.[6] Reading their paper today, it resonates with current concerns: technological developments and the increasing press prurience required a right to be 'left alone'. In the modern context of genetics, Allen proposes a broader typology of privacy: 'physical privacy', 'proprietary privacy', 'informational privacy', and 'decisional privacy'.[7] The first two, which seem strange 'privacies' today, are where Warren and Brandeis clearly see the Common Law as having reached in 1890. Law protects individuals' physical privacy through consent; private property law is equally well established. Warren and Brandeis identified 'informational privacy' and what might be described as 'reputational privacy' as the area where the law needed to develop in 1890. Allen pointed to the vast and compelling literature around the woman's right to choose in discussing the right to 'decisional privacy'. Legal theory, in part, responds to current privacy issues. Today, two major privacy issues in research are the protection of personal data protection and informed consent.

7.4 PRIVACY IN SPECIFIC LEGAL RESPONSES: PERSONAL DATA PROTECTION AND INFORMED CONSENT

The development of the automated processing of personal data has focussed privacy, at least in part, around 'informational privacy'.[8] The Organization for Economic Co-operation and Development *OECD Guidelines on the Protection of Privacy and Transborder Flows of*

---

[6] S. D. Warren and L. D. Brandeis, 'The Right to Privacy', (1890) *Harvard Law Review*, 4(5), 193–220.

[7] A. L. Allen, 'Genetic Privacy: Emerging Concepts and Values' in M. A. Rothstein (ed.) *Genetic Secrets: Protecting Privacy and Confidentiality in the Genetic Era* (New Haven: Yale University Press, 1997), pp. 31–60.

[8] See the tone, for example, of the Council of Europe website, where the focus is on privacy of personal data. 'Council of Europe Data Protection Website', (Council of Europe), www.coe.int/en/web/data-protection

Personal Data set an international standard in 1980 that remains at the core of data protection law.[9] The guidelines are transposed into regional and national laws.[10]

Data protection, as an expression of an area of privacy, seeks to balance a variety of interests in the processing of personal data within the non-absolute nature of privacy; the object of data protection is to create the legal conditions under which it is possible and appropriate to process personal data. Taking the European Union General Data Protection Regulation 2016/679 (GDPR) as an example, there are four elements in data protection law: data protection principles;[11] legal bases for processing personal data;[12] information that must be given to the data subject;[13] and, rights of the data subject.[14] Each element contains a balance of interests.

For stand-alone research with human participants directly contacted by the researcher, the route through the GDPR is clear. Security and data minimisation standards (i.e. only gathering, analysing and keeping data for such a time necessary for the purpose of the project) are clear; data subjects can be informed about the project fully, and data subjects rights can be respected. More complex data-sharing methodologies – perhaps the project envisaged in Section 7.2 above – are more difficult to negotiate through the GDPR. Are original consents valid for the new processing? Was the consent too broad for the new GDPR requirements? Might the processing be compatible with the original purpose for which the data were gathered? Could the new processing be in the public interest? How should data subjects be informed about the proposed new project? Each of these questions is open to debate in the GDPR. And the problem is how can the lack of definitional clarity in the rights be resolved in such a way that it accommodates all the positions on the spectrum of interests indicated in Section 7.2 above? One could say, law must produce a working definition and in a democracy, all differences cannot be accommodated, so some will be disappointed. However, the sensitivity of the data in the example above shows that there is a danger that those who are not within the working definition of privacy will be alienated from participating in key areas of social life, perhaps even avoiding interaction with, say, health research or medical services to their detriment.

If one of the current legal discussions is around informational privacy, the other is around decisional privacy. Informed consent is a legal mechanism to protect decisional privacy, not just in research, but across consumer society. The right of individual adults to make their own choices is largely unchallenged.[15] The choices must be free and informed. The question is, how informed must a choice be to qualify as a valid choice from an individual? This, in many modern biomedical research methodologies, is contested. A biobank, where data are gathered for the purpose of providing datasets for future, as yet undefined, research projects, depends on creating the biobank at the outset through a 'broad' consent. How though can an individual be said to give 'informed consent' if the purposes for which the consent is asked cannot be explained in detail? How can a consent that is 'for research' be specific enough to be an

[9] Organization for Economic Co-operation and Development, 'OECD Guidelines on the Protection of Privacy and Transborder Flows of Personal Data', (OECD, 1980). See also, OECD, 'The OECD Privacy Framework', (OECD, 2013).
[10] See, for example, Council of Europe Convention 108; European Union Directive 95/46/EC replaced by the General Data Protection Regulation 2016/679.
[11] GDPR, Article 5.
[12] GDPR, Articles 6 and 9.
[13] GDPR, Articles 13 and 14.
[14] GDPR, Articles 15–22.
[15] Although it is not an absolute right. See, for example, A. Smith, *The Theory of Moral Sentiments* (1759) or J. S. Mill, *On Liberty* (1859).

adequate safeguard of privacy interests? (See, for example, Kaye and Prictor, Chapter 10, in this volume for a specific discussion of consent in this context.)

The privacy issue is: what constitutes sufficient information upon which a participant can base her choice? Two conditions have to be satisfied: the quality of the information that will be made available; and (who determines) the amount of information that is necessary to underpin a decision. A non-specialist participant is not necessarily in a position to judge the first of these conditions. That is the role of independent review boards, standing as a proxy for the participant to assess the quality and trustworthiness of the scientific and methodological information that will be offered to the participant. For the second condition, what is sufficient information upon which to make a decision and who determines that decision, is a matter for the individual participant, and should not be seen as part of the role of the ethics committee, researcher or other body. The purpose of informed consent is to protect the individual participant from, essentially, paternalism – the usurping of the participant's free choice of whether or not to participate (unless the decision is palpably to the detriment of an individual who is not deemed competent to make a choice). Therefore, in the general case, it is inappropriate to remove the determination of what is sufficient information to inform the particular person from that person, or to determine for them what are appropriate or inappropriate considerations to bring to the decision-making process. This would seem to be crucial in ensuring an individual's decisional privacy – the extent of the right to make decisions for oneself.

7.5 REALISING PRIVACY IN MODERN RESEARCH GOVERNANCE

So far I have made two claims about privacy. First, individuals hold a range of sensitivities about their privacy (and we could add that this is a dynamic balance depending also upon the relationships between individuals and the emotional setting or moment of the relationship.[16] Second, the law produces a mechanism for resolving conflicts that fall within its definition of privacy, but it does not provide a complete normative definition of privacy that meets all the social functions required of the concept (that will confront researchers negotiating privacy relationships with their participants). Two observations might help with locating our thinking at this point. First, there is not a complete, normative definition of privacy in any discipline that satisfactorily meets the dynamic nature of privacy. There are many different definitions and conceptualisations, but there is no granular agreement on the normative question – what ought I to understand as 'my permitted private life'.[17] Second, the presentation so far might appear to suggest that privacy is a matter of individual autonomy, in opposition to society. This, in the remainder of the chapter, I will argue is not the case, by exploring how privacy might be operationalised, in our case, in research. The question is: what tools can we use to understand our relationships as individuals in society?

To do this, I suggest that there are three areas that can usefully be considered by both researchers (and participants) in the particular circumstances of a research project and by society in trying to understand the conceptualisation of privacy in modern society: the public interest, confidentiality and discourse.

---

[16] See M. J. Taylor, *Genetic Data and the Law: A Critical Perspective on Privacy Protection* (Cambridge University Press, 2012).

[17] This can be seen in privacy debates in other academic disciplines. See, for example, J. DeCew, 'Privacy', *(The Stanford Encyclopedia of Philosophy*, Spring 2018 Edition), E. N. Zalta (ed) www.plato.stanford.edu/archives/spr2018/entries/privacy/; A. Westin, *Privacy and Freedom* (New York: Atheneum, 1967); A. Westin, 'Social and Political Dimensions of Privacy', (2003) *Journal of Social Issues*, 59(2), 431–453.

The public interest, the common good, as a measure of solidarity is very attractive. It addresses directly the range of sentiments problem to which I refer to throughout the chapter (see Kieslich and Prainsack, Chapter 5, and Sorbie, Chapter 6, in this section, and Taylor and Whitton, Chapter 24, later in this volume). Appealing to the public interest is a practical mechanism that answers the individual's privacy sensitivities with the following: whatever you believe to be your privacy, these are the supervening arguments why, for example, you should let me stand on your land or use your personal data (your privacy has to accommodate these broader needs of others). The difficulty with the public interest is that it seems itself to have no definition or internal rules. Appeals to the public interest seem to be constructed loosely through a utilitarian calculus: the greatest utility for the greatest number. Mill himself identifies the problem: the tyranny of the majority. The problem has two elements. The claim to 'supervening utility' could seem itself to be a subjective claim, so those in the minority, suffering the consequences of a loss of amenity (in this case the breach of their privacy), are not immediately convinced of the substance of the argument. The construction does not balance the magnitude of the loss to the individual with the benefit (or avoided loss) of another individual; rather, the one stands against the many. This is not particularly satisfactory, especially when one links this back to the fundamental breach and the sense of the loss of privacy cutting to the personhood of the individual. Adopting the arguments of Arendt, we might phrase this more strongly. Arendt identifies the individual as constituted in two parts: the physical and the legal. In her studies of totalitarianism, she finds that tyrannies occur where the two parts of the individual are separated by bureaucracies and the legal individual is forgotten. Left with only the physical individual, the human is reduced to an expendable commodity.[18] Simple appeals to the public interest could be in danger of overlooking the whole individual and producing an alienation of those whose rights are removed in the public interest or common good.

Another way of constructing the appeal to the public interest can be through deontological rather than ontological theories, particularly those of Kant and Rawls. Taking Kant,[19] a first step would be to consider the losses to individuals – the person who stands to lose their privacy rights, and a person who would suffer a loss if that privacy was not breached. A second step would be to require each of those individuals to consider their claim to their privacy through the lens of the second formulation of the Categorical Imperative – that one should treat others as ends in themselves, not merely as means to one's own ends.[20] Because privacy is not an absolute right, when making such a claim, we must each ask: do I merely instrumentalise the other person in the balance by making this privacy claim? This is a matter of fact: which of us will suffer most? The third element is to acknowledge that the law can require me to adopt that choice if I fail to make it for myself, as it is the choice I should have made unprompted (I can argue that the calculation on the facts is incorrect, but not that the calculation ought not to be made). Rawls might construct it slightly differently: whereas I might prefer a particular action preserving my privacy, I must accept the breach of my privacy as reasonable in the circumstances. Using his

---

[18] H. Arendt, *The Human Condition* (Chicago University Press, 1958).

[19] I have developed this idea previously: D. Townend, 'Privacy, Politeness and the Boundary Between Theory and Practice in Ethical Rationalism' in P. Capps and S. Pattinson (eds), *Ethical Rationalism and the Law* (Oxford: Hart Publishing, 2017), pp. 171–189.

[20] I. Kant, *Groundwork of the Metaphysics of Morals* (1785). See M. Rohlf, 'Immanuel Kant', (*The Stanford Encyclopedia of Philosophy*, Spring 2020 Edition), E. N. Zalta (ed.), www.plato.stanford.edu/archives/spr2020/entries/kant/ (section 5.4).

'veil of ignorance', when I do not know my potential status in society, I must adopt this measure to protect the least-well-off member of society when the decision is made.[21]

In the example raised in Section 7.2, using this public interest consideration helps to reconcile the range of sensitivities problem. As a researcher trying to design privacy safeguards, I can use the calculations to evaluate the risks and benefits identifiable in the research, and then present the evaluation to participants and regulators. The public interest creates a discourse that steps outside self-interest. However, this sets off a klaxon that the public interest is not antithetical to privacy, as presented here; public interest is part of privacy. And I agree. Here, I am suggesting that using public interest arguments is a mechanism for defining the relationship of the individual to others (that is, to other individuals). The result is not saying that the public interest 'breaches' the privacy of the participant, but that it helps to define the individual's privacy in relation to others (for the individual and for other people and institutions). It brings to the subjectivity of the dynamic range of sentiments (that I identified as an issue at the outset) the solidarity and community that is also part of one's privacy.[22] This holistic understanding of privacy as a private life lived in community not reducible to a simply autonomy-based claim is best explored by Laurie's 'spatial', psychological privacy.[23]

Confidentiality is a second legal tool to ensure participants' rights are safeguarded. Arguably, it is a more practical tool or concept for researchers than privacy. Confidentiality earths abstract privacy concepts in actionable relationships and duties. Taking Common Law confidentiality as an example, it is constructed either expressly, as a contractual term, or it is implied into the conduct of a contract or through equity into the relationship between individuals.[24] Confidentiality depends on concrete, known parameters of the relationship, or parameters that one ought, in good conscience, to have known. Like data protection, it does not prohibit behaviour; rather confidentiality creates an environment in which particular behaviours can occur. This is important in the context of health research regulation because many potential research participants will be recruited through the professional relationships they enjoy with healthcare professionals; it is a tool that can be extended into other researcher–participant relationships. Confidentiality and the trust-based nature of that relationship can both help with recruitment and provide a welcome degree of reassurance about privacy protection.

Finally, and implicit throughout the operationalisation of privacy, privacy is a negotiated space that requires public engagement through discourse. Discourse ethics has a modern iteration, but a long history. The virtue ethics of Aristotle and Ancient Greek philosophy is dependent on the identification of the extent and nature of the virtues and their application in human life; Shaftesbury's early enlightenment 'politeness'[25] and the salons of the Age of Reason again ground the discussion of the questions, 'who are we, and how ought we to behave?' in public, albeit intellectual, discourse; today, Habermas et al. advocate this inclusion as a part of a participative democracy, perhaps reiterating the central arguments of the early Frankfurt School

---

[21] J. Rawls, *A Theory of Justice* (Cambridge, MA: Belknap Press, 1971, Revised Edition 1999). See L. Wenar, 'John Rawls', (*The Stanford Encyclopedia of Philosophy*, Spring 2017 Edition), E. N. Zalta (ed), www.plato.stanford.edu/archives/spr2017/entries/rawls/.

[22] And this is what the court advocates in W v. *Egdell* [1989] EWCA Civ 13 and is arguably the purpose of the derogations in human rights law discussed above.

[23] G. Laurie, *Genetic Privacy: A Challenge to Medico-Legal Norms* (Cambridge University Press, 2002).

[24] *Francome* v. *Mirror Group Newspapers Ltd* [1984] 1 WLR 892 (UK); *Campbell* v. *MGN Ltd* [2004] UKHL 22. See Taylor and Whitton, Chapter 24, in this volume.

[25] A. A. Cooper, Third Earl of Shaftesbury, *Characteristics of Men, Manners, Opinions, Times*, L. E. Klein (ed.), (Cambridge University Press, 1999); L. Klein, *Shaftesbury and the Culture of Politeness: Moral Discourse and Cultural Politics in Early Eighteenth-Century England* (Cambridge University Press, 1994).

against the false consciousness of the Culture Industry.[26] The thrust of this whole chapter is that privacy must be debated and understood in the lives of individuals; universities, professional bodies, and ethics committees must facilitate conversations that empower individuals to realise their decisional privacy in making choices about the nature of their participation in society.[27]

7.6 CONCLUSION

This chapter has focused on different aspects of a conceptual problem raised in relation to a modern research dilemma: how do we negotiate privacy-protecting research where individuals hold a dynamic range of sensitivities about their relationships to others in society? We have seen that whereas human rights law does not present granular definitions of privacy and courts use privacy concepts to resolve disputes in the area, attempts in legal theory and specific areas of law (personal data protection and informed consent) do not fill the conceptual gaps. The argument I advance is that using the public interest, confidentiality and public engagement discourse in constructing research protocols will go some way to address those gaps. It will also strengthen the relationship between researchers and the public they seek to engage and to serve, and could facilitate a greater understanding of the methods and objectives of science.

---

[26] See, for example, J. Habermas, *Between Facts and Norms: Contributions to a Discourse Theory of Law and Democracy* (tr. W. Rehg) (Cambridge, US-MA: MIT Press, 1996) (originally published in German, 1992); M. Horkheimer and T. W. Adorno, *Dialectic of Enlightenment* (tr. J. Cumming) (New York: Herder and Herder, 1972) (original publication in German, 1944).

[27] Townend, 'Privacy'.

8

Trustworthy Institutions in Global Health Research Collaborations

Angeliki Kerasidou

8.1 INTRODUCTION

Trust is often cited as being fundamental in biomedical research and in research collaborations. However, despite its prominence, its specific meaning and role remain vague. What does trust mean, and is it the same whether directed towards individuals and research institutions? What is it about trust that makes it important in global health research, and how can we effectively promote it? This chapter analyses the meaning of trust and discusses its importance and relevance in the context of global health research collaborations.

In recent decades, biomedical research has moved away from a one-researcher-one-project model to adopt a more collaborative way of working that brings together researchers from different disciplines, institutions and countries. Global health research, a field that has emerged as a distinct area of biomedical research, exemplifies this trend towards collaborative partnerships. Global health, as '*collaborative* trans-national research ...',[1] often relies on collaborations between researchers and institutions from high-income countries (HIC) and low-and-middle income countries (LMIC). LMICs still carry the highest burden of disease globally and have a high prevalence of many illnesses that pose global threats (e.g. infectious diseases). This has motivated a number of national and international funders to support global health research. The redirection of funds towards global health has resulted in increased interest among HIC researchers in working on diseases such as malaria, tuberculosis, HIV–AIDS and conditions such as malnutrition, and also a new impetus in forming partnerships with colleagues from LMICs.

Global health research is seen as a natural field for collaborative work for two reasons. First, by definition, the problems that global health research is trying to answer are complex, multifaceted and transcend borders and boundaries; tackling such problems requires collaborations between disciplines, countries and institutions. Second, most of the issues global health is concerned with affect less affluent parts of the world. Building health research capacity in these countries, and strengthening their public health systems, is seen as the most effective and sustainable way to ensure the successful progress and implementation of global health research, and to meet global health priorities. The importance of trust and the role of institutions in establishing and promoting trust relationships are often noted in discussions regarding global health research

---

[1] R. Beaglehole and R. Bonita, 'What Is Global Health?', (2010) *Global Health Action*, 3(1), 5142.

collaborations. Trust is often presented as a foundational element of research participation,[2] data sharing[3] and sharing of samples and other resources.[4]

Here, I give an account of what it means for an institution to be trusted and be trustworthy in the context of global health research. I employ the example of data sharing to illustrate the importance and value of trustworthiness as an institutional moral characteristic. I use the term 'institution' to refer to groups or collectives that actively undertake research, such as universities and research centres. I conclude that trust is important in global health research collaboration because of the power imbalance between partners that often characterises such collaborations. In order to promote trust, institutions need to focus on being trustworthy by developing a behaviour that corresponds to the aims, principles and values they profess to uphold, and by demonstrating that they have incorporated into their functions, rules and regulations the particular needs of their partners and collaborators.

8.2 WHAT IS A COLLABORATION?

We use the term 'collaboration' in our everyday language to signify many different types of partnerships. Yet, not all ways of working together are collaborations. The term denotes a particular type of partnership where two or more partners come together to achieve a common aim or goal.[5] Collaborations are non-hierarchical structures, based on the sharing of decision-making and responsibility that rely more on capacity and expertise, rather than on functions or titles.[6] Consider, for example, a collaboration between a statistics unit and an epidemiology unit working together on a population health project investigating lung cancer. The two groups are committed to the aim of the project and share equal responsibility for its successful completion. They bring different expertise into the project, participate equally in decisions regarding its running and direction and share ownership for its outputs (e.g. authorship on academic publications). Collaborations are characterised by transparency, openness in communication, synergy and honest appreciation of each other's positions. Transparency facilitates a collective awareness of the project, its structure, strengths and weaknesses, and promotes collective ownership. Open communication allows for the free flow of information and exchange of ideas, but also for the expression of concerns. The easier it is for people to talk to each other and share their thoughts and viewpoints, the easier it is for a project to stay on track and reach its goals. Understanding each other's positions and particular circumstances is also important, as it helps with setting expectations at the right level, anticipating problems and foreseeing areas where conflict may arise. Finally, synergy, which describes the drive and desire to achieve the common goal and recognition of the partners' interdependence in fulfilling it, is what drives such partnerships.[7]

[2] M. Guillemin et al., 'Do Research Participants Trust Researchers or Their Institutions?', (2018) *JEEHRE*, 13(3), 285–294.

[3] R. Milne et al., 'Trust in Genomic Data Sharing among Members of the General Public in the UK, USA, Canada and Australia', (2019) *Human Genetics*, 138(11–12), 1237–1246.

[4] P. Tindana et al., 'Ethical Issues in the Export, Storage, and Reuse of Human Biological Samples in Biomedical Research: Perspectives of Key Stakeholders in Ghana and Kenya', (2014) *BMC Medical Ethics*, 15(76).

[5] E. A. Henneman et al., 'Collaboration: A Concept Analysis', (1995) *Journal of Advanced Nursing*, 21(1), 103–109; D. D'Amour et al., 'The Conceptual Basis for Interprofessional Collaboration: Core Concepts and Theoretical Frameworks', (2005) *Journal of Interprofessional Care*, 19(sup 1), 116–131.

[6] Henneman et al., 'Collaboration: A Concept Analysis'.

[7] D'Amour et al., 'The Conceptual Basis for Interprofessional Collaboration'.

Other types of partnerships or co-working include cooperation – which brings partners together who do not share the same goal, but who need each other's skills and expertise to reach their individual aims – and hiring or commissioning someone to do a specific job. For example, someone can be brought into a project to complete a very specific task, such as to conduct a systematic literature review, collect samples or develop informed consent forms for a clinical trial. Once the task is completed, the person's involvement in the project is ended. None of these types of partnerships can be described as collaborations as they lack the fundamental characteristics of non-hierarchical and synergistic co-labouring.

8.3 TRUST AND TRUSTWORTHINESS

Alongside synergy and horizontal organisation, trust is regularly cited as a fundamental characteristic of collaborations.[8] D'Amour notes that 'the term collaboration conveys the idea of sharing … in a spirit of harmony and *trust*'.[9] One empirical study that investigated what underpins successful collaborations in global health research from the perspective of scientists and other research actors, identified trust between partners as one of the major contributing factors.[10] But why is trust so crucial for collaborations? Trust is yet another term that is often used in our everyday language but not always to describe the same thing. A short analysis would help to better define this term and see how it applies to the context of global health research.

Trust is an attitude towards a person whom we hope, and have good reasons to believe, will behave in a way that confirms our trust. This attitude can take different forms. People can trust others wholeheartedly and perfectly (A trusts B), for example their mother or spouse. But most commonly, trust is perceived as a three-part relationship (A trusts B to *x*). There are three main attributes of a trust relationship: vulnerability, assumption of good will from the trustee towards the trustor and voluntariness.[11] Vulnerability stems from the fact that when trusting, the trustor becomes vulnerable to the trustee as they acknowledge and accept that the trustee can decisively affect the outcome of the entrusted action.[12] This is what justifies feelings of gratitude or of betrayal when trust is confirmed or broken.[13] Vulnerability is not, however, a personal characteristic of the trustor. Rather, it is a relational property that emerges from the act of trusting. Consider the following example: a researcher shares some potentially significant pre-publication findings with a colleague who also works in a similar area. She has previously worked with this colleague and trusts him. In her correspondence she stresses the importance of these findings and asks the colleague to keep them confidential. If the colleague confirms her trust, and keeps the findings confidential, she will feel her trust is confirmed; if, however, the colleague ignores her trust and publishes the findings or shares them with others (e.g. at a conference), she will justifiably feel betrayed. The feeling of betrayal is predicated on the fact that she has no assurances, other than her trust, to protect her from the colleague's decision and behaviour,

[8] A. W. Pike et al., 'A New Architecture for Quality Assurance: Nurse-Physician Collaboration', (1993) *Journal of Nursing Care Quality*, 7(3), 1–8; D'Amour et al., 'The Conceptual Basis for Interprofessional Collaboration', 116; M. Parker and P. Kingori, 'Good and Bad Research Collaborations: Researchers' Views on Science and Ethics in Global Health Research', (2016) *PLoS ONE* 11(10).
[9] D'Amour et al., 'The Conceptual Basis for Interprofessional Collaboration'.
[10] Parker and Kingori, 'Good and Bad Research Collaborations'.
[11] A. Kerasidou, 'Trust Me, I'm a Researcher!: The Role of Trust in Biomedical Research', (2017) *Med Health Care Philos*, 20(1), 43–50.
[12] R. Holton, 'Deciding to Trust, Coming to Believe', (1994) *Australasian Journal of Philosophy*, 72(1), 63–76; S. Wright, 'Trust and Trustworthiness', (2010) *Philosophia*, 38(3), 615–627.
[13] A. Baier, 'Trust and Antitrust', (1986) *Ethics*, 96(2), 231–260.

and this is what makes her vulnerable towards her trustor. Seeking assurances, by trying to constrain someone's behaviours as a way of limiting one's vulnerability would indicate that the trustor mistrusts, or lacks trust for that person. This is why trusting requires some level of optimism about the trustee or a normative attitude that the trustee ought to do what the trustor wills them to do.[14]

The second characteristic of trust is the belief that the trustee has good will towards the trustor.[15] It is this belief that counterbalances vulnerability and provides a reasonable justification for trusting someone. To return to the previous example, the researcher reveals the pre-publication findings to her colleague because she has good reasons to believe that he has good will towards her and will not intentionally harm or hurt her. If she did not have good reasons to believe this, then choosing to reveal her findings and make herself vulnerable towards him, all things being equal, would be unjustified. Some challenge the importance of good will in trust, by suggesting that trust may be warranted when we believe that those we trust (trustees) will conform to social constraints and norms, or that they will act in the ways we expect out of self-interest.[16] Yet, while social constrains and self-interest could increase people's reliability, it is questionable whether such motives can underpin trust. A belief in the good will of the trustee signifies that the trustor has good reason to assume that the trustee cares about her and/or about the things about which she also cares. Although this could be problematic in situations where one does not have insight into the 'psychology of the one-trusted',[17] one could still justify trust on the belief in the other's good will by adopting a wide notion of good will, which includes commitment to benevolence and conscientious moral attitude.[18]

The third characteristic is that trust is voluntary, insofar as it cannot be forced or demanded. As Baier notes: '"Trust me!" is for most of us an invitation which we cannot accept at will – either we do already trust the one who says it, in which case it serves at best as reassurance, or it is properly responded to with, "Why should and how can I, until I have cause to?"'[19] Trust takes time to establish, and requires an expectation that people will behave not only in the way we assume they *will*, but rather in the way we assume they *should*.[20] A consistent demonstration of good will, as well as capacity to perform the entrusted action, can provide a good reason for trust. And those who want to be trusted can help generate such relationships by fostering and increasing their trustworthiness.

Trustworthiness is a moral characteristic of the trustee and signifies that they have an attitude of good will towards the trustor by being responsive to the trustor's dependency upon them.[21] The motivation for behaving trustworthily also matters. Trustworthiness signifies something more than just the mere observation of rules and regulations out of self-interest or duty. It is not just a tactic to avoid punishment or penalties, or to fulfil one's sense of duty. Potter describes trustworthiness as a virtue. 'In evaluating someone's trustworthiness', she argues 'we need to know that she can be counted on, *as a matter of the sort of person she is*, to take care of those

[14] V. McGeer, 'Trust, Hope and Empowerment', (2008) *Australasian Journal of Philosophy*, 86(2), 237–254.

[15] Baier, 'Trust and Antitrust'.

[16] R. Hardin, Trust and Trustworthiness (Russell Sage Foundation, 2002); O. O'Neill, A Question of Trust (Cambridge: Cambridge University Press, 2002).

[17] S. Blackburn, *Ruling Passion: A Theory of Practical Reasoning* (Oxford: Oxford University Press, 1998).

[18] K. Jones, 'Trust as an Affective Attitude', (1996) *Ethics*, 107(1), 4–25.

[19] Baier, 'Trust and Antitrust', 244.

[20] M. Urban Walker, *Moral Repair: Reconstructing Moral Relations after Wrongdoing* (Cambridge University Press, 2006).

[21] K. Jones, 'Trustworthiness', (2012) *Ethics*, 123(1), 61–85.

things with which we are considering entrusting her' (emphasis added).[22] However, it is important to note that expectation for one to behave in a certain manner does not compel the trustee to behave in the expected way. The fact that one is being counted on forms an important consideration to be taken into account but does not force one to act in a certain way – otherwise one would be forced to act in a 'trustworthy' way even when the trust placed on one is unjustified or misguided.

So far, I have argued that trust is a relational mode predicated on a reasonable belief in the trustee's skill to perform the entrusted action and also good will towards the trustor. Trust cannot be forced or demanded, and by trusting, one makes oneself vulnerable toward the person they choose to trust. Trustworthiness is a moral characteristic that indicates that someone can be counted on. It is not necessary that a trustworthy person is automatically trusted, but trustworthy behaviour can illicit trust. In the context of these definitions we can explore this chapter's main questions: what is the role of trust in global health research collaborations and can institutions be trusted? If trust is commonly perceived as a characteristic of interpersonal relationships and trustworthiness as a personal quality or virtue, is it possible to talk meaningfully about trusting institutions, or to ascribe moral characteristics such as trustworthiness to collectives?

8.4 TRUST IN GLOBAL HEALTH RESEARCH

In 2013, the Council on Health Research for Development (COHRED) published a report on fair collaborations in global health.[23] It noted that relying on HIC collaborators' good will has not been sufficient to ensure fair and just collaborations between partners. What was needed instead, the report recommended, was to build LMIC institutions' capacity in contract negotiations. The implication seems to be here, that instead of just trusting people to behave fairly and justly and thus opening oneself up to having their trust betrayed, one needs to ensure that people will behave this way. This could be achieved by putting in place contracts that direct and set the parameters of right behaviour. One way of understanding this contractual relationship is as relationship of reliance. Relationships of reliance are based on proven capacity and clear systems of accountability. In such relationships, the expectation is that the partners will act based on self-interest. Collaborators can ensure successful partnerships by aligning their interests and by putting in place rules to secure against defection. What makes relationships of reliance preferable to relationships of trust is that the former do not require an assumption of good will, nor do they require the trustor to become vulnerable to the trustee.[24]

One important condition, however, must apply for relationships of reliance to work. Reliance requires power parity between partners.[25] This is because in relationships that operate on self-interest, it is far easier for the stronger partners to shift the balance to their favour. This is particularly relevant for global health research collaborations, which often bring together institutions from HICs and LMICs. Giving LMIC researchers and institutions the tools to defend and promote their own interests is one way of promoting reliable – rather than trusting – partnerships and addressing relationships of dependency, and COHRED's efforts are a valuable step towards this. However, there are a number of reasons why building trust relationships and

[22] N. Nyquist Potter, *How Can I Be Trusted?: A Virtue Theory of Trustworthiness* (Rowman & Littlefield, 2002), p. 7.
[23] COHRED, 'Where There Is No Lawyer: Guidance for Fairer Contract Negotiation in Collaborative Research Partnerships', (COHRED, 2013).
[24] This is not to say that in relationships of reliance things cannot go wrong. One can fail to accurately predict the other person's action, which can result in harm or loss.
[25] A. Kerasidou, 'The Role of Trust in Global Health Research Collaborations', (2019) *Bioethics*, 33(4), 495–501.

promoting trustworthiness remains important in this context. First, in situations where power parity between partners is lacking, trust can be an essential foundation on which to build a good and fair collaboration. For example, attitudes of good will – a crucial feature of trust – can counterbalance self-interested motivations. Second, trust could facilitate good collaborative partnerships, by creating a safe environment in which partners can focus on achieving the common goal rather than on protecting their own interests. Finally, it is common sense that everyone, given the option, would prefer to work with partners they trust and not only with those they can reliably predict their behaviour.

If we accept that trust remains relevant in global health research, what we need to consider next is how it could be promoted. One of the reasons that could justify and encourage a trust relationship is trustworthiness. Although trustworthiness cannot always and *de facto* guarantee trust, moral agents who want to be trusted by their partners and collaborators could do worse than to try to cultivate and demonstrate their trustworthiness. However, while trustworthiness can be attributed to individual persons, collaborations in global health research however, are not just between individuals, but also between institutions. It is important, therefore, to examine whether it is reasonable to talk about 'trustworthy' institutions.

8.5 TRUSTWORTHY INSTITUTIONS

Prior to ascertaining whether institutions (e.g. universities, research centres) can be trustworthy, we must establish whether it is reasonable to talk about groups and collectives possessing such moral characteristics as trustworthiness.[26] In other words, can institutions involved in global health research collaborations be moral agents? There are two main reasons, which, I believe, give support to the view that collectives are entities that could be treated as moral agents: first, such a position chimes with the way we think about the role of collectives in public life and also the way we treat them in practical terms. For example, we expect universities to adhere to ethical principles when conducting research and we hold them responsible when they fail to do so.[27] In law, collectives are treated as bearers of rights and responsibilities and can be penalised for wrongdoing and for failing to meet their duties and obligations. Second, the view that institutions are moral agents reflects a growing realisation that many issues require the action of collectives in order to be resolved. Actions such as conducting large-scale research aimed at halting pandemics or reversing climate change are unavailable to individuals but possible to groups and institutions. If we accept that these actions reflect duties that *ought* to be met, then these duties will have to be ascribed to actors that *can* meet them.[28]

Being trustworthy means that a given individual (or institution) acts not only as they are expected to, but in a way that demonstrates that they have taken into account the fact that someone is counting on them.[29] Trustworthiness is a characteristic or moral attitude that is revealed through one's actions and also in one's 'values, commitments and loyalties'.[30] When it comes to institutions, their trustworthy character is revealed in their professed goals and aims, at

---

[26] The types of groups or collectives I have in mind are those who submit to a common goal, can act as one body and present organisational structures and rules, e.g. universities, research bodies and international agencies, and not those based merely on the sharing of a common characteristic (e.g. a disease).

[27] J. Couzin-Frankel, 'A Lonely Crusade', (2014) *Science*, 344(6186), 793–797; C. Elliot, 'Guinea-pigging', *The New Yorker* (31 December 2007).

[28] For a comprehensive defence of institutions as moral agents see: C. List and P. Pettit, *Group Agency: The Possibility, Design, and Status of Corporate Agents* (Oxford University Press, 2011).

[29] Jones, 'Trustworthiness*'; Wright, 'Trust and Trustworthiness'.

[30] Potter, *How Can I Be Trusted?*, p. 7.

their institutional structures, internal rules and regulations as indicators for their moral motivations,[31] and in their reputation and track record as indicator for their skill and commitment to right action.[32] Researchers and groups in global health who are looking for collaborations would perceive institutions that declare to care about things they also care as more trustworthy, rather than institutions that do not profess such interests. An institution's track record and proven capacity in their ability to reach these shared goals would add to its trustworthiness. Importantly, being trustworthy is not about following rules but acquiring a disposition of trustworthiness. Behaving in a certain way only for fear of penalty demonstrates a self-interested orientation, rather than concern for others or about what others value. Therefore, institutions would need to demonstrate that their commitment to trustworthy behaviour is principled and corresponds with their aims and purpose, rather than motivated by a desire to avoid sanctions and penalties, including loss of future collaborations. For institutions participating in global health research, this will mean demonstrating that they have incorporated into their structures, rules and regulations central aims of global health such as addressing health inequalities, improving health through rigorous research and promoting research capacity in countries that lack it. Using the example of data sharing and open access may help to illustrate this point.

8.6 TRUSTWORTHINESS IN DATA SHARING COLLABORATIONS

Data sharing is often presented as foundational to global health research.[33] Health and health-related data (e.g. genomic, phenotypic or clinical data) are an inexhaustible resource that could be used repeatedly to address multiple research questions, provide answers to a plethora of global health issues and thereby help reduce the global burden of disease. For example, data sharing between countries and institutions is essential in the attempt to understand and respond to epidemics and pandemics, as the cases of the H5N1 avian flu in 2007 and the outbreaks of Ebola in 2014 and ZIKA in 2015 have demonstrated.[34] Recognition that health data offer valuable resources with multiple applications has led to a position where data sharing is seen as both a scientific and moral imperative in biomedical research,[35] while failure to share has been variously described as being unscientific, contrary to research integrity, wasteful and unjust.[36] In recent years, a lot of effort has been put into facilitating and promoting the open sharing of data.[37] Progress in data sharing tools, methods and policies is seen as the 'innovation with the farthest-reaching impacts among the global medical community'.[38] This has led to the wide endorsement of data sharing and open access policies by many international research bodies,

---

[31] P. A. French, 'Types of Collectivities and Blame', (1975) *The Personalist*, 56(2), 65–85; R. Bachmann and A. Inkpen, 'Understanding Institutional-Based Trust Building Processes in Inter-Organizational Relationships', (2011) *Organizaition Studies*, 32(2), 281–301.

[32] Hardin, *Trust and Trustworthiness*.

[33] E. Pisani et al., 'Beyond Open Data: Realising the Health Benefits of Sharing Data', (2016) *BMJ*, 355.

[34] K. Littler et al., 'Progress in Promoting Data Sharing in Public Health Emergencies', (2017) *Bulletin World Health Organisation*, 95(4), 243–243A.

[35] H. Bauchner et al., 'Data Sharing: An Ethical and Scientific Imperative', (2016) *JAMA*, 315(12), 1238–1240.

[36] M. Munafo et al., 'Open Science Prevents Mindless Science', (2018) *BMJ*, 363; P. Langat et al., 'Is There a Duty to Share? Ethics of Sharing Research Data in the Context of Public Health Emergencies', (2011) *Public Health Ethics*, 4 (1), 4–11; P. C. Gotzsche, 'Why We Need Easy Access to All Data from All Clinical Trials and How to Accomplish It', (2011) *Trials*, 12(1), 249.

[37] M. Wilkinson et al., 'The FAIR Guiding Principles for Scientific Data Management and Stewardship', (2016) *Scientific Data*, 3.

[38] 'Is Data Sharing a Path to Global Health?', (*WIRED*, 5 February 2018), www.datamakespossible.westerndigital.com/data-sharing-panacea-global-health.

funding organisations, academic publishers and policymakers.[39] It seems that adopting and promoting open access of data and the implementation of a robust open data sharing policy would signal an institution's moral character as being one dedicated to open, transparent and robust science, and to maximising research benefits for all. But would this mean that such an institution is trustworthy?

In the context of global health research, data sharing can be ethically and practically complex. Despite its potential benefits, there are significant ethical and societal barriers to the wide implementation of open data sharing policies and practices.[40] Leaving aside confidentiality and consent, a significant issue in global health stems from the uneven ability of institutions in different parts of the world to utilise data.[41] As Serwadda and others note, advancements in technology that make data collection, storage and sharing easier, and the shift in the social and scientific norms to support openness and sharing, is undermining equitable collaborations between HIC and LMIC.[42] This has led to 'a landscape, often characterised by limited capacity and deep mistrust, for acceptance and implementation of open data policies'.[43] Furthermore, despite claims that open data sharing could lead to advancements that would be beneficial to all, including to the communities of origin, this is not always the case. Often, the new therapeutics developed are either too expensive for LMICs to purchase, or these countries lack adequate public health structures to make use of any new actionable knowledge. For example, in 2007, Indonesia refused to share its H5N1 avian flu data and samples unless their country was guaranteed affordable access to vaccines – and researchers from other LMIC seem to think that this was a fair response to an unfair situation.[44] Although data sharing could accelerate the production of new and useful knowledge, it can also contribute to the perpetuation of global injustices and undercut the stated goals of global health research.

Adopting an open access policy to data sharing could make an institution reliable, in the sense that its partners would know what to expect and would be able to predict its behaviour and actions with a certain degree of accuracy. Would recognising this institution's reliability in this domain, however, amount to it been perceived as being trustworthy by its partners?[45] Although adopting certain (moral) rules and acting consistently is an indication of a certain (moral) character, trustworthiness requires more than that; it requires an attitude of good will and

[39] European Medicines Agency, 'European Medicines Agency Policy on Publication of Clinical Data for Medicinal Products for Human Use', (European Medicines Agency, 2014); F. Godlee and T. Groves, 'The New BMJ Policy on Sharing Data from Drug and Device Trials', (2012) *BMJ*, 345(7884), 10; The Wellcome Trust, *Policy on Data Management and Sharing* (London, England: The Wellcome Trust, 2009); National Institutes of Health, *Final NIH Statement on Sharing Research Data* (Bethesda: National Institutes of Health, 2003).

[40] S. Bull and M. Parker, 'Sharing Public Health Research Data: Towards the Development of Ethical Data-Sharing Practice in Low- and Middle-Income Settings', (2015) *Journal of Empirical Research on Human Research Ethics*, 10(3), 217–224.

[41] I. Jao et al., 'Research Stakeholders' Views on Benefits and Challenges for Public Health Research Data Sharing in Kenya: The Importance of Trust and Social Relations', (2015) *PLoS ONE*, 10(9).

[42] D. Serwadda et al., 'Open Data Sharing and the Global South – Who Benefits?', (2018) *Science*, 359(6376), 642–643.

[43] Ibid., 642.

[44] K. T. Emerson and M. C. Murphy, 'A Company I Can Trust? Organizational Lay Theories Moderate Stereotype Threat for Women', (2015) *Personality and Social Psychology Bulletin*, 41(2), 295–307.

[45] Hawley argues that drawing a distinction between reliability and trustworthiness of institutions is not useful because 'we can require of our institutions that they are reliable in the respects that matter to us' see: K. J. Hawley, 'Trustworthy Groups and Organisations' in P. Faulkner and T. Simpson (eds), *The Philosophy of Trust* (Oxford University Press, (2017), p. 20. In her case, Hawley has in mind public institutions with whom 'we' as citizens have a special kind of relationship, meaning that these institutions have a duty to be responsive to our needs and particular circumstances. Whether research institutions have the same duty towards researchers in other countries or to the global research community is not immediately clear. An argument will have to be made to demonstrate that research institutions fall within this special category. However, this investigation falls outside the remit of this chapter.

responsiveness to the other's needs. Hence, a trustworthy institution in global health research would not blankly endorse an open data sharing policy, but would retain a flexible stance, leaving room for adapting its policies with the specificities of its collaborators in mind. Such adaptations might include time-specific embargoes on data release to give partners a fair head start on using their data, restrictions of use to protect the stated research aims of such partners and embedding contextually meaningful capacity building activities into their collaboration.[46] Although an open data sharing policy could ensure that maximum value and utility is extracted from data, allowing for the negotiation of a managed access policy would signal an institution with good will towards its collaborators and 'a direct responsiveness to the fact that the other is counting on [it]'.[47]

8.7 CONCLUSION

Trust is and will remain important in global health research collaborations, at least until the power imbalance between LMIC- and HIC-based researchers and their institutions is addressed. Institutions committed to advancing the aims of global health, including helping build research capacity in LMIC, should aim to promote fair and trusting collaborations. The best way of achieving this is by cultivating and demonstrating their trustworthiness as a way of eliciting justified trust. Being trustworthy requires more than just the observation of rules or the incorporation of moral principles in policies and structures. Although such moral attitudes would likely increase an institution's reliability, trustworthiness also demands attention to the relational aspect of trust. Trustworthiness requires that the institution is concerned with its partners and what its partners value, acknowledges its partners' vulnerability and demonstrates 'a direct responsiveness to the fact that the other is counting on [it]'.[48] The practice of data sharing provides a useful case to examine what being a trustworthy institution might look like in practice. Moving forward, more research will be required to fully examine the relationship between rules, regulations and policies and the moral character of institutions in global health.

[46] M. Parker et al., 'Ethical Data Release in Genome-Wide Association Studies in Developing Countries', (2009) *PLOS Medicine*, 6(11), e1000143.
[47] Jones, 'Trustworthiness', 62.
[48] Ibid.

9

Vulnerabilities and Power

The Political Side of Health Research

Iain Brassington

9.1 INTRODUCTION

In this chapter, I will argue that there is a political dimension to research, and that accounts of health research regulation that ignore political relations between stakeholders are therefore incomplete. The concept of vulnerability – particularly vulnerability to exploitation – provides the grit around which the claims are built. This is because vulnerability is an inescapable part of human life; because research participation may magnify vulnerability, even while health research itself promises to mitigate certain vulnerabilities (most directly vulnerability to illness, but indirectly vulnerability to economic hardships that may follow therefrom); and because vulnerability is manifested in, exacerbated by, or mitigated through, inherently political relationships with others, the groups and communities of which we are a part, and in the context of which all research takes place. I shall not be making any normative claims about research regulation here, save for the suggestion that decision-makers ought to take account of latent political aspects in their deliberations. For the most part, I shall simply attempt to sketch out some of those political aspects.

9.2 SETTING THE SCENE

Certain key terms ought to be defined at the offset.

By *vulnerability*, I understand a susceptibility to harm or wrong arising from a physical or social contingency above and beyond that found in a recognisably decent human life.

By *the vulnerable*, I understand those who are at an elevated risk of harm or wrong arising from such contingencies.

By *power*, I understand the capacity to act, or to resist being acted upon.[1]

By *power relations*, I understand the interplay of agents' relative power.

By *the political*, I understand the domain in which power relations are manifested.[2]

By *exploitation*, I understand the use of some thing or person to serve one's ends.

---

[1] This falls within a tradition that goes back at least as far as Hobbes: 'The POWER *of a Man*, (to take it Universally,) is his present means, to obtain some future apparent Good'. T. Hobbes, *Leviathan* (Cambridge University Press, 1999), p. 62. More recently, Miranda Fricker has defined 'social power' as 'a practically socially situated capacity to control others' actions'. M. Fricker, *Epistemic Injustice* (Oxford: Clarendon, 2007), p. 13; I take this to be related.

[2] H. Lasswell, *Politics: Who Gets What, When, How* (New York: McGraw-Hill, 1936), p. 3: 'The study of politics is the study of influence and the influential'. Combining this with Fricker's account above, gives us reason to think that social

Some elaboration is in order. At its most basic, vulnerability is any susceptibility to harms or wrongs; but such an understanding is generally unhelpful, because (*per* Rogers) 'it obscures rather than enables the identification of the context-specific needs of particular groups'[3] – plausibly, one may read this as 'individuals and groups' – and because (*per* Wrigley and Dawson) 'if everyone is vulnerable, then no one is'.[4] A more nuanced and useful conceptualisation of vulnerability would relate it to a susceptibility to harms or wrongs greater than is normally found in a recognisably decent human life. Correspondingly, in stating that '[s]ome groups and individuals are *particularly* vulnerable and may have an *increased likelihood* of being wronged or of incurring additional harm',[5] the Declaration of Helsinki is plainly referring to the ways in which persons may be *further* vulnerable above a universal baseline. That said, I will indicate below that, and how, the more basic understanding is not without utility.

Wendy Rogers provides an account of some of the difficulties of conceptualising vulnerability in this volume (see Chapter 1) and taxonomies of different kinds of vulnerability have been offered elsewhere.[6] I will neither rehearse nor assess those accounts here, save to highlight the idea of *pathogenic vulnerability*, the sources of which include morally dysfunctional or abusive interpersonal and social relationships, and sociopolitical oppression or injustice,[7] and which thereby illustrates plainly one of the political aspects of vulnerability. However, we conceptualise or parse it, though, vulnerability invites politically-informed responses. Wrigley and Dawson assert that vulnerability 'implies an ethical duty to safeguard [the vulnerable person's or group's] well-being because the person or group is unable to do so adequately themselves'.[8] For his part, ten Have claims that '[w]hat makes vulnerability problematic is the possibility of abuse and exploitation'; for him, vulnerability need not be eliminated, so long as it can be 'compensated, diminished, and transformed'.[9] Putative duties to safeguard the vulnerable, or to militate against abuse, could be discharged by individuals in some cases, and by the state in others. Venturing claims one way or the other implies a political position, because it speaks to decisions about how and by whom power may be exerted over, and on behalf of, another.

Exploitation, as defined above, implies the exercise of power over another: the exploiter is *in this context* more powerful than the exploited. As a manifestation of the power relations between agents, it is therefore a political phenomenon; and if exploitation violates a right of the exploited, it may be wrongful. Insofar as that vulnerability is susceptibility to certain harms or wrongs, it

---

power and politics are inseparable, that we therefore cannot talk about politics without talking about power, and that talking about power will at least often be talking about the political.

[3] W. Rogers, 'Vulnerability and Bioethics' in C. Mackenzie et al. (eds), *Vulnerability: New Essays in Ethics and Feminist Philosophy* (Oxford University Press, 2014), p. 69.

[4] A. Wrigley and A. Dawson, 'Vulnerability and Marginalized Populations' in D. Barrett et al. (eds), *Public Health Ethics: Cases Spanning the Globe* (Dordrecht: Springer, 2016), p. 204

[5] World Medical Association, 'WMA Declaration of Helsinki – Ethical Principles for Medical Research Involving Human Subjects', (1964), §19, www.wma.net/policies-post/wma-declaration-of-helsinki-ethical-principles-for-medical-research-involving-human-subjects/. Emphasis added.

[6] C. Mckenzie et al., 'Introduction: What Is Vulnerability, and Why does It Matter for Moral Theory?' in C. Mckenzie et al. (eds), *Vulnerability: New Essays in Ethics and Feminist Philosophy* (Oxford University Press, 2014), p. 7ff; F. Luna, 'Elucidating the Concept of Vulnerability: Layers Not Labels', (2009) *International Journal of Feminist Approaches to Bioethics*, 2(1), 121–139; F. Luna, 'Identifying and Evaluating Layers of Vulnerability – A Way Forward', (2019) *Developing World Bioethics*, 19(2), 86–95.

[7] W. Rogers et al., 'Why Bioethics Needs a Concept of Vulnerability', (2012) *International Journal of Feminist Approaches to Bioethics*, 5(2), 25.

[8] Wrigley and Dawson, 'Vulnerability and Marginalized Populations', p. 203

[9] Indeed, he goes so far as to entertain the (for my money, implausible) suggestion that 'Love would be impossible if we [did] not make ourselves vulnerable to another person.' H. ten Have, *Vulnerability: Challenging Bioethics* (Abingdon: Routledge, 2016), pp. 112–113.

includes susceptibility to wrongful exploitation; and since exploitation is a political phenomenon, vulnerability to wrongful exploitation will therefore also be political. The relevance of this will become clear as we proceed.

9.3 INDIVIDUALS' VULNERABILITY IN RESEARCH

It is in the nature of research that outcomes are uncertain; this means that healthy volunteers in medical trials might be susceptible to unexpected harms. If research concerns a treatment's effectiveness, it will often be necessary to recruit patients into a trial; but such a cohort will, by definition, be of people with medical needs, some of which may be otherwise unmet. The prospect of a health benefit, especially if there are few other extant or affordable treatment options, may mean that this somatic vulnerability is accompanied by vulnerability to exploitation: the patient may allow herself to be enrolled into a trial into which she would not have allowed herself to be enrolled otherwise. Moreover, participants' ability to control their exposure to risk may be limited: even without perfect knowledge, researchers are likely to have greater insight into the risks, and are able to control information in a way that participants, who rely on researchers for information, are not. This is a form of epistemic power held by researchers. Indeed, researchers may be *perceived* as having control over information even when they do not; and this perception may give them a 'credibility excess'[10] that is itself a source of epistemic power, insofar as that it can influence the decisions that participants make, perhaps to the extent of inhibiting their making them at all. How researchers and research managers handle the power disparity between them and participants is a political problem writ small.

Even putting the political aspect of this relationship to one side, it would be reasonable to expect that researchers address questions about the broader political context of their programme and protocol. After all, if someone enrols as a research subject because it is the only way they can access treatment, or because it is the only way they can afford it or other necessities, this tells us something about the characteristics of the state in which they live – notably, how just it is. Correspondingly, acknowledged political injustice may alter the likelihood that a person would act as a participant, how they behave as a participant, and whether their participation is voluntary. The political questions are clear. Does the political environment in which a person lives provide adequate protection against exploitation? What should be done if it does not?

At times, it may be that political circumstances make ethically acceptable research impossible. *In extremis*, this might be because certain people are forced to participate by an overweening government: prisoners, say, may be particularly vulnerable to this kind of pressure in some regimes. But participatory voluntariness may also be eroded by the lure of medical treatment that participants would not otherwise have, perhaps because it is not normally within the state's abilities to provide it. On the other hand, refusing to carry out research because the context in which it is proposed creates vulnerabilities or militates against their mitigation, may simply mean that would-be participants are deprived of benefits that they might have had – Ganguli-Mitra and Hunt touch on this problem when they consider the use of experimental interventions during the 2013–2016 Ebola outbreak in Chapter 32 of this volume – and that scientific opportunities are lost as well. A further problem is that some illnesses are illnesses of poverty; it may not be possible to carry out research on those illnesses without recruiting people who are socioeconomically vulnerable, because less socioeconomically vulnerable people would be less susceptible to the

---

[10] Fricker, 'Epistemic Injustice', p. 17; I. Kidd and H. Carel, 'Epistemic Injustice and Illness', (2017) *Journal of Applied Philosophy*, 34(2), 172–173.

illness in question. (That said, one may wonder whether prioritising poverty alleviation would dilute any imperative to research the illnesses that it causes.)[11]

There is unlikely to be an easy way to determine whether a given political situation is conducive to ethically sound research. Possibly the best that could be said is that good research practice may require an awareness of, and sensitivity to, the prevailing political dispensation as it applies to certain individuals.

9.4 INDIVIDUALS AND GROUPS

How well do these considerations translate to groups?

For the moment, I shall assume that groups are aggregates of individuals, and that groups' vulnerabilities are aggregates of individuals' vulnerabilities. Admittedly, this is a simplification: something might be good or bad for the group as a whole without being good or bad for each and every member thereof; a group's integrity, say, may be vulnerable in a way irreducible to its members' vulnerabilities. But, for the time being, and given space constraints, I think that the simplification is not gross.

Granted that groups are aggregates of individuals, discriminatory or otherwise unjust political arrangements may exacerbate or even generate vulnerabilities in those individuals *qua* group-members. Most obviously, individuals may be at increased susceptibility to harm or wrong if they lack legal or political representation, education, and so on, because of their membership of a particular group. This kind of powerlessness to resist injustice is a political product generating a pathogenic vulnerability – and a state in which injustice is not addressed is itself unjust, or vicious in some other way.[12] Further, the legacy of historic injustices may linger even if the unjust policies were ditched long ago.

But even having been identified as a member of a group at all may generate vulnerabilities in individuals, irrespective of the political circumstances. To give a simple example, a public health research programme may require population-level data-gathering. Any given individual may feature in such research by dint of having been identified as belonging to a target group – but they might not be aware that the research is taking place. Already, then, we will be confronted with the possible wrong of individuals not being treated as ends in themselves. This wrong has a political dimension in that the power of research subjects is a consideration: one is powerless to withdraw from a study in which one does not know that one is a subject.[13] We might say that researchers who think their work is worth the effort ought to approach those persons who may be captured by it – something that is in principle in their power to do – to give them a chance to opt out, and that research without this opt-out would be impermissible. This would restore to individuals some power. Yet giving people the chance to opt out of a large cohort study would be very difficult in practice, and – perhaps more importantly – would risk undermining the study's scientific integrity, which is itself a criterion of its moral permissibility. There is no clear solution to this sort of problem, though awareness of it is an important precursor to formulating best practice.

---

[11] I have nodded towards this point elsewhere, though without making it explicitly: see I. Brassington, 'John Harris' Argument for a Duty to Research', (2007), *Bioethics*, 21(3), 160–168, esp. at 165. Again, it is hard to see how there is *not* a political aspect to such arguments.

[12] Here, I follow John Rawls's opening gambit: 'Justice is the first virtue of social institutions'. J. Rawls, *A Theory of Justice (Revised edition)* (Oxford University Press, 1999), p. 3.

[13] I use 'subjects' rather than 'participants' here, since to say that one might *participate* in research about which one is unaware is oxymoronic.

Even if that problem is solved, others present themselves. A group might be characterised by an elevated occurrence of certain characteristics. Imagine that members of group *A* typically have an unusually high susceptibility to a given disease, and that members of *B* typically have an elevated inherited resistance to it. Facts like this would generate legitimate questions that would be worth investigating: by learning about how it is that some human bodies are more resilient or susceptible to an illness than others, we could glean insight that would help us prevent it or treat it when it occurs. Yet both groups would also be vulnerable to injustice and exploitation. Thinking about the distribution of the eventual benefits of the research will help show how.

Clearly, medical research contributes to the development of new treatments, at least some of which provide profits for the manufacturers; and the profit motive may drive socially-desirable research. However, the line between just profit and profiteering, which is by definition unjust, is crossed if the benefits of the research are not fairly distributed between researchers – and their backers – and participants. Thus, for example, if any drugs arising from research dependent on the participation of members of *A* are profitmaking, and those participants derive no benefit – perhaps because socioeconomic deprivation makes the drugs unaffordable – that would be a paradigmatic example of injustice. Even if *A* is a reasonably well-represented and educated group, it or its members might be exploited in other ways, perhaps by being targeted specifically for expensive medical interventions. *B* would be less vulnerable on these fronts, since its members' need for any drugs is, by stipulation, reduced. However, again, if members of *B* received no benefit at all from research into which their contribution was crucial, they might still have been exploited. After all, exploitation does not always imply harm – but to have contributed to something that benefits others is to have been exploited; and if this was without recompense, or at least without the opportunity to waive recompense, it is arguably to have been treated wholly as a means to their end, and therefore to have been wronged.

That groups can be exploited or treated unjustly – such as in the ways illustrated by *A* and *B* – is sufficient to show that there is a power differential in play; and because the political domain is that in which power relations are manifested, it is also straightforward to point out that this has a political dimension. As such, a full assessment of the ethics of a given piece of research, and a convincing regulatory policy, would take into account the political situation, both locally and globally.

On the local scale, it would be important to keep in mind questions such as whether the group's vulnerability to exploitation is exacerbated by things like systemic discrimination or economic disadvantage, which may make it difficult for members of a community to assert moral rights that themselves may not be fully reflected in law. The better protected a group is in law, the better able it and its members will be to avoid or resist exploitation in other contexts.

Globally, if research is carried out on people from low-income countries, and the benefits of that research flow overwhelmingly towards high-income countries, what we see is, in effect, a transfer of benefits from the least-wealthy to the most. In this context, the Swiss NGO Public Eye estimates that

> [a]lthough most clinical trials are conducted in the United States and Europe, over the last 20 years there has been a strong tendency towards offshoring to developing and emergent countries. The proportion of testing in emerging markets increased from 10 percent to 40 percent. This continued to increase between 2006 and 2010, while the proportion of clinical trials conducted in Western Europe and the United States fell from 55 percent to 38 percent.[14]

---

[14] Public Eye, 'Ethical Violation', www.publiceye.ch/en/topics/medicines/ethical-violation.

Such a transfer is facilitated and guaranteed by a system of domestic and international laws through the framing of which power becomes visible; and keeping those laws in place, or altering them to reduce the chance of exploitation, is correspondingly a matter of the political will of the powerful. And though individual researchers are powerless to do much about laws that facilitate unjust exploitation derived from research, they are able to do something about the design of individual research programmes, and whether or not they go ahead to begin with.

Yet this is not the most difficult problem in the way of handling group vulnerabilities in health research: that concerns how researchers and regulators should respond when the interests, wishes, and vulnerabilities of different members of a group are in tension. It is this problem to which I turn my attention now.

9.5 GROUP MEMBERSHIP AND GROUP VULNERABILITIES

Return to groups *A* and *B* from the example above. Suppose that researchers are particularly interested in a gene that is common in *A* but not in *B*; they hypothesise that this gene is relevant to understanding the medical condition they are studying. This presents a problem for consent: because genes are not confined to one member of the group, any individual's participation in the programme automatically recruits other members as what we might call 'indirect participants'. It might therefore be argued that every member of the group is vulnerable to having been wronged, even if the 'direct participant' – the person, say, whose blood is drawn – has given full consent. How might we take account of this vulnerability in other members of the group?

It is a commonplace that full, informed consent is at least a part of protecting the rights of research participants; from that we can infer that it would be part of mitigating their vulnerability. But obtaining the consent of each member of the group before beginning the research would be wildly impractical for any but the smallest groups in the most confined geographical areas. More, we would have to decide whether assent to participation must be unanimous: whether, that is, the permission of a person who would presumably not be a direct participant in the research should be a requirement to secure the participation of those who would be. Inasmuch as that this is a question about the relationship of individuals to each other, it is political.

Another layer of complication is added if we deny that a group's vulnerabilities are reducible to those of the aggregate of its members – and it seems as though this may sometimes be the case. Plausibly, there will be situations in which the vulnerabilities of individuals and of groups do not map onto each other particularly closely, if at all: groups can be vulnerable in their own right. For example, the size of a tribe of hunter-gatherers may fall as its members urbanise; we might therefore want to say that the group is increasingly vulnerable even as individual members, thanks to better access to things like health care and education, become less so. But if this is correct, then even addressing every individual's vulnerability may not address wholly the vulnerabilities of the group in the abstract, and so even unanimous consent may be insufficient to prevent impersonal harms or wrongs. Yet it does not seem plausible to say that a research programme should not go ahead because it is impossible to guarantee that the vulnerabilities of the group as a whole will not be exploited. Partly, this is because it seems to sacrifice the (probably admirable) willingness to participate of identifiable members of the community on the altar of concerns about everyone and no one in particular. And partly it is because, though the vulnerabilities of identifiable other members of the community and of the community itself may be important, they are not likely to be the only relevant moral consideration. After all: *everyone* who stands to benefit – directly or indirectly, tangibly or intangibly – from the research has an interest in its going ahead. These are political problems: to echo Bernard Crick, 'conflicts of interest, when public, create political activity'.[15]

Having a representative or representative body that can speak on behalf of the group broadly understood may be suggested as a way forward. For example, Charles Weijer argues that, although some groups and communities 'do not possess a legitimate political authority empowered to make binding decisions on behalf of members', which means that 'it would be both impossible and inappropriate to seek community consent for research participation', they 'may nonetheless have representative groups, and researchers ought to engage these groups in a dialogue concerning study design, conduct, and research results'.[16] Yet we may still wonder how we determine who represents the community and in what way, and what we should do if and when the views of members of the group or community broadly understood diverge from the views of its notional representatives. We should not forget the possibility that would-be research participants may be vulnerable to peer pressure, either to participate or not to, from the group of which they are a part. In this light, it is not obvious what should happen if one member of group A or B from the example above is willing to volunteer as a research participant when those representatives are opposed, or *vice versa*. Again: since these problems concern how individuals and groups interact, they are plainly political.

Neither should we forget that individuals may be members of several communities or groups simultaneously. As such, referring to membership of *a* community is likely to mask other problems. Accordingly, when, in the context of genetic research, Jones et al. state that, 'depending on the research focus',

> a community may include a group sharing a common geographic location, ethnicity, disease, occupation, etc. as well as virtual communities linked regionally, nationally or internationally[17]

they leave open questions about whether one must specify just one of these, and which – if any – takes priority over the other. At some point, someone would have to stipulate that the 'kind' of community in question is *this* or *that*; but such stipulations would appear to be always disputable, and likely politicised to boot.

When considering research involving vulnerable groups then, the relative power of the researcher (and the researcher's backers) and the participant is not the only consideration. Researchers' power relative to that of the group as a whole would also be important to keep in mind; at the same time, so would the power of the group as a whole in relation to the individual participant. Finally, even if we think that the interests of the community are significant, there is a lingering question of where the boundaries of the community should be drawn: sufficiently cosmopolitan politics may deny that the boundaries of this or that group are significant.[18] These questions are inescapably political given the understanding of the political as that domain in which power is manifested, but also political in a more everyday sense, because they speak to problems of how individuals relate to the groups and communities of which they are a part, and how we define group or community membership.

---

[15] B. Crick, *In Defence of Politics* (London: Bloomsbury Academic, 2013), 10.

[16] See, for example, C. Weijer, 'Community Consent for Genetic Research', (2006) *eLS*, 3.

[17] D. Jones et al., 'Beyond Consent: Respect for Community in Genetic Research', (2014) *eLS*, 4.

[18] I am conflating 'group' and 'community' here – but they may not be quite the same. We can arrange people or things into groups notwithstanding that they have no sense of community. A community is a *kind* of group: one that recognises, self-identifies as a community under the auspices of, *and endorses the importance of* some common feature. I do not think that this distinction makes much difference for the points I am making.

9.6 POLITICS AND PROTECTION

I noted earlier in this chapter Wrigley and Dawson's claim that there is an imperative to mitigate vulnerability. Allowing that there is such an imperative, it speaks to the obligations individuals have to each other, but also to the responsibilities of the community, as expressed through the state. Either way, there is a political dimension to it. More, it is reasonable to suppose that health research is one of the things that might be enlisted as a means of mitigating universally-shared human vulnerabilities, and it is likely that a functioning state of some sort is necessary to facilitate such research. Indeed, the idea that political existence is in one way or another crucial to human flourishing has been a touchstone of western philosophy since Aristotle.[19]

It should also be remembered that, as well as facilitating research, protecting research subjects from harms and wrongs more generally – notably, through regulation – falls within the state's demesne. It is in this light that we might consider moves such as the reforms to the Mexican General Health Law approved in 2008, which made 'the sampling of genetic material and its transport outside of Mexico without prior approval [...] illegal'.[20] The Genomic Sovereignty amendment states that Mexican-derived human genome data are the property of Mexico's government, and prohibits and penalises their collection and use in research without prior government approval.[21] This may be seen as an attempt by the Mexican state to protect vulnerable groups within it from the depredations of large and wealthy biotech companies. Such moves may be seen as particularly called-for when, for example, the results of genetic research might be patentable. In such circumstances, a national government can shield minority groups that might not be able to resist unjust exploitation on their own, and can work to give them authority over what happens to data derived from their members.

This is not the only way to see things, though. Cooperation with commercial research institutions could provide vulnerable groups – think again of groups A and B above – with a way to capitalise on their own genetic resources, by entering into benefit-sharing agreements that guarantee them a portion of any proceeds. Such cooperation may also provide a way for research attention to be paid to conditions that may be more prevalent in that community than elsewhere. On this basis, legislative moves such as Mexico's may be seen as an appropriation, however well-meaning, of the rights of some of its people(s) to decide for themselves how to handle data derived from their genes. Alternatively, it may be national governments that are best able to persuade biotech companies to research certain conditions at all; and the state may be able to use its power not to prevent a group exploiting its genetic resources, or to coopt them, but to ensure that the group in question it is able to exploit them effectively, since only national governments have the heft to ensure that the exploitation is not of the objectionable sort.

9.7 CONCLUSION

Research promises us a way to address human vulnerabilities, but it may exacerbate others in the process. Ensuring informed consent from participants may be a means of mitigating some of these, but not others. Those that it might mitigate often have a political genesis; but the relationship between researcher and participant can only really be understood when its own

---

[19] Aristotle, *The Politics* (London: Penguin, 1992).
[20] B. Séguin et al., 'Genomics, Public Health and Developing Countries: The Case of the Mexican National Institute of Genomic Medicine (INMEGEN)', (2008) *Nature Reviews Genetics*, 9(S1), S5–S9, S6. Slightly modified.
[21] R. Benjamin, 'A Lab of Their Own: Genomic Sovereignty as Postcolonial Science Policy', (2009) *Policy and Society*, 28(4), 341–355.

inherent political dynamic is acknowledged, too. More, the complications of the political aspects of research are magnified when we are dealing with vulnerable groups and communities, and with their members.

It has not been the aim of this chapter to offer any normative suggestions; nevertheless, fully to account for individuals' vulnerability, and reliably to avoid exacerbating or exploiting it unjustly, researchers should probably take account not just of the familiar ethical norms of health research, such as informed consent, but also of the political context in which such norms are applied.

Tools, Processes and Actors

Introduction

Edward Dove and Nayha Sethi

This section of the volume explores the tools, processes and actors at play in regulating health research. Regulators rely on a number of tools or regulatory devices to strike a balance between promoting sound research and protecting participants. Some of the paradigmatic examples are (informed) consent and research ethics review of proposed projects; both are explored in this section. Other examples include intellectual property (especially patents), data access governance models, and benefit-sharing mechanisms. Much of the contemporary scholarship on and practice of health research regulation relies on, and criticises, these tools. Relatedly, and arguably, regulation itself is processual; it is about guiding human practices towards desirable endpoints while avoiding undesirable consequences. There has been little discussion of this processual aspect of regulation to date and the specific processes at play in health research. Contributors in this section explore some of the most crucial processes, including risk–benefit analysis, research ethics review and data access governance mechanisms. Further, as becomes apparent, processes can themselves become tools or mechanisms for regulation. Finally, one cannot robustly explore the contours of health research regulation without a consideration of the roles regulatory actors play. Here, several contributors look at the institutional dimension of regulatory authorities and the crucial role experts and science advisory bodies play in constructing health research regulation.

Despite the breadth of topics explored within this section, an overarching theme emerges across the thirteen chapters: that technological change forces us to reassess the suitability of pre-existing tools, processes, and regulatory/governance ecosystems. While a number of tools and processes are long-standing features of health research regulation and are practised by a variety of long-standing actors, they are coming under increasing pressure in twenty-first-century research, driven by pluralistic societal values, learning healthcare systems, Big Data-driven analysis, artificial intelligence and international research collaboration across geographic borders that thrives on harmonised regulation. As considered by the authors, in some cases, new tools, processes or actors are advocated; in other cases, it may be more beneficial to reform them to ensure remain they fit for purpose and provide meaningful value to health research regulation.

Much of the discussion focusses therefore not only on the nature of these long-standing tools, processes, and actors, but also on how they might be sustained – if at all – well into the twenty-first century. For example, the digital-based data turn necessitates reconsidering fundamental principles like consent and developing new digital-based mechanisms to put participants at the heart of decision-making, as discussed by Kaye and Prictor (Chapter 10). Shabani, Thorogood and Murtagh (Chapter 19) also speak to the challenges that data intensive research is presenting for governance and in particular the challenges of balancing the need to grant (open) access to databases with the need to protect the rights and interests of patients and participants.

This leads to another related theme emerging within this section: the need to examine more closely the participatory turn in health research regulation. Public and participant involvement is becoming an increasingly emphasised component of health research, as illustrated by public engagement exercises becoming mandatory within many research funding schemes. But, as Aitken and Cunningham-Burley (Chapter 11) note, many different forms of public engagement exist and we need to ask 'why' publics are engaged, rather than simply 'how' they are engaged. They suggest that framing public engagement as a political exercise can help us to answer this question. For Chuong and O'Doherty (Chapter 12), the process of participatory governance also necessitates unpacking, particularly due to the varied approaches taken towards embedding deliberative practices and including patients and participants as *partners* within health research initiatives. Both of these chapters help set up discussion and analysis to come later in this book, specifically the contribution from Burgess (Chapter 25), who makes a case for mobilising public expertise in the design of health research regulation.

Beyond the inclusion of publics and participants in decision-making, many authors in this section raise additional questions about decision-making tools and processes involving other regulatory actors. For example, Dove (Chapter 18) notes how research ethics committees have evolved into regulatory entities in their own right, suggesting that they can play an important role in stewarding projects towards an ethical endpoint. Similarly, McMahon (Chapter 21) explores the ways in which institutions (and their scaffolding) can shape and influence decision-making in health research and argues that this ought to be reflected when drafting legal provisions and guidance. On the question of guidance, Sethi (Chapter 17) lays out different implications that rules, principles and best-practice-based approaches can carry for health research, including the importance of capturing previous lessons learned within regulatory approaches. Sethi's discussion of principles-based regulation helps round out the discussion to come later in this book, specifically Vayena and Blassime's contribution (Chapter 26) on Big Data and a proposed model of adaptive governance. Sethi's chapter also engages with another key theme emerging within this section: the construction of knowledge-bases and expertise. For example, Flear (Chapter 16) suggests that basing current framings of regulatory harm as technological risk marginalises critical stakeholder knowledges of harm, in turn limits knowledge-bases. Indeed, in considering how governments make use of expertise to inform health research regulation, Meslin (Chapter 22) concludes that it will be best served when different stakeholders are empowered to contribute to the process of regulation, and when governments are open to advice from the expertise of experts and non-experts alike.

Many of the authors highlight the need to analyse how we anticipate and manage the outputs (beneficial and harmful) of health research. For example, Coleman (Chapter 13) questions the robustness and objectivity attributed to risk–benefit analysis, despite the heavy reliance placed upon it within health research. Similarly, benefit sharing has become a key requirement for many research projects but, as discussed by Simm (Chapter 15), there are practical challenges to deploying such a complex tool to distinct concrete projects. Patents are also a standard feature of health research and innovation. As considered by Nicol and Nielsen (Chapter 14), these can be used both as a positive incentive to foster innovation and, paradoxically, as a means to stifle collaboration and resource sharing.

Three final cross-cutting themes must be kept in mind as we continue to attempt to improve health research regulation. First, in closing this section, Nicholls (Chapter 20) reminds us that we must be mindful of the constant need to evaluate and adapt our approaches to the varying contexts and ongoing developments in health research regulation. Second, in recognition of the fragility of public trust and the necessity of public confidence for health research initiatives to succeed, we must continue to strive for transparency, fairness and inclusivity within our practices. Finally, as we seek to refine and develop new approaches to health research regulation, we must acknowledge that no one tool or process can provide a panacea for the complex array of values and interests at stake. All must be kept under constant review as part of a well-functioning learning system, as Laurie argues in the Afterword to this volume.

10

Consent

Jane Kaye and Megan Prictor

10.1 INTRODUCTION

Informed consent is regarded as the cornerstone of medical research; a mechanism that respects human dignity and enables research participants to exercise their autonomy and self-determination. It is a widely accepted legal, ethical and regulatory requirement for most health research. Nonetheless, the practice of informed consent varies by context, is subject to exceptions, and, in reality, often falls short of the theoretical ideal.[1] The widespread use of digital technologies this century has revolutionised the collection, management and analysis of data for health research, and has also challenged fundamental principles such as informed consent. The previously clear boundaries between health research and clinical care are becoming blurred in practice, with implications for implementation and regulation. Through our analysis we have identified the key components of consent for research articulated consistently in international legal instruments. This chapter will: (1) describe the new uses of data and other changes in health research; (2) discuss the legal requirements for informed consent for research found in international instruments; and (3) discuss the challenges in meeting these requirements in the context of emerging research data practices.

10.2 THE CHANGING NATURE OF RESEARCH

Health research is no longer a case simply of the physical measurement and intimate observation of patients. Rather, it increasingly depends upon the generation and use of data, and new analysis tools such as Artificial Intelligence (AI). Health research has been transformed by innovations in digital technologies enabling the collection, curation and management of large quantities of diverse data from multiple sources. The intangible nature of digital data means that it can be perfectly replicated indefinitely, instantly shared with others across geographical borders and used for multiple purposes, such as clinical care and research. The information revolution enables data to be pulled from different sources such as electronic medical records; wearables and smart phones monitoring chronic conditions; and datasets outside the health care system yielding inferences about an individual's health. These developments have significant implications for informed consent.

---

[1] C. Grady, 'Enduring and Emerging Challenges of Informed Consent', (2015) *New England Journal of Medicine*, 372(9), 855–862.

New technologies have enabled the development of ambitious scientific agendas, new types of infrastructure such as biobanks and genomic sequencing platforms and international collaborations involving datasets of thousands of research participants. Much innovation is driven by collaborations between clinical and research partners that provide practical need and clinical data, and companies offering technical expertise and resources. Examples are: national genomic initiatives including Genomics England (UK), All of Us (USA), Aviesan (France), Precision Medicine Initiative (China); international research collaborations like the Human Genome Project, Global Alliance for Genomics and Health, the Personal Genome Project; and mission-orientated collaborations such as Digital Technology Supercluster (Canada) and the UK Health Data Research Alliance.

The greatest challenges emerge around informed consent in these new contexts where already-collected data can be used in ways not anticipated at the time of collection and data can be sent across jurisdictional borders. When data and tissue samples are being collected for multiple unknown future research uses, explicit informed consent to the research aims and methods may not be possible. In response to this practical challenge, the World Medical Association (WMA) adopted the *Declaration of Taipei on Ethical Considerations regarding Health Databases and Biobanks* (2002, revised 2016). It stipulates that instead of consenting to individual research, individuals may validly consent to the purpose of the biobank, the governance arrangements, privacy protections, risks associated with their contribution and so on. This form of 'broad consent' is really an agreement that others will govern the research, since determinations about appropriate uses of the data and biomaterials are decided by researchers with approval by research ethics committees or similar bodies.[2]

10.3 THE BASIS FOR INFORMED CONSENT

The moral force of consent is not unique to health research; it is integral to many interpersonal interactions, as well as being entrenched in societal values. The key moral values at play in medical research are: **autonomy** – the right for an individual to make his or her own choice; **beneficence** – the principle of acting with the participant's best interests in mind; **non-maleficence** – the principle that 'above all, do no harm'; and **justice** – emphasising fairness and equality among individuals.[3] The concepts of voluntariness and transparency embedded in informed consent speak to the ethical value of respect for human beings, their autonomy, their dignity as free moral agents and their welfare. This respect for individuals has resulted in special protections for those who are not legally competent to provide informed consent. Beneficence requires that the probable benefits of the research project outweigh the harms. In the context of informed consent, non-maleficence demands that harm is minimised by researchers being attuned to participant welfare and fully disclosing likely benefits and risks to permit adequately informed choice. The principle of justice in the research setting requires that potential participants are equally provided with adequate information to make a knowledgeable decision, helping to avoid participant exploitation. Consideration of the ethical principles underpinning informed consent also requires reflection on cultural values, such as those pertaining to specific indigenous communities or ethnic groups. Cultural values may lead researchers to consider, for example, whether unique harms to cultural integrity and heritage could accrue to certain groups through specific research projects, and whether respect for human beings should be seen

---

[2] S. Boers et al., 'Broad Consent Is Consent for Governance', (2015) *American Journal of Bioethics*, 15(9), 53–55.
[3] T. Beauchamp and J. Childress, *Principles of Biomedical Ethics*, 4th Edition (Oxford University Press, 1994).

through a lens of collective, as well as individual, autonomy and well-being.[4] These ethical principles underpin informed consent in health research practice, but not all of them have been implemented into law.

10.4 LEGAL REQUIREMENTS FOR INFORMED CONSENT

The requirements for informed consent emerged from a range of egregious examples of physical experimentation on humans. Among the most notable examples were the Nuremberg trials following World War II, although concern about harmful research practices internationally had surfaced decades earlier.[5] The trial of Nazi doctors produced a ten-point Code that became the foundation of modern health research ethics. Voluntary consent was its first and arguably most emphasised principle.[6] It has since been espoused in declarations by international and non-governmental organisations. A key instrument is the WMA's Declaration of Helsinki (1964, as amended) setting out the basic requirements for informed consent for research.

> In medical research involving human subjects capable of giving informed consent, each potential subject must be adequately informed of the aims, methods, sources of funding, any possible conflicts of interest, institutional affiliations of the researcher, the anticipated benefits and potential risks of the study and the discomfort it may entail, post-study provisions and any other relevant aspects of the study. The potential subject must be informed of the right to refuse to participate in the study or to withdraw consent to participate at any time without reprisal.[7]

Crucial to this formulation is the need to communicate and provide detailed information to the 'human subject'. While this information should be comprehensive enough for participants to make an informed decision, it positions the researcher as the information provider and the subject as a passive recipient. Yet, the Declaration also posits ongoing engagement as an essential requirement as the participant can withdraw consent at any time.

The principle of free consent also forms part of the United Nations' International Covenant on Civil and Political Rights (Article 7). Further guidelines and conventions promulgated by international organisations such as the International Council for Harmonisation of Technical Requirements for Pharmaceuticals for Human Use (ICH),[8] the Council for International Organizations of Medical Sciences,[9] and the Council of Europe,[10] endorse and explain these principles. The ICH Good Clinical Practice Guideline considers consent in the context of human clinical trials; it establishes a unified quality standard for the European Union, Japan and

---

[4] For instance: National Health and Medical Research Council [Australia], 'Ethical Conduct in Research with Aboriginal and Torres Strait Islander Peoples and Communities', (NHMRC, 2018); L. Jamieson et al., 'Ten Principles Relevant to Health Research among Indigenous Australian Populations', (2012) *Medical Journal of Australia*, 197(1), 16–18.

[5] A. Dhai, 'The Research Ethics Evolution: From Nuremberg to Helsinki', (2014) *South African Medical Journal*, 104 (3), 178–180.

[6] Trials of War Criminals before the Nuremberg Military Tribunals under Control Council Law No. 10 [Nuremberg Code] (1949) para. 1.

[7] World Medical Association, 'Declaration of Helsinki – Ethical Principles for Medical Research Involving Human Subjects', (World Medical Association, 1964, 2013 version), para. 26. [hereafter 'Declaration of Helsinki']

[8] International Council for Harmonisation of Technical Requirements for Pharmaceuticals for Human Use (ICH), 'Guideline for Good Clinical Practice', (ICH, 1996).

[9] Council for International Organizations of Medical Sciences, 'International Ethical Guidelines for Biomedical Research Involving Human Subjects', (CIOMS, 2002, updated 2016).

[10] Convention for the Protection of Human Rights and Dignity of the Human Being with regard to the Application of Biology and Medicine: Convention on Human Rights and Biomedicine, Oviedo, 04/04/1997, in force 01/12/1999, ETS No. 164.

the USA. The Oviedo Convention and the 2005 Additional Protocol relating to biomedical research similarly foreground consent, stipulating that it be 'informed, free, express, specific and documented'.[11] The European General Data Protection Regulation (GDPR) has raised the bar for informed consent for data use worldwide. In Australia, the National Health and Medical Research Council's *National Statement on Ethical Conduct in Human Research* (2007, updated 2018) is the principle guiding document for health research. From these documents, several key components can be discerned, such as competence, transparency and voluntariness, and that consent must be informed.

Only 'human subjects capable of giving informed consent' are the subject of the Helsinki Declaration statement about consent. Ethicists have described competent people as those who have 'the capacity to understand the material information, to make a judgement about the information in light of his or her values, to intend a certain outcome, and to freely communicate his or her wish to caregivers or investigators'.[12] Special protections pertain to those not competent to give consent, such as some young children and some people who are physically, mentally or intellectually incapacitated. These protections centre upon authorisation by a research ethics committee and consent provided by a legal representative. The potential participant may still be asked to assent to the research.

Assessing competence represents a challenge in relation to biobanks and other longitudinal research endeavours where people contributing data or tissue samples may have shifting competence over time; for instance people who were enrolled into research as children will become competent to provide consent for themselves as they reach adulthood.[13] People impacted by cognitive decline or mental illness may lose competence to provide consent, either temporarily or indefinitely. Periodically revisiting consent for participants is an ethically appropriate, yet logistically demanding, response.

As indicated above, the Nuremberg Code and the Declaration of Helsinki outline a range of information that potential research participants are to be given to enable them to be informed before making a choice about enrolment. The ICH Guideline goes into further detail regarding clinical trials, stating that the information should be conveyed orally and in writing (4.8.10), and that that the explanation should include:

- Whether the expected benefits of the research pertain to the individual participants;
- What compensation is available if harm results;
- The extent to which the participant's identity will be disclosed;
- The expected duration of participation;
- How many participants are likely to be involved in the research.

National or regional statutes and guidelines stipulate the required informational elements for consent to health research in their jurisdictions, mirroring the elements contained in the international instruments to varying degrees.[14]

---

[11] Additional Protocol to the Convention on Human Rights and Biomedicine, Concerning Biomedical Research,' Strasbourg, 21/05/2005, in force 01/09/2007, CETS No. 195, Article 14.1.

[12] Beauchamp and Childress, *Principles of Biomedical Ethics*, p. 135.

[13] M. Taylor et al., 'When Can the Child Speak for Herself?', (2018) *Medical Law Review*, 26(3), 369–391.

[14] For example, Human Biomedical Research Act 2015, sec. 12 (Singapore); National Health and Medical Research Council, Australian Research Council, and Universities Australia, 'National Statement on Ethical Conduct in Human Research', (NHMRC, 2007), ch 2.2. [hereafter 'NHMRC National Statement']; Health Research Authority, 'Consent and Participant Information Guidance', (HRA) (UK) ; Federal Policy for the Protection of Human Subjects ('Common Rule'), 45 CFR part 46, para. 46.114, (1991); The Medicines for Human Use (Clinical Trials) Regulations 2004 No. 1031, Schedule 1 (UK).

Limited disclosure of information may sometimes be permitted, for instance in a study of human behaviour where the research aims would be frustrated by full disclosure to participants.[15] It may also be a necessary consequence of the difficulty of comprehensive disclosure in the context of Big Data science, where not all the uses of the data (that may not be collected directly from the individual) can be anticipated when the data are collected.

The Declaration of Helsinki requirement that research participants must be 'adequately informed' points to further consideration of how best to communicate the complex information described above. This is the focus of much recent law and guidance.[16] Research has shown repeatedly that participants often do not understand the investigative purpose of clinical trials, key concepts such as randomisation and the risks and benefits of participation.[17] Using simple language and providing enough time to consider the information can help, as well as tailoring information to participant age and educational level. Researchers have evaluated tools to assist with communicating information in ways that support understanding.[18] Complex, heterogenous and changing research endeavours that cross geographic boundaries and blur the lines between clinical care, daily life and research pose an additional challenge to the requirement for transparency.

A consistent requirement of international conventions, law and guidelines for ethical research is that for consent to be valid, it must be voluntary.[19] The Nuremberg Code obliges researchers to avoid 'any element of force, fraud, deceit, duress, over-reaching, or other ulterior form of constraint or coercion'.[20] Beyond the problem of overt coercion by another person, other considerations in evaluating voluntariness include: deference to the perceived power of the researcher or institution;[21] the mere existence of a power imbalance;[22] the existence of a dependent relationship with the researcher;[23] and the amount paid to participants.[24] On power and vulnerabilities, see further Brassington, Chapter 9, this volume.

These concerns are largely associated with duress as a result of specific relationships developed through personal interactions. In Big Data or AI analysis, the concept of voluntariness must be reconsidered, as often the data users are not known to the data subject and the nature of the duress may not be straightforwardly attributed to particular relationships. An example is companies that provide direct-to-consumer genetic tests, where the provision of test results also enables the companies to use the data for purposes including marketing and research. This is a

---

[15] NHMRC National Statement, chap. 2.3.

[16] NHMRC National Statement, para. 5.2.17; Regulation (EU) 2016/679 of the European Parliament and of the Council of 27 April 2016 on the protection of natural persons with regard to the processing of personal data and on the free movement of such data, and repealing Directive 95/46/EC (General Data Protection Regulation), OJ 2016 L 119/1 Recital 58.

[17] M. Falagas et al., 'Informed Consent: How Much and What Do Patients Understand?', (2009) *American Journal of Surgery*, 198(3), 420–435; On risk-benefit analysis, see also Coleman, Chapter 13 in this volume.

[18] For example: A. Synnot et al., 'Audio-Visual Presentation of Information for Informed Consent for Participation in Clinical Trials', (2014) *Cochrane Database of Systematic Reviews*, (5); J. Flory and E. Emanuel, 'Interventions to Improve Research Participants' Understanding in Informed Consent for Research: A Systematic Review', (2004) *JAMA*, 292(13), 1593–1601.

[19] Additional Protocol to the Convention on Human rights and Biomedicine, Article 14.1; ICH, 'Guideline for Good Clinical Practice', paras 2.9 and 3.1.8; NHMRC National Statement, para. 2.2.9; General Data Protection Regulation, Article 4(11); Declaration of Helsinki, para. 25.

[20] Nuremberg Code, para. 1.

[21] NHMRC National Statement, para. 2.2.9.

[22] General Data Protection Regulation, Recital 43.

[23] Declaration of Helsinki, para. 27.

[24] ICH, 'Guideline for Good Clinical Practice', para. 3.1.8; NHMRC National Statement, para. 2.2.10.

different kind of duress as people lured through the fine print in click-wrap contracts are then enrolled into research.[25]

Traditionally, valid informed consent occurs before the participant's involvement in the research;[26] no specific timing is recommended as long as there is time for the person to acquire sufficient understanding of the research. In selected circumstances, 'deferred' consent – where individuals do not know they are enrolled in a clinical trial so that the sample is not biased and they are asked for consent later on[27] – a waiver of consent or an opt-out approach might be justifiable. These are typically addressed within relevant guidance.[28] Once-off informed consent before a project starts may, however, be insufficient to acquit researchers' responsibilities in the context of longitudinal data-intense research infrastructures. Modalities that permit ongoing or at least repeated opportunities to refresh consent, such as staged consent and Dynamic Consent, considered below, are a developing response to this issue.

It is a key principle of health research, traceable back to the Declaration of Helsinki, that potential research participants have a right to decline the invitation to participate without giving a reason and should not incur any disadvantage or discrimination as a consequence.[29] Further, people who have consented must be free to withdraw consent at any time without incurring disadvantage. The GDPR stipulation that 'It shall be as easy to withdraw as to give consent',[30] has energised research into technology-based tools to facilitate seamless execution of a withdrawal decision, or even to support shifting levels of participation over time.[31]

Newer research methods and infrastructures characterised by open-ended research activities and widespread data sharing add complexity to the interpretation of 'withdrawing consent'. International guidelines have acknowledged that withdrawal in this context might equate to no new data collection while raising a question over whether existing samples and data must be destroyed or remain available for research.[32]

10.5 THE LIMITATIONS OF CONSENT

In research involving human participants, the informed consent process is foregrounded.

As a legal mechanism intended to protect human subjects in the way envisaged by international instruments, it is also recognised that consent may be insufficient. People often do not understand what they have agreed to participate in, retain the information about the research or even recall that they agreed to be involved.[33] Consent is not the only legal basis for conducting

[25] A. Phillips, *Buying your Self on the Internet: Wrap Contracts and Personal Genomics* (Edinburgh University Press, 2019).

[26] ICH, 'Guideline for Good Clinical Practice', para. 2.9.

[27] L. Johnson and S. Rangaswamy, 'Use of Deferred Consent for Enrolment in Trials is Fraught with Problems', (2015) *BMJ*, 351.

[28] NHMRC National Statement, chap. 2.3; The paper N. Songstad et al. and on behalf of the HIPSTER trial investigators, 'Retrospective Consent in a Neonatal Randomized Controlled Trial', (2018) *Pediatrics*, 141(1), e20172092 presents an example of deferred consent.

[29] Declaration of Helsinki, paras 26, 31.

[30] General Data Protection Regulation, Article 7(3).

[31] K. Melham et al., 'The Evolution of Withdrawal: Negotiating Research Relationships in Biobanking' (2014) *Life Sciences, Society and Policy*, 10(1), 1–13.

[32] Council for International Organizations of Medical Sciences and World Health Organization, 'International Ethical Guidelines for Epidemiological Studies', (CIOMS, 2009) p. 48.

[33] J. Sugarman et al., 'Getting Meaningful Informed Consent From Older Adults: A Structured Literature Review of Empirical Research', (1998) *Journal of the American Geriatrics Society*, 46(4), 517–524; P. Fortun et al., 'Recall of Informed Consent Information by Healthy Volunteers in Clinical Trials', (2008) *QJM: An International Journal of Medicine*, 101(8) 625–629; R. Broekstra et al., 'Written Informed Consent in Health Research Is Outdated', (2017)

health research. While there is variation between jurisdictions, broadly speaking research involving data or tissue may be able to proceed without consent in certain circumstances. These include if: there is an overriding public interest and consent is impracticable; there is a serious public health threat; the participant is not reasonably identifiable; or the research carries low or negligible risk. Many researchers have sought to augment traditional modes of consent at the point of entry to research, to support informed decision-making by potential participants. New consent processes seek to enable truly informed consent rather than doing away with this fundamental requirement.

Traditionally, consent is operationalised as a written document prepared by the researcher setting out the information described above. The participant's agreement is indicated by their signature and date on the document. Concerns about participant problems with reading and understanding the form have led to initiatives including simplified written materials, extra time and the incorporation of multimedia tools.[34] More nuanced consent modalities might encompass different tiers of information – with simple, minimally compliant information presented first, linking to more comprehensive explanation – and different staging of information, for instance with new choices being presented to participants at a later time.[35]

Scholars have also considered when and how it might be appropriate to diverge from the notion of the individual human subject as the autonomous decision-maker for health research participation, towards a communitarian approach informed by ethical considerations pertaining to culture and relationships. The concept of informed consent must, in this context, expand to incorporate the possibility of family and community members at least being consulted, perhaps even deciding jointly. Osuji's work on relational autonomy in informed consent points to decisions 'made not just in relation to others but with them, that is, involving them: family members, friends, relations, and others'.[36] This approach might particularly suit some groups, with extensive examples deriving from Australian aboriginal and other Indigenous communities,[37] family members with shared genetic heritage[38] and some Asian and African cultures.[39] Communitarian-based consent processes may not meet legal requirements for informed consent to research, but may nevertheless be a beneficial adjunct to standard processes in some instances.

European Journal of Public Health, 27(2), 194–195; Falagas et al., 'Informed Consent'; H. Teare et al., 'Towards "Engagement 2.0": Insights From a Study of Dynamic Consent with Biobank Participants', (2015) *Digital Health*, 1, 1–13.

[34] A. Nishimura et al., 'Improving Understanding in the Research Informed Consent Process', (2013) *BMC Medical Ethics*, 14(1), 1–15; Synnot et al., 'Audio-Visual Presentation'; B. Palmer et al., 'Effectiveness of Multimedia Aids to Enhance Comprehension of Research Consent Information: A Systematic Review', (2012) *IRB: Ethics & Human Research*, 34(6), 1–15; S. McGraw et al., 'Clarity and Appeal of a Multimedia Informed Consent Tool for Biobanking', (2012) *IRB: Ethics & Human Research*, 34(1), 9–19; C. Simon et al., 'Interactive Multimedia Consent for Biobanking: A Randomized Trial', (2016) *Genetics in Medicine*, 18(1), 57–64.

[35] E. Bunnik et al., 'A Tiered-Layered-Staged Model for Informed Consent in Personal Genome Testing', (2013) *European Journal of Human Genetics*, 21(6), 596–601.

[36] P. Osuji, 'Relational Autonomy in Informed Consent (RAIC) as an Ethics of Care Approach to the Concept of Informed Consent', (2017) *Medicine, Health Care and Philosophy*, 21(1), 101–111, 109.

[37] F. Russell et al., 'A Pilot Study of the Quality of Informed Consent Materials for Aboriginal Participants in Clinical Trials', (2005) *Journal of Medical Ethics*, 31(8), 490–494; P. McGrath and E. Phillips, 'Western Notions of Informed Consent and Indigenous Cultures: Australian Findings at the Interface', (2008) *Journal of Bioethical Inquiry*, 5(1), 21–31.

[38] J. Minari et al., 'The Emerging Need for Family-Centric Initiatives for Obtaining Consent in Personal Genome Research', (2014) *Genome Medicine*, 6(12), 118.

[39] H3Africa Working Group on Ethics, 'Ethics and Governance Framework for Best Practice in Genomic Research and Biobanking in Africa', (H3Africa, 2017).

10.6 NEW DIGITAL CONSENT MECHANISMS

The pervasion of technology into all aspects of human endeavour has transformed health research activities and the consent processes which support them. Electronic consent may mean simply transferring the paper form to a computerised version. Internationally, electronic signatures are becoming generally accepted as legally valid in various contexts.[40] These may comprise typewritten or handwritten signatures on an electronic form, digital representations such as fingerprints or cryptographic signatures. Progress is being made on so-called digital, qualified or advanced electronic signatures which can authenticate the identity of the person signing, as well as the date and location.[41]

Semi-autonomous consent is emerging in computer science; it refers to an approach in which participants record their consent preferences up-front, a computer enacts these preferences in response to requests – for instance, invitations to participate in research – and the participants review the decisions, refine their expressed preferences and provide additional information.[42] This could be a way to address consent fatigue by freeing participants from the need to make numerous disaggregated consent decisions. It is a promising development at a time when increasing uses of people's health data for research may overwhelm traditional tick-box consent.

Dynamic Consent is an approach to consent developed to accommodate the changes in the way that medical research is conducted. It is a personalised, digital communication interface that connects researchers and participants, placing participants at the heart of decision-making. The interface facilitates two-way communication to stimulate a more engaged, informed and scientifically-literate participant population where individuals can tailor and manage their own consent preferences.[43] In this way it meets many of the requirements of informed consent as stipulated in legal instruments[44] but also allows for the complexity of data flows characterising health research and clinical care. The approach has been used in the PEER project,[45] CHRIS,[46] the Australian Genomics Health Alliance[47] and the RUDY project.[48] It seems appropriate to have digital consent forms for a digital world that allow for greater flexibility and engagement with patients when the uses of data for research purposes cannot be predicted at the time of collection.

[40] United Nations, 'United Nations Convention on the Use of Electronic Communications in International Contracts', (UNCITRAL, 2005) Article 9(3); Electronic Transactions Act 1999 (Cth) sec. 8(1); Regulation (EU) No 910/2014 of the European Parliament and of the Council of 23 July 2014 on electronic identification and trust services for electronic transactions in the internal market and repealing Directive 1999/93/EC (2014); CFR Code of Federal Regulations Title 21 Part 11, (1997) (USA); Electronic Signatures in Global and National Commerce Act 2000, Pub. L. No. 106-229, 114 Stat. 464 (2000) (USA).

[41] Health Research Authority and Medicines and Healthcare Products Regulatory Agency, 'Joint Statement on Seeking Consent by Electronic Means', (HRA and MHPRA, 2018) p. 5.

[42] R. Gomer et al., 'Consenting Agents: Semi-Autonomous Interactions for Ubiquitous Consent', Proceedings of the 2014 ACM International Joint Conference on Pervasive and Ubiquitous Computing (Seattle, Washington: ACM Press, 2014), pp. 653–58.

[43] J. Kaye et al., 'Dynamic Consent: A Patient Interface for Twenty-First Century Research Networks', (2015) European Journal of Human Genetics, 23, 141–146.

[44] M. Prictor et al. 'Consent for Data Processing Under the General Data Protection Regulation: Could "Dynamic Consent" be a Useful Tool for Researchers?', (2019) Journal of Data Protection and Privacy, 3(1), 93–112.

[45] Genetic Alliance, 'Platform for Engaging Everyone Responsibly', www.geneticalliance.org/programs/biotrust/peer.

[46] CHRIS eurac research, 'Welcome to the CHRIS study!', (CHRIS), www.de.chris.eurac.edu.

[47] Australian Genomics Health Alliance, 'Introducing CTRL', www.australiangenomics.org.au/introducing-ctrl-a-new-online-research-consent-and-engagement-platform.

[48] H. Teare et al., 'The RUDY Study: Using Digital Technologies to Enable a Research Partnership', (2017) European Journal of Human Genetics, 25, 816–822.

10.7 CONCLUSION

The organisation and execution of health research has undergone considerable change due to technological innovations that have escalated in the twenty-first century. Despite this, the requirements of informed consent enshrined in the Nuremberg Code are still the basic standard for health research. These requirements were formulated specifically in response to atrocities that occurred through physical experimentation. They continue to be applied to data-based research that is very different in its scope and nature, and in the issues it raises for individuals compared to physically-based research, that was the template for the consent requirements found in international instruments. The process for obtaining and recording consent has undergone little change over time and is still recorded through paper-based systems, reliant on one-to-one interactions. While this works well for single projects with a focus on the prevention of physical, rather than informational harm, it is less suitable when data are used in multiple settings for diverse purposes.

Paper-based systems are not flexible and responsive and cannot provide people with the information that is needed in a changing research environment. Digital systems such as Dynamic Consent provide the tools for people to be given information as the research evolves and to be able to change their mind and withdraw their consent. However, given the complexity and scale of research, when data are collected from a number of remote data points it is difficult for consent to effectively respond to all of the issues associated with data-intensive research. The use of collective datasets that concern communal or public interests are difficult to govern through individual decision-making mechanisms such as consent.[49]

Consent is only one of the many governance mechanisms that should be brought into play to protect people involved in health research. Additionally, attention should be given to the ecosystem of research and informational governance that consist of legal requirements, regulatory bodies and best practice that provide the protective framework that is wrapped around health research. Despite its shortcomings, informed consent is still fundamental to health research, but we should recognise its strengths and limitations. More consideration is needed on how to develop better ways to enable the basic requirements of informed consent to be enacted through digital mechanisms that are responsive to the characteristics of data-intensive research. Further research needs to be directed to how the governance of health research should adapt to this new complexity.

[49] J. Allen, 'Group Consent and the Nature of Group Belonging: Genomics, Race and Indigenous Rights', (2009) *Journal of Law, Information and Science*, 20(2), 28–59.

11

Forms of Engagement

Mhairi Aitken and Sarah Cunningham-Burley

11.1 INTRODUCTION

Public engagement (PE) is part of the contemporary landscape of health research and innovation and considered a panacea for what is often characterised as a problem of trust in science or scientific research, as well as a way to ward off actual or potential opposition to new developments. This is quite a weight for those engaging in engagement to carry, and all the more so since PE is often underspecified in terms of purpose. PE can mean and involve different things but such flexibility can come at the price of clarity. It may allow productive creativity but can limit PE's traction.[1]

In this chapter we provide a synthesis of current conceptualisations of PE. We then consider what kinds of publics are 'engaged with' and what this means for the kinds of information exchanges and dialogues that are undertaken. Different forms of PE 'make up' different kinds of publics: engagements do not, indeed cannot, start with a clean sheet – neither with a pure public nor through a pure engagement.[2] As Irwin,[3] among others, has noted, PE is a political exercise and this wider context serves to frame what is engaged about. It is therefore all the more important to reflect on the practice of PE and what it is hoped will be achieved. We argue that clarity and transparency about the intention, practice and impact of PE are required if PE is to provide an authentic and meaningful tool within health research governance.

11.2 ENGAGING WITH CRITIQUE

PE has been a subject of debate for many years, particularly in the Science and Technology Studies literature, through what is termed critical public understanding of science. From Wynne's[4] seminal work onwards, this critique has championed the range of expertise that can come to bear on matters scientific and has provided analytical verve to critiques of the institutional arrangements of both science and PE. Criticisms of top-down models of PE were dominant throughout the 1990s and the 'deficit model of public understanding' was roundly

[1] S. Parry et al., 'Heterogeneous Agendas Around Public Engagement in Stem Cell Research: The Case for Maintaining Plasticity', (2012) *Science and Technology Studies*, 12(2), 61–80.

[2] K. Braun and S. Schultz, '"... A Certain Amount of Engineering Involved": Constructing the Public in Participatory Governance Arrangements', (2010) *Public Understanding of Science*, 19(4), 403–419.

[3] A. Irwin, 'The Politics of Talk: Coming to Terms with the 'New' Scientific Governance', (2012) *Social Studies of Science*, 36(2), 299–320.

[4] B. Wynne 'May the Sheep Safely Graze? A Reflexive View of the Expert–Lay Knowledge Divide' in S. Lash et al. (eds), *Risk, Environment and Modernity: Towards a New Ecology* (London: Sage, 1998).

debunked not least for suggesting that public ignorance of science was a fundamental cause of loss of trust. This critique played an important role in bringing about a new emphasis on two-way processes of PE that went beyond 'educating the public'.[5] New commitments to dialogue and engagement – 'the participatory turn' – have become more commonplace and mainstream.[6] However, as Stilgoe and colleagues have commented, the shift from deficit model approaches to dialogic PE, has been only partially successful:

> It has been relatively easy to make the first part of the argument that monologues should become conversations. It has been harder to convince the institutions of science that the public are not the problem. The rapid move from doing communication to doing dialogue has obscured an unfinished conversation about the broader meaning of this activity.[7]

Herein lies further threats to the integrity of PE.

PE is now a component of much health research where engagement or patient and public involvement is often a funding requirement. This is particularly pronounced in the UK where public understanding and engagement in science has gained increasing institutional traction since the House of Lords report in 2000. For some, the deficit model of public understanding has simply been replaced with a deficit model of public trust, to which 'more understanding' and, even, 'more dialogue' remain a solution.[8] So, on the one hand the deficit model of publics in need of education about science lingers on, sometimes under the guise of trust. Yet, on the other, we see PE being taken up across sectors – and there is evidence of PE, sometimes, bringing science and its governance to account.

PE can be productive as many commentators have posited.[9] The task for health research governance is to ensure that participatory practices are not skewed towards institutional ends but allow diverse voices into the policy making process so that they can make a difference to how health research is conducted, regulated and held accountable to the very publics it purports to serve. As Braun and Schultz note:

> The question that is increasingly discussed in public understanding of science (PUS) today is not so much whether there is a trend towards participation but what we are to make of it, how to assess it, how to understand the dynamics propelling it, how to systematise and interpret the different forms and trajectories it takes, what the benefits, pitfalls or unintended side-effects of these forms and trajectories are and for whom.[10]

We now turn to consider some of these questions.

11.3 FORMS OF PUBLIC ENGAGEMENT

Enthusiasm for, and professed commitment to, PE does not easily translate into meaningful engagement in practice. This is in no small part due to the fact that the term 'public engagement' can be interpreted in many different ways and PE is undertaken for a variety of reasons.

[5] M. Kurath and P. Gisler, 'Informing, Involving or Engaging? Science Communication, in the Ages of Atom-, Bio-and Nanotechnology', (2009) *Public Understanding of Science*, 18(5), 559–573.

[6] Irwin, 'The Politics of Talk'.

[7] J. Stilgoe et al., 'Why Should We Promote Public Engagement with Science?', (2014) *Public Understanding of Science*, 23(1), 4–15, 8.

[8] S. Cunningham-Burley, 'Public Knowledge and Public Trust', (2006) *Community Genetics*, 9(3), 204–210; B. Wynne, 'Public Engagement as a Means of Restoring Public Trust in Science – Hitting the Notes, but Missing the Music?', (2006) *Community Genetics*, 9(3), 211–220.

[9] A. Irwin and M. Michael, *Science, Social Theory and Public Knowledge* (Berkshire: Open University Press, 2003).

[10] Braun and Schultz, '... A Certain Amount of Engineering', 404.

Key challenges around PE are that the different ideas about its role and value manifest in a variety of purposes and rationales, whether implicit or explicit. PE can be underpinned by normative, substantive or instrumental rationales.[11] A normative position suggests that PE should be conducted as it is 'the right thing to do' – something that is part and parcel of both public and institutional expectations. An instrumental position regards PE as a means to particular ends. For example, PE might be conducted to secure particular outcomes such as greater public support for a policy or project. Such a position aligns PE closely with institutional aims and objectives: it promotes public support through understanding and addressing public concerns. A substantive position suggests that the goal of PE is to lead to benefits for participants or wider publics: this can include empowering members of the public, enhancing skills or building social capital.[12] While these varying rationales are not mutually exclusive, they lead to different understandings and expectations regarding the objectives and role of PE, as well as different ideas of what it means for such processes to be 'successful'.

Rowe and Frewer argue that public involvement 'as widely understood and imprecisely defined can take many forms, in many different situations (contexts), with many different types of participants, requirements, and aims (and so on), for which different mechanisms may be required to maximize effectiveness (howsoever this is defined)'.[13] Choosing between different forms requires consideration of purpose and an awareness of the wider context within which engagement is taking place; its effectiveness is more than a matter of method. Academic and practitioner literatures on PE contain many different typologies and classifications of forms of engagement. These often take as their starting point Arnstein's[14] ladder of public participation. This sets out eight levels of participation, in the form of a hierarchy of engagement. On the bottom rung of the ladder (non-participation), engagement is viewed instrumentally as an opportunity to educate the public and/or engineer support, a common effort when seeking to fill a knowledge deficit or garner social support for a new development. In the middle of the ladder, tokenistic forms of participation include informing and consulting members of the public, where consultation does not involve a two way process, but rather positions the public as having views and attitudes that might be helpful to seek as part of policy development. Again, this is not an unusual mode of engagement in the context of health research. Arnstein suggested that both of these could be valuable first steps towards participation but that they are limited by the lack of influence that participants have. Consultation is described as being a cosmetic 'window-dressing ritual' with little impact, although the extent of impact would depend on how the results of any consultation are subsequently used, rather than being intrinsic to the method itself. The top rungs of the ladder, which move towards empowerment and ownership of process, require redistribution of power to members of the public; while the participatory turn gestures towards such an approach, institutional practices often militate against its enactment.

Arnstein's model has been adapted by a large number of individuals and organisations in developing alternative classification systems and models. This has resulted in a proliferation of typologies, tool kits and models which can be referred to in designing and/or evaluating PE

[11] D. J. Fiorino, 'Citizen Participation and Environmental Risk: A Survey of Institutional Mechanisms', (1990) *Science, Technology, & Human Values*, 15(2), 226–243.

[12] J. Wilsdon and R. Willis, *See-Through Science: Why Public Engagement Needs to Move Upstream* (London: Demos, 2004).

[13] G. Rowe and L. J. Frewer, 'Evaluating Public-Participation Exercises: A Research Agenda', (2004) *Science, Technology, and Human Values*, 29(4), 512–556, 252.

[14] S. R. Arnstein, 'A Ladder of Citizen Participation', (1969) *Journal of the American Planning Association*, 35(4), 216–224.

approaches. Aitken has observed that these models, whilst adopting varying terminology and structures, typically follow common patterns:

> Each starts with a 'bottom' layer of engagement which is essentially concerned with information provision [...] They then have one (or more) layer(s) with limited forms of public feedback into decision-making processes (consultation), and finally they each have a 'top' layer with more participatory forms of PE which give greater control to participants.[15]

Forms of engagement classified as 'awareness raising' are essentially concerned with the dissemination of information. Where awareness raising is conducted on its own (i.e. where this represents the entirety of a PE approach) this represents a minimal form of PE. It may even be argued that awareness raising on its own – as one-sided and unidirectional information provision – should not be considered PE. Rowe and Frewer note that at this level, 'information flow is one-way: there is no involvement of the public per se in the sense that public feedback is not required or specifically sought'.[16] Awareness raising is limited in what it can achieve, but the focus on increasing understanding of particular issues may be a prerequisite for the deliberative approaches discussed below.

Examples of PE activities focussed on awareness raising include campaigns by national public health bodies such as Public Health England's 'Value of Vaccines',[17] or the creation and dissemination of videos and animations to explain the ways that people's health data is used in research.[18]

Consultation aims to gather insights into the views, attitudes or knowledge of members of the public in order to inform decisions. It can involve – to varying degrees – two-way flows of information. Wilcox contends that: 'Consultation is appropriate when you can offer some choices on what you are going to do – but not the opportunity [for the public] to develop their own ideas or participate in putting plans into action'.[19] Consultation provides the means for public views to be captured and taken into consideration, but does not necessarily mean that these views, or public preferences and/or concerns will be acted on or addressed.

Consultation can be either a one-way or two-way process. In a one-way process, public opinion is sought on pre-defined topics or questions, whereas a two-way process can include opportunities for respondents to reflect on and/or question information provided by those running engagement exercises.[20] Such two-way processes can ensure the questions asked, and subsequently the responses given, reflect the interests and priorities of those being engaged. It can also facilitate dialogue and 'deeper' forms of engagement with the aim of characterising, in all their complexity, public attitudes and perspectives.

It is widely recognised that consultation will be best received and most effective when it is perceived to be meaningful. This means that participants want to know how their views are taken into account and what impact the consultation has had (i.e. how has this informed decision-making). Davidson and colleagues caution that: 'Consultation can be a valuable

[15] M. Aitken, 'E-Planning and Public Participation: Addressing or Aggravating the Challenges of Public Participation in Planning?', (2014) *International Journal of E-Planning Research (IJEPR)*, 3, 38–53, 42.
[16] G. Rowe and L. J. Frewer, 'A Typology of Public Engagement Mechanisms', (2005) *Science, Technology, & Human Values*, 30(2), 251–290, 255.
[17] Public Health England, 'Campaign Resource Cente', (Public Health England), www.campaignresources.phe.gov.uk/resources/campaigns.
[18] For example those produced by Understanding Patient Data, 'Data Saves Lives Animations', (*Understanding Patient Data*), www.understandingpatientdata.org.uk/animations.
[19] D. Wilcox, 'The Guide to Effective Participation', (Brighton: Partnerships, 1994), 11.
[20] Rowe and Frewer, 'Typology of Public Engagement'.

mechanism for reflecting public interests, but can also lead to disappointment and frustrations if participants feel that their views are not being taken seriously or that the exercise is used to legitimise decisions that have already been made'.[21] Again, we see that choice of method is no guarantee of meaningful engagement in terms of influence on the practices of research and its governance.

Approaches taken to consultation include: public consultations where any member of the public is able to submit a written response; surveys and questionnaires with a sample which aims to be representative of the wider population (or key groups within it); and, focus groups, deliberative engagement or community-based participatory methods to engage more deeply with communities to shape both research processes and outcomes.

Approaches to PE that can be classified under the heading of empowerment are those that would be positioned at the top of Arnstein's ladder of participation. These approaches involve the devolution of power to participants and the creation of benefits for participants and/or wider society. This can be achieved through public-led forms of engagement where public members themselves design the process and determine its objectives, topics of relevance and scope or through partnership approaches.[22] It might also be achieved through engagement approaches that bring together public members in ways that build relationships and social capital that will continue after the engagement process ends.[23] Both invited and uninvited[24] forms of engagement can involve empowerment, so it is possible to engineer a flattening of hierarchies of knowledge and expertise as well as respond to efforts of publics to come together to define and debate issues of concern.

Empowering forms of engagement can lead to outcomes of increased relevance to communities and that most accurately reflect public interests and values. However, they can also be more expensive than traditional forms of engagement, given that they necessitate more open and flexible timeframes and may require extra skills related to facilitation and negotiation. Certainly, they may confront the more uncomfortable social, political and economic consequences and drivers of health research.

One example of engaging with some of the wider issues raised by health data and research is the dialogue commissioned by the Scottish Government to deliberate about private and third sector involvement in data sharing.[25]

While a hierarchical classification, such as Arnstein's, serves to highlight the importance of how the public are positioned in different modes of engagement, each broad approach described above can add different value and play important roles in PE. In practice it may be most appropriate for PE to use a range of methods reflecting different rationales and objectives. Rather than conceptualising them hierarchically, it is more helpful to think of these methods as overlapping and often working alongside each other within any PE practice or strategy.

[21] S. Davidson et al., 'Public Acceptability of Data Sharing between the Public, Private and Third Sectors for Research Purposes', (2013) *Social Research Series* (Edinburgh: Scottish Government), 4.30.

[22] L. Belone et al., 'Community-Based Participatory Research Conceptual Model: Community Partner Consultation and Face Validity', (2016) *Qualitative Health Research*, 26(1), 117–135.

[23] INVOLVE, 'People and Participation: How to Put Citizens at the Heart of Decision-Making' (*INVOLVE*, 2005), www.involve.org.uk/sites/default/files/field/attachemnt/People-and-Participation.pdf.

[24] P. Wehling, 'From Invited to Uninvited Participation (and Back?): Rethinking Civil Society Engagement in Technology Assessment and Development', (2012) *Poiesis & Praxis*, 9(1), 43–60.

[25] Davidson et al., 'Public Acceptability'.

11.4 TYPES OF PUBLICS

PE and involvement professionals, policy documents and critical scholars increasingly refer to 'publics' as a way to problematise and differentiate within and between different kinds of public. The adoption of such a term signifies that publics are diverse and that we cannot talk of a homogeneous public. However, beyond that, the term may obscure more than it reveals: what kinds of publics are we talking about when we talk about PE, and how are these related to particular forms of engagement? As Braun and Schultz note '"The public," we argue, is never immediately given but inevitably the outcome of processes of naming and framing, staging, selection and priority setting, attribution, interpellation, categorisation and classification'.[26] How members of 'the public' are recruited is more than a practical matter: the process embodies the assumptions, aims and priorities of those designing the engagement.

On the whole, publics are constructed or 'come into being' within PE practices rather than being self-forming. As with types of PE, different categorisations of publics have been developed. Degeling and colleagues highlight three different types: citizens (ordinary people who are unfamiliar with the issues, a kind of pure public); consumers (those with relevant personal experience, a kind of affected public) and advocates (those with technical expertise or partisan interests).[27] And each of these was linked to different types of PE. Citizens were treated as a resource to increase democratic legitimacy; consumers were directed to focus on personal preferences; advocates were most commonly used as expert witnesses in juries – directly linked to policy processes. However, overall the 'type' of public sought was often not explicit, and their role not specified.

Braun and Schultz[28] elaborate a four-fold distinction: the general public, the pure public, the affected public and the partisan public. Different PE methods serve to construct different kinds of publics. The general public is a construct required for opinion polls and surveys; pure publics for citizen conferences and juries; affected publics for consultative panels; partisan publics for stakeholder consultations. However, as with the different types of PE, in practice there will be overlaps across these dimensions and subject positions will shift as expertise is crafted through the processes of engagement and facilitation.[29] Different types of expertise are presumed here too: the general public gives policy makers knowledge about people's attitudes; the pure public creates a 'mature' citizen who becomes knowledgeable and can develop sophisticated arguments; affected publics bring expertise to 'educate' the expert – very common in health research regulation; and a partisan public may be deliberately configured to elicit viewpoints 'out there' in society to assess the 'landscape of possible argument'.[30]

Types of PE and the categorisation of different publics involve processes of inclusion and exclusion and the legitimacy of PE can easily be challenged because of who participates: some voices may be prioritised over others, and challenges may be made to participants' expertise. We turn now to a case study of how PE is being enacted in one area of health research to explore how we might deal with these problematics of how and who.

---

[26] Braun and Schultz, '. . . A Certain Amount of Engineering', 406
[27] C. Degeling et al., 'Which Public and Why Deliberate? – A Scoping Review of Public Deliberation in Public Health and Health Policy Research', (2015) *Social Science & Medicine*, 131, 114–121.
[28] Braun and Schultz, '. . . A Certain Amount of Engineering'.
[29] A. Kerr et al., 'Shifting Subject Positions: Experts and Lay People in Public Dialogue', (2007) *Social Studies of Science*, 37(3), 385–411.
[30] Braun and Schultz, '. . . A Certain Amount of Engineering', 414.

11.5 PUBLIC ENGAGEMENT IN DATA INTENSIVE HEALTH RESEARCH:
PRINCIPLES FOR AN INCLUSIVE APPROACH

The digitisation of society has led to an explosion of interest in the potential uses of more and more population data in research; this is particularly true in relation to health research.[31] However, recent years have also brought a number of public controversies, particularly regarding proposed uses of health data. Two high profile examples from England are the failed introduction of the care.data scheme to link hospital and GP records[32] and Google Deep Minds' involvement in processing health data at an NHS Trust in London.[33] The introduction of Australia's National Electronic Health Record Systems (NEHRS) also floundered, demonstrating the importance of taking account how such programmes reflect, or jar, with public values.[34] Such controversies have drawn attention to the importance of engaging with members of the public and stakeholders to ensure that data are used in ways which align with public values and interests and to ensure that public concerns are adequately addressed.

The growing interest in potential uses of population data, and the increasing recognition of the importance of ensuring a social licence for their use, have resulted in considerable interest in understanding public attitudes and views on these topics.[35] With the expansion of research uses of (health) data there has been a growing interest in public acceptability. As Bradwell and Gallagher have suggested, 'personal information use needs to be far more democratic, open and transparent' and this means 'giving people the opportunity to negotiate how others use their personal information in the various and many contexts in which this happens'.[36] PE is seen as key to the successful gathering and use of health data for research purposes.

As a recent consensus statement on PE in data intensive health research posits, there are particular reasons to promote PE in data intensive health research[37] including its scale – here the wider public is an 'affected' public and the distance is increased between researchers and those from whom data are gathered, thus requiring a new kind of social licence.[38] This requires novel thinking about how best to engage publics in shaping acceptable practices and their effects.

As well as recognising diverse practices, aims and effects, and building reflexive critique into PE for health research regulation and governance, we need to articulate some common commitments that can help steer a useful path through this diversity and thereby challenge criticisms of institutional capture and tokenism.[39] These commitments must include clarity of purpose and transparency, which will help deal with the challenges of multiple but often

---

[31] K. McGrail et al., 'A Position Statement on Population Data Science: The Science of Data about People', (2018) *International Journal of Population Data Science*, 3(1), 1–11.

[32] P. Carter et al., 'The Social Licence for Research: Why care.data Ran into Trouble', (2015) *Journal of Medical Ethics*, 41(5), 404–409.

[33] J. Powles and H. Hodson, 'Google DeepMind and Healthcare in an Age of Algorithms', (2017) *Health and Technology*, 7, 351–367.

[34] K. Garrety et al., 'National Electronic Health Records and the Digital Disruption of Moral Orders', (2014) *Social Science & Medicine*, 101, 70–77.

[35] M. Aitken et al., 'Public Responses to the Sharing and Linkage of Health Data for Research Purposes: A Systematic Review and Thematic Synthesis of Qualitative Studies', (2016) *BMC Medical Ethics*, 17(1), 73; Social Research Institute, 'The One-Way Mirror: Public Attitudes to Commercial Access to Health Data', (Wellcome Trust, 2016).

[36] P. Bradwell and N. Gallagher, *We No Longer Control What Others Know about Us, But We Don't Yet Understand the Consequences . . .The New Politics of Personal Information* (London: Demos, 2007), pp. 18–19.

[37] M. Aitken et al., 'Consensus Statement on Public Involvement and Engagement with Data Intensive Health Research', (2019) *International Journal of Population Data Science*, 4(1), 1–11.

[38] Carter et al., 'The Social Licence for Research'.

[39] Aitken et al., 'Consensus Statement'.

implicit purposes and goals. Inclusion and accessibility will broaden reach and two way communication – dialogue – is a necessary but not sufficient condition for impact. The latter can only be achieved if there is institutional buy-in, a commitment to respond to and utilise PE in governance and research. Given the challenges of assessing whether or not PE is impactful, something we discuss in the conclusion below, PE should be designed with impact in mind and be evaluated throughout. It is clear that you cannot straightforwardly get the right public and the right mechanism and be assured of meaningful and impactful PE. The choices are complicated and inflected with norms and goals that need to be explicitly stated and indeed challenged.

We now turn, as a conclusion, to review some of the outstanding issues that a critical approach to PE brings and make the case for robust evaluation.

11.6 CONCLUSION

The prominent emphasis on PE in relation to health research can be seen as a reflection of a wider resurgence of interest in PE in diverse policy areas.[40] For example, Coleman and Gotzehave pointed to a widespread commitment to PE, conceived of as a mechanism for addressing problems in democratic societies.[41] For Wilsdon and Willis, the emphasis on engagement represents a wider pattern whereby the 'standard response' of government to public ambivalence or hostility towards technological, social or political innovation is 'a promise to listen harder'.[42]

PE is not straightforward, and fulfilling the commitments of PE presents challenges and dilemmas in practice. There are many different ways of approaching PE, and these lead to different ideas of what constitutes success. There is no agreed best practice in evaluation; different rationales lead to different approaches to evaluation. Approaches underpinned by normative rationales will evaluate the quality of PE processes (Was it done well?); instrumental rationales lead to a focus on outcomes (Was it useful? Did it achieve the objectives?); and substantive rationales will assess the value added for participants or wider society (Did participants benefit from the process? Were there wider positive impacts?). Evaluation following substantive rationales is typically focussed on longer term outcomes, compared to evaluation following normative or instrumental rationales. Such longer term outcomes may be indirect and difficult to quantify or measure.

While the literature on methods of doing PE continues to proliferate, evaluation of PE remains under-theorised and underreported. The current evidence base is limited, but existing approaches to evaluating PE tend to reflect instrumental rationales and focus on direct outcomes of PE rather than substantive rationales and indirect, less tangible outcomes or impacts.[43] Wilson and colleagues[44] have observed that there is a tendency to focus on 'good news' in evaluating PE and that positivist paradigms shaping research projects or programmes can limit the opportunities to fully or adequately evaluate the complexities of PE as a social process.

[40] M. Pieczka and O. Escobar, 'Dialogue and Science: Innovation in Policy-Making and the Discourse of Public Engagement in the UK', (2013) *Science and Public Policy*, 40(1), 113–126.

[41] J. Gotze and S. Coleman, *Bowling Together: Online Public Engagement in Policy Deliberation* (London: Hansard Society, 2010).

[42] Wilsdon and Willis, *See-Through Science*, p. 16.

[43] J. P. Domecq et al., 'Patient Engagement in Research: A Systematic Review', (2014). *BMC Health Services Research*, 14(1), 89.

[44] P. Wilson et al., (2015) 'ReseArch with Patient and Public invOlvement: A RealisT evaluation – the RAPPORT study', (2015) *Health Services and Delivery Research*, 3(38), 1–9.

This is significant as it means that while a variety of rationales and purposes are acknowledged in relation to PE, there is very limited evidence of the extent to which these are realised. This in turn has negative implications for the recognition – and consequently, the institutional support – that PE receives. By providing evidence only of narrow and direct outcomes, instrumental approaches to evaluation obscure the varied and multiple benefits that can result from PE. While 'the move from "deficit to dialogue" is now recognised and repeated by scientists, funders and policymakers [. . .] for all of the changing currents on the surface, the deeper tidal rhythms of science and its governance remain resistant'.[45] Despite growing emphasis on dialogue and co-inquiry, simplistic views of the relationship between science and the public persist[46] and PE is often conducted in instrumental ways which seek to manufacture trust in science rather than foster meaningful dialogue. Greater reflection is required on the question of *why* publics are engaged rather than *how* they are engaged.

Finally, in designing, conducting and using PE in health research, we need to be reflective and critical, asking ourselves whether the issues are being narrowly defined and interpreted within existing frameworks (that often focus on privacy and consent). Does this preclude wider discussions of public benefit and the political economy of Big Data research for health? PE can and should improve health research and its regulation by questioning institutional practices and societal norms and using publics' contributions to help shape solutions.

---

[45] Stilgoe et al., 'Why Should We Promote Public Engagement with Science?', 4.
[46] Kurath and Gisler, 'Informing, Involving or Engaging?'.

12

Participatory Governance in Health Research

Patients and Publics as Stewards of Health Research Systems

Kim H. Chuong and Kieran C. O'Doherty

12.1 INTRODUCTION

This chapter discusses participatory governance as a conceptual framework for engaging patients and members of the public in health research governance, with particular emphasis on deliberative practices. We consider the involvement of patients and members of the public in institutional mechanisms to enhance responsibility and accountability in collective decision-making regarding health research. We illustrate key principles using discussion of precision medicine, as this demonstrates many of the challenges and tensions inherent in developing participatory governance in health research more generally. Precision medicine aims to advance healthcare and health research through the development of treatments that are more precisely targeted to patient characteristics.

Our central argument in this chapter is that patients and broader publics should be recognised as having a legitimate role in health research governance. As such, there need to be institutional mechanisms for patients and publics to be represented among stewards of health research systems, with a role in articulating vision, identifying research priorities, setting ethical standards, and evaluation. We begin by reviewing relevant scholarship on patient and public engagement in health research, particularly in the context of the development and use of Big Data for precision medicine. We then examine conceptualisations of participatory governance and outline stewardship as a key function of governance in a health research system. Thereafter, we propose the involvement of patients and publics as stewards who share leadership and oversight responsibilities in health research, and consider the challenges that may occur, most notably owing to professional resistance. Finally, we discuss the conditions and institutional design elements that enable participatory governance in health research.

12.2 PATIENT AND PUBLIC ENGAGEMENT IN HEALTH RESEARCH

Beresford identifies two broad approaches that have predominated in public engagement in health and social research since the 1990s.[1] *Consumerist approaches* reflect a broad interest in the market and seek consumer feedback to improve products or enhance services; in contrast, *democratic approaches* are concerned with people having more say in institutions or

[1] P. Beresford, 'User Involvement in Research and Evaluation: Liberation or Regulation?', (2002) *Social Policy & Society*, 1(2), 95–105.

organisations that have an impact on their lives. Unlike consumerist approaches, democratic approaches are explicit about issues of power, the (re)distribution of power and a commitment to personal and collective empowerment. Well-known examples of democratic approaches include the social movements initiated by people living with disability and HIV/AIDS, where these communities demanded greater inclusion in the development of scientific knowledge and health policy decisions.[2] Moral and ethical reasons based on democratic notions of patient empowerment and redistribution of power, and consequentialist arguments that patient and public engagement can improve research credibility and social acceptance, are also offered by health researchers.[3] It should be noted that patient and public engagement does not, in and of itself, constitute an active role for members of the public in health research and policy decision-making. Conceptual models have often highlighted the multiple forms that engagement can take, which vary in the degree to which members of the public are empowered to participate in an active role (see Aitken and Cunningham-Burley, Chapter 11).

In recent years, the potential to link large data sources and harness the breadth and depth of such Big Data has been hailed as bringing 'a massive transformation' to healthcare.[4] Data sources include those collected for health services (e.g. electronic health records), health research (e.g. clinical trials, biobanks, genomic databases), public health (e.g. immunisation registries, vital statistics), and other innovative sources (e.g. social media). Achieving the aims of precision medicine relies on the creation of networks of diverse data sources and scientific disciplines to capture a more holistic understanding of health and disease.[5] Conducting research using such infrastructure represents a shift from individual and isolated projects to research enterprises that span multiple institutions and jurisdictions. While the challenges of doing patient and public engagement *well* have been widely recognised, the emergence of precision medicine highlights the stakes and urgency of involving patients and publics in meaningful ways.

Biomedical research initiatives that involve large, networked research infrastructure rely on public support and cooperation. Rhetorical appeals to democratising scientific research, empowerment and public benefits, have been employed in government-sponsored initiatives in the USA and UK in attempts to foster to public trust and cultivate a sense of collective investment and civic duty to participate, notably to agree to data collection and sharing.[6] Such appeals have been explicit in the US Precision Medicine Initiative (PMI)[7] since its inception, whereas they have been used post hoc in the NHS England care.data programme after public backlash. The failure of care.data illustrates the importance of effective and meaningful public engagement – rather than tokenistic appeals – to secure public trust and confidence in its oversight for large-scale, networked research. Established to be a centralised data sharing system that linked vast amounts of

[2] C. Barnes, 'What a Difference a Decade Makes: Reflections on Doing 'Emancipatory' Disability Research', (2003) *Disability & Society*, 18(1), 3–17; S. Epstein, 'The Construction of Lay Expertise: AIDS Activism and the Forging of Credibility in the Reform of Clinical Trials', (1995) *Science, Technology, & Human Values*, 20(4), 408–437.
[3] J. Thompson et al., 'Health Researchers' Attitudes towards Public Involvement in Health Research', (2009) *Health Expectations*, 12(2), 209–220.
[4] E. Vayena and A. Blassimme, 'Health Research with Big Data: Time for Systemic Oversight', (2018) *Journal of Law, Medicine & Ethics*, 46(1), 119–129.
[5] Ibid., 120.
[6] J. P. Woolley et al., 'Citizen Science or Scientific Citizenship? Disentangling the Uses of Public Engagement Rhetoric in National Research Initiatives', (2016) *BMC Medical Ethics*, 17(33), 1–17.
[7] The US PMI was launched in 2015 with the aims of advancing precision medicine in health and healthcare. A cornerstone of the initiative is the All of Us Research Program, a longitudinal project aiming to enroll 1 million volunteers to contribute their genetic data, biospecimens and other health data to a centralised national database. 'National Institutes of Health', www.allofus.nih.gov/.

patient data including electronic health records from general practitioners, care.data was suspended and eventually closed in 2016 after widespread public and professional concerns, including around its 'opt-out' consent scheme, transparency, patient confidentiality and privacy, and potential for commercialisation.[8] See further, Burgess, Chapter 25, this volume.

Research using Big Data raises many unprecedented social, ethical, and legal challenges. Data are often collected without clear indication of their uses in research (e.g. electronic health records) or under vague terms regarding their future research uses (e.g. biobanks). Challenges arise with regard to informed consent about future research that may not yet be conceived; privacy and confidentiality; potential for harms from misuses; return of results and incidental findings; and ownership and benefit sharing, which have implications for social justice.[9] As cross-border sharing of data raises the challenges of marked differences in regulatory approaches and social norms to privacy, there have been calls for an international comparative analysis of how data privacy laws might have affected biobank practices and the development of a global privacy governance framework that could be used as foundational principles.[10] Arguments have been made that relying on informed consent – which was developed primarily for individual studies – is insufficient to resolve many of the social and ethical challenges in the context of large-scale, networked research; rather, the focus should be on the level of systemic oversight or governance.[11] Laurie proposes an 'Ethics+' governance approach that appraises biobank management in processual terms.[12] This approach focuses on the dynamics and interactions of stakeholders in deliberative processes towards the management of a biobank, and allows for adaptation to changes in circumstances, ways of thinking, and personnel.

12.3 PARTICIPATORY GOVERNANCE IN HEALTH RESEARCH SYSTEMS

The concept of governance has theoretical roots in diverse disciplines and has been used in a variety of ways, with a variety of meanings.[13] In the health sector, the concept of governance has been informed by a systems perspective, notably the World Health Organization's framework for health systems.[14] In their review, Barbazza and Tello claim that: 'Despite the complexities and multidimensionality inherent to governance, there does however appear to be general consensus that the governance function characterizes a set of processes (customs, policies or laws) that are formally or informally applied to distribute responsibility or accountability among actors of a

---

[8] S. Sterckx et al., '"You Hoped We Would Sleep Walk into Accepting the Collection of Our Data": Controversies Surrounding the UK care.data Scheme and Their Wider Relevance for Biomedical Research', (2016) *Medicine, Health Care, and Philosophy*, 19(2), 177–190.

[9] W. Burke et al., 'Informed Consent in Translational Genomics: Insufficient without Trustworthy Governance', (2018) *Journal of Law, Medicine & Ethics*, 46(1), 79–86; A. Cambon-Thomsen et al., 'Trends in the Ethical and Legal Frameworks for the Use of Human Biobanks', (2007) *European Respiratory Journal*, 30(2), 373–382; E. Wright Clayton and A. L. McGuire, 'The Legal Risks of Returning Results of Genomic Research', (2012) *Genetics in Medicine*, 14(4), 473–477

[10] E. S. Dove, 'Biobanks, Data Sharing, and the Drive for a Global Privacy Governance Framework', (2015) *Journal of Law, Medicine & Ethics*, 43(4), 675–689.

[11] Burke et al., 'Informed Consent', 83–85; K. C. O'Doherty et al., 'From Consent to Institutions: Designing Adaptive Governance for Genomic Biobanks', (2011) *Social Science & Medicine*, 73(3), 367–374; Vayena and Blasimme, 'Health Research with Big Data', 123–127.

[12] G. Laurie, 'What Does It Mean to Take an Ethics+ Approach to Global Biobank Governance?', (2017) *Asian Bioethics Review*, 9(4), 285–300.

[13] G. Stoker, 'Governance as Theory: Five Propositions', (1998) *International Social Science Journal*, 50(155), 17–28.

[14] E. Barbazza and J. E. Tello, 'A Review of Health Governance: Definitions, Dimensions and Tools to Govern', (2014) *Health Policy*, 116(1), 1–11; F. A. Miller et al., 'Public Involvement in Health Research Systems: A Governance Framework', (2018) *Health Research Policy and Systems*, 16(1), 1–15.

given [health] system'.[15] Common values, such as 'good' or 'democratic,' and descriptions of the type of accountability arrangement, such as 'hierarchical' or 'networked,' may be used to denote how governance should be defined. The notion of distributed responsibility or accountability relates to the assertion that governance is about collective decision-making and involves various forms of partnership and self-governing networks of actors.[16]

A systems perspective allows for a more integrated and coordinated view of health research activities that may be highly fragmented, specialised and competitive.[17] Strengthening the coordination of research activities promotes more effective use of resources and dissemination of scientific knowledge in the advancement of healthcare. The vision of a learning healthcare system, which was first proposed by the US Institute of Medicine (IOM), illustrates a cycle of continuous learning and care improvement that bridges research and clinical practice.[18] The engagement of patients, their families and other relevant stakeholders is identified as a fundamental element of a learning healthcare system.[19] Engaging patients as active partners in the cycle is argued to both secure the materials required for research (i.e. data and samples) and enhance patient trust.[20]

Pang and colleagues propose stewardship as a key function within a health research system that has four components: defining a vision for the health research system; identifying research priorities and coordinating adherence to them; setting and monitoring ethical standards; and monitoring and evaluating the system.[21] Other key functions of a health research system include: financing, which involves securing and allocating research funds accountably; creating and sustaining resources including human and physical capacity; and producing and using research. An important question is therefore how to engage and incorporate the perspectives and values of patients and publics in governance, particularly in terms of stewardship.

Internationally, participatory governance has been explored in multiple reforms in social, economic, and environmental planning and development that varied in design, issue areas and scope.[22] Fung and Wright use the term 'empowered participatory governance' to describe how such reforms are 'participatory because they rely upon the commitment and capacities of ordinary people to make sensible decisions through reasoned deliberation and empowered because they attempt to tie action to discussion'.[23] They outline three general principles: (1) a focus on solving practical problems that creates situations for participants to cooperate and build congenial relationships; (2) bottom-up participation, with laypeople being engaged in decision-making while experts facilitate the process by leveraging professional and citizen insights; and (3) deliberative solution generation, wherein participants listen to and consider each other's

[15] Barbazza and Tello, 'Health Governance', 3.
[16] Stoker, 'Governance as Theory', 21–24.
[17] T. Pang et al., 'Knowledge for Better Health – A Conceptual Framework and Foundation for Health Research Systems', (2003) *Bulletin of the World Health Organization*, 81(11), 815–820.
[18] Institute of Medicine, *Best Care at Lower Cost: The Path to Continuously Learning Health Care in America* (Washington, DC: National Academies Press, 2013).
[19] K. H. Chuong et al., 'Human Microbiome and Learning Healthcare Systems: Integrating Research and Precision Medicine for Inflammatory Bowel Disease', (2018) OMICS: *A Journal of Integrative Biology*, 22(20), 119–126; S. M. Greene et al., 'Implementing the Learning Health System: From Concept to Action', (2012) *Annals of Internal Medicine*, 157(3), 207–210; W. Psek et al., 'Operationalizing the Learning Health Care System in an Integrated Delivery System', (2015) *eGEMs*, 3(1), 1–11.
[20] Psek et al., 'Learning Health Care System'.
[21] Pang et al., 'Health Research Systems', 816–818.
[22] A. Fung and E. O. Wright (eds), *Deepening Democracy: Institutional Innovations in Empowered Participatory Governance* (New York, NY: Verso, 2003).
[23] Ibid., p. 5.

positions and offer reasons for their own positions. A similar concept is collaborative governance, which is defined by Ansell and Gash as 'a governing arrangement where one or more public agencies directly engage non-state stakeholders in a collective decision-making process that is formal, consensus-oriented, and deliberative and that aims to make or implement public policy or manage public programs or assets'.[24] The criterion of formal collaboration implies established arrangements to engage publics. Participatory governance is advocated to contribute to citizen empowerment, build local communities' capacity, address the gap in political representation and power distribution, and increase the efficiency and equity of public services. Unfortunately, however, successful implementation of participatory governance ideals is 'a story of mixed outcomes' with the failures still outnumbering the successful cases.[25]

Yishai argues that the health sector has remained impervious to the practice of participatory governance: patients have not had a substantial voice in health policy decisions, even though they may enjoy the power to choose from different health services and providers as consumers.[26] Professional resistance to non-expert views and marginalisation of public interests by commercial interests are cited as some of the reasons for the limited involvement of patients. Similarly, there are concerns that public voices are not given the same weight as those of professionals in health research decision-making. Tokenism, engaging patients as merely a 'tick-box exercise' – for funding or regulatory requirements – and devaluing patient input in comparison to expert input are common concerns.[27] Furthermore, most engagement efforts are limited to preliminary activities and not sustained across the research cycle; the vast majority of biomedical research initiatives do not engage publics beyond informed consent for data collection and sharing.[28]

Deliberative practices, such as community advisory boards and citizens' forums, have been suggested as mechanisms to allow public input in the governance of research with Big Data.[29] Public deliberation has been used to engage diverse members of the public to explore, discuss and reach collective decisions regarding the institutional practices and governance of biobanks, and the use and sharing of linked data for research.[30] However, in many instances, public input is limited to the point in time at which the deliberative forum is convened. One example of ongoing input is provided by the Mayo Clinic Biobank deliberation, which was used as a seeding mechanism for the establishment of a standing Community Advisory Board. To address

---

[24] C. Ansell and A. Gash, 'Collaborative Governance in Theory and Practice', (2008) *Journal of Public Administration Research and Theory*, 18(4), 543–571, 544.

[25] F. Fischer, 'Participatory Governance: From Theory to Practice' in D. Levi-Faur (ed.), *The Oxford Handbook of Governance* (New York, NY: Oxford University Press, 2012), pp. 458–471.

[26] Y. Yishai, 'Participatory Governance in Public Health: Choice, but No Voice' in D. Levi-Faur (ed.), *The Oxford Handbook of Governance* (New York, NY: Oxford University Press, 2012), pp. 527–539.

[27] J. P. Domecq et al., 'Patient Engagement in Research: A Systematic Review', (2014) *Health Services Research*, 14(89), 1–9; G. Green, 'Power to the People: To What Extent has Public Involvement in Applied Health Research Achieved This?', (2016) *Research Involvement and Engagement*, 2(28), 1–13; P. R. Ward et al., 'Critical Perspectives on 'Consumer Involvement' in Health Research: Epistemological Dissonance and the Know-Do Gap', (2009) *Journal of Sociology*, 46(1), 63–82.

[28] E. Manafo et al., 'Patient Engagement in Canada: A Scoping Review of the 'How' and 'What' of Patient Engagement in Health Research', (2018) *Health Research Policy and Systems*, 16(1), 1–11; Woolley et al., 'Citizen Science', 5.

[29] Burke et al., 'Translational Genomics', 84; Vayena and Blasimme, 'Health Research with Big Data', 125.

[30] S. M. Dry et al., 'Community Recommendations on Biobank Governance: Results from a Deliberative Community Engagement in California', (2017) *PLoS ONE*, 12(2), e0172582; K. C. O'Doherty et al., 'Involving Citizens in the Ethics of Biobank Research: Informing Institutional Policy through Structured Public Deliberation', (2012) *Social Science & Medicine*, 75(9), 1604–1611; J. E. Olson and others, 'The Mayo Clinic Biobank: A Building Block for Individualized Medicine', (2013) *Mayo Clinic Proceedings*, 88(9), 952–962; J. Teng et al., 'Sharing Linked Data Sets for Research: Results from A Deliberative Public Engagement Event in British Columbia, Canada', (2019) *International Journal of Population Data Science*, 4(1), 13.

the challenge of moving from one-time input to ongoing, institutionalised public engagement, O'Doherty and colleagues propose four principles to guide adaptive biobank governance: (1) recognition of participants as a collective body, as opposed to just an aggregation of individuals; (2) trustworthiness of the biobank, with a reflexive focus of biobank leaders and managers on its practices and governance arrangements, as opposed to a focus on the trust of participants divorced from considerations of how such trust is earned; (3) adaptive management that is capable of drawing on appropriate public input for decisions that substantively affect collective patient or public expectations and relationships; and (4) fit between the particular biobank and specific structural elements of governance that are implemented.[31]

A few cases of multi-agency research networks that engage patients or research participants in governance are also available. For instance, the Patient-Centered Outcomes Research Institute (PCORI) in the USA established multiple patient-powered research networks, each focusing on a particular health condition (www.pcori.org). In the UK, the Managing Ethico-social, Technical and Administrative issues in Data ACcess (METADAC) was established as a multi-study governance infrastructure to provide ethics and policy oversight to data and sample access for multiple major population cohort studies. Murtagh and colleagues identify three key structural features: (1) independence and transparency, with an independent governing body that promotes fair, consistent and transparent practices; (2) interdisciplinarity, with the METADAC Access Committee comprising individuals with social, biomedical, ethical, legal and clinical expertise, and individuals with personal experience participating in cohort studies; and (3) patient-centred decision-making, which means respecting study participants' expectations, involving them in decision-making roles and communicating in a format that is clear and accessible.[32]

12.4 ENABLING CONDITIONS AND INSTITUTIONAL DESIGNS

12.4.1 Enabling Conditions: Power/Resource Imbalances and Representativeness

Fung and Wright propose that an enabling condition to facilitate participatory governance is 'a rough equality of power, for the purposes of deliberative decision-making, between participants'.[33] Nonetheless, power and resource imbalances are a common problem in many cases of patient and public engagement. Patients and publics bring different forms of knowledge that could be seen as challenging traditional scientific knowledge production and the legitimacy of professional skills and knowledge. Such knowledge could be constructed positively by researchers, but it could also be constructed in ways that question its validity compared to professional/academic knowledge.[34] Furthermore, patients and publics may not always be capable of articulating their needs as researchable questions, which limits the uptake of their ideas in research prioritisation, or a perceived mismatch may lead to resistance from researchers to act upon priorities identified by patients and publics.[35]

Articulating a vision for advancing patient and public engagement in a health research system is important, whether it is at an organisational or broader level.[36] We further propose recognition

[31] O'Doherty et al., 'Adaptive Governance', 368.
[32] M. J. Murtagh et al., 'Better Governance, Better Access: Practising Responsible Data Sharing in the METADAC Governance Infrastructure', (2018) *Human Genomics*, 12(1), 1–12.
[33] Fung and Wright, *Deepening Democracy*, p. 24.
[34] Thompson et al., 'Health Researchers' Attitudes'; Ward et al., 'Critical Perspectives'.
[35] F. A. Miller et al., 'Public Involvement and Health Research System Governance: Qualitative Study', (2018) *Health Research Policy and Systems*, 16(1), 1–15.
[36] Miller et al., 'Health Research Systems', 4–5.

of patients and publics as having legitimate representation as stewards or governors, with a role in articulating vision, identifying research priorities, setting ethical standards, and evaluation. Moreover, we suggest that formal arrangements are required to enable patients and publics in their role as stewards and governors within institutional architecture. A range of innovative mechanisms have been explored and implemented. For instance, ArthritisPower, which is a patient-powered research network within PCORI, established a governance structure in which patients have representation and overlapping membership across the Executive Board, Patient Governor Group and Research Advisory Board. Clear communication of expectations, provision of well-prepared tools for engagement (e.g. work groups organised around particular tasks or topics, online platform for patient governors to connect) and regular assessments of patient governors' viewpoints are found to be necessary to support and build patients' capacity within a multi-stakeholder governance structure.[37]

It should be recognised that members of the public vary in their capacity to participate, deliberate and influence decision-making. Those who are advantaged in terms of education, wealth or membership in dominant racial/ethnic groups often participate more frequently and effectively in deliberative decision-making.[38] Power and resource imbalances can result in the problem of co-optation whereby stronger stakeholders are able to generate support for their own agendas. The lack of representation of certain groups – i.e. youth, Indigenous, Black and ethnic minority groups – has been noted in many efforts of patient and public engagement in health research,[39] which reflects structural barriers and/or historical discrimination and mistrust due to past ethical violations. This raises challenges of how to promote and support inclusion and equity in decision-making. This also serves as a valuable counterpoint on power dynamics as discussed by Brassington, chapter 9.

There are also concerns that patients may risk becoming less able to represent broader patient perspectives as they become more trained and educated in research and more involved in the governance of research activities. For instance, Epstein documented the use of 'credibility tactics', such as the acquisition of the language of biomedical science by HIV/AIDS activists to gain acceptance in the scientific community, and Thompson and colleagues identified the emergence of professionalised lay experts who demonstrated considerable support for dominant scientific paradigms and privileged professional or certified forms of expertise among patients and caregiver participants in cancer research settings in England.[40] To guard against this, the governance structure of ArthritisPower maintains a mix of veteran and new members by limiting patient governors' memberships to three years.[41]

12.4.2 Institutional Designs: Relationships, Trust and Leadership Support

Fung and Wright outline three institutional design elements that are necessary for participatory governance: (1) devolution of decision-making power to local units that are charged and held accountable with implementing solutions; (2) centralised supervision and coordination to

---

[37] W. B. Nowell et al., 'Patient Governance in a Patient-Powered Research Network for Adult Rheumatologic Conditions', (2018) *Medical Care*, 56(10 Suppl 1), S16–S21.

[38] Fung and Wright, *Deepening Democracy*, p. 34.

[39] Miller et al., 'Health Research System Governance', 7; Green, 'Power to the People', 10.

[40] Epstein, 'The Construction of Lay Expertise', 417–426; J. Thompson et al., 'Credibility and the 'Professionalized' Lay Expert: Reflections on the Dilemmas and Opportunities of Public Involvement in Health Research', (2012) *Health*, 16 (6), 602–618.

[41] Nowell et al., 'Patient Governance', S21.

connect the local units, coordinate and distribute resources, reinforce quality of local decision-making, and diffuse learning and innovation; and (3) transformation of formal governance procedures to institutionalise the ongoing participation of laypeople.[42] At a national level, devolution of power implies that the state solicits local units, such as community organisations and local councils, to devise and implement solutions. Members of the public are engaged at a local level through these organisations as stakeholders who are affected by the targeted problems. Within a health research system, network or organisation, patients and publics may serve on advisory boards and committees as members within a multi-stakeholder governance structure.

In this section, we discuss factors that may facilitate or impede the participation of patients and publics in the governance structures of health research systems, networks or organisations. It is important to consider multilevel engagement strategies for matching participation opportunities to varying interests, capacities and goals of patients and publics.[43] These strategies may range from patients and publics having one-time input into a targeted issue, to serving in leadership roles as members of a research team or governing body. Involving patients and publics in governance structures in an ongoing manner requires relationship building over much longer periods of time.

Clarity of roles and purposes of patient and public engagement is needed for relationship building, as well as for developing and maintaining trust. Participatory forms of governance are more feasible when stakeholders have opportunities to identify mutual gains in collaboration. However, pre-existing relationships can discourage stakeholders from seeing the value of collaboration. In health research that spans multiple sites, approaches and willingness to engage patients and publics may differ considerably across the participating sites.[44] Establishing new relationships with patients as partners may be considered too risky and jeopardising to current relationships by some sites.

Additionally, engagement activities that focus on 'patients', 'citizens' or 'members of a community', may each carry different sets of assumptions. Patients often have a personal connection to the health issue in question, whereas community members are selected to represent a collective experience and perspective. In national biomedical research initiatives, engagement as 'citizens' may lead to the exclusion of certain groups, such as advocacy groups and charities, from governing committees to avoid 'special interests'.[45] While people may be able to navigate and draw on different aspects of their lives to inform research and policy, further exploration is needed to understand the common and distinctive aspects between different types of roles that people occupy.[46] In any case, clarity regarding roles and responsibilities, and transparency in the aims of engagement are necessary for relationship and trust building.

Fung and Wright assert that centralised supervision and coordination is needed to stabilise and deepen the practice of participatory governance among local units.[47] At a national level, centralised coordination is a component of leadership capacity to ensure accountability, distribute resources, and facilitate communication and information sharing across local units. According to Ansell and Gash, facilitative leadership is important for bringing together stakeholders, promoting the representation of disadvantaged groups, and facilitating dialogue and

[42] Fung and Wright, *Deepening Democracy*, pp. 20–24.
[43] For an example, see A. P. Boyer et al., 'Multilevel Approach to Stakeholder Engagement in the Formulation of a Clinical Data Research Network', (2018) *Medical Care*, 56(10 Suppl 1), S22–S26.
[44] K. S. Kimminau et al., 'Patient vs. Community Engagement: Emerging Issues', (2018) *Medical Care*, 56(10 Suppl 1), S53–S57.
[45] Woolley et al., 'Citizen Science or Scientific Citizenship', 11.
[46] See Kimminau et al., 'Patient vs. Community Engagement', for a comparison of the two.
[47] Fung and Wright, *Deepening Democracy*, pp. 21–22.

trust-building in the collaborative processes.[48] Trust-building requires commitment and mutual recognition of interdependence, shared understanding of the problem in question and common values, and face-to-face dialogue. Senior leadership and supportive policy and infrastructure are recognised as building blocks for embedding patient and public engagement in a health research system.[49]

12.5 CONCLUSION

In this chapter, we have discussed the potentials and challenges of involving patients and publics as stewards or governors of health research, whether within a broad health system, a research network, or a specific organisation. We have also outlined some of the conditions and institutional design elements that may impede or facilitate the engagement of patients and publics in governance structures, focusing on issues of power/resource imbalances, representativeness, relationships, trust and leadership support. Some conditions and institutional design elements are necessary for the implementation of participatory governance, but our discussion is not intended to be comprehensive or prescriptive. In particular, we are not proposing a specific governance structure or body as an ideal. Governance structures can vary in their purposes and constituencies. With rapid scientific advances and potential for unanticipated ethical and social issues, a multi-stakeholder governance structure needs to contain an element of reflexivity and adaptivity to evolve in ways that are respectful of diverse needs and interests while responding to changes. Moreover, the literature on patient and public engagement has documented the need for rigorous evaluation of the impact of engagement on healthcare and health research, especially given the problems of inconsistent terminology and lack of validated frameworks and tools to evaluate patient and public engagement.[50] Stronger evidence of the impact and outcomes, both intended and unintended, of patient and public engagement may help normalise the role of patients and publics as partners in health research regulation.

---

[48] Ansell and Gash, 'Collaborative Governance', 554–555.
[49] Miller et al., 'Health Research System Governance', 6–7.
[50] Manafo et al., 'Patient Engagement', 4–7. Also, Aitken and Cunningham-Burley, Chapter 11, this volume.

13

Risk-Benefit Analysis

Carl H. Coleman

13.1 INTRODUCTION

This chapter explores the concept of risk-benefit analysis in health research regulation, as well as ethical and practical questions raised by identifying, quantifying, and weighing risks and benefits. It argues that the pursuit of objectivity in risk-benefit analysis is ultimately futile, as the very concepts of risk and benefit depend on attitudes and preferences about which reasonable people disagree. Building on the work of previous authors, the discussion draws on contemporary examples to show how entities reviewing proposed research can improve the process of risk-benefit assessment by incorporating diverse perspectives into their decision-making and engaging in a systematic analytical approach.

13.2 IDENTIFYING RISKS

The term 'risk' refers to the possibility of experiencing a harm. The concept incorporates two different dimensions: (1) the magnitude or severity of the potential harm; and (2) the likelihood that this harm will occur. The significance of a risk depends on the interaction of these two considerations. Thus, a low chance of a serious harm, such as death, would be considered significant, as would a high chance of a lesser harm, such as temporary pain.

In the context of research, the assessment of risk focuses on the *additional* risks participants will experience as a result of participating in a study, which will often be less than the total level of risks to which participants are exposed. For example, a study might involve the administration of various standard-of-care procedures, such as biopsies or CT scans. If the participants would have received these same procedures even if they were not participating in the study, the risks of those interventions would not be taken into account in the risk-benefit analysis. As a result, it is possible that a study comparing two interventions that are routinely used in clinical practice could be considered low risk, even if the interventions themselves are associated with a significant potential for harm. This is the case with a significant proportion of research conducted in 'learning health systems', which seek to integrate research into the delivery of healthcare. Because many of the research activities in such systems involve the evaluation of interventions patients would be undergoing anyway, the risks of the research are often minimal, even when the risks of the interventions themselves may be high.[1]

---

[1] J. Lantos et al., 'Considerations in the Evaluation and Determination of Minimal Risk in Pragmatic Clinical Trials', (2015) *Clinical Trials*, 12(5), 485–493.

The risks associated with health-related research are not limited to potential physical injuries. For example, in some studies, participants may be asked to engage in discussions of emotionally sensitive topics, such as a history of previous trauma. Such discussions entail a risk of psychological distress. In other studies, a primary risk is the potential for unauthorised disclosure of sensitive personal information, such as information about criminal activity, or stigmatised conditions such as HIV, or mental disorders. If such disclosures occur, participants could suffer adverse social, legal, or economic consequences.

Research-related risks can extend beyond the individuals participating in a study. For example, studies of novel interventions for preventing or treating infectious diseases could affect the likelihood that participants will transmit the disease to third parties.[2] Similarly, studies in which psychiatric patients are taken off their medications could increase the risk that participants will engage in violent behaviour.[3] Third-party risks are an inherent feature of research on genetic characteristics, given that information about individuals' genomes necessarily has implications for their blood relatives.[4] Thus, if a genetic study results in the discovery that a participant is genetically predisposed to a serious disease, other persons who did not consent to participate in the study might be confronted with distressing, and potentially stigmatising, information that they never wanted to know.

In some cases, third-party risks extend beyond individuals to broader social groups. As the Council for International Organizations of Medical Sciences (CIOMS) has recognised, research on particular racial or ethnic groups 'could indicate – rightly or wrongly – that a group has a higher than average prevalence of alcoholism, mental illness or sexually transmitted disease, or that it is particularly susceptible to certain genetic disorders',[5] thereby exposing the group to potential stigma or discrimination. One example was a study in which researchers took blood samples from members of the Havasupai tribe in an effort to identify a genetic link to type 2 diabetes. After the study was completed, the researchers used the blood samples for a variety of unrelated studies without the tribe members' informed consent, including research related to schizophrenia, inbreeding and migration patterns. Tribe members claimed that the schizophrenia and inbreeding studies were stigmatising, and that they never would have agreed to participate in the migration research because it conflicted with the tribe's origin story, which maintained that the tribe had originated in the Grand Canyon. The researcher institution reached a settlement with the tribe that included monetary compensation and a formal apology.[6]

Despite the prevalence of third-party risks in research, most ethics codes and regulations do not mention risks to anyone other than research participants. This omission is striking given that some of these same sources explicitly state that benefits to non-participants should be factored into the risk-benefit analysis. A notable exception is the EU Clinical Trials Regulation, which states that the anticipated benefits of the study must be justified by 'the foreseeable risks and

---

[2] N. Eyal et al., 'Risk to Study Nonparticipants: A Procedural Approach', (2018) *Proceedings of the National Academy of Sciences*, 115(32), 8051–8053.

[3] G. DuVal, 'Ethics in Psychiatric Research: Study Design Issues', (2004) *Canadian Journal of Psychiatry*, 49(1), 55–59.

[4] A. McGuire et al., 'Research Ethics and the Challenge of Whole-Genome Sequencing', (2008) *Nature Reviews Genetics*, 9(2), 152–156.

[5] Council for International Organizations of Medical Sciences, 'International Ethical Guidelines for Health-Related Research Involving Humans', (CIOMS, 2016), p. 13.

[6] M. Mello and L. Wolf, 'The Havasupai Indian Tribe Case: Lessons for Research Involving Stored Biologic Samples', (2010) *New England Journal of Medicine*, 363(3), 204–207.

inconveniences',[7] without specifying that those risks and inconveniences must be experienced by the participants themselves.

In addition to omitting any reference to third-party risks, the US Federal Regulations on Research With Human Participants state that entities reviewing proposed research 'should not consider possible long-range effects of applying knowledge gained in the research (e.g. the possible effects of the research on public policy) as among those research risks that fall within the purview of its responsibility'.[8] This provision is intended 'to prevent scientifically valuable research from being stifled because of how sensitive or controversial findings might be used at a social level'.[9]

13.3 IDENTIFYING BENEFITS

The primary potential benefit of research is the production of generalisable knowledge – i.e. knowledge that has relevance beyond the specific individuals participating in the study. For example, in a clinical trial of an investigational drug, data sufficient to establish the drug's safety and efficacy would be a benefit of research. Data showing that an intervention is *not* safe or effective – or that it is inferior to the existing standard of care – would also count as a benefit of research, as such knowledge can protect future patients from potentially harmful and/or ineffective treatments they might otherwise undergo.

Whether a study has the potential to produce generalisable knowledge depends in part on how it is designed. The randomised controlled clinical trial (RCT) is often described as the 'gold standard' of research, as it includes methodological features designed to eliminate bias and control for potential confounding variables.[10] However, in some types of research, conducting an RCT may not be a realistic option. For example, if researchers want to understand the impact of different lifestyle factors on health, it might not be feasible to randomly assign participants to engage in different behaviours, particularly over a long period of time.[11] In addition, ethical considerations may sometimes preclude the use of RCTs. For example, researchers investigating the impact of smoking on health could not ethically conduct a study in which non-smokers are asked to take up smoking.[12] In these situations, alternative study designs may be used, such as cohort or case-control studies. These alternative designs can provide valuable scientific information, but the results may be prone to various biases, a factor that should be considered in assessing the potential benefits of the research.[13]

A recent example of ethical challenges to RCTs arose during the Ebola outbreak of 2013–2016, when the international relief organisation Médicins Sans Frontières refused to participate in any RCTs of experimental Ebola treatments. The group argued that it would be unethical to withhold the experimental interventions from persons in a control group when 'conventional

---

[7] Article 28 of the European Union Clinical Trials Regulation 536/2014, OJ 2014 No. L 158/1.

[8] The Federal Policy for the Protection of Human Subjects ('Common Rule'), 45 C.F.R. § 46.111(a)(2) (1991).

[9] A. London et al., 'Beyond Access vs. Protection in Trials of Innovative Therapies', (2010) *Science*, 328(5980), 829–830, 830.

[10] J. Grossman and F. Mackenzie, 'The Randomized Controlled Trial: Gold Standard, or Merely Standard?', (2005) *Perspectives in Biology & Medicine*, 48(4), 516–534.

[11] J. Younge et al., 'Randomized Study Designs for Lifestyle Interventions: A Tutorial', (2015) *International Journal of Epidemiology*, 44(6), 2006–2019.

[12] C. J. Mann, 'Observational Research Methods. Research Design II: Cohort, Cross Sectional, and Case-Control Studies', (2003) *Emergency Medicine Journal*, 20(1), 54–60.

[13] D. Grimes and K. Schulz, 'Bias and Causal Associations in Observational Research', (2002) *Lancet*, 359(9302), 248–252.

care offers little benefit and mortality is extremely high'.[14] The difficulty with this argument was that, in the context of a rapidly evolving epidemic, the results of studies conducted without concurrent control groups would be difficult to interpret, meaning that an ineffective or even harmful intervention could erroneously be deemed effective. Some deviations from the 'methodologically ideal approach', such as the use of adaptive trial designs, could have been justified by the need 'to accommodate the expectations of participants and to promote community trust'.[15] However, any alternative methodologies would need to offer a reasonable likelihood of producing scientifically valid information, or else it would not have been ethical to expose participants to any risk at all.

The potential benefit of scientific knowledge also depends on the size of a study, as studies with very small sample sizes may lack sufficient statistical power to produce reliable information. Some commentators maintain that underpowered studies lack any potential benefit, making them inherently unethical.[16] Others point out that small studies might be unavoidable in certain situations, such as research on rare diseases, and that their results can still be useful, particularly when they are aggregated using Bayesian techniques.[17]

Often, choices about study design can require trade-offs between internal and external validity. While an RCT with tightly controlled inclusion and exclusion requirements is the most reliable way to establish whether an experimental intervention is causally linked to an observable result – thereby producing a high level of internal validity – if the study population does not reflect the diversity of patients in the real world, the results might have little relevance to clinical practice – thereby producing a low level of external validity.[18] In assessing the potential benefits of a study, decision-makers should take both of these considerations into account.

In addition to the potential benefit of generalisable knowledge, some research also offers potential benefits to the individuals participating in the study. Benefits to study participants can be divided into 'direct' and 'indirect' (or 'collateral') benefits.[19] Direct benefits refer to those that result directly from the interventions being studied, such as an improvement in symptoms that results from taking an investigational drug. In some studies, there is no realistic possibility that participants will directly benefit from the study interventions; this would be the case in a Phase I drug study involving healthy volunteers, where the purpose is simply to identify the highest dose humans can tolerate without serious side effects. Indirect benefits include those that result from ancillary features of the study, such as access to free health screenings, as well as the psychological benefits that some participants receive from engaging in altruistic activities. Study participants may also consider any payments or other remuneration they receive in exchange for their participation as a type of research-related benefit.

Most commentators take the position that only potential direct benefits to participants and potential contributions to generalisable knowledge should be factored into the risk-benefit analysis. The concern is that, otherwise, 'simply increasing payment or adding more unrelated

[14] C. Adebamowo et al., 'Randomised Controlled Trials for Ebola: Practical and Ethical Issues', (2014) *Lancet*, 384 (9952), 1423–1424, 1423.

[15] C. Coleman, 'Control Groups on Trial: The Ethics of Testing Experimental Ebola Treatments', (2016) *Journal of Biosecurity, Biosafety and Biodefense Law*, 7(1), 3–24, 8.

[16] E. Emanuel et al., 'What Makes Clinical Research Ethical?', (2000) *JAMA*, 283(20), 2701–2711.

[17] R. Lilford and A. Stevens, 'Underpowered Studies', (2002) *British Journal of Surgery*, 89(2), 129–131.

[18] B. Freedman and S. Shapiro, 'Ethics and Statistics in Clinical Research: Towards a More Comprehensive Examination', (1994) *Journal of Statistical Planning and Inference*, 42(1), 223–240.

[19] N. King, 'Defining and Describing Benefit Appropriately in Clinical Trials', (2000) *Journal of Law, Medicine & Ethics*, 28(4), 332–343.

services could make the benefits outweigh even the riskiest research'.[20] Other commentators reject this position on the ground that it is not consistent with the ethical imperative to respect participants' autonomy, and that it could preclude studies that would advance the interests of participants, investigators, and society.[21] The US Food and Drug Administration has stated that payments to participants should not be considered in the context of risk-benefit assessment,[22] but it has not taken a position on consideration of other indirect benefits, such as access to free health screenings.

13.4 QUANTIFYING RISKS AND BENEFITS

Once the risks and benefits of a proposed study have been identified, the next step is to quantify them. Doing this is complicated by the fact that the significance of a particular risk or benefit is highly subjective. For example, a common risk in health-related research is the potential for unauthorised disclosure of participants' medical records. This risk could be very troubling to individuals who place a high degree of value on personal privacy, but for persons who share intimate information freely, the risk of unauthorised disclosure might be a minor concern. In fact, in some studies, the same experience might be perceived by some participants as a harm and by others as a benefit. For example, in a study in which participants are asked to discuss prior traumatic experiences, some participants might experience psychological distress, while others might welcome the opportunity to process past experiences with a sympathetic listener.[23]

In addition to differing attitudes about the potential outcomes of research, individuals differ in their perceptions about risk-taking itself. Many people are risk averse, meaning that they would prefer to forego a higher potential benefit if it enables them to reduce the potential for harm. Others are risk neutral, or even risk preferring. Similarly, individuals exhibit different levels of willingness to trade harmful outcomes for good ones.[24] For example, some people are willing to tolerate medical treatments with significant side effects, such as chemotherapy, because they place greater value on the potential therapeutic benefits. Others place greater weight on avoiding pain or discomfort and would be disinclined to accept high-risk interventions even when the potential benefits are substantial.

Another challenge in attempting to quantify risks and benefits is that the way that risks and benefits are perceived can be influenced by a variety of cognitive biases. For example, one study asked subjects to imagine that they had lung cancer and had to decide between surgery and radiation. One group was told that 68 per cent of surgical patients survived after one year, while a second group was told that 32 per cent of surgical patients died after one year. Even though the information being conveyed was identical, framing the information in terms of a risk of death increased the number of subjects who chose radiation from 18 per cent to 44 per cent.[25] Another common cognitive bias is the 'availability heuristic', which leads people to attach greater weight

---

[20] Emanuel et al., 'What Makes Clinical Research Ethical?', 2705.
[21] See, e.g. A. Wertheimer, 'Is Payment a Benefit?', (2013) *Bioethics*, 27(2), 105–116.
[22] US Food and Drug Administration, 'Payment and Reimbursement to Research Subjects', (US Food and Drug Administration, 2018), www.fda.gov/regulatory-information/search-fda-guidance-documents/payment-and-reimburse ment-research-subjects.
[23] T. Opsal et al., '"There Are No Known Benefits …" Considering the Risk/Benefit Ratio of Qualitative Research', (2016) *Qualitative Health Research*, 26(8), 1137–1150.
[24] C. Troche et al., 'Evaluation of Therapeutic Strategies: A New Method for Balancing Risk and Benefit', (2000) *Value in Health*, 3(1), 12–22.
[25] P. Slovic, 'Trust, Emotion, Sex, Politics, and Science: Surveying the Risk-Assessment Battlefield', (1999) *Risk Analysis*, 19(4), 689–701.

to information that is readily called to mind.[26] For example, if a well-known celebrity recently died after being implanted with a pacemaker, the risk of pacemaker-related deaths may be perceived as greater than it actually is.

Individuals' perceptions of risks and benefits can also be influenced by their level of social trust, which has been defined as 'the willingness to rely on those who have the responsibility for making decisions and taking actions related to the management of technology, the environment, medicine, or other realms of public health and safety'.[27] In particular, research suggests that, when individuals are considering the risks and benefits of new technologies, their level of social trust has 'a positive influence on perceived benefits and a negative influence on perceived risks'.[28] This is not surprising: those who trust that decision-makers will act in their best interests are less likely to be fearful of changes, while those who lack such trust are more likely to be worried about the potential for harm (see Aitken and Cunningham-Burley, Chapter 11, in this volume).

Compounding these subjective variables is the fact that risk-benefit analysis typically takes place against a backdrop of scientific uncertainty. This is true for all risk-benefit assessments, but it is especially pronounced in research, as the very reason research is conducted is to fill an evidentiary gap. While evaluators can sometimes rely on prior research, including animal studies, to identify the potential harms and benefits of proposed studies, most health-related research takes place in highly controlled environments, over short periods of time. As a result, prior research results are unlikely to provide much information about rare safety risks, long-term dangers or harms and benefits that are limited to discrete population subgroups.

13.5 WEIGHING RISKS AND BENEFITS

Those responsible for reviewing proposed research must ultimately weigh the risks and benefits to determine whether the relationship between them is acceptable. This process is complicated by the fact that risks and benefits often cannot be measured on a uniform scale. First, 'risks and benefits for subjects may affect different domains of health status',[29] as when a risk of physical injury is incurred in an effort to achieve a potential psychological benefit. Second, 'risks and benefits may affect different people';[30] risks are typically borne by the participants in the research, but most of the benefits will be experienced by patients in the future.

Several approaches have been suggested for systematising the process of risk-benefit analysis in research. The first, and most influential, approach is known as 'component analysis'. This approach calls on decision-makers to independently assess the risks and potential benefits of each intervention or procedure to be used in a study, distinguishing those that have the potential to provide direct benefits to participants ('therapeutic') from those that are administered solely for the purpose of developing generalisable knowledge ('non-therapeutic'). For therapeutic interventions, there must be genuine uncertainty regarding the relative therapeutic benefits of the intervention as compared to those of the standard of care for treating the participants'

---

[26] T. Pachur et al., 'How Do People Judge Risks: Availability Heuristic, Affect Heuristic, or Both?', (2012) *Journal of Experimental Psychology: Applied*, 18(3), 314–330.

[27] M. Siegrist et al., 'Salient Value Similarity, Social Trust, and Risk/Benefit Perception', (2000) *Risk Analysis*, 20(3), 353–362, 354.

[28] Ibid., 358.

[29] D. Martin et al., 'The Incommensurability of Research Risks and Benefits: Practical Help for Research Ethics Committees', (1995) *IRB: Ethics & Human Research*, 17(2), 8–10, 9.

[30] Ibid., 8.

condition or disorder (a standard known as 'clinical equipoise'[31]). For non-therapeutic interventions, the risks must be minimised to the extent consistent with sound scientific design, and the remaining risks must be reasonable in relation to the knowledge that is expected to result. In addition, when a study involves a vulnerable population, such as children or adults who lack decision-making capacity, the risks posed by nontherapeutic procedures may not exceed a 'minor increase above minimal risk'.[32]

Component analysis has been influential, but it is not universally supported. Some critics maintain that the distinction between therapeutic and non-therapeutic procedures is inherently ambiguous, as 'all interventions offer at least some very low chance of clinical benefit'.[33] Others argue that the approach's reliance on clinical equipoise rests on the mistaken assumption that researchers have a duty to promote each participant's medical best interests, which conflates the ethics of research with those of clinical care.[34]

One alternative to component analysis is known as the 'net risk test', which is based on the principle that the fundamental ethical requirement of research is 'to protect research participants from being exposed to excessive risks of harm for the benefit of others'.[35] The approach has four elements. First, for each procedure involved in a study, the risks to participants should be minimised and the potential clinical benefits to participants enhanced, to the extent doing so is consistent with the study's scientific design. Second, instead of clinical equipoise, the approach requires that, 'when compared to the available alternatives, a research procedure must not present an excessive increase in risk, or an excessive decrease in potential benefit, for the participant'.[36] Third, to the extent particular procedures involve greater risks than benefits, those net risks 'must be justified by the expected knowledge gained from using that procedure in the study'.[37] Finally, the cumulative net risks of all of the procedures in a study must not be excessive.[38]

Both component analysis and the net risk test can add structure to the process of risk-benefit analysis by focusing attention on the risks and potential benefits of each intervention in a study. The advantage of this approach is that it reduces the likelihood that potential direct benefits from one intervention will be used as a justification for exposing participants to risks from unrelated interventions that offer no direct benefits. However, neither approach eliminates the need for subjective determinations. Under component analysis, the principle of clinical equipoise offers a benchmark for judging the risks and potential benefits of therapeutic procedures, but for non-therapeutic procedures, the only guidance offered is that the risks must be 'reasonable' in relation to the knowledge expected to result. The net benefit test dispenses with clinical equipoise entirely, instead relying on a general principle of avoiding 'excessive risk'. Whether a

---

[31] B. Freedman, 'Equipoise and the Ethics of Clinical Research', (1987) *New England Journal of Medicine*, 317(3), 141–145.

[32] C. Weijer, 'The Ethical Analysis of Risks and Potential Benefits in Human Subjects Research: History, Theory, and Implications for US Regulation' in National Bioethics Advisory Commission, *Ethical and Policy Issues in Research Involving Human Participants. Volume II – Commissioned Papers and Staff Analysis* (Bethesda, MD: National Bioethics Advisory Commission), pp. 1–29, p. 24.

[33] A. Rid and D. Wendler, 'Risk-Benefit Assessment in Medical Research – Critical Review and Open Questions', (2010) *Law, Probability and Risk*, 9(3–4), 151–177, 157.

[34] Ibid., 158.

[35] Ibid., 164.

[36] Ibid.

[37] Ibid.

[38] D. Wendler and F. Miller, 'Assessing Research Risks Systematically: The Net Risks Test', (2007) *Journal of Medical Ethics*, 33(8), 481–486.

particular mix of risks and potential benefits is 'reasonable' or 'excessive' is ultimately left to the judgment of those charged with reviewing the study.

Most regulations and ethics codes provide little guidance on the process of weighing the risks and potential benefits of research. The primary exception is the CIOMS guidelines, which adopts what it describes as a 'middle ground' between component analysis and the net risk test. In most respects, the CIOMS approach reflects component analysis, including its reliance on clinical equipoise as a standard for evaluating interventions or procedures that have the potential to provide direct benefits to participants. However, the guidelines also call for a judgment that 'the aggregate risks of all research interventions or procedures ... must be considered appropriate in light of the potential individual benefits to participants and the scientific social value of the research',[39] a requirement that mirrors the final step of the net risk test.

Neither component analysis nor the net risk test explicitly sets an upper limit on permissible risk, at least in studies involving competent adults. However, one of the developers of component analysis has stated that 'the notion of excessive net risks, and the underlying ethical principle of non-exploitation, clearly impose a cap on the risks that individuals are allowed to assume for the benefit of others'.[40] The notion of an upper limit on risk also appears in several ethical guidelines. For example, the CIOMS guidelines state that 'some risks cannot be justified, even when the research has great social and scientific value and adults who are capable of giving informed consent would give their voluntary, informed consent to participate in the study'.[41] Similarly, the European Commission has suggested that certain 'threats to human dignity and shared values' should never be traded against the potential scientific benefits of research, including 'commonly shared values like privacy or free movement ... certain perceptions of the integrity of a person (e.g. cloning, technological modifications) ... [and] widely shared view[s] of our place in the world (e.g. inhumane treatment of animals or threat to biodiversity)'.[42]

In light of the inherent ambiguities involved in weighing the risks and benefits of research, the results of risk-benefit assessments can be heavily influenced by the type of decision-making process used. The next section looks at these procedural issues more closely.

13.6 PROCEDURAL ISSUES IN RISK-BENEFIT ANALYSIS

In most health-related research, the process of risk-benefit assessment is undertaken by interdisciplinary bodies known as research ethics committees (RECs), research ethics boards (REBs), or institutional review boards (IRBs). These committees make judgments based on predictions about the preferences and attitudes of typical research participants, which do not necessarily reflect how the actual participants would react to particular risk-benefit trade-offs.[43] In addition, because few committees rely on formal methods of risk-benefit analysis, decisions are likely to be influenced by individual members' personal attitudes and cognitive biases.[44] For this reason, it is

---

[39] Council for International Organizations of Medical Sciences, 'International Ethical Guidelines', xi, 9.
[40] Wendler and Miller, 'Assessing Research Risks Systematically', 165.
[41] Council for International Organizations of Medical Sciences, 'International Ethical Guidelines', 10.
[42] European Commission Directorate-General for Research and Innovation, 'Research and Innovation, Research, Risk-Benefit Analyses, and Ethical Issues', (European Union, 2013).
[43] M. Meyer, 'Regulating the Production of Knowledge: Research Risk-Benefit Analysis and the Heterogeneity Problem', (2013) *Administrative Law Review*, 65(2), 241–242.
[44] C. Coleman, 'Rationalizing Risk Assessment in Human Subject Research', (2004) *Arizona Law Review*, 46(1), 1–51.

not surprising that different committees' assessments of the risks and potential benefits of identical situations exhibit widespread variation.[45]

Some commentators have proposed techniques to promote greater consistency in risk-benefit assessments. For example, it has been suggested that committees issue written assessments that could be entered into searchable databases.[46] Others have called on committees to engage in a formal process of 'evidence-based research ethics review', in which judgments about risks and potential benefits would be informed by a systematic retrieval and critical appraisal of the best available evidence.[47]

Outside of research ethics, a variety of techniques have been developed to systematise the process of risk-benefit analysis. For example, several quantitative approaches to risk-benefit assessment exist, such as the Quality-Adjusted Time Without Symptoms and Toxicity (Q-TWIST) test, which 'compares therapies in terms of achieved survival and quality-of-life outcomes',[48] or the 'standard gamble', which assigns utility values to health outcomes based on individuals' stated choice between hypothetical health risks.[49] Committees reviewing proposed studies can draw on these quantitative analyses when relevant ones exist.

In some cases, formal consultation with the community from which participants will be drawn can be an important component of assessing risks and benefits. For example, in the study of Havasupai tribe members discussed above, prior consultation with the community could have alerted researchers to the fact that research on migration patterns was threatening to the tribe's cultural beliefs. In cancer research, consultation with patient advocacy groups may help identify concerns about potential adverse effects that might not have been sufficiently considered by the researchers.[50] Further lessons might be learned from the the analysis by Chuong and O'Doherty, Chapter 12, this volume.

13.7 CONCLUSION

Risk-benefit analysis is a critical part of the process of evaluating the ethical acceptability of health-related research. The primary challenge in risk-benefit assessment arises from the fact that perceptions about risks and potential benefits are inherently subjective. Those charged with assessing the ethical acceptability of research should make efforts to incorporate as many different perspectives into the process as possible, to ensure that their decisions do not simply reflect their own idiosyncratic views.

---

[45] T. Caulfield, 'Variation in Ethics Review of Multi-Site Research Initiatives', (2011) *Amsterdam Law Forum*, 3(1), 85–100.

[46] Coleman, 'Rationalizing Risk Assessment', 1176–1179.

[47] E. Anderson and J. DuBois, 'Decision-Making with Imperfect Knowledge: A Framework for Evidence-Based Research Ethics', (2012) *Journal of Law, Medicine and Ethics*, 40(4), 951–966.

[48] Troche et al., 'Evaluation of Therapeutic Strategies', 13.

[49] S. van Osch and A. Stiggelbout, 'The Construction of Standard Gamble Utilities', (2008) *Health Economics*, 17(1), 31–40.

[50] N. Dickert and J. Sugarman, 'Ethical Goals of Community Consultation in Research', (2005) *American Journal of Public Health*, 95(7), 1123–1127.

14

The Regulatory Role of Patents in Innovative Health Research and Its Translation from the Laboratory to the Clinic

Dianne Nicol and Jane Nielsen

14.1 INTRODUCTION

Regulators must ensure that innovative health research is safe and undertaken in accordance with laws, ethical norms and social values, and that it is translated into clinical outcomes that are safe, effective and ethically appropriate. But they must also ensure that innovative health research and translation (IHRT) is directed towards the most important health needs of society. Through the patent system, regulators provide an incentive-based architecture for this to occur by granting a temporary zone of exclusivity around patented products and processes. Patents thus have the effect of devolving control over IHRT pathways to patentees and to those to whom patentees choose to license their patent rights.

The sage words of Stephen Hilgartner set the backdrop for this chapter: 'Patents do not just allocate economic benefits; they also allocate leverage in negotiations that shape the technological and social orders that govern our lives'.[1] Patents have been granted for many – if not all – of the major recent innovations in health research, from the earliest breakthroughs like recombinant DNA technology, the polymerase chain reaction, the Harvard Oncomouse and the BRCA gene sequences, through to a whole variety of viruses, monoclonal antibodies, receptors and vectors, thousands of DNA sequences, embryonic stem cell technology, intron sequence analysis, genome editing technologies and many more.[2] These innovations have laid the foundations for whole new health research pathways, from basic research, through applied research, to diagnostic and therapeutic end points.[3] Broad patent rights over these fundamental innovations give patentees the freedom to choose how these research pathways will be progressed. Essentially then, the patent grant puts patentees in a position to assert significant private regulatory control over IHRT.

The first part of this chapter outlines this regulatory role of patents in IHRT. The chapter then considers the ways in which patentees choose to use their patent rights in IHRT, and the scope for government intervention. The chapter then explores recent actions by patentees that indicate

---

[1] S. Hilgartner, 'Foundational Technologies and Accountability', (2018) *American Journal of Bioethics*, 18(12), 63–65.

[2] Organisation for Economic Cooperation and Development, 'Key Biotechnology Indicators', (OECD, 2019), www .oecd.org/innovation/inno/keybiotechnologyindicators.htm; Nuffield Council on Bioethics, 'The Ethics of Patenting DNA', (Nuffield Council on Bioethics, 2002), 39–44; D. Nicol, 'Implications of DNA Patenting: Reviewing the Evidence', (2011) *Journal of Law, Information and Science* 7, 21(1).

[3] J. P. Walsh et al., 'Effects of Research Tool Patents and Licensing on Biomedical Innovation' in W. M. Cohen and S. A. Merrill (eds), *Patents in the Knowledge-Based Economy* (The National Academies Press, 2003), pp. 285–340, see particularly pp. 332–335.

a willingness to moderate the use of their patent rights by engaging in self-regulation and other forms of collaborative regulation. Finally, the chapter concludes with a call for greater government oversight of patent use in IHRT. Although self-regulation has merit in the absence of clear governmental direction, it is argued that private organisations should not have absolute discretion in deciding how to employ their patents in areas such as health, but that they must be held to account in exercising their state-sanctioned monopoly rights.

14.2 PATENTS AS A FORM OF PRIVATE REGULATION

In many markets, the regulation of market entry, prices, product availability and development is left to the market to varying degrees, there being at least some general consensus that competitive decision-making is a hallmark of market efficiency.[4] At the same time, granting patent rights removes an element of competition from a market in order to induce innovation and disclosure.[5] While it is unclear how much innovation is optimal, it has been suggested that there is unlikely to ever be too much from an economic welfare perspective.[6]

Although primary innovators are arguably best placed to organise and control follow-on innovation,[7] vesting decision-making power in a single private entity has the potential to scuttle efficiency in much the same way as absolute government control. Nonetheless, conferring this power on individual entities through the grant of patents – and accompanying Intellectual property (IP) rights – is generally justified on efficiency grounds.[8] However, non-efficiency goals such as distributive fairness may also be important drivers of private regulatory arrangements and may be incorporated either consciously or unconsciously in regulatory schemes.[9]

Granting a patent gives a property right in an invention. As Mark Lemley observes, IP constitutes both a form of government regulation and a property right around which parties can contract,[10] and its confused identity partly explains why policy makers have grappled with exactly how to manage the delicate innovation balance. Studies have provided mixed evidence as to the necessity to grant IP rights: in some technology areas, patents are viewed as necessary in order to recoup research and development investment, but this is by no means universal.[11]

---

[4] F. M. Scherer and D. Ross, *Industrial Market Structure and Economic Performance* (Boston: Houghton Mifflin, 1990), p. 660; K. J. Arrow, 'Economic Welfare and the Allocation of Resources for Invention' in The National Bureau of Economic Research (eds), *The Rate and Direction of Inventive Activity: Economic and Social Factors* (Princeton University Press, 1962),pp. 609–626.

[5] R. P. Merges, *Justifying Intellectual Property* (Cambridge, MA: Harvard University Press, 2011), p. 27; R. Mazzoleni and R. R. Nelson, 'Economic Theories about the Benefits and Costs of Patents', (1998) *Journal of Economic Issues*, 32(4), 1031–1052, 1039.

[6] Federal Trade Commission, 'To Promote Innovation: The Proper Balance of Competition and Patent Law and Policy', (FTC, 2003), ch 2, their n30.

[7] E. Kitch 'The Nature and Functions of the Patent System', (1977) *Journal of Law and Economics*, 20(2), 265–290; R. P. Merges, 'Of Property Rules, Coase, and Intellectual Property', (1994) *Columbia Law Review*, 94(8), 2655–2673, 2661; M. A. Lemley, 'Ex Ante versus Ex Post Justifications for Intellectual Property', (2004) *University of Chicago Law Review*, 71(1), 129–149.

[8] R. Feldman, 'Regulatory Property: The New IP,' (2016) *Columbia Journal of Law & the Arts*, 40(1), 53–103; F. K. Hadfield, 'Privatising Commercial Law', (2001) *Regulation*, 24(1), 40–45, 44; O. Feeney et al., 'Patenting Foundational Technologies: Lessons from CRISPR and Other Core Biotechnologies', (2018) *The American Journal of Bioethics*, 18(12), 36–48.

[9] S. L. Schwarcz, 'Private Ordering', (2002) *Northwestern University Law Review*, 91(1) 319–350.

[10] M. Lemley, 'The Regulatory Turn in IP', (2013) *Harvard Journal of Law and Public Policy*, 36(1), 109–115.

[11] R. Levin et al., 'Appropriating the Returns From Industrial Research and Development', (1987) *Brookings Papers on Economic Activity: Microeconomics*, 3, 783–831; W. Cohen et al., 'Protecting Their Intellectual Assets: Appropriability Conditions and Why US Manufacturing Firms Patent (or Not)', (2000), *Working Paper No. 7552, National Bureau of*

The value of patents in IHRT has not been unequivocally established, although there is some evidence to suggest they are crucial for signalling purposes.[12] Patent law can be said to form a 'corrective' function in the health context, particularly in relation to pharmaceuticals and biotechnology, where the development of clinical products is subject to substantial regulation.[13] Without patents, it is argued that researchers would not commit the considerable investment required to conduct research with the ultimate aim of a clinical outcome.

14.3 USE OF PATENT RIGHTS IN INNOVATIVE HEALTH RESEARCH AND TRANSLATION

Patentees can limit who enters a field by choosing who, if anyone, they will authorise to use their patents. This can create problems for broad breakthrough technologies, where insistence on exclusivity gives patentees and their licensees control over whole research pathways, allowing them to dictate how those pathways develop. Patentees and their licensees could choose to block others completely from using the technology, or restrict access, or charge excessive prices for use. Conversely, they could allow their patented technology to be used widely for minimal costs. The tragedy of the anticommons posited by Michael Heller and Rebecca Eisenberg, adds further complexity, speculating that a proliferation of patents in particular areas of technology exacerbates the problem because no one party has an effective privilege of use.[14] Rather, agreement with multiple patentees would be required in order to utilise a particular resource.

Fortunately, empirical studies have revealed little evidence of blocking or anticommons effects in IHRT,[15] suggesting that, on the whole, working solutions employed by researchers have allowed them to work around 'problematic' patents so that research and development may progress. 'Working solutions' mean strategies such as entering into licence agreements or other collaborative arrangements; inventing around problematic patents; relying on research exemptions; or challenging the validity of patents.[16] These working solutions can be viewed as facets of the regulatory scheme that encompasses the grant of patent rights. However, solutions that involve entering into a licence agreement or other collaborative arrangement also involve a degree of conformity on the part of a patentee. It may be fruitless to approach a patentee unless they are willing to negotiate, which takes time and effort on their part, as well as on the part of the licensee. Unless these processes can be streamlined, the incentive to license is low.

14.4 SCOPE FOR GOVERNMENT INTERVENTION

Arguably, the fruits of all health-related research should be distributed openly, because of its vital social function of improving healthcare. However, this is hardly a realistic option for aspects such as drug development, where the enormous cost of satisfying regulatory requirements for

---

Economic Research. See also E. Mansfield, 'Patents and Innovation: An Empirical Study', (1986) *Management Science*, 32(2), 173–181.

[12] E. Burrone, 'Patents at the Core: The Biotech Business', (WIPO, 2006), www.wipo.int/sme/en/documents/patents_biotech_fulltext.html.

[13] Lemley, 'The Regulatory Turn in IP'.

[14] M. A. Heller and R. S. Eisenberg, 'Can Patents Deter Innovation? The Anticommons in Biomedical Research', (1998) *Science*, 280(5364), 698–701.

[15] Walsh et al., 'Effects of Research Tool Patents and Licensing', pp. 285, 335; D. Nicol and J. Nielsen, 'Patents and Medical Biotechnology: An Empirical Analysis of Issues Facing the Australian Industry', (2003) *Occasional Paper Series (6) Centre for Law and Genetics*, 174–193; but note R. S. Eisenberg, 'Noncompliance, Nonenforcement, Nonproblem? Rethinking the Anticommons in Biomedical Research', (2008) *Houston Law Review*, 45(4), 1059–1099.

[16] Nicol and Nielsen, 'Patents and Medical Biotechnology', 208–225.

marketing approval must be recoverable. For other aspects of IHRT, however, the case for more open access is compelling, particularly since it generally originates in public research laboratories, funded by governments from the public purse.[17] Yet the ensuing patents may ultimately be controlled by private parties, whether spin-offs or more established firms. This phenomenon has been referred to by Jorge Contreras and Jacob Sherkow as 'surrogate licensing'.[18]

Given the public contribution made to IHRT, the argument for open access, at least for research purposes, is appealing. Public funders are within their rights to insist on some form of open dissemination in such circumstances.[19] But what are the options when patentees or their licensees insist on exclusivity, even for the most fundamental research tools? If governments see patents as providing a broader social function beyond giving monopoly rights to patentees – albeit temporary in nature – they must ensure that, along with incentives to innovate, the patent system provides appropriate incentives to disseminate innovative outputs, or other regulatory mechanisms to compel the provision of access where needed.[20] Patents provide patentees with significant freedom to decide who can enter a particular field of research, and what they can do. Some jurisdictions do have legislative provisions allowing government or private providers to step in should patentees fail to work the invention.[21] Most countries exempt from infringement the steps needed for regulatory approval of generic pharmaceuticals and other chemicals.[22] Some also exempt use of the patent for experimental purposes, although the scope of protected experimental use remains unclear.[23] However, the reality is that the role of governments in regulating patent use is limited.

14.5 EMERGENT SELF-REGULATORY MODELS FOR USE OF PATENT RIGHTS IN INNOVATIVE HEALTH RESEARCH AND TRANSLATION

Recognising these limitations on government control of patent use, some promising developments are emerging in IHRT that indicate that patentees and their licensees are willing to consider a range of self-regulatory models in ensuring optimal patent utilisation. Some of the more prominent examples are discussed below.

14.5.1 Non-exclusive Research Tool Licensing

Because foundational research tools are just that – foundational to whole new areas of research – best practice dictates they should be licensed non-exclusively. US funding agencies and

[17] L. Pressman et al., 'The Licensing of DNA Patents by US Academic Institutions: An Empirical Study', (2006) *Nature Biotechnology*, 24(1), 31.

[18] J. L. Contreras and J. S. Sherkow, 'CRISPR, Surrogate Licensing, and Scientific Discovery', (2017) *Science*, 355(6326), 698–700; J. S. Sherkow, 'Patent Protection for CRISPR: An ELSI Review', (2017) *Journal of Law and the Biosciences*, 4(3), 565–576, 570–571.

[19] A. K. Rai and B. N. Sampat, 'Accountability in Patenting of Federally Funded Research', (2012) *Nature Biotechnology*, 30(10), 953–956; K. J. Egelie et al., 'The Ethics of Access to Patented Biotech Research Tools from Universities and Other Research Institutions,' (2018) *Nature Biotechnology*, 36(6), 495.

[20] Referred to by some commentators as 'carrots' and 'sticks'; see e.g. I. Ayres and A. Kapczynski, 'Innovation Sticks: The Limited Case for Penalizing Failures to Innovate', (2015) *University of Chicago Law Review*, 82(4), 1781–1852.

[21] For example, US: 28 USC § 1498(a) (government use) (2011); Australia: *Patents Act 1990* (Cth) section 133 (compulsory licensing), section 163 (government use).

[22] For example, US: *Roche Products Inc. v. Bolar Pharmaceuticals Co.*, 733 F.2d 858 (Fed. Cir. 1984), 35 USC § 271(e)(1)); *Patents Act 1990* (Cth) sections 119A and 119B.

[23] R. Dreyfuss, 'Protecting the Public Domain of Science: Has the Time for an Experimental Use Defense Arrived?', (2004) *Arizona Law Review*, 946(3), 457–472; K. J. Strandburg, 'What Does the Public Get? Experimental Use and the Patent Bargain', (2004) *Wisconsin Law Review*, 2004(1), 81–155.

universities agree; for example, the US National Institutes of Health released guidance to this effect in 1999 and 2005.[24] In 2007, the Association of University Technology Managers, recognising that 'universities share certain core values that can and should be maintained to the fullest extent possible in all technology transfer agreements', provided nine key points to consider in licensing university patents. Point 5 recommends 'a blend of field-exclusive and non-exclusive licenses'.[25]

Yet non-exclusive licensing is not cost-free. The problem that it presents to users is that it imposes a fee in return for not being sued for infringement, with little or no additional benefit for the user.[26] Inclusion of reach through rights to future uses adds to the burden on follow-on researchers.[27] If governments were really concerned about the toll of research tool patent claims on IHRT they could choose to exclude them, or to require them to be exchanged through some form of statutory licensing scheme, with minimal or no licensing fees and no other restrictive terms. For now, however, governments seem content to leave such decisions to patentees.

We are witnessing some interesting developments in this area, illustrating that government intervention may not yet be necessary. Companies like Addgene and the Biobricks Foundation have been established as intermediaries to facilitate no-cost, non-exclusive patent licensing and sharing of research materials for genome editing and synthetic biology research, respectively.[28] There are also other examples of these types of intermediary arrangements, or 'clearinghouses' as they are sometimes called, in IHRT. Such arrangements appear to provide a valuable social function provided that fees are not excessive and that technology that is of real value to IHRT is included, so that the clearinghouse does not become a 'market for lemons'.[29]

14.5.2 *Mixed Licensing Models*

Realistically, a more nuanced approach over the simple choice of exclusive or non-exclusive licensing is needed, involving a mix of licensing strategies for a single patented technology. Licensing of the clustered regularly interspersed palindromic repeats (CRISPR) patents illustrates this point. CRISPR, as explained in Chapter 34, is a genome editing technology that has captivated the research world because of its ease of use and enhanced safety, owing to reduced incidence of off-target effects.[30]

---

[24] US Department of Health and Human Services, National Institutes of Health, 'Principles and Guidelines for Recipients of NIH Research Grants and Contracts on Obtaining and Disseminating Biomedical Research Resources: Final Notice', (1999) *Federal Register* 72090, 64(246); US Department of Health and Human Services, National Institutes of Health, 'Best Practices for the Licensing of Genomic Inventions: Final Notice', (2005) *Federal Register 18413*, 70(68); see also Organisation for Economic Co-Operation and Development, 'Guidelines for the Licensing of Genetic Inventions', (OECD, 2006).

[25] Association of University Technology Managers, 'In the Public Interest: Nine Points to Consider in Licensing University Technology', (Association of University Technology Managers, 2007), www.autm.net/AUTMMain/media/Advocacy/Documents/Points_to_Consider.pdf.

[26] A. D. So et al., 'Is Bayh-Dole Good for Developing Countries? Lessons from the US Experience', (2008) *PLoS Biology*, 6(10), e262.

[27] J. Nielsen, 'Reach-Through Rights in Biomedical Patent Licensing: A Comparative Analysis of their Anti-Competitive Reach', (2004) *Federal Law Review*, 32(2), 169–204.

[28] J. Nielsen et al., 'Provenance and Risk in Transfer of Biological Materials', (2018) *PLoS Biology*, 16(8), e2006031

[29] E. van Zimmeren et al., 'Patent Pools and Clearinghouses in the Life Sciences', (2011) *Trends in Biotechnology*, 29(11), 569–576; see also D. Nicol et al., 'The Innovation Pool in Biotechnology: The Role of Patents in Facilitating Innovation', (2014) *Centre for Law and Genetics Occasional Paper No. 8*. 249–250.

[30] V. Iyer et al., 'No Unexpected CRISPR-Cas9 Off-target Activity Revealed by Trio Sequencing of Gene-edited Mice', (2018) *PLoS Genetics*, 14(7), p. e1007503.

Already, we are witnessing the adoption of nuanced approaches for licensing CRISPR patents. For example, the Broad Institute, one of the giants of CRISPR technology, non-exclusively licences CRISPR constructs freely for public sector research through Addgene, and charges a fee for use in more commercially-oriented research. Broad exclusively licences to its own spin-off company, Editas, for therapeutic product development. Broad describes this as an 'inclusive innovation model'.[31] However, this model has been criticised by Oliver Feeney and colleagues on the basis that the decision whether to allow other uses for therapeutic purposes is left to Editas.[32] They see this as a 'significant moral hazard', because of the potential restrictions it imposes on therapeutic development. While Feeney and colleagues propose government-imposed time limitations on exclusivity as a means of addressing such hazards, it is doubtful, given past history, that governments would be persuaded to incorporate this level of post-grant regulatory intervention within the patent system.

Knut Egelie and colleagues, equally concerned about CRISPR patent licensing, argue that public research organisations should commit more fully to a self-regulatory model that balances social responsibilities with commercial activity.[33] Their 'transparent licensing model' would minimise fees and other restrictions for uses of patented subject matter as research tools, and narrow field-of-use exclusive licences for commercial development. They suggest government intervention as an alternative to this self-regulatory model, referring to some of the recently emerging contractual funding strategies in Europe. However, they themselves criticise both options, the former for lacking public control and the latter for over-regulation and unnecessary bureaucracy. More cooperative and collaborative strategies, involving both public sector and private sector organisations, might provide alternative models.

14.5.3 *Collaborative Licensing*

A greater commitment to social responsibility might be achieved by patentees and their licensees through entry into collaborative IP arrangements.[34] Patent pools have been used in some high technology areas – particularly information technology – to overcome patent thickets and cluttered patent landscapes.[35] In IHRT, however, complex arrangements such as patent pools have gained limited traction,[36] primarily because of the lack of need to date. Simpler strategies

---

[31] Broad Institute, 'Information About Licensing CRISPR Genome Editing Systems', (Broad Institute, 2017), www .broadinstitute.org/partnerships/office-strategic-alliances-and-partnering/information-about-licensing-crispr-genome-edi.

[32] Feeney et al., 'Patenting Foundational Technologies', 40.

[33] K. J. Egelie et al., 'The Emerging Patent Landscape of CRISPR–Cas9 Gene Editing Technology', (2016) *Nature Biotechnology*, 3(10), 1025.

[34] A. Krattiger and S. Kowalski, 'Facilitating Assembly of and Access to Intellectual Property: Focus on Patent Pools and a Review of other Mechanisms' in A. Krattiger et al. (eds), *Intellectual Property Management in Health and Agricultural Innovation: A Handbook of Best Practices* (MIHR, Oxford UK and PIPRA Davis California, US, 2007) p. 131; P. Gaulé, 'Towards Patents Pools in Biotechnology?', (2006) *Innovation Strategy Today*, 2, 123; G. Van Overwalle et al., 'Models for Facilitating Access to Patents on Genetic Inventions', (2006) *Nature Reviews Genetics*, 7(2), 143; van Zimmeren et al., 'Patent Pools and Clearinghouses'; Organisation for Economic Cooperation and Development, 'Collaborative Mechanisms for Intellectual Property Management in the Life Sciences', (OECD, 2011); Nicol et al., 'The Innovation Pool'.

[35] R. P. Merges, 'Institutions for Intellectual Property Transactions: The Case of Patent Pools' in R. C. Dreyfuss et al. (eds), *Expanding the Boundaries of Intellectual Property: Innovation Policy for the Knowledge Society* (Oxford University Press; 2001), ch 6.

[36] E. van Zimmeren et al., *Patent Licensing in Medical Biotechnology in Europe: A Role for Collaborative Licensing Strategies?* (Catholic University of Leuven Centre for Intellectual Property Rights; 2011), 82; Nicol et al., 'The Innovation Pool', 238–239, 250.

such as non-exclusive licensing and clearinghouses appear to be adequate at the present time, and predicted anticommons effects have not yet emerged.[37]

Where patentees are reluctant to engage in collaborative strategies, there is some scope to mandate engagement. Patent pools, for example, have in some instances – especially in the USA – been established by government regulators in order to ease innovative burdens and address competition law concerns.[38] Mandatory arrangements are rarely optimistically embraced, and prospects for the sustainability of collaborative arrangements is probably significantly greater where they are voluntary. Patent pools are complex structures and involve many legal considerations. Although there has been some success in establishing patent pooling-type arrangements in public health emergencies like HIV/AIDS and other epidemics,[39] it is difficult to see what would motivate patentees to come together to create such complex structures in IHRT at the present time, particularly given the rapid pace of technological development and change.

Patent aggregation is another increasingly popular strategy, referring to the process of collecting suites of IP required to conduct research and development within a particular field of use. The process of patent aggregation has brought with it some negative press, because of concerns that aggregators could be 'patent trolls', whose sole motivation is extracting licensing revenue.[40] However, not all aggregators have this trolling motivation, but rather license out entire bundles of patents on a non-exclusive basis. To this extent, their role in advancing the research agendas in IHRT can be seen as broadly facilitative.[41]

14.5.4 *Ethical Licensing*

Aside from the social good associated with self-regulatory models of patent use discussed above, there are other ethical and social considerations that could be addressed through more public-focused approaches to licensing. For example, even where public sector organisations exclusively license to private partners, whether spin-offs or established firms, it is common practice for the license terms to reserve rights for the organisation's researchers to continue to conduct research using patented subject matter.[42] Reservation of the right to engage in broader sharing of patented subject matter for non-commercial research purposes might also be included in such agreements, effectively circumventing the lack of a statutory or common law research exemption in some jurisdictions.

Patent pledges and non-assertion covenants can be used to serve essentially the same purpose.[43] The role of reservation of rights could also extend to humanitarian uses, which has been

[37] Gaulé, 'Towards Patents Pools in Biotechnology?', 123, 129; Nicol et al., 'The Innovation Pool', 238.

[38] D. Serafino, 'Survey of Patent Pools Demonstrates Variety of Purposes and Management Structures', (2007) *KEI Research Note 6*, www.keionline.org/book/survey-of-patent-pools-demonstrates-variety-of-purposes-and-management-structures.

[39] UNITAID, 'The Medicines Patent Pool', (UNITAID), www.unitaid.org/project/medicines-patent-pool/#en.

[40] M. A. Lemley, 'Are Universities Patent Trolls?', (2008) *Fordham Intellectual Property, Media and Entertainment Law Journal*, 18(3), 611–631; A. Layne-Farrar and K. M. Schmidt, 'Licensing Complementary Patents: "Patent Trolls", Market Structure, and "Excessive" Royalties', (2010) *Berkeley Technology Law Journal*, 25(2), 1121.

[41] A. Wang, 'Rise of the Patent Intermediaries', (2010) *Berkeley Technology Law Journal*, 25(1), 159, 167, 173.

[42] A. B. Bennett, 'Reservation of Rights for Humanitarian Uses' in A. Krattiger et al. (eds), *Intellectual Property Management in Health and Agricultural Innovation: A Handbook of Best Practices* (Oxford, UK: MIHR; and Davis, USA: PIPRA; 2007), p. 41.

[43] J. Contreras, 'Patent Pledges', (2015) *Arizona State Law Journal*, 47(3), 543–608; A. Krattiger, 'The Use of Nonassertion Covenants: A Tool to Facilitate Humanitarian Licensing, Manage Liability, and Foster Global Access' in A. Krattiger et al. (eds), *Intellectual Property Management in Health and Agricultural Innovation: A Handbook of Best Practices*, (Oxford, UK: MIHR; and Davis, USA: PIPRA; 2007), p. 739.

mooted specifically in the context of agricultural biotechnology. As Alan Bennett notes, these voluntary measures can serve the purpose of meeting the humanitarian and commercial needs of developing countries in the absence of national policies to this effect.[44] Such measures could be equally effective in the context of humanitarian uses of innovative health technologies, an area which likewise suffers from a lack of clear government policy direction.

There has been recent discussion on the efficacy of introducing ethical terms into patent licences for the new genome editing technologies, particularly CRISPR. The emergence of this technology triggered a range of ethical debates in relation to its applications in agriculture, the natural environment – for example, in pest eradication through a combination of CRISPR and gene drives – and humans – for example, in genetic enhancement, germline genome modification and gene editing research using human embryos.[45]

The Broad Institute, through Editas, and other public research organisations and their licensees, are already using licences that exclude these types of ethically questionable uses, whether in human or non-human contexts. As Christi Guerrini and colleagues note, there are some obvious advantages with this approach, including that: licence terms are enforceable; they can be tailored; and they are negotiated, leading to better buy in.[46] Given that the regulation of genome editing varies widely across jurisdictions,[47] the introduction of ethical licensing terms also has the advantage of creating enforceable obligations across the jurisdictions where the patent has been granted and where the licence applies. Potentially, then, ethical licences could impose global standards on uses of CRISPR technology, which is otherwise considerably conjectural if relying on agreement between countries.

Despite the apparent attractiveness of ethical licensing, however, there is likely to be some unease with the notion of devolving decisions about what is or is not ethical to patentees.[48] In areas such as this, which are highly contentious, community consensus would usually be a precursor to government regulation. Is regulatory failure in this area significant enough to justify private action? Is this a step too far when it amounts to ceding regulation to private entities?

14.6 CONCLUSION

Patents play a key role in the progress of IHRT. By granting patents, governments devolve to patentees considerable decision-making power about who can enter particular fields of IHRT and what they can do. This chapter has shown that patentees can and do choose to exercise this power wisely, by engaging in open and collaborative models for patent use. However, not all choose do so, and governments currently have limited regulatory tools with which to compel such engagement.

Patentees can decide to work collaboratively with other interested parties, or not. They can decide whether to share broadly, or not. They can even decide what types of uses are ethical or unethical. This is a significant set of delegated powers. Regulators have at their disposal various policy levers that could provide them with broad discretion to specify criteria for patent

---

[44] Bennett, 'Reservation of Rights'.

[45] Sherkow, 'Patent Protection for CRISPR', 565–576, 572–573.

[46] C. J. Guerrini et al., 'The Rise of the Ethical License', (2017) *Nature Biotechnology*, 25(1), 22; Sherkow, 'Patent Protection for CRISPR'.

[47] R. Isasi et al., 'Editing Policy to Fit the Genome?', (2016) *Science*, 351(6271), 337–339.

[48] N. de Graeff et al., 'Fair Governance of Biotechnology: Patents, Private Governance, and Procedural Justice', (2018) *American Journal of Bioethics*, 18(12), 57–59, 58.

eligibility, periods of exclusivity and access.[49] Regulatory control can be asserted by governments both pre-grant, influencing the ways in which patents are granted, and post-grant, on the ways in which patents are used. Governments can use these regulatory tools to impose limits on these delegated powers, but these are not being fully utilised at present.

The current situation is that non-enforceable guidelines have been issued in some jurisdictions to assist patentees in deciding how to exercise their powers, but not in others. Internationally, although the OECD has issued licensing guidelines,[50] for the most part there is no jurisdictional consensus on how best to set limits on the exercise of patent rights. This is not surprising in view of the diversity of technologies and actors involved and given jurisdictional discrepancies. More research is needed to assist governments in finding optimal ways to support, guide and regulate public research organisations and private companies in their use of the patent system in IHRT.

ACKNOWLEDGEMENTS

The research presented in this chapter is supported by funding from the Australian Research Council, DP180101262.

[49] D. L. Burk and M. A. Lemley, 'Policy Levers in Patent Law', (2003) *Virginia Law Review*, 89(7), 1575–1696.
[50] OECD, 'Recommendation of the Council on the Licensing of Genetic Inventions', (OECD/LEGAL/0342, 2007).

15

Benefit Sharing

From Compensation to Collaboration

Kadri Simm

15.1 INTRODUCTION

Benefit sharing pertains to the distribution of benefits and burdens arising from research. More specifically, it concerns what, if anything, is owed to individuals, communities or even populations that participate in research (benefits to investors, to other populations or the social value of research more generally understood are not the focus of benefit sharing).

Traditionally, health research has been concerned with compensating those participants who have been more or less directly involved. The practice of benefit sharing, especially in agriculture, introduced a perspective that recognised the contributions of communities and populations in safeguarding biological resources.[1] The issue is further complicated in human genetics as genetic information is by nature shared, and thus implicates individuals and communities who might not have participated in research in the traditional sense. At the same time, contemporary global research activities have increasingly been associated with for-profit companies. Some of their practices – 'helicopter research', ethics dumping – have given credence to broader political and social worries that have now been harnessed to the concept of benefit sharing, which was initially used within more limited research settings.

Framing benefit-sharing debates are several central concepts – the duty to avoid exploitation, the rights and interests of all research stakeholders, the requirements of fairness and compensation, and the various principles of distributive justice. In many ways, benefit sharing as an ethics and governance framework attempts to deal with most of those concerns and anxieties. Thus, responses to the question, 'why is benefit sharing a duty?' vary. In practical terms, benefit sharing is a thoroughly context-sensitive topic. It matters which risks and harms are involved in research (if any), who the investigators and funders are (for-profit, local, NGOs etc), where research takes place (developed or low- or middle-income countries), who is involved (e.g. vulnerable groups), what local needs are, and whether research is successful.

In what follows, I will give a brief overview of the ethical arguments and historical dynamics behind benefit-sharing practices, then outline major governance frameworks and discuss the potential problems around applying this concept in health research. The overall aim of this chapter is to highlight the complexity of benefit sharing and argue that success hinges on the careful balancing of universal research ethics duties with the particularities of concrete research

---

[1] Well-known examples of problematic research that motivated the international community to formulate benefit-sharing framework were the Neem tree and Canavan-disease controversies.

projects taking place in distinct locations. Benefit sharing is no panacea for solving the inequalities of access and opportunities associated with global health research. Yet it can be a profoundly empowering tool, especially as the framework is shifting from compensation
to collaboration.

15.2 HISTORY AND RATIONALE OF BENEFIT SHARING

Looking back, the rationale behind access and benefit-sharing justifications has been dynamic.
It was originally employed in the context of agriculture and non-human biological resources
(plants, animals, microorganisms). The 1992 UN Convention on Biological Diversity (CBD)
acknowledged national sovereignty in all genetic resources and requested 'fair and equitable
sharing of the benefits arising out of the utilization of genetic resources'.[2] As the majority of the
world's biological diversity is found in developing countries, benefit sharing was seen as a
necessary instrument in guaranteeing these countries' continuing interest in safeguarding this
heritage and curbing biopiracy (when indigenous knowledge and resources are patented or
otherwise exploited by third parties with no permission or compensation for the locals). The
supplementary Nagoya Protocol on Access and Benefit-sharing (2010) is a legal framework that
supports the implementation of the objectives of CBD.[3]

Since the 1990s, benefit sharing emerged as an important component of health research and
made its appearance in various international documents (in the rest of the chapter, I will focus
on benefit sharing in health research only, excluding research on non-human materials and
populations). The Human Genome Organisation (HUGO) Ethics Committee Statement on
benefit sharing formulates:

> A benefit is a good that contributes to the well-being of an individual and/or a given community
> (e.g. by region, tribe, disease-group …). Benefits transcend avoidance of harm (non-
> maleficence) in so far as they promote the welfare of an individual and/or of a community.
> Thus, a benefit is not identical with profit in the monetary or economic sense. Determining a
> benefit depends on needs, values, priorities and cultural expectations.[4]

Benefits put forward by scientists, as well as the pharmaceutical industry, patients, investors and
public health officials, span a wide array of potential valued 'goods', from improved health and
science to financial gains and wider social benefits.[5] A fixed definition of what would constitute
a benefit would be quite useless, or worse, unfair (an informative list of possible benefits
regarding non-human research is available from the annex of the Nagoya Protocol). Potential
benefits and harms arising from clinical trials would be rather different from those associated
with population biobanks, for example. Benefits can be related to healthcare, but they could also
encompass other socially important goals, such as support for infrastructure, development of
local research capacities and build-up of community resilience. The kind and scope of potential

---

[2] United Nations 'Convention on Biological Diversity', (United Nations, 1992).
[3] Secretariat of the Convention on Biological Diversity, 'Nagoya Protocol on Access to Genetic Resources and the Fair
and Equitable Sharing of Benefits Arising from their Utilization to the Convention on Biological Diversity', (United
Nations Secretariat of the Convention on Biological Diversity, 2011).
[4] Human Genome Organization Ethics Committee, 'Genetic Benefit-Sharing', (2000) *Science*, 290(5489), 49.
[5] K. Simm, 'Benefit-Sharing: An Inquiry Regarding the Meaning and Limits of the Concept in Human Genetic
Research', (2005) *Genomics, Society and Policy*, 1(2), 29–40.

benefits has few limits, although the minimum threshold for satisfying the 'reasonable availability' should surpass the simple licensing of drugs or interventions with market prices.[6]

When is an appropriate time for benefit sharing? These issues deserve consideration from the very earliest phases of research design. It is necessary to find out the characteristics and needs of the potential research sites to ensure that the planned investigations, as well as potential benefits, respond to those needs. Equally, benefit sharing could involve long-term follow up of participants or training and employment of community members that continues for years after research has ended.

The HUGO statement on benefit sharing mapped the following justifications for the concept in human genetic research:

1. Descriptive argument: There is an 'emerging international consensus'[7] that benefits should be shared with participants.
2. Common heritage argument – we all share (in one sense) the same genome, so there is a shared interest in genetic heritage of humankind; thus, the Human Genome Project should benefit all humanity.
3. Justice-based arguments – compensatory (compensation in return for contribution), procedural (procedural justice should be adhered to in benefit-sharing) and distributive (equitable allocation and access to resources and goods) justice as important aspects to consider.
4. Solidarity argument on two levels: first, as a potential basis for benefit sharing among a group of research participants (communities, host populations); second, to foster health for wider communities and eventually the whole of humanity, thus benefits should not be limited strictly to those participating in research.[8]

Of these various justifications, the overall concern fuelling benefit-sharing debates has been justice, and the concept itself has been likened to a device in the toolbox of justice.[9] Yet, justice is notoriously difficult to pin down given that the principles of justice vary – one can refer to equality as fundamental, or point at the importance of merit, and in healthcare contexts the principle of need has often served as central. Decisions about what justice requires (i.e. what principles are important in a particular context) can result in divergent benefit-sharing patterns and practices – how benefits are defined and by whom, as well as with whom the sharing is foreseen.[10] Certain justifications necessarily exclude or include specific groups or communities. For example, the compensatory logic associated with the principles of merit and desert would benefit those directly involved but could leave out those who did not directly participate but are nevertheless part of the community. Focus on a shared human heritage of genetic resources tends to disregard the needs and deserts of particular communities where research is undertaken. This is why, for example, in the agricultural and plant genetics context, the early employment of the global heritage model was quickly replaced by the nationalisation and property model of genetic resources.[11] The patenting practices through which the 'shared free resources' were

[6] E. J. Emanuel, 'Benefits to Host Countries' in E. J. Emanuel et al. (eds), *The Oxford Textbook of Clinical Research Ethics* (Oxford University Press, 2008), p. 722.
[7] HUGO Ethics Committee, 'Statement on Benefit-Sharing', (Human Genome Organisation, 2000).
[8] K. Simm, *Benefit-Sharing: An Inquiry into Justification*, PhD thesis, Tartu University, (2005).
[9] D. Schroeder, 'Benefit-Sharing: It's Time for a Definition', (2007) *Journal of Medical Ethics*, 33(4), 205–209.
[10] K. Simm, 'Benefit-Sharing: A Look at the History of an Ethics Concern', (2007) *Nature Reviews Genetics*, 8(7), 496.
[11] E. Tsioumani, 'Beyond Access and Benefit-Sharing: Lessons from the Law and Governance of Agricultural Biodiversity', (2018) *The Journal of World Intellectual Property*, 21(3–4), 106–122.

turned into private profits and property were eventually rejected and the nationalisation of biological resources took over as the dominant framework.

To conclude, benefit-sharing negotiations always entail choices between some publics over others and upholding of certain principles before others. The above considerations about what justice requires have historically played a role in benefit-sharing discussions and none of them may be discounted as irrational or irrelevant. So how have these justice-related concerns been framed, operationalised, and translated into regulation and governance?

15.3 REGULATION AND GOVERNANCE FRAMEWORKS

Ethically sound and respectful research practices do not only benefit researchers, participants and science but also support public trust towards research in general.[12] All approaches to benefit sharing assume the baseline of the usual ethics requirements for research (thus benefit sharing does not substitute some or all ethics principles but is to be considered an additional one). In 1993, the Council for International Organisations of Medical Sciences (CIOMS) argued that 'any product developed will be made reasonably available to inhabitants of the underdeveloped community in which the research was carried out'.[13] In the latest updated Guidelines from 2016, exploitative research was defined as the kind of research that did not respond to the health needs of the community where it took place or who would later not be able to access or afford the resulting product.[14]

The prominence of benefit sharing as an ethics requirement in global health research is exemplified by the existence of many national[15] and international documents, statements and opinions. Both national and international health research organisations, policy think tanks and research funders have thought it important to discuss and state their views on the matter. Most discuss benefit sharing in the context of research in developing countries: the European Group on Ethics in Science and New Technologies to the European Commission's *Opinion on Ethical Aspects of Clinical Research in Developing Countries* (2003), the Nuffield Council on Bioethics' *The Ethics of Research Related to Healthcare in Developing Countries* (first paper in 2002), the US National Bioethics Advisory Commission's *Ethical and Policy Issues in International Research* (2001), and the Wellcome Trust's Statement on *Research Involving People Living in Developing Countries: Position Statement and Guidance Notes for Applicants*.[16] Even general health research frameworks have included references to benefit

[12] C. D. DeAnglis, 'Conflict of Interest and the Public Trust', (2000) *JAMA*, 284(17), 2237–2238.
[13] Council for International Organizations of Medical Sciences (CIOMS), 'International Ethical Guidelines for Biomedical Research Involving Human Subjects', (CIOMS, 1993), 2nd version.
[14] CIOMS, 'International Ethical Guidelines for Health-Related Research Involving Humans', (CIOMS, 2016), 4th edition.
[15] An early example of national regulation on benefit-sharing comes from the Canadian provinces of Newfoundland and Labrador. E.g. D. Pullman and A. Latus, 'Benefit-Sharing in Smaller Markets: The Case of Newfoundland and Labrador', (2003) *Community Genetics*, 6(3), 178–181.
[16] European Group on Ethics in Science and New Technologies to the European Commission (2003), 'Opinion on Ethical Aspects of Clinical Research in Developing Countries', (European Group on Ethics in Science and New Technologies to the European Commission, 2003); Nuffield Council on Bioethics, 'The Ethics of Research Related to Healthcare in Developing Countries', (Nuffield Council on Bioethics, 2002); US National Bioethics Advisory Commission (NBCA), 'Ethical and Policy Issues in International Research: Clinical Trials in Developing Countries: Report and Recommendations of the National Bioethics Advisory Commission', (Rockville, MD: NBAC, 2001), Vol. 1; Wellcome Trust, 'Research Involving People Living in Developing Countries: Position Statement and Guidance Notes for Applicants', (Wellcome), www.wellcome.ac.uk/funding/guidance/guidance-notes-research-involving-people-low-and-middle-income-countries.

sharing in their more recent drafts – for example the WHO's *Good Clinical Practice*, the World Medical Association's Declaration of Helsinki (2013), and the UNESCO Universal Declaration on Bioethics and Human Rights (2005).[17]

All of the above documents constitute what may be called soft law (i.e. non-binding instruments), yet a number of them have been influential in regulating health research practices (especially the WHO, CIOMS and funders' guidelines). When applied routinely, such ethics regulations could be considered customary international law,[18] but there have also been calls to formulate dedicated legal instruments to provide stronger support for benefit-sharing negotiations.[19] The latest attempt to ensure that benefit sharing constitutes an important normative aspect of research is the *Global Code of Conduct for Research in Resource-Poor Settings* (2018), which the European Commission endorsed as a reference document for its research funding programme Horizon 2020.[20]

While declarations and guidelines can highlight important principles and values for research, their interpretation and implementation are less straightforward. Over time, the developments in health research practices and the pressures from various stakeholders have resulted in a repeated re-framing of benefit sharing as various competing accounts have been promoted.

The earliest versions advanced a duty to benefit the particular people participating in research or a somewhat wider circle of beneficiaries (communities or populations in the case of Low and Middle Income Countries (LMICs)). This is the *'reasonable availability model'* espoused by CIOMS, which has traditionally tied the benefits to products or interventions resulting from a particular research project. An ethical prerequisite here is that research should respond to the health needs of the community and therefore any positive results of research are directly relevant to those needs.

A somewhat overlapping concept of post-trial obligations has also been argued for and applied in the context of health research, especially clinical trials. The language of post-trial obligations has its roots in the 2000 edition of the Declaration of Helsinki (§30: 'At the conclusion of the study, every patient entered into the study should be assured of access to the best proven prophylactic, diagnostic and therapeutic methods identified by the study.').[21] Later versions of the Declaration specify this duty further. Post-trial obligations are often formulated as prior agreements that are signed between stakeholders before research is begun and there exist a number of successful examples of post-trial access agreements globally.[22]

The reasonable availability model has been roundly criticised for a variety of reasons.[23] Most importantly, it is said that the focus on types of benefits arising from particular research projects

[17] World Health Organization, 'Handbook for Good Clinical Research Practice', (WHO, 2002).; WMA, 'Declaration of Helsinki', (WMA, 2000); UNESCO, 'Universal Declaration on Bioethics and Human Rights', (UNESCO, 2005).

[18] P. Andanda et al., 'Legal Frameworks for Benefit-Sharing: From Biodiversity to Human Genomics' in D. Schroeder and J. Cook Lucas (eds), *Benefit-sharing. From Biodiversity to Human Genetics* (Springer, 2013), pp. 33–64.

[19] B. Dauda and K. Dierickx, 'Benefit-Sharing: An Exploration on the Contextual Discourse of a Changing Concept', (2013) *BMC Medical Ethics*, 14(1), 36.

[20] D. Schroeder et al., 'Global Code of Conduct for Research in Resource-Poor Settings' (*GlobalCodeofConduct*), www .globalcodeofconduct.org/.

[21] WMA, 'Declaration of Helsinki'.

[22] E.g. J. M. Lavery, 'The Obligation to Ensure Access to Beneficial Treatments for Research Participants at the Conclusion of Clinical Trials' in E. J. Emanuel et al. (eds), *The Oxford Textbook of Clinical Research Ethics* (Oxford University Press, 2008), pp. 697–708; A. K. Page, 'Prior Agreements in International Clinical Trials: Ensuring the Benefits of Research to Developing Countries', (2002) *Yale Journal of Health Policy, Law and Ethics*, 3(1), 35–66.

[23] Participants in the 2001 Conference on Ethical Aspects of Research in Developing Countries, 'Moral Standards for Research in Developing Countries: From 'Reasonable availability' to 'Fair Benefits'', (2004) *Hastings Center Report*, 34(3), 17–27; Emanuel, 'Benefits to Host Countries', p. 723

does not adequately remove the dangers of exploitation and it unnecessarily limits the scope of potential benefits. Thus, the alternative *'fair benefits' model* was proposed, widening the scope of potential benefits as well as beneficiaries.[24] Benefits should not be limited to the results of particular research projects, and the distribution of benefits could take place both during as well as after research. Yet, while the increased flexibility in benefit-sharing discussions is a pragmatically useful development, it might also involve adverse side-effects. For example, a community might agree to participate in research that will not target their health needs at all, but will provide other benefits that they need.[25] This means that some of the fundamental ethical premises of research in LMICs have been effectively replaced. Perhaps this is acceptable – after all, such flexibility can be construed as less paternalistic and respectful of local needs. But it could also hint at the problematic infiltration of commercial bargaining rules into health research, which I discuss further below.

The latest re-framing, driven largely by funders, construes benefit sharing as a comprehensive *cooperative tool for capacity-building* that is justified via the larger framework of global health research and justice concerns.[26] In 2002, the Nuffield Council on Bioethics suggested that healthcare-related research in developing countries should proceed through genuine partnerships that provide transfer of knowledge and technology to strengthen the expertise of local partners. More recently, a group of influential research funders (NIH, Wellcome and the African Society of Human Genetics) have launched an H3Africa benefit-sharing vision where the more established avenues of 'reasonable availability' and 'fair benefits' have been replaced by straightforward requests for capacity building as the objective of collaborative research.[27] Such activities thus no longer constitute simply one of the options in the extensive list of potential benefits that parties to the benefit-sharing arrangement should consult and pick from. Benefit sharing is here no longer a positive side-effect or even an intended externality to a successful research project. Rather, it has been moved to the very core – it is one of the most important reasons the research collaboration should take place at all. In many ways, this is a welcome development, as benefit sharing has often been misunderstood as disbursement of tangible research 'results'.

15.4. WHAT, WHEN AND HOW: THE PRACTICALITIES OF BENEFIT SHARING

Much of the rationale for benefit sharing is articulated in the language of principles and values. Somewhat less guidance is given on the procedural aspects – how these principles and values are to be negotiated, prioritised and enforced. In most cases, a variety of potential benefits and beneficiaries can realistically be considered based on diverse justificatory reasons and local needs. Obviously, the host population needs to be the judge of the value of benefits to itself.[28] An answer to a practical question of whom does one talk to when negotiating with communities should look for engagement with those who might bear burdens for research, but are not given a voice (this concerns especially the voice of women in LMICs – their meaningful participation in

[24] Participants, 'Moral Standards', 2004.
[25] A. J. London and K. J. S. Zollmann, 'Research at the Auction Block: Problems for the Fair Benefits Approach to International Research', (2010) *Hastings Center Report*, 40(4), 36.
[26] E.g. F. Mutapi, 'Africa Should Set Its Own Health-Research Agenda', (2019) *Nature*, 575(7784), 567.
[27] B. Dauda and S. Joffe, 'The Benefit-Sharing Vision of H3Africa', (2018) *Developing World Bioethics*, 18(2), 165–170.
[28] Participants, 'Moral Standards', 2004.

all phases of benefit-sharing negotiations should be required[29]). At the same time, one needs to be conscious – and transparent – of the fact that defining and refining participant categories or negotiation partners is already a highly selective, political act.[30]

While community involvement is a crucial part of the benefit-sharing process, the mere fact of participation and consent does not necessarily guarantee the fairness of the agreement.[31] To ensure transparency and that involved communities and populations do have a fair chance to make up their minds about research participation, an influential statement recommended that publicly accessible repositories of previous benefit-sharing agreements be created.[32] This would provide a chance for stakeholders to assess the fairness of what they are offered and would support the procedural side of benefit sharing. Critics, however, have claimed that the principles and structures of transparency and fairness that the fair benefits approach supports might turn out to be an 'ethical Trojan horse'.[33] The proposed auction-like model could make host communities compete against each other in offering services to global research contract organisations, turning benefit-sharing negotiations into 'a race to the bottom'.[34] While the funders of non-profit research or even public–private partnerships could be held accountable for checking the fairness of the reached deals, much of for-profit research lacks such oversight structures.

15.5 WORRIES AND FUTURE CHALLENGES

While benefit sharing is by now a relatively standard and well-established requirement regarding ethical research practices (especially in LMICs), I would like to draw attention to several critical points that problematise the appropriateness and scope of benefit sharing in research settings.

Some of the most discussed worries associated with sharing benefits with research participants concern the dangers of therapeutic misconception and undue inducement. Research has traditionally been about serving future generations and producing generalisable knowledge. Focus on benefiting research participants introduces the risk that they might volunteer because they expect research to benefit them directly. While research participants are often well cared for, this should not be mistaken for therapy.

Undue inducement concerns instances where benefit-sharing negotiations result in overly generous and disproportionate advantages to participants such that their ability to rationally weigh the benefits and harms of participation might be jeopardised. In the LMIC context, the local public health infrastructure might be minimal or lacking; clinical trials and other types of research often offer services that are not otherwise available. Access to medical services might motivate research participation and raise the potential of undue inducement. In these situations, a proper balance between potential risks and benefits is crucial to ensure fairness and to distinguish undue inducement from fair compensation.

A different kind of unease about the extensive employment of benefit-sharing language and practices in health research was voiced already decades ago. Debates then revolved

[29] J. Cook Lucas and F. A Castillo, 'Fair for Women? A Gender Analysis of Benefit-Sharing' in D. Schroeder and J. Cook Lucas (eds), *Benefit-Sharing. From Biodiversity to Human Genetics* (Springer, 2013), pp. 129–152.
[30] C. Hayden, 'Taking as Giving: Bioscience, Exchange, and the Politics of Benefit-Sharing', (2007) *Social Studies of Science*, 37(5), 729–758.
[31] S. Gbadegesin and D. Wendler, 'Protecting Communities in Health Research from Exploitation', (2006) *Bioethics* 20 (5), 252.
[32] Participants, 'Moral Standards', 2004.
[33] London and Zollmann, 'Research at the Auction Block', 44.
[34] Ibid., 41.

around benefit sharing as a side-effect of unwelcome commercialisation of health research. Often focused on the patenting of the human genome,[35] the arguments ranged from the consequentialist (threats to scientific progress as it changes the altruistic motivation for scientific research) to the deontological (metaphysical dangers to the 'ethical self-understanding of the species'[36]). The worry was that benefit sharing as a conceptual framework had opened health research up to the vagaries of global commercial markets and had turned it into a shameless profit-driven activity, where the services of the participants were nothing but tradable commodities.

Over the past decade, we have grown used to the increasing prominence of for-profit health research. The noble idea of volunteering for research to support the project of science that may benefit humankind is no longer easily applicable nor ethically acceptable in the context of global biomedical research where powerful for-profit companies choose to do their research among possibly vulnerable populations in LMICs. While altruistic volunteering and even a gift-relationship dynamic might still be possible for health research within affluent and more sheltered communities, it would be distinctly unfair to insist on this rationale for other contexts. Even in developed countries, fierce battles regarding patenting and access to screening tests have taken place between those who contributed to research and those who were granted a patent (e.g. the Canavan disease controversy in the USA).

A different kind of worry is that if benefit sharing is motivated by the wider concerns of global justice ('an effective way of helping people in LMIC'[37]), then benefit-sharing practices and procedures are not well-equipped to deal with these much larger and complex challenges arising from global (and local) political, social and economic inequalities. Indeed, numerous funders have explicitly stated that too wide a scope for post-trial or benefit-sharing obligations (bordering on aid) is not to be required of investigators; some of the funders are, in fact, prohibited from funding healthcare provision. Furthermore, while it is clear that in many cases research is undertaken by for-profit companies who may go on to earn substantial benefits, there are also numerous trials and projects that do not translate into profits and may prove unsuccessful. Yet even such research constitutes valuable knowledge that is crucial to guide further research. The framework of benefit sharing as capacity-building gets around that challenge because it no longer focuses nor depends on the tangible results but on the cooperative aspects of research where 'negative' results are also valuable for involved local researchers.

Benefit sharing is an attempt to offer the vulnerable and the burdened communities a fair and well-earned chance to improve their situation. This means that benefit sharing can sometimes rightfully be associated with the tendencies to commodify relations and objects that, in a different world, would perhaps be guided by other, more altruistic and less monetised motives. Yet, from the perspective of LMICs, the dynamic of benefit-sharing logic over the past decades has enabled those countries themselves to increasingly have a say in steering benefit sharing. It should no longer be constrained by a particular research project or be seen as contributing towards the local scarcities in a haphazard way of plugging the holes in responding to the most desperate needs. Rather, benefit sharing is increasingly construed as a

---

[35] R. Chadwick and A. Hedgecoe, 'Commercialisation of the Human Genome' in J. Burley and J. Harris (eds), *A Companion to Genethics* (Oxford: Blackwell, 2004), pp. 334–345.

[36] J. Habermas, *The Future of Human Nature* (Cambridge: Polity, 2003), p. 71.

[37] London and Zollmann, 'Research at the Auction Block', 37.

systematic tool within the wider project of collaboration, of taking control of one's resources and setting one's own research and health policies and priorities. In short, it is coming to be seen as crafting a space for a 'lab of their own'.[38]

Such an interpretation of benefit sharing frames it as part of a more general tendency of rethinking the function and practice of research and science in society. This has been visible, for example, in the European Commission's funding guidelines. The requirement of transparency in setting research priorities, the democratising of science through involvement of various stakeholder groups (e.g. patients) in the early stages of research, and the rhetoric of responsible research and innovation are all instances of opening up research as a social practice, shifting away from a view of research as a boxed-up end-product. Perhaps some benefit-sharing partnerships might already be viewed as examples of such 'power sharing',[39] although one should remain cautious in terms of the concept's ability to revolutionise health research around the globe.

Benefit sharing is not immune to the many changes happening in health research: learning healthcare systems are doing away with the once central distinctions between clinical and research ethics; multi-site research makes it difficult to assess the contributions of distinct locations and partners; and it is unclear what the relationship will be between benefit sharing and data sharing in the context of open data and the increased role of health-related data in health research. Certain flexibility that has always been necessary for a successful implementation of benefit-sharing frameworks – the integration of universal ethical principles with the particular research partnerships – needs to continue to ensure that, at least as long as we live in an imperfect world of great inequalities, benefit sharing can successfully be integrated into the evolving practices of health research. Yet we need to be cautious about pinning too many hopes on that one framework.

15.6 CONCLUSION

Benefit sharing in health research is by now a well-established ethical requirement. There are a plethora of documents and established best practices to guide the researchers, funders and regulators, as well as communities and other stakeholders. The rationale for benefit sharing has evolved and continues to do so. Starting from the idea that individuals and communities taking certain risks and accepting potential harms deserve compensation and should not be exploited, we have now reached frameworks that view capacity-building and development support as one of the primary goals of research cooperation.

Benefit sharing is an activity that is grounded in potentially conflicting sets of justifications. While that might seem philosophically problematic (leading to e.g. various inconsistencies, potentially contradictory duties), in pragmatic terms, detailed global agreements are not necessary. It is best to regard benefit sharing as a mandatory ethics frame(work) that is to be applied to all international research collaborations as it highlights certain moral concerns and provides conceptual and governance resources for dealing with those. But the actual agreements need to be contracted by particular stakeholders and the details of the planned

---

[38] R. Benjamin, 'A Lab of Their Own: Genomic Sovereignty as Postcolonial Science Policy', (2009) *Policy and Society,* 28(4), 341–355.
[39] D. E. Winickoff, 'From Benefit-Sharing to Power Sharing: Partnership Governance in Population Genomics Research' in J. Kaye and M. Stranger (eds), *Principles and Practice in Biobank Governance* (Routledge, 2016), pp. 53–65.

research and the distinct context will determine which sets of concerns are paramount, which justifications make sense, what benefits are realistic, and who should be involved. There is a danger of potential relativism involved in such a governance framework, but only combining universal research norms with unique contextual components provides the sensitivity and flexibility that is needed for ethical health research as a collaborative enterprise.

16

Taking Failure Seriously

Health Research Regulation for Medical Devices, Technological Risk and Preventing Future Harm

*Mark L. Flear**

16.1 INTRODUCTION

Failure in health research regulation is nothing new. Indeed, the regulation of clinical trials was developed in response to the Thalidomide scandal, which occurred some fifty years ago.[1] Yet, health research regulation is at the centre of recent failures.[2] Metal-on-metal hip replacements,[3] and, more recently, mesh implants for urinary incontinence and pelvic organ prolapse in women – often referred to as 'vaginal mesh' – have been the subject of intense controversy.[4] Some have even called the latter controversy 'the new Thalidomide'.[5] In these cases, previously licensed medical devices were used to demonstrate the safety of supposedly analogous new medical devices, and obviate the need for health research involving humans.[6]

In this chapter, I use health research regulation for medical devices to look at the regulatory framing of harm through the language of technological risk, i.e. relating to safety. My overall argument is that reliance on this narrow discourse of technological risk in the regulatory framing of harm may marginalise stakeholder knowledges of harm to produce a limited knowledge base. The latter may underlie harm, and in turn lead to the construction of failure.

I understand failure itself in terms of this framing of harm.[7] Failure is taken to be ontologically and normatively distinct from harm, and as implicating the design and functioning of the system

* Many thanks to all those with whom I have discussed the ideas set out in this chapter, especially the editors and Ivanka Antova, Richard Ashcroft, Daithi Mac Sithigh, Katharina Paul and Barbara Prainsack. The discussion in this chapter is developed further in: Mark L Flear, 'Epistemic Injustice as a Basis for Failure? Health Research Regulation, Technological Risk and the Epistemic Foundations of Harm and Its Prevention', (2019) *European Journal of Risk Regulation* 10(4), 693–721.

[1] In the United Kingdom, the scandal resulted in the Medicines Act 1968 and its related licensing authority. See E. Jackson, *Law and the Regulation of Medicines* (London: Hart Publishing, 2012), pp. 4–5.
[2] Relatedly, see S. Macleod and S. Chakraborty, *Pharmaceutical and Medical Device Safety* (London: Hart Publishing, 2019).
[3] C. Heneghan et al., 'Ongoing Problems with Metal-On-Metal Hip Implants', (2012) *BMJ*, 344(7846), 23–24.
[4] See the articles comprising 'The Implant Files', (*The Guardian*), www.theguardian.com/society/series/the-implant-files.
[5] H. Marsden, 'Vaginal Mesh to Treat Organ Prolapse Should Be Suspended, Says UK Health Watchdog', (*The Independent*, 15 December 2017).
[6] The famous Poly Implant Prothése silicone breast implants scandal concerned fraud rather than the kinds of problems with health research regulation discussed in this chapter – see generally C. Greco, 'The Poly Implant Prothése Breast Prostheses Scandal: Embodied Risk and Social Suffering', (2015) *Social Science and Medicine*, 147, 150–157; M. Latham, '"If It Ain't Broke Don't Fix It": Scandals, Risk and Cosmetic Surgery', (2014) *Medical Law Review*, 22(3), 384–408.
[7] This may extend beyond physical harm to social harm, environmental harm 'and so on' – see R. Brownsword, *Rights, Regulation and the Technological Revolution* (Oxford University Press, 2008), p. 119. Also see pp. 102–105.

or regime itself. Failure is understood as arising when harm is deemed to thwart expectations of safety built into technological framings of regulation. This usually occurs from stakeholder perspectives. Stakeholders include research participants, patients and other interested parties. However, the new force of failure in public discourse and regulation,[8] apparent in the way it 'now saturates public life',[9] ensures that the language of failure provides a means to integrate stakeholder knowledges of harm with scientific-technical knowledges.

In the next section, I use health research relating to medical devices to reflect on the role of expectations and harm in constructing failure. This sets the scene for the third section, where I outline the roots of failure in the knowledge base for regulation. Subsequently, I explain how the normative power of failure may be used to impel the integration of expert and stakeholder knowledges, improving the knowledge base and, in turn, providing a better basis on which to anticipate and prevent future failures. The chapter thus appreciates how failure can amount to a 'failure of foresight', which may mean it is possible to 'organise' failure and the harm it describes out of existence.[10]

16.2 EXPECTATIONS AND FAILURE IN HEALTH RESEARCH

Failure has long been understood, principally though not exclusively, in Kurunmäki and Miller's words, '*as arising from risk* rather than sin'.[11] Put differently, failure can be understood in principally consequentialist, rather than deontological, terms.[12] This understanding does not exclude legal conceptualisations of failure in tort law and criminal law, in which the conventional idea of liability is one premised on 'sin' or causal contribution.[13] However, within contemporary society and regulation, such deontological understandings are often overlaid with a consequentialist view of failure.[14]

This is apparent in recent work by Carroll and co-authors. Through their study of material objects and failure, they describe failure as 'a situation or thing as [sic] *not being in accord with expectation*'.[15] According to van Lente and Rip, expectations amount to 'prospective structures' that inform 'statements, brief stories and scenarios'.[16] It is expectation, rather than anticipation or hope, then, that is central to failure. Unlike expectation, anticipation and hope do not provide a

---

[8] For definition of 'regulation' see the Introduction to this volume.

[9] L. Kurunmäki and P. Miller, 'Calculating Failure: The Making of a Calculative Infrastructure for Forgiving and Forecasting Failure', (2013) *Business History*, 55(7), 1100–1118, 1100. Emphasis added. More broadly, for comment on the 'stream of failures' since the 1990s, see M. Power, *Organised Uncertainty* (Oxford University Press, 2007), p. 5.

[10] B. Turner, *Man-Made Disasters* (Wykeham 1978). For application to organisations, see B. Hutter and M. Power (eds), *Organisational Encounters with Risk* (Cambridge University Press, 2005), p. 1. Some failures are 'normal accidents' and cannot be organised out of existence – see C. Perrow, *Normal Accidents: Living with High-Risk Technologies* (New York: Basic Books, 1984).

[11] Kurunmäki and Miller, 'Calculating Failure', 1101. Emphasis added.

[12] For discussion, see R. Brownsword and M. Goodwin, *Law and the Technologies of the Twenty-First Century: Text and Materials* (Cambridge University Press, 2012), p. 208.

[13] Indeed, Poly Implant Prothése silicone breast implants and vaginal mesh have been the subject of litigation – for discussion of each see, Macleod and Chakraborty, *Pharmaceutical and Medical Device Safety*, pp. 232–234 and pp. 259–263, respectively. For a recent case on vaginal mesh involving a class action against members of the Johnson & Johnson group in which the court found in favour of the claimants, see *Gill v. Ethicon Sarl (No. 5)* [2019] FCA 1905.

[14] A. Appadurai, '"Introduction" to Special Issue on "Failure"', (2016) *Social Research*, 83(3), xx–xxvii.

[15] T. Carroll et al., 'Introduction: Towards a General Theory of Failure' in T. Carroll et al. (eds), *The Material Culture of Failure: When Things Go Wrong* (Bloomsbury, 2018), pp. 1–20, p.15. Emphasis added.

[16] H. van Lente and A. Rip, 'Expectations in Technological Developments: An Example of Prospective Structures to be Filled in by Agency' in C. Disco and B. van der Meulen (eds), *Getting New Technologies Together: Studies in Making Sociotechnical Order* (Berlin: De Gruyter, 1998), p. 205.

sense of how things *ought* to be, so much as how they could be or an individual or group would like them to be.[17] Indeed, as Bryant and Knight explain: 'We expect because of what the past has taught us to expect ... [Expectation] *awakens a sense of how things ought to be, given particular conditions*'.[18]

This normative dimension distinguishes expectation from other future-oriented concepts and furnishes 'a standard for evaluation', for whether a situation is 'good or bad, desirable or undesirable',[19] and, relatedly, a failure. Indeed, for Appadurai '*[t]he most important thing about failure is that it is not a fact but a judgment*'.[20] Expectations rely on the past to inform a normative view of some future situation or thing, such as that it will be safe. When, through the application of calculative techniques that determine compliance with the standard for evaluation, this comes to be seen as thwarted, there is a judgment of failure.[21] Expectations, and hence a key ground for establishing failure, are built into regulatory framings[22] and the targets of regulation.[23]

These insights can be applied and developed through the example of health research regulation for medical devices. In this instance, technological risk, i.e. safety, provides the framing for medical devices within the applicable legislation and engenders an expectation of safety.[24] However, in respect of metal-on-metal hips and vaginal mesh, harm occurred, and the expectation of safety was thwarted downstream once these medical devices were in use.

Harm was consequent, seemingly in large part, on the classification of metal-on-metal hips and vaginal mesh as Class IIb devices. IIb devices are medium to high-risk devices, which are usually devices installed within the body for thirty days or longer. This meant that it was possible for manufacturers to rely on substantial equivalence to existing products to demonstrate conformity with general safety and performance requirements. These requirements set expectations for manufacturers and regulators to demonstrate safety, both for the device and the person within which it was implanted. Substantial equivalence obviates the need for health research involving humans via a clinical investigation.

It is noted in one *BMJ* editorial that this route '*failed to protect* patients from substantial harm'.[25] Heneghan et al. point out that in respect of approvals by the Food and Drug Administration in the USA, which are largely mirrored in the European Union (EU): 'Transvaginal mesh products for pelvic organ prolapse have been *approved on the basis of weak*

---

[17] R. Bryant and D. Knight, *The Anthropology of the Future* (Cambridge University Press, 2019), p. 28 for anticipation and p. 134 for hope.

[18] Ibid., p. 58. Emphasis added.

[19] Ibid., p. 63.

[20] Appadurai, 'Introduction', p. xxi. Emphasis added. Also see A. Appadurai, *Banking on Words: The Failure of Language in the Age of Derivative Finance* (University of Chicago Press, 2016).

[21] Beckert lists past experience among the social influences on expectations – see J. Beckert, *Imagined Futures: Fictional Expectations and Capitalist Dynamics* (Cambridge, MA: Harvard University Press, 2016), p. 91.

[22] Brownsword, *Rights, Regulation and the Technological Revolution*; K. Yeung, 'Towards an Understanding of Regulation by Design' in R. Brownsword and K. Yeung (eds), *Regulating Technologies: Legal Futures, Regulatory Frames and Technological Fixes* (London: Hart Publishing, 2008), pp. 79–107.

[23] T. Dant, *Materiality and Society* (Open University Press, 2005); D. MacKenzie and J. Wajcman (eds), *The Social Shaping of Technology*, 2nd Edition (Buckingham: Open University Press, 1999); L. Winner, 'Do Artefacts Have Politics?', (1980) *Daedalus*, 109(1), 121–136.

[24] Medical devices are defined by their intended function, as determined by the manufacturer, for medical purposes – see Article 2(1) of the Medical Devices Regulation (EU) 2017/745 of the European Parliament and of the Council of 5 April 2017 on medical devices, amending Directive 2001/83/EC, Regulation (EC) No. 178/2002 and Regulation (EC) No. 1223/2009 and repealing Council Directives 90/385/EEC and 93/42/EEC OJ 2017 L 117/1. On the classification of medical devices, see Point 1.3, Annex VIII.

[25] C. Allan et al., 'Europe's New Device Regulations Fail to Protect the Public', (2018) *BMJ*, 363, k4205, 1.

evidence over the last 20 years'.[26] This study traced the origins of sixty-one surgical mesh implants to just two original devices approved in the USA in 1985 and 1996. The reliance on substantial equivalence meant that safety and performance data came from implants that were already on the market, sometimes for decades, and that were no longer an accurate predicate. In other words, on the basis of past experience – specifically, of 'substantially equivalent' medical devices – there was an unrealistic expectation that safety would be ensured through this route, and that further research involving human participants was unnecessary.

Stakeholders reported adverse events including: 'Pain, impaired mobility, recurrent infections, incontinence/urinary frequency, prolapse, fistula formation, sexual and relationship difficulties, depression, social withdrawal or exclusion/loneliness and lethargy'.[27] On this basis, stakeholders, including patient groups, demanded regulatory change. Within the EU, new legislation was introduced, largely in response to these events. The specific legislation applicable to the examples considered in this chapter, the Medical Devices Regulation (MDR),[28] came into force on 26 May 2020 (Article 123(2) MDR).

In respect of metal-on-metal hips and vaginal mesh, the legislation reclassifies them as Class III. Class III devices are high risk and invasive long-term devices. Future manufacturers of these devices will, in general, have to carry out clinical investigations to demonstrate conformity with regulatory requirements (Recital 63 MDR). The EU's new legislation takes up a whole chapter on clinical investigations and thus safety. The legislation is deemed to provide a 'fundamental revision' to 'establish a robust, transparent, predictable and sustainable regulatory framework for medical devices which ensures a high level of safety and health whilst supporting innovation' (Recital 1 MDR). One interpretation of the legislation is that it is a direct response to problems in health research for medical devices, and intended to provide '*a better guarantee for the safety of medical devices*, and to restore the loss of confidence that followed high profile scandals around widely used hip, breast, and vaginal mesh devices'.[29]

As regards metal-on-metal hips and vaginal mesh, however, there has been little or no suggestion of failure by those formally responsible, and who might be held accountable if there were – perhaps especially if it could be said there were any plausible causal contribution by them towards harm. Instead, the example of medical devices demonstrates how the construction of failure does not necessarily hinge on official accounts of harm as amounting to 'failure'. This is apparent in the various quotations from non-regulators noted above. As Hutter and Lloyd-Bostock put it, these are 'terms in which events are construed or described in the media or in political discourse or by those involved in the event'. As they continue, what matters is an 'event's construction, interpretation and categorisation'.[30]

Failure is an interpretation and judgment of harm. Put differently, 'failure' arises through an assessment of harm undertaken through calculative techniques and judgments. Harm becomes refracted through these. At a certain point, the expectations of safety built into framing are

---

[26] Carl J. Heneghan et al., 'Trials of Transvaginal Mesh Devices for Pelvic Organ Prolapse: A Systematic Database Review of the US FDA Approval Process', (2017) *BMJ Open*, 7(12), e017125, 1. Emphasis added.

[27] Macleod and Chakraborty, *Pharmaceutical and Medical Device Safety*, p. 238.

[28] Medicine Devices Regulation (EU) 2017/745. Implementation of this legislation is left to national competent authorities.

[29] Allan et al., 'Europe's New Device Regulations', 1. Emphasis added.

[30] B. Hutter and S. Lloyd-Bostock, *Regulatory Crisis: Negotiating the Consequences of Risk, Disasters and Crises* (Cambridge University Press, 2017), p. 3. On understandings of failure, see S. Firestein, *Failure. Why Science Is So Successful* (Oxford University Press, 2016), pp. 8–9.

understood by stakeholders as thwarted, and the harm becomes understood as a failure.[31] Official discourses are significant, not least because they help to set expectations of safety. But these discourses do not necessarily control stakeholder interpretations and knowledge of harm, or how they thwart expectations of safety, and lead to the construction of failure[32]

In what follows, I shift attention to the *lacunae* and blind spots in the knowledge base for the regulation of medical devices, which are made apparent by the harm and failure just described. I outline these missing elements before turning to discuss the significance of failure for improving health research regulation.

16.3 USING FAILURE TO ADDRESS THE SYSTEMIC CAUSES OF HARM

Failure, at its root, emerges from the limited knowledge base for health research regulation: for medical devices, and other areas framed by technological risk, it is derived from an archive of past experience and scientific-technical knowledge. The focus on performance (i.e. the device performs as designed and intended, in line with a predicate) marginalised attention to effectiveness (i.e. producing a therapeutic benefit) and patient knowledge on this issue. Moreover, in relation to vaginal mesh implants, female knowledges and lived experiences of the devices implanted within them have tended to be sidelined or even overlooked. The centrality of the male body within research and models of pain, and gender-based presumptions about pain,[33] help to explain the time taken to recognise a safety problem in respect of medical devices, and the gaping hole in research and knowledge.

Another part of the explanation for the latter problem is that there was a lengthy delay in embodied knowledge and experiences of pain being reported and recognised – effectively sidelining and ignoring those experiences. New guidance on vaginal mesh in the United Kingdom (UK) has faced criticism on gender-based lines. Safety concerns are cited and it is recommended that vaginal mesh should not be used to treat vaginal prolapse. However, as the UK Parliament's All Party Parliamentary Group on Surgical Mesh Implants said, the guidelines: 'disregard mesh-injured women's *experiences* by stating that there is no long-term evidence of adverse effects'.[34]

The latter may amount to epistemic injustice, what Fricker describes as a 'wrong done to someone specifically in their *capacity as a knower*'.[35] More than a harm in itself, epistemic

[31] Kurunmäki and Miller, 'Calculating Failure', 1101. Cf I. Hacking, *Historical Ontology* (Cambridge, MA: Harvard University Press, 2002) – applied in e.g. B. Allen, 'Foucault's Nominalism' in S. Tremain (ed.), *Foucault and the Government of Disability* (University of Michigan Press, 2018); D. Haraway, *The Haraway Reader* (New York: Routledge, 2004); D. Roberts, 'The Social Immorality of Health in the Gene Age: Race, Disability and Inequality' in J. Metzl and A. Kirkland (eds), *Against Health* (New York University Press, 2010), pp. 61–71.

[32] Kurunmäki and Miller, 'Calculating Failure', 1101. Cf Hutter and Lloyd-Bostock, *Regulatory Crisis*, pp. 9–18 and pp. 19–21 for framing and routines.

[33] See, for example, R. Hurley and M. Adams, 'Sex, Gender and Pain: An Overview of a Complex Field', (2008) *Anesthesia & Analgesia*, 107(1), 309–317. Also see M. Fox and T. Murphy, 'The Body, Bodies, Embodiment: Feminist Legal Engagement with Health' in M. Davies and V. E. Munro (eds), *The Ashgate Research Companion to Feminist Legal Theory* (London: Ashgate, 2013), pp. 249–265.

[34] National Institute for Health and Care Excellence (NICE), 'Urinary Incontinence and Pelvic Organ Prolapse in Women: Management, NICE Guideline [NG123]', (NICE, 2019). This guidance was issued in response to the NHS England Mesh Working Group – see 'Mesh Oversight Group Report', (NHS England, 2017). Also see 'Mesh Working Group', (NHS), www.england.nhs.uk/mesh/.For criticism, see H. Pike, 'NICE Guidance Overlooks Serious Risks of Mesh Surgery', (2019) *BMJ*, 365, l1537.

[35] M. Fricker, *Epistemic Injustice: Power and the Ethics of Knowing* (Oxford University Press, 2007), p. 1. Emphasis added. Also see I. J. Kidd and H. Carel, 'Epistemic Injustice and Illness', (2017) *Journal of Applied Philosophy*, 34(2), 172–190.

injustice may limit stakeholder ability to contribute towards regulation, leading to other kinds of harm and failure. This is especially true in the case of health research regulation, where stakeholders may be directly or indirectly harmed by practices and decisions that are grounded on a limited knowledge base. Moreover, even in respect of the EU's new legislation on medical devices, doubts remain whether these will prevent future harms and thus failures similar to those mentioned above. Indeed, the only medical devices that are required to evidence therapeutic benefit or efficacy in controlled conditions before marketing are those that incorporate medicinal products.[36]

A deeper explanation for the marginalisation of stakeholder knowledges of harm, and a key underpinning for failure, lies in the organisation of knowledge production. Hurlbut describes how: 'Framed as epistemic matters – that is, as problems of properly assessing the risks of novel technological constructions – problems of governance become questions for experts'.[37] This framing constructs a hierarchy of knowledge that privileges credentialised knowledge and expertise, while marginalising those deemed inexpert or 'lay'. Bioethics plays a key role here. As a field, bioethics tends to focus on technological development within biomedicine and principles of individual ethical conduct or so-called 'quandary ethics', rather than systemic issues related to epistemic – or social – justice. Consequently, bioethics often privileges and bolsters scientific–technical knowledge, erases social context and renders 'social' elements as little more than 'epiphenomena'.[38] In this setting, stakeholder knowledges and forms of expertise relating to harm are, as Foucault explained, 'disqualified . . . [as] naïve knowledges, hierarchically inferior knowledges, knowledges that are below the required level of erudition or scientificity'.[39]

The specific contemporary cultural resonance of the language of failure means that it can be used as a prompt to overcome this marginalisation and improve the knowledge base for regulation. Specifically, the language of failure can be used to generate a risk to organisational standing and reputation. Adverse public perceptions may cast failure as regulatory failure, effectively framing regulators as *'part of the cause of disasters and crises'*.[40] A perception of regulatory failure thus has key implications for the accountability and legitimacy of regulation and regulators – and such perception is therefore to be avoided by them. Relatedly, regulators want to avoid the shaming and blaming that often accompany talk of failure. Blaming can even amplify[41] or extend the duration of an institutional risk to standing and reputation. This may

---

[36] For discussion, see C. J. Heneghan et al., 'Transvaginal Mesh Failure: Lessons for Regulation of Implantable Devices', (2017) *BMJ*, 359, j5515.

[37] J. B. Hurlbut, 'Remembering the Future: Science, Law, and the Legacy of Asilomar' in S. Jasanoff and S. Kim, *Dreamscapes of Modernity: Sociotechnical Imaginaries and the Fabrication of Power* (University of Chicago Press, 2015), p. 129. Original emphasis.

[38] On 'quandary ethics', see P. Farmer, *Pathologies of Power: Health, Human Rights, and the New War on the Poor* (University of California, 2003), pp. 204–205. Also see D. Callaghan, 'The Social Sciences and the Task of Bioethics', (1999) *Daedalus*, 128(4), 275–294, 276. On bioethics and social context, see J. Garrett, 'Two Agendas for Bioethics: Critique and Integration', (2015) *Bioethics*, 29(6), 440–447; A. Hedgecoe, 'Critical Bioethics: Beyond the Social Science Critique of Applied Ethics', (2004) *Bioethics*, 18(2), 120–143, 125. Also see B. Hoffmaster (ed.), *Bioethics in Social Context* (Philadelphia: Temple University Press, 2001).

[39] M. Foucault, *Society Must Be Defended* (London: Penguin Books, 2004), p. 7.

[40] Hutter and Lloyd-Bostock, 'Regulatory Crisis', p. 8. Emphasis added. For discussion, see M. Lodge, 'The Wrong Type of Regulation? Regulatory Failure and the Railways in Britain and Germany', (2002) *Journal of Public Policy*, 22(3), 271–297; R. Schwartz and A. McConnell, 'Do Crises Help Remedy Regulatory Failure? A Comparative Study of the Walkerton Water and Jerusalem Banquet Hall Disasters', (2009) *Canadian Public Administration*, 52(1), 91–112.

[41] For discussion, see A. Boin et al. (eds), *The Politics of Crisis Management: Public Leadership Under Pressure* (Cambridge University Press, 2005); C. Hood, *The Blame Game: Spin, Bureaucracy, and Self-Preservation in*

produce a crisis for regulation, including for its legitimacy, quite apart from any interpretation and judgment of failure or regulatory failure.

The risk posed by failure to standing and reputation may prompt the integration of stakeholder knowledges with the scientific–technical knowledges that currently underpin regulation. The potential to use failure in this way is already apparent in the examples above, and perhaps especially vaginal mesh. Stakeholders have been largely successful in presenting their knowledges of harm, placing a spotlight on health research regulation and demanding change to prevent future failure.

Despite the limitations within much bioethics scholarship, there is a growing plethora of approaches to injustice, most recently and notably vulnerability, within which embodied risk and experiential knowledge are central.[42] These approaches are buttressed by a developing scientific understanding of the significance of environmental factors to genetic predisposition to vulnerability and embodied risk.[43] Further, within such approaches, the centrality of the human body and experience is foregrounded precisely to recast the objects of bioethical concern. The goal: to prompt a response from the state to fulfil its responsibilities in respect of rights.[44] In the context of health research, this research can be leveraged to counter the lack of alertness and communicative failures for which institutions and powerful people must take responsibility,[45] and expand the knowledges that count in regulation.

There are mechanisms to facilitate the integration of stakeholder with scientific–technical knowledges and improve health research for medical devices. Further attention to effectiveness could yield important additional data (i.e. on producing a therapeutic benefit) on top of performance (i.e. the device performs as designed and intended). Similar to clinical trials for medicines, which produce data to demonstrate safety, quality and efficacy, this would require far more involvement and data from device recipients. Recipient involvement and data could come pre- or post-marketing – or both. Involvement pre-marketing seems both desirable and possible:

> The manufacturers' argument that [randomised controlled trials] are often infeasible and do not represent the gold standard for [medical device] research is clearly refuted. As high-quality evidence is increasingly common for pre-market studies, *it is obviously worthwhile to secure these standards through the [Medical Devices Regulation] in Europe and similar regulations in other countries.*[46]

Government (Princeton University Press, 2011); N. Pidgeon et al., *The Social Amplification of Risk* (Cambridge University Press, 2003).

[42] M. Fineman, 'The Vulnerable Subject and the Responsive State', (2010) *Emory Law Journal*, 60(2), 251–275. Also see work on: precarity (J. Butler, *Precarious Life: The Power of Mourning and Violence* (London: Verso, 2005)); the capabilities approach (M. Nussbaum, *Creating Capabilities* (Cambridge, MA: Harvard University Press, 2011); A. Sen, 'Equality of What?' in S. McMurrin (ed.), *Tanner Lectures on Human Values, Volume 1* (Cambridge University Press, 1980), pp. 195–220); and a feminist approach to flesh (C. Beasley and C. Bacchi, 'Envisaging a New Politics for an Ethical Future: Beyond Trust, Care and Generosity – Towards an Ethic of Social Flesh', (2007) *Feminist Theory*, 8(3), 279–298).

[43] This includes understanding in epigenetics and neuroscience – see N. Rose and J. Abi-Rached, *Neuro: The New Brain Sciences and the Management of the Mind* (Princeton University Press, 2013); D. Wastell and S. White, *Blinded by Science: The Social Implications of Epigenetics and Neuroscience* (Bristol: Policy Press, 2017).

[44] Most notably, see Fineman, 'The Vulnerable Subject'. For application to bioethics, see M. Thomson, 'Bioethics & Vulnerability: Recasting the Objects of Ethical Concern', (2018) *Emory Law Journal*, 67(6), 1207–1233.

[45] For discussion, see A. Boin et al. (eds), *The Politics of Crisis Management*, especially p. 215 and p. 218. This responsibility is grounded in virtue theory. For discussion see Fricker, *Epistemic Injustice*.

[46] S. Sauerland et al., 'Premarket Evaluation of Medical Devices: A Cross-Sectional Analysis of Clinical Studies Submitted to a German Ethics Committee', (2019) *BMJ Open*, 9(2), 6. Emphasis added. For a review of approaches to the collection of data, see D. B. Kramer et al., 'Ensuring Medical Device Effectiveness and Safety: A Cross-National Comparison of Approaches to Regulation', (2014) *Food Drug Law Journal*, 69(1), 1–23. The EU's new legislation on

One proposed model for long-term implantable devices, such as those discussed in this chapter, involves providing limited access to them through temporary licences that restrict use to within clinical evaluations, with long follow-up at a minimum of five years. Wider access could be provided once safety, performance *and* efficacy have been adequately demonstrated. In addition, wider public access to medical device patient registries, including the EU's Eudamed database, could be provided so as to ensure transparency, open up public discourse around safety and tackle epistemic injustice.[47]

16.4 CONCLUSION

In this chapter, I described how failure is constructed and becomes recognised through processes that determine whether harm has thwarted the expectation of safety built into technological framings of regulation. Laurie is one of the few scholars to illuminate, not only how health research regulation transforms its participants into instruments, but how this may underlie failure:

> if we fail to see involvement in health research as an essentially transformative experience, then we blind ourselves to many of the human dimensions of health research. More worryingly, *we run the risk of overlooking deeper explanations about why some projects fail and why the entire enterprise continues to operate sub-optimally.*[48]

By looking at the organisation of knowledge that supports regulatory framings of medical devices, it becomes clear how the marginalisation of stakeholder knowledge may provide a deeper explanation for harm and failure. Failure can be used to prompt the take-up of stakeholder knowledges of harm in regulation, by recasting regulation or using its mechanisms differently in light of those knowledges, so as to better anticipate and prevent future harm and failure, and enable success. See further on users' experiences, Harmon, Chapter 39, this volume.

Why, then, has more not been done to ensure epistemic integration as a way to enhance regulatory capacities to anticipate and prevent failure? Epistemic integration would involve bringing stakeholders within regulation via their knowledges. As such, epistemic integration would seem to undermine the dominant position of those deemed expert within extant processes. Knowledge of harm becomes re-problematised: what knowledges from across society are required by regulation in order to ensure its practices are ethical and legitimate? Integration of diverse knowledges might reveal to society at large the limits of current regulation to deal with risk and uncertainty. More deeply, epistemic integration would challenge modernist values on the import of empirically derived knowledge, and the efficacy

medical devices has sought to improve *inter alia* post-marketing data collection, such as through take-up of the Unique Device Identification. This is used to mark and identify medical devices within the supply chain. For discussion of this and other aspects of the EU's new legislation, see A. G. Fraser et al., 'The Need for Transparency of Clinical Evidence for Medical Devices in Europe', (2018) *Lancet*, 392(10146), 521–530.

[47] On licensing, see Heneghan et al., 'Transvaginal Mesh Failure'. Also see B. Campbell et al., 'How Can We Get High Quality Routine Data to Monitor the Safety of Devices and Procedures?', (2013) *BMJ*, 346(7907), 21–22. On access to data, see M. Eikermann et al., 'Signatories of Our Open Letter to the European Union. Europe Needs a Central, Transparent, and Evidence Based Regulation Process for Devices', (2013) *BMJ*, 346, f2771; Fraser et al., 'The Need for Transparency'.

[48] G. Laurie, 'Liminality and the Limits of Law in Health Research Regulation: What Are We Missing in the Spaces In-Between?' (2016) *Medical Law Review*, 25(1), 47–72, 71. Emphasis added.

of society's technological 'fixes' in addressing its problems. However, scientific–technical knowledge and expertise would still be necessary in order to discipline 'lay' knowledges and ensure their integration within the epistemic foundations of decision-making. To resist epistemic integration is, therefore, essentially to bolster extant power relations. As the analysis in this chapter suggests, these relations are actually antithetical to addressing failure and maintaining the protections that are central to ethical and legitimate health research and regulation more generally.

Rules, Principles and the Added Value of Best Practice in Health Research Regulation

Nayha Sethi

17.1 INTRODUCTION

In this chapter I consider some important implications of adopting rules, principles and supplementary guidance-based approaches to the regulation and governance of health research. This is a topic that has not yet received sufficient attention given how impactful different regulatory approaches can be on health research. I suggest that each approach has strengths and limitations to be factored-in when considering how we shape health research practices. I argue that while principles-based approaches can be well suited to typically complex health research landscapes, additional guidance is often required. I explore why this is so, highlighting in particular the added value of best practice and noting that incorporating additional guidance within regulatory approaches demands its own important considerations, which are laid out in the final section.

17.2 THE SIGNIFICANCE OF REGULATORY APPROACHES

Determining which regulatory/governance approach (RGA)[1] to adopt is a recurring predicament spanning the diverse spectrum of health research activities. For example, a key challenge concerning emerging technologies is regulatory lapse – law's inability to keep up with the fast pace of technological development and adoption.[2] Novel practices/technologies may be subsumed under pre-existing frameworks through processes of commensuration such as legislative analogy. Alternatively, it may be determined that entirely new frameworks are required.[3] Pre-existing frameworks may be too rigid and restrictive, or conversely, overly flexible and permissive.[4] Content included within RGAs may deviate substantially from what takes place 'on the ground', raising problems for those charged with interpreting regulation, leading to theory-

---

[1] I collectively refer to regulatory and governance approaches (RGA) in recognition of the fact that rules, principles and other guidance may manifest as legislation typically associated with regulation as well as other forms of guidance associated with governance. For more discussion on the relationships between regulation and governance, see the Introduction of this volume.

[2] L. B. Moses, 'Recurring Dilemmas: the Law's Race to Keep Up with Technological Change', (2007) *University of Illinois Journal of Law, Technology and Policy*, 2007(2), 239– 285; R. Brownsword and M. Goodwin, *Law and the Technologies of the Twenty-First Century* (Cambridge University Press, 2012).

[3] A. Faulkner and L. Poort, 'Stretching and Challenging the Boundaries of Law: Varieties of Knowledge in Biotechnologies Regulation', (2017) *Minerva*, 55(2), 209–228.

[4] Multiple examples are offered throughout this volume. See, for example, Kaye and Prictor's (Chapter 10) discussion on the challenges of digital transformation for consent.

practice gaps.[5] Similarly, current approaches may fail to reflect embodied experiences of the subjects affected by regulation.[6]

Universally, then, important questions arise relating to what form RGAs should take. Should they manifest as specific prescriptive norms, which often appear in the form of rules? Or would high-level and more abstract norms, such as those typically communicated through principles be more effective? Is additional guidance needed alongside rules and principles? If so, what form should this take? Each approach can have repercussions for the patients, researchers, regulators, developers, manufacturers, technologies and other key actors, subjects and objects constituting health research ecosystems. It is imperative that prior to adopting a particular RGA, the respective benefits and limitations of different potential approaches are granted due regard.

Many spheres of health research are widely populated by rules, principles and supplementary guidance. These manifest in diverse forms including: international instruments, primary and secondary legislation, ethical frameworks, professional guidance, codes of conduct, best practice instantiations, recommendations and standards. Consider, for instance, use of patient health data for research purposes. UK-based researchers wishing to access such data must consider the requirements laid out within (among others) the General Data Protection Regulation (Regulation (EU) 2016/679) (GDPR), the UK Data Protection Act 2018 and the NHS Act 2006. Additionally, they must consider guidance from the Information Commissioner's Office, and adhere to the Caldicott Principles and applicable professional guidance. Technical standards such as those set out by the International Organization for Standardization (ISO) must also be observed. Researchers may be required to obtain research ethics committee approval, and demonstrate due consideration and mitigation of the risks and privacy impacts of data uses and whether or not such uses carry social value and are in the public interest. Many additional spheres of health research can prove similarly labyrinthine.

Navigating such complex regulatory frameworks and interpreting provisions included within them is challenging. A balance must be sought between offering clear articulations of what is required, permitted and prohibited, while retaining sufficient flexibility to guarantee applicability across a wide array of contexts. This tension between specificity and flexibility is a recurring dilemma for regulation. Regardless of the technologies/activities under consideration, an additional balance must be sought of providing adequate coverage of the range of pre-existing activities associated with a specific type of research and simultaneously avoiding the risk of becoming obsolete when new applications of, or progressions in, those research practices/technologies appear. For example, one of the driving factors for the introduction of the GDPR was the drastic transformations in how data are used today as compared to when its predecessor, the European Directive 95/46/EC, was drafted. Given the challenges of navigating health research regulation, we ought also to consider how best to communicate norms, while supporting decision-makers in the inevitable exercise of discretion. These concerns lead us to engage with two dominant RGA approaches in health research: rules-based and principles-based approaches. The next section considers these, placing an emphasis on principles-based approaches, which, I argue, can be especially helpful for complex regulatory landscapes.

---

[5] N. Sethi, 'Research and Global Health Emergencies: On the Essential Role of Best Practice', (2018) *Public Health Ethics*, 11(3), 237–250.

[6] As explored by Flear in this volume (see Chapter 16).

17.3 RULES AND PRINCIPLES-BASED APPROACHES

To understand what rules-based and principle-based approaches are, we should briefly define rules and principles. It may be more meaningful to talk of 'rule and principle-type features' than attempt to provide hard and fast definitions of rules and principles. These mean different things to different people in different contexts[7] and grey areas exist where differentiation based on any sole 'typical' characteristic is unhelpful. For instance, reliance upon high specificity of language as the identifying feature of rules is problematic because rules can be articulated in general terms. Conversely, principles are not always communicated through abstract language, despite frequently being described as such. Consider the Nuremberg Code 1947, the norms included within it, referred to as principles, are articulated through prescriptive language: 'The experiment should be conducted only by scientifically qualified persons. The highest degree of skill and care should be required through all stages of the experiment of those who conduct or engage in the experiment'.[8] Another example is the CIOMS International Ethical Guidelines for Health-related Research Involving Humans, its content being collectively referred to as 'rules and principles'. Upon closer inspection, it is unclear which of the guidelines are rules and which principles.[9] Given these definitional challenges, reference is made here to typical *but not unequivocal* features of rule and principle-like norms.

Rules are typically specific, prescriptive and fixed iterations of what to do.[10] They may be conceptualised in terms of rigidity, enforceability and whether they carry legal obligations. They can be characterised according to their pedigree or the manner in which they were adopted or developed.[11] Examples include rules contained within the GDPR and the UK Data Protection Act 2018. According to legal theorist Alexy, 'rules are norms which are always either fulfilled or not. If a rule validly applies, then the requirement is to do exactly what it says, neither more nor less. In this way rules contain fixed points in the field of the factually and legally possible'.[12] Rules can be considered as norms that are applicable in an all-or-nothing fashion, i.e. barring an exception to a rule, they either apply to a scenario or they do not.[13] A rule-based approach (RBA) to regulation is dominated by such rule-like norms.

In contrast, principles are frequently characterised as high-level, general and abstract norms.[14] These may be ethical and/or legal, conceptualised as broad iterations of individual or sets of ethical values, such as those included within Beauchamp and Childress' Four Principles

[7] K. Wildes, 'Principles, Rules, Duties and Babel: Bioethics in the Face of Postmodernity', (1992) *Journal of Medicine and Philosophy*, 17(5), 483–485.

[8] Nuremberg Code, 1949.

[9] CIOMS, 'International Ethical Guidelines for Health-related Research Involving Humans', (Council for the International Organization of Medical Sciences, 2016), xii.

[10] S. Arjoon, 'Striking a Balance Between Rules and Principles-based Approaches for Effective Governance: A Risk-based Approach', (2006) *Journal of Business Ethics*, 68(1), 53–82; J. Braithwaite, 'Rules and Principles: A Theory of Legal Certainty', (2002) *Australian Journal of Legal Philosophy*, 27, 47–82; T. Beauchamp and J. Childress, *Principles of Biomedical Ethics*, 7th Edition (Oxford University Press, 2013).

[11] As considered in the longstanding Hart-Dworkin debate on legal positivism. See H. Hart, *The Concept of Law*, 2nd Edition, P. Bulloch (ed.), (Oxford: Clarendon Press, 1994) and R. Dworkin, 'The Model of Rules', (1967) *University of Chicago Law Review*, 35(1), 14–46.

[12] R. Alexy, *A Theory of Constitutional Rights* (Oxford University Press, 2002), p. 4.

[13] Dworkin, 'Model'; M. Redondo, 'Legal Reasons: Between Universalism and Particularism', (2005) *Journal of Moral Philosophy*, 2(1), 47–68.

[14] D. Clouser and B. Gert, 'A Critique of Principlism', (1990) *The Journal of Medicine and Philosophy*, 5(2), 219–236; J. Raz, 'Legal Principles and the Limits of the Law', (1972) *Yale Law Journal*, 81(5), 823–854; Beauchamp and Childress, *Principles*; Dworkin, 'Model'.

(Principlism).[15] Accordingly, respect for beneficence and non-maleficence implies that health research should aim to provide benefit and to minimise foreseeable harm. The Four Principles are considered prima facie in nature, implying that they must be satisfied barring conflict between the principles. Within legal theory, it has been suggested that principles are optimisation requirements, i.e. norms that can be satisfied to varying degrees[16] as opposed to the 'valid or not' quality of rule-like norms. Principles can be articulated in more general and less legally enforceable terms but equally, breach of principles can lead to legal repercussions. For example, infringement of any of the seven principles included within the GDPR renders organisations subject to fines of up to €20 million or 4 per cent total worldwide annual turnover. Regardless of whether enshrined within legislative provisions or guidance documents, principles have the potential to shape behaviour within the health research setting, given the various commitments – legal, moral, political and other – to which they can give rise. A principle-based approach (PBA) is dominated by such principle-like norms.

Choosing to adopt a PBA dominant path in preference to RBA or vice versa carries important repercussions for health research landscapes and actors navigating them. The specific question of RBA v PBA received attention within the context of financial market regulation during the shift from RBA to PBA in the 1990s.[17] Different categories of PBA were identified, including full, polycentric,[18] formal and substantive.[19] Their commonality lies in a preference towards broad principle-like standards over detailed prescriptive and specific rules for setting standards of behaviour.[20] In contrast, discussions within bioethics have centred on: (1) specific content of particular rules/principles; (2) how principles ought to be balanced against each other when conflict arises – including whether certain principles ought to take priority over others; (3) which particular rules/principles ought to be in/excluded from ethical frameworks; and (4) how to extract action-guiding content from abstract principles.[21]

Within health research regulation more specifically, consideration of PBA in contrast to RBA has been more limited. Some contributions exist in the contexts of regulating the use of stem cells[22] and health data[23] in health research. In those arenas, PBA has been preferred over RBA in recognition of the value of principles and limitations of rules but without concluding that approaches dominated by principles obviate the need for rules. Rules play pivotal roles in delineating 'boundaries beyond which research ought not to stray and therefore over which society requires closer regulatory oversight'.[24] Rule-like norms may provide certainty to decision-makers given their typically detailed and prescriptive nature. The value of hard and fast rules, particularly manifested via legislation is not under dispute. Rather, as will become apparent, I suggest that their rigidity can leave RBA-dominated frameworks ill-suited to the demands of

[15] Beauchamp and Childress, *Principles*.
[16] Alexy, *Theory*.
[17] J. Black et al., 'Making a Success of Principles-Based Regulation', (2007) *Law and Financial Markets Review*, 1(3), 191–206.
[18] J. Black, *The Rise, Fall and Fate of Principles Based Regulation*, (2010), LSE Law Society and Economy Working Papers (17/2010).
[19] K. Alexander and N. Moloney, *Law Reform and Financial Markets* (Cheltenham: Edward Elgar Publishing, 2011).
[20] Black et al., 'Making a Success'.
[21] H. Richardson, 'Specifying, Balancing and Interpreting Bioethical Principles', (2000) *Journal of Medicine and Philosophy*, 25(3), 285–307.
[22] S. Devaney, 'Regulate to Innovate: Principles-Based Regulation of Stem Cell Research', (2011) *Medical Law International*, 11(1), 53–68.
[23] G. Laurie and N. Sethi, 'Towards Principles-Based Approaches to Governance of Health-Related Research Using Personal Data', (2013) *European Journal of Risk Regulation*, 4(1), 43–57.
[24] Devaney, 'Innovate', 60.

complex regulatory landscapes, especially rapidly-evolving health research terrains. In contrast, PBA affords the flexibility fast-paced technological change often necessitates. Principles can create and leave space for interpretation and the exercise of discretion, essential when dealing with difficult decisions and ensuring applicability across a variety of contexts.[25] The remainder of this section therefore focuses on PBA and, through exploration of several key functions principles can perform (often in contrast to rules), explains how they may be especially useful to health research regulation.

Principles can protect against over/under-inclusiveness of activities or subjects of regulation, in contrast to rules. Where specific and prescriptive rules are employed, there is a risk that by virtue of their rigidity, rules either fail to capture relevant activities within them, or are applied to activities that ought not to fall under their purview. It is impossible to legislate for every eventuality, particularly at the cutting edge of scientific research. Consider the proliferation of data-driven technologies that revealed the European Data Protection Directive 95/46/EC was no longer fit for purpose. Its replacement, the GDPR, seeks to better reflect the status quo viz potential data use and applications, albeit that ongoing and rapid developments in Artificial Intelligence (AI), computing and analytics are generating new regulatory concerns.[26] The GDPR is, however, underpinned by seven high-level 'principles' to be factored-in to all interpretations of activities falling within its scope. These may have more longevity and reach than prescriptive rules because principles are less likely to be as detailed and technology-specific as rules tend to be. Of course, principles may also necessitate revision, for example to reflect changes in consensus around what the overarching principles ought to be, but they are more likely to outlast the technological changes that can frequently make prescriptively drafted rules obsolete.

A further strength of principles is their interpretive/guiding function, in communicating the spirit with which more specific norms – including rules – ought to be applied, especially where tensions exist within law, e.g. simultaneously restricting, banning and promoting behaviour. This function can be observed in the approach of the UK Human Fertilisation and Embryology Authority (HFEA). It includes within its Code of Practice a series of regulatory principles to be adhered to when licensed activities are carried out under the Human Fertilisation and Embryology Act 1990, as per S8(1(ca). For example, the first principle states licensed centres must 'treat prospective and current patients and donors fairly, and ensure that all licensed activities are conducted in a non-discriminatory way'.[27] Such overarching principles can guide and assist decision-makers in all of their related activities.

The paramountcy of stakeholder engagement is a strong theme within this volume.[28] High-level principles can provide an effective dialogical tool for engagement with different stakeholders, enabling ongoing moral debate, and identifying interests and values at stake. As I argue elsewhere, PBA are more conducive to fostering meaningful dialogue because they avoid prescribing specifically (as rules often do), what ought to be done. Further, they 'promote reflection precisely on this point ... and in particular, they offer us the opportunity to lay out the core values which matter to us in the specific context. Rules, in contrast, can do the opposite, they can either prohibit something that might not be problematic or ... grant licence where

---

[25] Black et al., 'Success'.

[26] For example, discussion within House of Lords Select Committee on Artificial Intelligence 2017–2019, 'AI in the UK; Ready, Willing and Able?,' 16 April 2018 HL Paper 100; G. Hinton, 'Deep Learning – A Technology with the Potential to Transform Health Care', (2018) *JAMA*, 320(11), 1101–1102.

[27] HFEA Code of Practice, Edition 9.0 (2019).

[28] See, for example, Choung and O'Doherty, Chapter 12, this volume.

there is little'.[29] The UK care.data debacle illustrates the danger of overreliance on rules and failure to effectively engage in discussion of core principles of concern to stakeholders.[30]

Relatedly, the legitimacy of RGAs that are *not* co-produced alongside individuals/groups affected by them is problematic. For example, dominant policy framings can tend to portray innovation *per se* in a positive light,[31] but some innovation is high risk. Appropriate frameworks must be developed as simultaneously responsive to potential value *and* dangers of innovation.[32] Key to this is explicit acknowledgement from the outset of the imperative to shape the trajectory of research and innovation alongside and for society. Lipworth and Rexler call for development of a bioethics of innovation, which necessitates dialogue and engagement with stakeholders. The framework for Responsible Research and Innovation[33] advanced by Stilgoe and colleagues contains four 'dimensions' (anticipation, reflexivity, inclusion and responsiveness) that are akin to high-level principles and can serve as a helpful framing device through which to engage in dialogue. Indeed, as Devaney notes, PBA may have the capacity and potential to reflect, encompass and be facilitative of the process of innovation itself.[34]

In this section, I have laid out some key strengths of PBA, illustrating why they may be better suited to complex health research landscapes than RBA. The discussion now advances to consider an equally important aspect of developing appropriate RGAs: the need for additional tools to support decision-making.

17.4 RULES AND PRINCIPLES: NECESSARY BUT NOT SUFFICIENT

PBA and RBA have limitations and additional tools are necessary to guide decision-makers. For instance, resolving conflict between principles is challenging. Balancing, whereby each principle is assigned a weight, is a methodology through which it is suggested competing principles may be prioritised. However, balancing implies commensurability (that each principle can be assigned a weight),[35] which is obviously problematic at a practical level. Further, balancing can give rise to subjectivism, decisionism and intuitionism where decision-makers justify weighting according to preconceived prejudices.[36] For instance, Principlism has been criticised for prioritising respect for autonomy over other principles. Balancing is also a challenge for RBA: inter-rule conflict may equally arise as '[r]ules look more certain when they stand alone; uncertainty is crafted in the juxtaposition with other rules'.[37] Conflict between competing norms may be inevitable and balancing requires both judgement and justification. Opting for one interpretation/resolution to a decision is only legitimate if it comes with well-

[29] N. Sethi, 'Reimagining Regulatory Approaches: On the Essential Role of Principles in Health Research Regulation', (2015) *SCRIPTed*, 12 (2), 91–116, 110.
[30] P. Carter et al., 'The Social Licence for Research: Why care.data Ran into Trouble', (2015) *Journal of Medical Ethics*, 41(5), 404–409; M. Quiroz-Aitken et al., 'Consensus Statement on Public Involvement and Engagement with Data-Intensive Health Research', (2019) *International Journal of Population Data Science*, 4(1). See also Burgess in Chapter 25 of this volume.
[31] W. Lipworth, and R. Axler, 'Towards a Bioethics of Innovation', (2016) *Journal of Medical Ethics*, 42(7), 445–449.
[32] Special issue, 'Regulating Innovative Treatments: Information, Risk Allocation and Redress', (2019) *Law Innovation and Technology*, 11(1).
[33] J. Stilgoe et al., 'Developing a Framework for Responsible Innovation', (2013) *Research Policy*, 42(9), 1568–1580.
[34] Devaney, 'Innovate'.
[35] H. Richardson, 'Specifying, Balancing, and Interpreting Bioethical Principles', (2000) *Journal of Medicine and Philosophy*, 25(3), 285–307.
[36] R. Veatch, 'Resolving Conflicts among Principles: Ranking, Balancing and Specifying', (1995) *Kennedy Institute of Ethics Journal*, 5(3), 199–218.
[37] Braithwaite, 'Rules and Principles'.

reasoned and justifiable bases. Nonetheless, decision-makers require support in determining how to approach balancing. Likewise, concrete examples are required to elucidate how conflict between principles and rules ought to be addressed in practice.

Another criticism of high-level norms is that their abstract nature leaves too much interpretative space; extracting meaningful, action-guiding content becomes challenging. For instance, the Declaration of Helsinki states: 'Groups that are underrepresented in medical research should be provided appropriate access to participation in research'.[38] This does not suggest what 'appropriate access' entails. Even rules, – particularly when articulated broadly – are open to challenges of interpretative uncertainty, given the 'open texture' of language.[39] Content included within rules/principles may be interpreted overly-cautiously in fear of potential regulatory repercussions, stifling important research which may actually be legally and ethically permissible, as has been the case in some data sharing contexts.[40] Alternatively, interpretative latitude can leave room for creative compliance and exploitation of excessively abstract norms. Further, any potential certainty derived by RBA or PBA articulated in prescriptive language still necessitates shared understandings of the content – especially key terminology – of rules/principles and the overall objectives to be pursued, again suggesting the need for supplementary guidance to aid interpretation.

Theory-practice gaps and the need for context-sensitivity are also significant. Failure to adequately reflect the practical realities of conducting research 'on the ground' risks rendering norms ineffective. For instance, health research activities during global health emergencies have revealed disparities between what regulations demand and what is practically feasible or context-appropriate. Requirements to obtain timely ethical approval and adhere to randomised control trial protocols are not always possible/appropriate in time-sensitive settings and where proven therapeutics are lacking. Traditional distinctions between medical care/practice/treatment and research/innovation activities are blurred.[41] Discerning between practices primarily seeking to benefit the individual patient (treatment) and those aimed towards generating generalisable knowledge (research) is difficult, particularly regarding 'innovative' practices.

It is apparent from this discussion that rules and principles, while indispensable as regulatory tools, possess weaknesses that can limit their effectiveness in decision-making. As considered next, more is often required to support decision-makers, particularly in interpreting relevant norms, offering context-sensitivity and reflecting practical realities.

17.5 SUPPLEMENTARY GUIDANCE AND THE ADDED VALUE OF BEST PRACTICE

Additional supplementary guidance alongside RBA and PBA exists across many health research domains. This appears in myriad forms including: standards; guidelines; codes of practice; good practice; and, as will receive special attention below, best practice.

---

[38] WMA General Assembly, 'Declaration of Helsinki – Ethical Principles for Medical Research Involving Human Subjects', (WMA, 1964, as amended).

[39] Hart, *Concept*.

[40] N. Sethi and G. Laurie, 'Delivering Proportionate Governance in the Era of eHealth: Making Linkage and Privacy Work Together', (2013) *Medical Law International*, 13(2–3), 168–204.

[41] A. Ganguli-Mitra and N. Sethi, 'Conducting Research in the Context of GHEs: Identifying Key Ethical and Governance Issues', (Nuffield Council on Bioethics, 2016); N. Sethi, 'Regulating for Uncertainty: Bridging Blurred Boundaries in Medical Innovation, Research and Treatment', (2019) *Law, Innovation and Technology* 11(1), 112–133.

Clinical guidance in the form of guidelines proliferated from the 1970s onwards in the UK to achieve technical, procedural and administrative standardisation in medical practice and to maintain professional autonomy.[42] Good Clinical Practice Guidelines and evidence-based guidance from the National Institute for Health and Care Excellence are UK-based examples. Internationally, the World Health Organization (WHO) continually issues and updates guidelines on a variety of health topics, each designed to ensure the appropriate use of evidence in health policies and interventions, in accordance with the standards set out for guideline development.[43]

The role of guidance and its role in shaping *health research* practices also necessitates attention. Numerous international guidance documents exist, including the CIOMS Guidelines, which are supported by a 'commentary'. Likewise, the fourteen guidelines included within WHO Guidance for Managing Ethical Issues in Infectious Disease Outbreaks are each accompanied by questions illustrating the scope of ethical issues and 'a more detailed discussion that articulates the rights and obligations of relevant stakeholders'.[44]

Given that regulatory landscapes are frequently complex and pre-saturated with rules and principles (in their various forms), it is paramount that the introduction of supplementary guidance is approached with caution. Arguably, more guidance alone could suffer from the same criticisms as PBR and RBR; simply more norms requiring interpretation. Important considerations must be factored-in to the design and implementation of guidance to ensure its effectiveness. For example, the legitimacy of guidance is interlinked with the sources from which it has been generated and we must ask who this gives power to. As mentioned above, guidance may represent a means for actors to preserve autonomy and freedom from external interference/control. This raises questions of fairness, justice and transparency. Consider legal and ethical issues associated with current/anticipated uses of data and AI within health research. Concerns have been raised of technology firms developing their own guidance, facilitating creative compliance and self-regulation.[45] Even where guidance is drafted by independent committees, diverse interests must be balanced. For example, in the contexts of Big Data and AI, trade-offs are apparent in the UK between protection of privacy rights of individual citizens and national ambitions of economic growth, international competitiveness and participation in the fourth industrial revolution.

Additionally, the form guidance takes carries important ramifications for legitimacy and uptake. Distinct categories of guidance exist, at times operating at different levels. Guidance comprises a broad category ranging from anecdotal to evidence-based guidelines. At the time of writing, COVID-19 is causing a global pandemic. As health systems around the world struggle to treat patients, a plethora of new guidance documents are emerging from multiple sources, based on varying degrees of evidence. These include guidelines to support decision-making around public health responses to containment and guidance to support frontline health workers in resource allocation. Where guidance diverges across different sources, challenges arise as to which guidance to follow and why. Due regard must be given to how guidelines interact with pre-existing regulation. Another important consideration is what the repercussions, if any, might be for non-observation of, or derogation from guidance.

---

[42] G. Weisz et al., 'The Emergence of Clinical Practice Guidelines', (2007) *Milbank Quarterly*, 85(4), 691–727.

[43] World Health Organization, *Handbook for Guideline Development*, 2nd Edition (WHO, 2014).

[44] World Health Organization, 'Guidance for Managing Ethical Issues in Infectious Disease Outbreaks', (WHO, 2016).

[45] P. Nemitz, 'Constitutional Democracy and Technology in the Age of Artificial Intelligence', (2018) *Philosophical Transactions of the Royal Society*, Series A 376(2133), 1–14.

Relatedly, this generates considerations around how compliance with guidance is to be measured, incentivised or even enforced. It also follows that fundamental questions arise around the processes involved in drafting, endorsing, disseminating and implementing guidance. Central to these, is the question of where public voices are in each of these processes, as considered elsewhere in this volume.[46]

I have argued previously that best practice (BP), a form of supplementary guidance, can be particularly helpful for decision-makers within health research[47] and in ways that are distinct from other forms of guidance. For example, BP, as co-produced through inclusive and consultative processes, can provide a platform for inclusion of public(s) and additional stakeholder perspectives.[48]

One example of such an approach can be viewed again in the context of AI. In recognition of the widespread development and adoption of AI applications, the European Commission has developed Guidelines for Trustworthy AI. The Commission has taken a phased approach to piloting these, including wide consultation with various stakeholders. It is notable that alongside the guidelines, explanatory notes are offered and the Commission has established a Community of Best Practices for Trustworthy AI.[49] Through the European AI Alliance, registered participants could share their own best practices on achieving trustworthy AI.

Further, a recent report on the regulation and governance of health research lamented the 'disconnect between those making high-level decisions on how regulations should be applied and those implementing them on the ground'.[50] BP instantiations also offer concrete examples of principles or rules 'in action', based on lessons learned from those experienced in interpreting and applying the relevant norms. BP can thus serve an important function of helping to bridge such problematic policy-practice divides.

BP instantiations can also support decision-makers in interpreting relevant legislative provisions and/or ethical frameworks and related obligations. They provide more detailed explanatory notes on the legislative or normative intent behind overarching principles/rules. They provide a mechanism through which to make explicit to intended users of guidance what the status of such guidance is, and how it relates to pre-existing rules, principles and additionally relevant guidance. BP also guides decision-makers in approaching resolution of conflicting principles or rules, which I identified earlier as a key challenge for PBA and RBA.

Finally, I noted earlier that rules can close down conversations. BP also carries such risks; reference to BP may decontextualise and thwart discussion/use of other practices. Arguably, use of the term 'best practice' or 'good practice' suggests a superlative and that derogations from BP interpretations are suboptimal. But best practices (as construed here) are subject to constant review and revision, thus by definition, always seeking what is 'best' in a given context. In turn, in order to remain fit for purpose, best practices require us to constantly revisit the underlying rules/principles to which they correspond. In this regard they drive a symbiotic relationship between all of the norms in play towards an optimal system of regulatory and governance approaches.

---

[46] See Aitken and Cunningham-Burley (Chapter 11) and Burgess (Chapter 25) in this volume.
[47] N. Sethi, 'Research and Global Health Emergencies: On the Essential Role of Best Practice', (2018) *Public Health Ethics* 11(3), 237–250.
[48] Laurie and Sethi 'Approaches'.
[49] High-Level Expert Group on Artificial Intelligence, 'Ethics Guidelines for Trustworthy AI', (European Commission, 2019), www.ec.europa.eu/digital-single-market/en/news/ethics-guidelines-trustworthy-ai
[50] Academy of Medical Sciences, 'Regulation and Governance of Health Research: Five Years on', (The Academy of Medical Sciences, 2016).

17.6 CONCLUSION

In this chapter I have outlined key considerations in adopting rules, principles and supplementary guidance-based approaches to health research regulation. In particular, I have laid out the suitability of principles for guiding decision-makers across complex regulatory landscapes. I suggested that the introduction of supplementary guidance could tend to limitations of PBA and RBA. But, in turn, I stressed that generating new guidance must be approached with caution and due regard to additional concerns. Finally, I highlighted the added value that best practice – to be distinguished from other forms of supplementary guidance – can bring to complex regulatory landscapes.

18

Research Ethics Review

Edward Dove

18.1 INTRODUCTION

Across most jurisdictions today, researchers who propose to involve humans, their tissue and/or their data in a health research project must first submit an application form, which includes the research protocol and attendant documents (e.g. information sheets and consent forms), to one or several committees of experts and lay persons, who then assess the ethics of the proposed research. In some jurisdictions, this review, known as research ethics review, is mandated by law. In these cases, the law may be general[1] or it may apply to specific kinds of health research, such as clinical trials of an investigational medicinal product[2] or health research involving adults lacking capacity.[3] In other jurisdictions, and depending on the type of research project, research ethics review may be required or expected by 'softer' forms of regulation, such as guidelines, policy or custom, with the processes for the review consequently less standardised – and more flexible – than in a rules-based regime.

The principal aim of these research ethics committees (RECs), also known as institutional review boards (IRBs) and research ethics boards (REBs),[4] is to protect the welfare and interests of prospective (and current) participants and to minimise risk of harm to them. Another aim is to promote ethical and socially valuable research. This phenomenon of evaluating the ethics of proposed health research and determining whether the research may proceed – and on what grounds – has been in existence largely since the 1960s.[5] Originally designed for review of clinical research involving healthy human volunteers, research ethics review has since expanded to cover all fields of health research, including social science-driven health research such as qualitative studies investigating patient experiences with a disease or treatments that they receive. Given their central role in determining the bounds of ethical research, it is unsurprising to learn that RECs have been subject to sustained scrutiny; in many quarters, this has resulted in criticism within the health research and academic community that, among other things, the process of research ethics review is not fit for purpose. The cumulative charge is that research

---

[1] See e.g. CC 810.30 Federal Act of 30 September 2011 on Research involving Human Beings (Switzerland).
[2] See e.g. The Medicines for Human Use (Clinical Trials) Regulations 2004 No. 1031 (UK); Food and Drug Regulations (CRC, c 870), C.05 (Division 5 – Drugs for Clinical Trials Involving Human Subjects) (Canada).
[3] See e.g. Mental Capacity Act 2005 (England and Wales) and Adults with Incapacity (Scotland) Act 2000.
[4] Henceforth in this chapter I will use the terminology 'REC' as shorthand.
[5] L. Stark, *Behind Closed Doors: IRBs and the Making of Ethical Research* (University of Chicago Press, 2012); A. Hedgecoe, *Trust in the System: Research Ethics Committees and the Regulation of Biomedical Research* (Manchester University Press, 2020).

ethics review by committees promotes a wicked combination of inexpert review, inconsistent opinions, duplicative work, mission creep and heavy-handed regulation of health research.

This chapter places this charge at the focal point. In what follows, I chart the process of research ethics review with a view towards arguing that RECs have become regulatory entities in their own right and very much are a form of social control of science. As I detail, while RECs are far from perfect in terms of regulatory design and performance, they do perform, at least in principle, a valuable role in helping to steward research projects towards an ethical endpoint. In what follows, I analyse the nature and aims of research ethics review and the body of academic research regarding research ethics review. In so doing, this chapter also offers a critique of existing work and suggests some future directions for both the regulatory design of research ethics review and also researching the field itself.

18.2 RESEARCH ETHICS REVIEW AS A REGULATORY PROCESS

Many scholars have long viewed the notion of *evaluation* of the *ethics issues* of a *proposed* research project by a *committee* of people qualified in some way to assess the project's ethics as necessary, but not necessarily sufficient, for the successful functioning of, and securing of public trust in, health research. RECs, it is said, reflect a pragmatic system of 'social control' by researchers' academic and community peers. As William May opined in 1975: 'The primary guarantee of protection of subjects against needless risk and abuse is in the review before the work is undertaken. [...] [I]t is the only stage at which the subject can be protected against needless risk of injury, discomfort, or inconvenience'.[6] John Robertson similarly concluded in 1979: 'The [REC] is an important structural innovation in the social control of science, and similar forms are likely to be developed for other such controversial areas'.[7] By influencing research in an event-licensing capacity – that is, by offering opinion on and approval (or rejection) of a research project *before* it commences – RECs are seen to mitigate risks to researchers, participants and society. To this extent, research ethics review can be cast as a regulatory process.

As RECs have become more entrenched in the regulatory apparatus of health research over the past half-century, they have come to hold tremendous power over how research is shaped – and thus, influence over what knowledge is produced – as well as how the relationship between a researcher and a research participant is circumscribed. As Laura Stark observes, ethics committees 'are empowered to turn a hypothetical situation (this project *may be* acceptable) into shared reality (this project *is* acceptable). [...] [T]hey change what is knowable'.[8]

But it remains unclear what exactly constitutes research ethics review. Indeed, we might ask whether RECs engage in ethics deliberation at all – and, just as critically, whether this matters to fulfilling their putative *regulatory* role of assessing the relevant ethics issues in a project. Perhaps the challenge lies with the term 'research ethics review'. This suggests less of a focus on formulaic, bureaucratic – arguably synonymous with 'regulatory' – answers to questions (e.g. 'Is there informed consent?'; 'Have they used our consent form template?') and more of a focus on seeking deeper, more philosophically engaged answers to penetrating questions, such as: 'Do we really need informed consent here?'; 'What sort of alternative and preferable safeguards

---

[6] W. May, 'The Composition and Function of Ethical Committees', (1975) *Journal of Medical Ethics*, 1(1), 23–29, 24.
[7] J. Robertson, 'Ten Ways to Improve IRBs', (1979) *Hastings Center Report*, 9(1), 29–33, 29.
[8] Stark, *Behind Closed Doors*, p. 5.

might there be and why?'; 'Is this research in the public interest?'; or 'What public good might come from this research and is the financial and social cost commensurate?'.

What is reasonably clear is that a REC provides a favourable opinion only if it is assured that the ethics issues in the proposed research are appropriately addressed by the researcher – and sponsor – before the project proceeds. As the issues will vary depending on the research in question, REC members receive training and guidance about the issues they should consider, both in general and in particular cases. For example, according to the Governance Arrangements for Research Ethics Committees (GAfREC), which is a formal governance document for National Health Service (NHS) RECs in the UK: 'The training and guidance reflect recognised standards for ethical research, such as the Declaration of Helsinki, and take account of applicable legal requirements'.[9] If REC members learn about what research ethics is supposed to entail according to 'recognised standards' and take account of 'applicable legal requirements', we might reasonably ask whether the REC meetings themselves reflect a kind of instantiated deliberative decision-making ethics – that is, ethics as input, process, and outcome – where members individually and collectively evaluate and come to decide on the ethical acceptability of research proposals by invoking and deliberating on standards and requirements more than (ethical) norms or principles. If this is so, the REC, as a form of a decision-making body, need not necessarily 'do ethics' at all.

Some evidence of this comes when we shift our gaze from theory to practice. As Mary Dixon-Woods and colleagues have found in their empirical investigation of REC opinion letters to researchers:

> Though clearly RECs are making firm recommendations to researchers in these [previously discussed] examples of both inconsistent and consistent advice, the source of ethical authority for the REC in coming to their conclusions is rarely explicit in the letters. GAfREC – which provides the framework within which RECs are expected to work – is not referred to in any of the letters in our sample. Specific ethical principles or even guidelines are rarely invoked explicitly, and when they are, it is to authenticate or legitimise the decisions of the committee [...].[10]

If the REC opinion letter is a reasonably accurate reflection of the contents of a REC meeting's discussion, then there is some doubt as to whether ethical rules, norms or principles are openly discussed. Other empirical research has affirmed this doubt.[11]

Yet, the names bestowed upon these bodies by many jurisdictions ('ethics committees' or 'ethics boards'), and the related expectation that they should engage in research ethics review – and related criticism that they do not do enough of this – may, in fact, be somewhat misplaced. I have suggested through my own empirical research that as RECs become institutionalised and professionalised, acting as multi-faceted and multidisciplinary micro-*regulators* of health research, and as further national and international regulations come into force that impact health research, RECs might be expected to act more as risk-assessing 'health research regulatory committees' writ large.[12] Somewhat similarly, based on her own recent empirical research, Sarah Babb makes the case that IRBs in the USA have transformed from academic committees to

[9] Health Research Authority, 'Governance Arrangements for Research Ethics Committees', (2020), para 5.3.1.

[10] M. Dixon-Woods et al., 'Written Work: The Social Functions of Research Ethics Committee Letters', (2007) *Social Science & Medicine*, 65(4), 792–802, 796.

[11] M. Fitzgerald et al., 'The Research Ethics Review Process and Ethics Review Narratives', (2006) *Ethics & Behavior*, 16(4), 377–395.

[12] E. Dove, *Regulatory Stewardship of Health Research: Navigating Participant Protection and Research Promotion* (Cheltenham: Edward Elgar, 2020).

'compliance bureaucracies', where specialised administrative staff members define and apply federal regulations.[13] Even if RECs do not engage in something approaching truly substantive ethics deliberation, and this is (partly) accepted as an outcome of practical constraints (e.g. limited resources and pressed time), might they still be able to fulfil their aim of targeting areas of health research that pose moral concern, and might they still be able to mitigate the manifestation of those concerns?

Indeed, I would argue that it is not necessarily problematic to acknowledge that RECs rarely engage in deep ethics deliberation. RECs *are* a valuable regulator in health research, and so if there are criticisms of them, we should look to those criticisms that speak to their regulatory functions – procedures, performance and so on – more than the absence or presence of ethics deliberation per se. By focusing here, we may come to see that concerns about efficiency, effectiveness, proportionality, reduced burden and so on, must be addressed more directly. Acknowledging this is not to say that RECs cannot spot and deal with thorny ethics issues when or if they arise, but it does allow us to be arguably more accurate and honest to cast them for what they are: regulators with a gatekeeping and promotional role about getting safe and good science done.

Let us, then, look at some of the persistent criticism of the research ethics review process that speak to the regulatory functions of RECs.

18.3 REC CRITICISMS: POOR DESIGN AND PERFORMANCE AND THE FETISHISATION OF CONSENT

For as long as they have existed, RECs have been subject to opprobrium from the research community and academic commentators, mainly because they are seen as under-, over- or simply mis-regulated bureaucratic bulwarks against otherwise ethical, minimally risky or non-risky research. For years, research into RECs has revealed a high level of variation of decision-making processes in RECs[14] and dissatisfaction from various stakeholders.[15] These criticisms can be grouped into concerns about (a) design and performance and (b) the fetishisation of consent.

18.3.1 Poor Design and Performance

Many of the problems scholars have identified with research ethics review have been due both to weak regulation – which contributes to procedural and substantive inconsistency of decision-making – and also over-regulation – which contributes to duplicative review and cumbersome and complex thickets of disproportionate regulation for research that presents minimal risk.[16] In their review of US IRBs, Emanuel and colleagues identified fifteen 'problems' and grouped them into three broad categories: (1) structural problems deriving from the organisation of the

[13] S. Babb, *Regulating Human Research: IRBs from Peer Review to Compliance Bureaucracy* (Palo Alto, CA: Stanford University Press, 2020).
[14] See e.g. B. Barber et al., *Research on Human Subjects: Problems of Social Control in Medical Experimentation* (New York: Russell Sage Foundation, 1973). See also Dixon-Woods et al., 'Written Work', 796.
[15] See e.g. G. Alberti, 'Local Research Ethics Committees: Time to Grab Several Bulls by the Horns', (1995) *BMJ*, 311(7006), 639–640; K. Jamrozik, 'The Case for a New System for Oversight of Research on Human Subjects', (2000) *Journal of Medical Ethics*, 26(5), 334–339; C. Warlow, 'Clinical Research Under the Cosh Again', (2004) *BMJ*, 329(7460), 241–242.
[16] G. Laurie and S. Harmon, 'Through the Thicket and Across the Divide: Successfully Navigating the Regulatory Landscape in Life Sciences Research', in E. Cloatre and M. Pickersgill (eds), *Knowledge, Technology and Law* (London: Routledge, 2014), pp. 121–136.

system as established by the US federal regulations, (2) procedural problems stemming from the ways in which individual IRBs operate, and (3) performance assessment problems resulting from the absence of systemic assessment of current protections.[17] Arguably, many of these structural, procedural and performance assessment problems also could be identified in RECs in other jurisdictions.

Indeed, the main design and performance concerns with the research ethics review process commentators have identified over the years include:

- inconsistency in procedures and substantive decisions within and across committees;
- delays or impediments to research due to slow-moving RECs that have no built-in efficiency incentive;
- cumbersome bureaucratisation and standardisation of application forms that are ill-suited to different types of research, that slow and muddy the process of ethics review and that lead to heavy administrative burdens for researchers;
- distortion of research methods imposed by RECs who may not be trained in research methods and are not qualified (or expected) to judge the scientific merit of applications;
- over- or exclusive reliance on prior (*ex ante*) review that inadequately assures that the *actual* conduct of research is in accordance with ethical standards;
- imposition of inappropriate consent requirements in certain types of research projects (e.g. surveys, behavioural intervention studies) that can lead to potential selection bias in participation and responses; and
- increased risk of unethical research, in part due to the ever-growing length of information sheets that participants do not bother to read, and also in part due to lengthy application forms that researchers and REC members alike either may not adequately read or quickly complete – in other words, the insidious growth of a 'tick-box mentality'.[18]

The cumulative account of these concerns suggests that better regulation is needed to improve the efficiency and effectiveness of research ethics review by RECs, and this may entail, among other things, streamlining existing regulation, enacting robust standard operating procedures (SOPs), designing templates tailored to the specific type of research project, and embedding in regulation and policy the emerging notion of stewardship. But before I address these ways forward, I now turn to a second persistent criticism of RECs, namely their fetishisation of consent.

18.3.2 *The Fetishisation of Consent*

Another major criticism of RECs centres on their putatively over-bearing emphasis on consent forms and information sheets, and minute wordsmithing of both, that leads to inevitable elongation of the documents and thereby increased risk of non- or miscomprehension by participants, which ironically may lead to other harms not related to the research, such as stigmatisation or disrespect. Since at least the 1960s,[19] commentators have argued that consent cannot and should not act as a stand-alone rampart to prevent unethical research. Yet many consider that RECs disproportionately fixate on consent as a locus for determining and setting

[17] E. Emanuel et al., 'Oversight of Human Participants Research: Identifying Problems to Evaluate Reform Proposals', (2004) *Annals of Internal Medicine*, 141(4), 282–291.
[18] Many of these criticisms are explored in R. Klitzman, *The Ethics Police? The Struggle to Make Human Research Safe* (Oxford University Press, 2015).
[19] H. Beecher, 'Ethics and Clinical Research', (1966) *New England Journal of Medicine*, 274(24), 1354–1360.

researchers' ethical behaviour, demonstrating 'the acme of self-defeating ritual compliance'.[20] Perhaps it is because 'these [consent] documents constitute one of the few aspects of researcher interactions with subjects – a very downstream process – that committees feel they can control.'[21]

This bureaucratic addiction to procedure and process, coupled no doubt with an uptick in legal – albeit siloed – regulation of health research, has led to a *legalisation* in the workings of RECs, which is to say: a fetishisation for more forms, longer forms and ongoing insistence on boiler-plate language tacked on to information sheets and consent forms so that RECs and institutions protect themselves and others from liability. Consent is treated as a panacea for all ethical concerns,[22] a kind of Pollyanna-ish hope that, 'If only we can inject all possible risks and relevant information into the form, then participants can truly exercise their autonomy'. This is not the 'good kind' of REC legalisation William Curran envisioned in 1969, replete with a common law-like generalisable body of precedents and principles of procedure and substance that allow the process of deliberation to flourish.[23] Instead, it is the troubling kind: rigid and overly standardised, treating ethics as a tick-box, form-ridden, technocratically structured event. Once again, this militates against ethics committees actually 'doing ethics' in the genuine sense that is understood of that discipline.

Given the groundswell of criticisms over the years, what, then, might be the future directions for research ethics review as a core process in health research regulation, and what might be the future directions for researching research ethics review to assess what is working well and not so well?

18.4 FUTURE DIRECTIONS FOR RESEARCH ETHICS REVIEW

While many support the underlying idea of *ex ante* ethics review by a competent committee as a means to protect and promote the rights, interests and welfare of participants, as this chapter has observed, many also have expressed dissatisfaction with the structure and function of the ethics review system and the individual processes of RECs. Multiple regulatory techniques and instruments have been employed over the years in the hopes of remedying the myriad problems attributed to RECs, foremost the concerns of inefficiency and ineffectiveness.

Scholars have proposed a number of changes to the regulatory design of research ethics review. For the purposes of this chapter, I want to focus on three that have gained attention recently and may be among the most promising: streamlining, standardisation and stewardship.

18.4.1 *Future Directions for Regulatory Design*

A number of jurisdictions are now *streamlining* the process of research ethics review in at least two ways. First, they have introduced proportionate review systems, whereby a research project that is deemed by assessors to present no (or limited) material ethics issues undergoes a lighter-touch review. In the UK, for example, under the Health Research Authority's (HRA)

[20] S. Burris and J. Welsh, 'Regulatory Paradox in the Protection of Human Research Subjects: A Review of Enforcement Letters Issued by the Office for Human Research Protection', (2007) *Northwestern University Law Review*, 101(2), 643–685, 678.
[21] Klitzman, *The Ethics Police*, p. 139.
[22] See e.g. S. Burris and K. Moss, 'US Health Researchers Review Their Ethics Review Boards: A Qualitative Study', (2006) *Journal of Empirical Research on Human Research Ethics*, 1(2), 39–58.
[23] W. Curran, 'Governmental Regulation of the Use of Human Subjects in Medical Research: The Approach of Two Federal Agencies', (1969) *Daedalus*, 98(2), 542–594.

Proportionate Review Service, such projects are reviewed via email correspondence, teleconference or at a face-to-face meeting by a sub-committee – comprising experienced expert and lay members – rather than at a full meeting of a REC.[24] The final decision is notified to the applicant by email within twenty-one calendar days of receipt of a valid application, which is a faster turnaround time than an application that goes to a full-committee review. Second, a group of efforts are underway internationally to streamline multiple REC review of multi-site research projects, which is seen as duplicative and disproportionate.[25] Since 2004, the UK requires only one NHS REC opinion per research project, even if the project involves multiple sites in the country. In the USA, since 2020, a revised rule in the Federal Policy for the Protection of Human Subjects – better known as the 'Common Rule' – generally requires US-based institutions that receive federal funding and are engaged in cooperative research projects (i.e. projects covered by the Common Rule that involve more than one institution in the USA) to use a *single* IRB for that portion of the research that takes place within the USA if certain requirements are met.[26] This 'sIRB rule' reflects a growing effort by regulators and policymakers in countries around the world – including Uganda,[27] Canada[28] and Australia[29] – to reduce the procedural inefficiencies, redundancies, delays and research costs that have become synonymous with the absence of research ethics review mechanisms designed for multi-site health research projects.[30]

A number of jurisdictions are also working on *standardisation* of the processes involved in ethics reviews, with the aim of achieving more consistent outcomes in review and fairness to applicants. The Care Act 2014 in the UK, for example, requires the HRA to co-operate with several other regulatory authorities in the exercise of their respective functions relating to health or social care research, 'with a view to co-ordinating and standardising practice relating to the regulation of such research'.[31] Standardisation is accomplished through various means, including the introduction and maintenance of:

- SOPs to ensure procedural consistency across RECs;
- template research application forms – including information sheets, consent forms and research protocols – for researchers to devise more thorough and ethically robust applications;
- template review forms for REC members to complete when reviewing applications; and
- systems of accreditation, qualification or certification of RECs to encourage mutual trust in each REC's processes of review.

It must be said, though, that while many commentators support standardisation as a way to drive consistency and fairness in ethics review, others blame standardisation for the growth of an undesirable 'tick-box' approach that many see as defining REC work today. This, however,

---

[24] Health Research Authority, 'Proportionate Review: Information and Guidance for Applicants', www.hra.nhs.uk/documents/1022/proportionate-review-information-guidance-document.pdf.

[25] See e.g. E. Dove et al., 'Ethics Review for International Data-Intensive Research', (2016) *Science*, 351(6280), 1399–1400.

[26] The Federal Policy for the Protection of Human Subjects ('Common Rule'), 45 C.F.R. § 46, Subpart A; The Federal Policy for the Protection of Human Subjections, 82 FR 7149, at 7265 (19 January 2017).

[27] Uganda National Council for Science and Technology, 'National Guidelines for Research involving Humans as Research Participants', (UNCST, 2014), s. 4.5.5, para. c

[28] *Clinical Trials Ontario*, www.ctontario.ca/.

[29] Victoria State Government, 'National mutual acceptance', (health.vic, 2018) www2.health.vic.gov.au/about/clinical-trials-and-research/clinical-trial-research/national-mutual-acceptance.

[30] E. Dove, 'Requiring a Single IRB for Cooperative Research in the Revised Common Rule: What Lessons Can Be Learned from the UK and Elsewhere?', (2019) *Journal of Law, Medicine & Ethics*, 47(2), 264–282.

[31] Care Act 2014, s. 111(1).

might be a product of the continued confusion about whether we see RECs as philosophically attuned ethics deliberation entities rather than as regulatory assessors situated within a wider health research ecosystem. I have argued above that the latter view is more accurate.

Third, the emerging concept of *regulatory stewardship* may have resonance in reforming the regulatory design of research ethics review to better account for the network of actors involved in bringing an application through the various regulatory thresholds in the research lifecycle. A key finding from recent empirical investigation[32] is the ability of actors within the health research regulatory space to serve as 'regulatory stewards'. Research suggests that regulatory stewardship involves different actors – including RECs and others involved in the regulation of health research – helping researchers and sponsors navigate complex regulatory pathways and work through the thresholds of regulatory approvals. Collective responsibility, as a component of regulatory stewardship, requires relevant actors to work together to design and conduct research that is ethical and socially and scientifically valuable and that ultimately aims to improve human health. This can only be accomplished if a framework delineates how and when regulators and regulatees should communicate with one another and makes clear who has what responsibility and role to be played (if any) at each stage in the research lifecycle.

The regulatory environment for research ethics review could be designed to provide clearer channels for RECs – and members within them who may have closer contact with researchers and sponsors – and their own managing regulators (e.g. institutions, ministries, regulatory authorities) to engage with researchers and sponsors in improving the quality of research protocols and applications, and in working through law, regulation, and regulatory approvals. These communicative channels may include online toolkits and more personalised support via email, telephone, or digital meetings.

All of this would have the added advantage of engaging multiple actors in earlier stages of the research design process, including on the actual ethics issues (or not) that arise. Where these are considerable, the further downstream ethics review will still have a role to play; however, where these are minimal or negligible, they might be addressed sooner in the regulatory pathway, leaving the REC to undertake its regulatory role more efficiently and effectively.

18.4.2 *Future Directions for Researching Research Ethics Review*

Further empirical evidence is needed to investigate questions about extant research ethics review processes and to test new models that seek to improve REC efficiency and effectiveness. There have been few in-depth qualitative studies of RECs focusing on assessment of regulatory design. This undermines effective regulation, as policymakers and regulators – through state actors or otherwise – increasingly seek to develop regulation through intricately documented evidence of problems and the effects of regulation on society. There is a need for qualitative research that explores how and why RECs make the decisions they do, and how the nested dynamics of RECs and central 'managing' regulators play into decisions.[33]

---

[32] Dove, 'Regulatory Stewardship'; see also G. Laurie et al., 'Charting Regulatory Stewardship in Health Research: Making the Invisible Visible', (2018) *Cambridge Quarterly of Healthcare Ethics*, 27(2), 333–347.

[33] S. Nicholls et al., 'A Scoping Review of Empirical Research Relating to Quality and Effectiveness of Research Ethics Review', (2015) *PLOS ONE*, 10(7), e0133639; see also, for a US example of research in this area, *AEREO: The Consortium to Advance Effective Research Ethics Oversight*, www.med.upenn.edu/aereo/.

Documented problems of RECs have largely relied on evidence and anecdote proffered by researchers. While there is a welcome growing corpus of empirical literature on RECs,[34] more evidence is needed from regulatory scholars who can go inside RECs to test new models via pilot studies or randomised controlled trials; or who can examine how RECs, both as individual members and as a body, see themselves and their committee in a changing regulatory environment, and can go inside regulatory bodies to gather the regulators' perspective on the roles of a REC within the health research regulatory space. Research ethics review thus remains an area ripe for investigation.

18.5 CONCLUSION

In this chapter, I have argued that RECs have become regulatory entities in their own right, governed by – depending on the jurisdiction – institutions, central regulatory agencies, administrative staff and offices, standardised forms and communications, and lengthy governance arrangements and SOPs. Just as some legal scholars speak of 'juridification',[35] which is an encroachment of law into ever more aspects of our society, so too might we speak of ethics review increasingly 'colonising' the health research regulatory space, structured according to the logic of its codes and customs. When RECs were first coming into being in the 1960s, Harvard Law Professor Louis Jaffe opined that '[a] general statutory requirement requiring institutional committees in any "experiment" would raise monstrous problems of interpretation, would unduly complicate medical practice, and would add unnecessary steps to experiments where the risks to the subject or patient are trivial.'[36]

Yet this is where we stand today, with REC review required formally by law or informally by policy for an array of health research, from the trivial to the complex and risky, albeit with more proportionate review processes than occurred previously. Over time, like all of health research, the regulatory space in which RECs are situated has expanded, along with the paperwork and resources researchers must dedicate in order to pass over the 'ethics hurdle'.

At the same time, scholars remind us that: 'The role of the Research Ethics Committee is to advise. It does not itself authorise research. This is the responsibility of [another] body under whose auspices the research will take place'.[37] While technically accurate – at least in many jurisdictions – this fails to appreciate the power of a REC to control what knowledge can be produced and how that knowledge is shaped. RECs, as noted previously, are a form of social control of science. The 'advisory' role of a REC masks its profound ability to impact health research, which is precisely why RECs have faced such criticism and undergone reform. They are not minor actors in the health research regulatory space; on the contrary, they may be

---

[34] For empirical studies of IRBs in the USA, see e.g. Stark, *Behind Closed Doors*; Babb, *Regulating Human Research*; Klitzman, *The Ethics Police*; J. F. Jaeger, 'An Ethnographic Analysis of Institutional Review Board Decision-Making' (PhD thesis, University of Pennsylvania 2006). For empirical studies of RECs in the UK, see e.g. A. Hedgecoe et al., 'Research Ethics Committees in Europe: Implementing the Directive, Respecting Diversity', (2006) *Journal of Medical Ethics*, 32(8), 483–486; J. Neuberger, *Ethics and Health Care: The Role of Research Ethics Committees in the United Kingdom* (King's Fund Institute, 1992).

[35] G. Teubner, 'Juridification: Concepts, Aspects, Limits, Solutions' in G. Teubner (ed.), *Juridification of Social Spheres* (Berlin: Walter de Gruyter & Co, 1987).

[36] L. Jaffe, 'Law as a System of Control', (1969) *Daedalus*, 98(2), 406–426, 412.

[37] I. Kennedy and P. Bates, 'Research Ethics Committees and the Law' in S. Eckstein (ed.), *Manual for Research Ethics Committees*, 6th Edition (Cambridge University Press, 2003), pp. 15–17, p. 16.

among the most important. And, as I have stressed, the obligations imposed on RECs have only increased over time as myriad regulation is brought to bear on them. Ethics and regulation must go hand-in-hand – indeed, one might say that the process of research ethics review must be co-produced with regulation, and regulation and ethical judgement are co-dependent. It is crucial that we appreciate the respective roles of each when it comes to entities such as the REC. This chapter has sought to reveal how we can better understand and deliver these dual roles.

19

Data Access Governance

Mahsa Shabani, Adrian Thorogood and Madeleine Murtagh

19.1 INTRODUCTION

Enabling researchers' access to large volumes of health data collected in both research and healthcare settings can accelerate improvements in clinical practice and public health. Because the source and subject of those data are people, data access governance has been of concern to scientists, ethics and regulatory scholars, policymakers and citizens worldwide. While researchers have long provided colleagues access to data in an ad hoc fashion, many research funders – e.g. US National Institutes of Health, Wellcome, Bill and Melinda Gates Foundation, United Kingdom Research and Innovation, European Research Council – journals,[1] professional societies and associations,[2] and regulators now systematically promote the deposit of research data in repositories that aim to provide responsible and timely access to data. Data sharing aims to enable meta-analyses and creative (re)uses, reduce duplicative effort in data generation, and improve reproducibility through validation studies, so as to support data-intensive research and thereby improve human health. In many countries, routinely collected healthcare data is also increasingly being made available to researchers. In both research and healthcare contexts, technical and governance strategies for promoting responsible data sharing and access continue to evolve.

The broad sharing of health research data promises many benefits, but it can also involve risks. Health research data can reveal sensitive information about individuals (in legal terms, data subjects) and their relatives, posing risks to privacy and of discrimination and stigmatisation. Broad sharing of health research data can also raise professional concerns for the researchers or organisations who produce data in terms of receiving adequate credit and recognition for their efforts in collecting, curating and analysing data.[3] Likewise, commercial research companies may be concerned their data will be appropriated or misused by competitors. Data access governance aims to promote organisational, scientific and societal interests in data re-use, while protecting the rights and interests of the range of stakeholders with an interest in data. Data access governance manages who has access to data, for what purposes, and under what

---

[1] D. Taichman et al., 'Sharing Clinical Trial Data: A Proposal from the International Committee of Medical Journal Editors', (2016) *Annals of Internal Medicine*, 164(7), 505–506.

[2] ACMG Board of Directors, 'Laboratory and Clinical Genomic Data Sharing Is Crucial to Improving Genetic Health Care: A Position Statement of the American College of Medical Genetics and Genomics', (2017) *Genetics in Medicine*, 19(7), 721– 722.

[3] M. Murtagh et al., 'International Data Sharing in Practice: New Technologies Meet Old Governance', (2016) *Biopreservation and Biobanking*, 14(3), 231–240.

conditions. Governance mechanisms include policies, due diligence processes, data access agreements and monitoring. Data access governance is closely linked to the concept of data stewardship, where organisations aim to ensure data are shared widely in the interest of science and society, while also mitigating associated ethical, societal and privacy risks.[4]

In contemporary data-driven science, data access governance often involves Data Access Committees (DACs) as the key institutional setting in which access decisions are made. DACs are diverse and may be composed of individuals with a range of relevant expertise, including familiarity with the scientific area, privacy and security, and research ethics.[5] As Lowrance notes, '...[s]ome DACs are formally constituted and appointed, while some are more casual. Some publish their criteria, decisions and decision rationales, but most don't. Some directly advise the data custodians, who then make the yes/no (or revise-and-reapply) access decisions. But many DACs make binding decisions'.[6]

Against this backdrop, this chapter examines the topic of data access governance. We discuss the underlying values and goals of data access governance, focusing in particular on the scientific and social implications for open access and data sharing, on the rights and interests of data subjects as well as those of data producers, and on the ethical conduct of data sharing. We contrast the general structural and normative components of open and controlled data access. We then present existing data access arrangements of organisations and repositories that exemplify varying modes of good practice. We argue these models exemplify the tension between promoting open access to databases on the one hand, and, on the other, protecting the rights and interests of the parties involved, including data subjects, researchers, funding organizations and commercial entities. We suggest that principles of transparency, fairness and proportionality in consideration of all stakeholders' interests and values is key to achieving this balance. We conclude by discussing existing challenges in data access governance, including potential conflicts between various stakeholders' views and interests, resource issues, (mis) coordination between oversight bodies, and the need for better harmonisation of access policies and procedures.

19.2 GOALS OF DATA ACCESS GOVERNANCE

Key goals of data access governance aim to strike a balance between protecting data subjects and data producers' rights and interests, while also promoting broad access to data to advance scientific research in the public interest.

19.2.1 *Protecting Data Subject Rights and Interests and Promoting Research Integrity and Ethics*

Data access governance supports research ethics principles for research involving human subjects. Minimising privacy risks to participants, respecting participant autonomy, and holding researchers accountable for the scientific validity and ethical conduct of research through research ethics committee (REC) approval and oversight, are key goals of governance

[4] The Expert Panel on Timely Access to Health and Social Data for Health Research and Health System Innovation, 'Accessing Health and Health-Related Data in Canada', (Council of Canadian Academies, 2015).
[5] Murtagh et al., 'Better Governance, Better Access: Practising Responsible Data Sharing in the METADAC Governance Infrastructure', (2018) *Human Genomics*, 12(1), 24.
[6] W. W. Lowrance, *Privacy, Confidentiality, and Health Research* (Cambridge University Press, 2012).

of data access.[7] These goals are increasingly furthered by engaging communities in the design of governance.

Privacy and security: Data access governance can protect participant privacy in several ways. Data access agreements, which are signed by data custodians and data users, typically include requirements regarding protecting privacy and security. Privacy safeguards include restrictions on unauthorised individual-level linkage of datasets, which may increase the re-identifiability of data, or prohibitions on attempting to re-identify participants. The greater the combinations of individual-level data for any given individual, the more likely re-identification becomes. Privacy rules in access processes are therefore often designed to control the level of individual-level data linkage. Security safeguards may include general or specific requirements to adopt physical, organisational, and technical protections, as well as data breach reporting obligations.

Respect for the provisions of ethical approvals: Data access governance models often aim to ensure users respect high standards of scientific integrity, and meet the ethical requirements related to compatibility of downstream use of data with the original consent obtained from the participants at the time of enrolment to a study and data collection. Where researchers have stated that data will only be used for certain kinds of research – e.g. disease-specific – this condition will inform the review of an access proposal by the relevant oversight bodies, notably DACs. Data access review may be informed by the following questions:[8] Does the application violate – or potentially violate – any of the ethical permissions granted to the study or any of the consent forms signed by the study participants or their guardians? Does the application run a significant risk of upsetting or alienating study participants or thereby reducing their willingness to remain as active participants in the research? Does the application run a significant risk of bringing disrepute to study, repository or steward and thereby reducing participant trust and willingness to remain as active participants in the research?

Respect for communities and relevant stakeholders: Responding to relevant stakeholders including communities' concerns and seeking to strike a balance between the views of different groups is fundamental to respecting these communities. This may mean championing the rights of less powerful groups and taking steps to seek out their views and actively responding to those views. In the context of data access, stakeholders include study participants and communities who provide the data, study managers and the researchers who develop the data and related resources, researchers who wish to access those data, the funders who support the studies which produce the data and the public who are the ultimate funders as well as beneficiaries of research. Each of these groups has a legitimate and vested interest in the responsible and respectful uses of data and provide a unique perspective on how such governance can be achieved. For example, study participants and community representatives sitting on oversight committees such as DACs can provide a unique insight into what other study participants may view as acceptable uses of data.

19.2.2 *Data Producer Rights and Interests*

One goal of access controls is to protect the rights and interests of the researchers or institutions generating data. Academically, researchers compete for high-impact publications and, in turn, for academic positions and promotions. Commercially, researchers and research institutions

---

[7] M.Aitken et al., 'Consensus Statement on Public Involvement and Engagement with Data-Intensive Health Research,' (2019) *International Journal of Population Data Science*, 4(1).

[8] Murtagh et al., 'METADAC Governance Infrastructure', 24.

may compete to develop commercial applications from research findings. These considerations are often addressed through publication and commercialisation clauses in data access agreements.

Data access governance may include publication policies that seek to ensure that data producers are appropriately recognised for their contribution to science. Given that publication remains the major currency in academia, there may be a tendency for data producers to request co-authorship as a condition of access. This is discouraged for reasons of scientific freedom and accountability. Having independent DAC members adjudicating access is one remedy to the potential conflicts of interest in such practices. A compromise position is sometimes used whereby the data producer has a right to review manuscripts before publication, or to at least to be informed in advance of forthcoming publications based on (re)analysis of shared datasets. Commercialisation policies aim to ensure that the data producer benefits from, or at least does not have its competitive position harmed by, downstream use of data.

Finally, responsible data access governance requires transparency, fairness and proportionality towards participants and other stakeholders. Transparency can be improved by the publishing of policies and procedures, as well as publication of approved data recipients and plain language summaries or abstracts of approved uses. Moreover, ensuring timely and consistent access review without imposing unnecessary constraints on data access are of salient importance with regard to fairness. Where data governance seeks to achieve competing goals of openness and privacy protection, as well as meeting social and participant expectations of data use, a proportionate balance needs to be struck. Proportionality may call for different types of access controls to be applied to different types of data. Increasingly, there is emphasis that the balance between public benefit and individual risks be evidence-based.[9]

19.3 DATA ACCESS GOVERNANCE: POLICIES, PROCESSES, AGREEMENTS AND OVERSIGHT

The values and goals of data access governance are operationalised through the policies and practices of DACs and various models of data access.

19.3.1 *Controlled Versus Open Access Data*

The nature of data – and the associated ethical, policy and legal issues – largely determines the access model, which can range from open to controlled to closed. Open access models generally make data available to any user, anywhere, over the internet, without financial or technical constraints. The Human Genome Project, for example, which sequenced the entire human genome, shared the sequence data openly. Subsequent publicly-funded projects sequenced more individuals and combined these data with richer social, demographic and clinical data, prompting concerns about the privacy of data subjects. Controlled access models emerged to ensure data could still be shared broadly with qualified and trusted researchers, while also protecting the privacy of data subjects and sometimes also the interests of researchers producing data. In controlled access, access is managed by a REC or increasingly by a specialised DAC,

---

[9] M. Shabani et al., 'Who Should Have Access to Genomic Data and How Should They Be Held Accountable? Perspectives of Data Access Committee Members and Experts', (2016) *European Journal of Human Genetics*, 24(12), 1671–1675; P. Burton et al., 'Policies and Strategies to Facilitate Secondary Use of Research Data in the Health Sciences', (2017) *International Journal of Epidemiology*, 46(6), 1729–1733.

which reviews requests for data access. In this regard, DACs often carry out a due diligence review of access requests and may hold deliberations over the scientific, feasibility and ethical aspects of the request. This is in line with the recommendations issued by the Organisation for Economic Co-operation and Development's (OECD) Council on Health Data Governance that review and approval processes should involve an evidence-based assessment and adhere to principles of transparency, objectivity and fairness. In addition, the OECD's recommendations underline the importance of independent multi-disciplinary review with an ultimate aim of risk mitigation for individuals and society.[10]

19.3.2 *Data Access Agreements*

One component of both controlled and open access models is the data agreement (termed 'data transfer', 'data access' or 'data use' agreement), which establishes the conditions governing the accessing researcher's use of the data. The terms of data access agreements typically address data subject protections, including prohibition on unauthorised linkage of individual-level data and attempts to re-identify participants, respect for consent-based use conditions and ensuring appropriate security safeguards are in place. The terms may also include protections for the rights and interests of the researchers producing data, such as publication embargoes to allow data producers the first attempt at publication or intellectual property clauses governing owner-ship of downstream commercialisation. Benefit-sharing clauses are important in countries with emergent research infrastructures. Other clauses may serve multiple stakeholders, such as obligations to only use data for specified purposes. Still other clauses may address the interests of science and society, such as requirements for open access publication, or to share analysis code or derived datasets. While data access agreements are legally binding if designed properly, their practical enforceability, especially across borders, is largely untested and remains a concern.[11] Especially where terms are associated with open access data, they are typically meant more as a means of communicating community norms to users.

19.3.3 *Monitoring of Data Use*

DACs may additionally develop tools and mechanisms to maintain ongoing oversight of downstream data uses. For instance, data users may be required to provide periodic reports regarding the projects in which data are being used. In addition, data users may be asked to report to the DAC the publications resulting from the data use, or issues arising from special conditions of access, e.g. risk management strategies for sensitive or potentially 'sensational' research, or return of incidental findings. Such oversight may enable the DACs to check compliance of the data uses, but implementation requires infrastructure and human resources that may be burdensome for DACs that do not have dedicated funding. There may also be important burdens – e.g. reporting or transparency obligations – placed on data users that discourage frivolous use. Research teams releasing data or DACs may have little ability to monitor data users or to directly sanction them for misuse, except by withdrawing or refusing access in the future. Some level of accountability is available via community reporting and

---

[10] OECD, 'Recommendations on Health Data Governance', (OECD), www.oecd.org/els/health-systems/health-data-governance.htm.

[11] Burton et al., 'Policies and Strategies for Secondary Data Use of Data'; Global Alliance for Genomics & Health, 'GA4GH Accountability Policy 2016', (Global Alliance for Genomics & Health, 2016).

norms. Research institutions, funders, journals and databases themselves may have mechanisms to hold researchers accountable for respecting their commitments.[12]

19.3.4 *Maintaining Transparency*

The constitution of DACs shape how policies and governance mechanisms are implemented in practice. DACs are the site around which tensions between the competing interests of stakeholders may play out and therefore, examining how they do or do not maintain transparency allows scrutiny of those governance processes. DAC members may be part of the scientific team that generated the data, though the independence of members is often advocated in order to avoid conflicts of interest. Real or perceived conflicts of interest may arise where the researcher who collected the data restricts access to potential competitors, described as data 'hugging' or hoarding by those advocating data sharing.[13] And yet, data producers have important expertise: they know the affordances and limits of the data as well as its provenance. In some DACs, this expertise is recognised by including members of the study team in an advisory role.[14] Furthermore, all stakeholders should have some representation in governance of data access including as decision-making members of DACs. Stakeholder engagement may also comprise forms of transparency, for example through publication of high-quality plain language summaries to communicate how study data are, or will be, used.

19.4 BEST PRACTICE EXAMPLES

Depending on the organisation or its specific needs, data access governance can emphasise different governance-related values and goals.

19.4.1 *Multi-study Access: European Genome-Phenome Archive (EGA)*

An example of the local access management model is the collection of study DACs under the framework of the EGA. EGA is a database of all types of 'sequence and genotype experiments, including case-control, population, and family studies, hosted at the European Bioinformatics Institute'.[15] According to the EGA website: 'The EGA will serve as a permanent archive that will archive several levels of data including the raw data (which could, for example, be re-analysed in the future by other algorithms) as well as the genotype calls provided by the submitters.'[16] Data submitters via EGA maintain control over the downstream uses of datasets via DACs located in the original study or consortium. An advantage of local data access review is that data generators who are familiar with the dataset can stay involved in the process of review and inform the access review procedure. The disadvantage of this model is that the access control is entirely left to the local committees, making it hard if not impossible to track/audit whether all data access requests are being handled in a timely manner.

---

[12] Global Alliance for Genomics & Health, 'GA4GH Accountability Policy 2016'.
[13] Murtagh et al., 'International Data Sharing'.
[14] Murtagh et al., 'METADAC Governance Infrastructure'.
[15] European Genome-phenome Archive (EGA), 'Introduction', (EGA, 2019), www.ega-archive.org/about/introduction.
[16] Ibid.

19.4.2 Centralised Access: Database of Genotypes and Phenotypes (dbGaP)

In contrast, dbGaP exemplifies a centralised approach to managing data access requests. The dbGaP is designed by the National Institutes of Health (NIH) to archive and distribute the results of studies that have investigated the interaction of genotype and phenotype. Within this database, sixteen DACs 'review requests for consistency with any data use limitations and approve, disapprove or return requests for revision', except for large studies in which a local DAC leads access review.[17] The centralised access model seems advantageous for smaller research groups who lack resources to establish their own data access review infrastructure. However, the handling of data access requests centrally may lead to latency in data access, due to complex administrative arrangements.

19.4.3 Tiered Access: International Cancer Genome Consortium/25K Initiative

The International Cancer Genome Consortium (now called the 25K Initiative) was a large-scale genomics research initiative aiming to generate and share 25,000 whole genome sequences from fifteen jurisdictions to better understand the genetic changes occurring in different forms of cancer.[18] The International Cancer Genome Consortium (ICGC) adopted a tiered access approach, with open access for data unlikely to be linked to other data that could re-identify individual participants, and controlled access for more sensitive data such as raw sequence and genotype files – though the exact data types in these two categories evolved over time.[19] These more sensitive data can only be accessed through the Data Access Committee Office (DACO) to protect the privacy and reasonable expectations of study participants, uphold scientific community norms of attribution and publication priority, and ensure the impartiality of access decisions. The DACO reviews the purpose and relevance of research proposals, and the trustworthiness of applicants to protect participant privacy and data security. The ICGC adopted a plain language access agreement restricting users from establishing parasitic intellectual property on primary data or attempting to re-identify individual participants, with signatures from the principal investigator and institutional signing official. Recognising that requirements for ethics review vary from country to country, the DACO asks applicants to indicate if their study of ICGC data requires local ethics approval.

19.4.4 Independent, Interdisciplinary Access Involving Stakeholder Participation in Decisions: METADAC (Managing Ethical, Socio-Technical and Administrative Issues in Data Access)[20]

METADAC provides data access governance for only the most sensitive data and data combinations (as well as sample access). While separating access in this way produces a complex data governance setting for researchers, the devolvement to different degrees of scrutiny for differently risky data allows resources for human-mediated decision making, where this is necessary and

---

[17] D. Paltoo et al., 'Data Use Under the NIH GWAS Data Sharing Policy and Future Directions', (2014) *Nature Genetics*, 46(9), 934–938, 934.

[18] International Cancer Genome Consortium, 'About Us', (International Cancer Genome Consortium, 2018), www.icgc .org/about-us.

[19] Y. Joly et al., 'Data Sharing in the Post-Genomic World: The Experience of the International Cancer Genome Consortium (ICGC) Data Access Compliance Office (DACO)', (2012) *PLoS Computational Biology*, 8(7), e1002549.

[20] Murtagh et al., 'METADAC Governance Infrastructure'.

allows administrative or algorithm-based decisions for low risk data types. The human-mediated decisions made by METADAC include a proportionate review process for routine-but-sensitive data access applications and full committee decision-making for the remaining sensitive data access applications. The METADAC committee comprises a highly multidisciplinary committee, including study-facing members (currently drawn from the participants of longitudinal studies not regulated by METADAC), with non-voting representation from the studies (including their technical teams) and the funders of these studies. Data access under METADAC does not require additional ethical approval as data sharing is based on tissue bank approval under the Human Tissue Act 2004,[21] study ethical approval and/or explicit participant consent to sharing. METADAC's key criteria for access follow precisely the questions outlined in '*Respect for the provisions of ethical agreements*' above. The METADAC committee does not review the scientific merit of data access applications except in the case of finite resources (i.e. samples).

19.4.5 *Data Producers' Rights and Interests: ClinicalStudyDataRequest*

ClinicalStudyDataRequest.com is a portal facilitating access to patient-level data from clinical studies carried out by pharmaceutical companies and academic researchers.[22] The portal involves independent review of proposals as well as protections for participant privacy and confidentiality. A major differentiator of this access model from the publicly funded genomic research context is protection of commercial interests. For pharmaceutical company-sponsored trials, the data sharing agreement requires users to keep all information provided confidential, in part to protect commercially sensitive information.[23] The user must also agree to give the sponsor an exclusive licence to any new intellectual property generated from the study. The agreement also requires users to publish or otherwise publicly disclose their results, which helps to ensure research is pursued for verification rather than commercial purposes.

19.4.6 *Transparency and Reflexive Governance: UK Biobank*
(Ethics and Governance Framework)

In the late 2000s, in what would be an example of reflexive data access governance,[24] the UK Biobank revised its Ethics and Governance Framework (to address challenges that were current at the time). More specifically, the UK Biobank had originally committed to destroy the data of participants who chose to withdraw from the biobank. However, it soon realised that it could not uphold this commitment due to technical issues.[25] These issues included the establishment of IT systems that made it impossible to destroy data completely in order 'to protect the integrity and security of those people who have taken part'.[26] One year after identifying these issues, the

---

[21] This Act applies to England, Northern Ireland and Wales. With the exception of section 45, which regulates DNA analysis, the Act does not extend to Scotland.

[22] ClinicalStudyDataRequest, 'Home', (ClinicalStudyDataRequest), www.clinicalstudydatarequest.com/Default.aspx.

[23] ClinicalStudyDataRequest, 'Data Sharing Agreement', (ClinicalStudyDataRequest), www.clinicalstudydatarequest.com/Help/Help-Data-Sharing-Agreement.aspx.

[24] G. Laurie, 'Reflexive Governance in Biobanking: On the Value of Policy Led Approaches and the Need to Recognise the Limits of Law', (2011) *Human Genetics*, 130(3), 347–356.

[25] UK Biobank, 'Revision of the UK Biobank Ethics and Governance Framework: 'No Further Use' Withdrawal Option', (UK Biobank, 2007), www.egcukbiobank.org.uk/sites/default/files/Right%20to%20withdraw%20from%20UK%20Biobank.pdf.

[26] UK Biobank, '"NO FURTHER USE" Withdrawal Option – UK Biobank's Commitment to Your Wishes', (UK Biobank, 2011), www.ukbiobank.ac.uk/.

UK Biobank discussed and agreed with its Ethics and Governance Council to amend the scope of its commitment: rather than destroying participant data, the biobank would commit to ensure these data would be made completely unusable. UK Biobank subsequently revised both the participant information materials and governance frameworks not only to reflect this change, but to also describe the underlying reasons. In effect, such transparency and reflexiveness could increase participant trust, and ultimately, participation in biobanks.

19.5 CHALLENGES AND FUTURE DIRECTIONS

19.5.1 *Resources, Effectiveness and Efficiency of Data Access Governance*

Not all research teams or repositories have the guidance, resources or expertise to establish responsible data access governance. Adequate support from funding agencies and institutions is key. This support may include establishing community data repositories to store and manage access on behalf of researchers.

Concerns regarding the workload of DACs in manually reviewing data access requests are the basis for emerging innovations around automation of at least some parts of the data access review.[27] One example of such efforts has been to automate the review of the conformity of the proposed data use with any use restrictions attached to the dataset – e.g. a consent agreement restricting use to non-commercial or disease specific research. In this regard, a recent initiative supported by the Global Alliance for Genomics and Health (GA4GH) developed a matrix for machine-readable consent forms. While these technical approaches will support the work substantially, there will likely always be a need for human review of the most sensitive or disclosive data access requests.

19.5.2 *Coordination between Oversight Bodies*

Oversight of access to biomedical databases would benefit considerably from further coordination between the relevant oversight bodies, such as DACs and RECs.[28] A single data-intensive research project may require access to multiple resources governed by multiple DACs, meaning multiple forms, reviews and delays. Multi-study DACs, such as METADAC, address the problem of repeated and time-consuming access processes. Requirements for multiple approvals from both ethics committees and DACs are dealt with in different ways. In the UK, for example, ethics review under the Human Tissue Act 2004 provides for broad approval for data sharing at the biobank level if relevant consents and other ethical safeguards are in place; permission for specific data access requests then only needs approval from the relevant DAC. Where national legislation is not in place, local or consortia arrangements are possible. The ICGC have disentangled ethics review from data access request review. Indeed, the ICGC's DACO consistently maintains that its DAC is not an ethics review committee and that it should not evaluate the consent forms of users or their

---

[27] S. Dyke et al., 'Registered Access: Authorizing Data Access', (2018) *European Journal of Human Genetics*, 26(12), 1721–1731.

[28] E. Dove et al., 'Ethics Review for International Data-intensive Research', (2016) *Science*, 351(6280), 1399–1400; M. Shabani et al., 'Oversight of Genomic Data Sharing: What Roles for Ethics and Data Access Committees?', (2017) *Biopreservation and Biobanking*, 15(5), 469–474.

research protocols, relying instead 'on the local ethics processes of the data users without imposing another layer of ethics review requirements on them'.[29]

19.5.3 *Harmonisation of Access Policies and Processes*

Interoperability of data access governance supports an important goal of data science, which is to combine similar datasets together to increase statistical power and thereby produce greater scientific insight. Access arrangements are currently fragmented, differing across countries, institutions and databases. These fragmented access arrangements have the potential to undermine usability of databases and produce data silos as users battle to conform to a variety of – sometimes contradictory – access requirements and conditions. Undertaking multiple roughly similar access processes to access different databases is not only burdensome, it also does not necessarily improve participant/data subject protections. Different aspects of access review can be streamlined so that they do not have to be repeated every time a researcher seeks access. Interoperability and predictability can be improved where different data stewards adopt standard access criteria. Central access portals could accept single requests to multiple data resources. This may be possible even where there are differences between the access conditions applying to the datasets. A step further would be to delegate certain aspects of access review. A common authentication body, for example, could be responsible for establishing the identity and affiliation of researchers, who could then present a single set of credentials to different access bodies.[30]

19.6 CONCLUSION

Data access governance has an ultimate goal of taking into account and maintaining balance between the rights and interests of various stakeholders involved in data sharing. A central aim of data access governance, of course, is to promote broad access to data to advance knowledge and improve human health. In doing so, it is essential to have a comprehensive overview of the rights and interests of the involved parties that might be in contrast with each other when establishing rules for data access reviews and approvals.

In view of increasing data sharing among researchers, it is crucial to ensure the DACs and RECs have sufficient resources to achieve the ultimate goals of access review, namely transparency, fairness and proportionality. In doing so, adopting a number of already proposed approaches would be advantageous, including – partly – automating the process of access review and introducing light-touch forms of review when sharing non-sensitive data.

Technological advancements could lead to heightened risks of re-identification of individuals when sharing sensitive health related data. Therefore, it is important to ensure the adopted governance mechanisms include adequate safeguards when sharing data. In addition, in establishing governance mechanisms, attention should be paid to the social values underpinning data sharing. Thereby, the focus of data governance should not be limited to only protecting the individual rights and interests of the involved parties, but also to fostering social values that can arise from promoting responsible data sharing.

---

[29] Joly et al., 'Data Sharing in the Post-Genomic World'.
[30] Global Alliance for Genomics & Health, 'GA4GH Data Use and Researcher ID Work Stream', (GA4GH), https://ga4gh-duri.github.io/.

20

Is the Red Queen Sitting on the Throne?

Current Trends and Future Developments in Human Health Research Regulation

Stuart G. Nicholls

20.1 INTRODUCTION

Human health research is a vast enterprise; worldwide, hundreds of billions of dollars are spent annually on health research involving millions of research participants.[1] This research is guided by multiple regulations and guidance documents that commonly reflect several core principles: the protection of the rights and welfare of individual research participants; the promotion of justice in the practice and outcomes of research; and that human health research should be socially valuable (see van Delden and van der Graff, Chapter 4, in this volume). In addition, oft-cited goals of health research regulation include the development of a culture of ethical concern among researchers and institutions, and the maintenance of public trust in the research enterprise.[2]

However, these generally accepted principles belie an ongoing tension between the protection of individual participants through appropriate regulation, and the facilitation of health research.[3] Authors have, for example, written regarding the amount of waste in research, including inefficient research regulation and management.[4] Others have pointed to the variation in decisions[5] and time taken – with associated costs[6] – of obtaining ethics

[1] L. Shamseer et al., 'Improving the Reporting and Usability of Research Studies', (2013) *Canadian Journal of Anaesthesia*, 60(4), 337–339; M. R. Macleod et al., 'Biomedical Research: Increasing Value, Reducing Waste', (2014) *Lancet*, 383 (9912), 101–104; I. Chalmers and P. Glasziou, 'Avoidable Waste in the Production and Reporting of Research Evidence', (2009) *Lancet*, 374(9683), 86–89; Food and Drug Administration, '2015–2016 Global Participation in Clinical Trials Report', (FDA, 2017); R. A. English et al., *Transforming Clinical Research in the Unites States. Challenges and Opportunities 2010 Workshop Summary* (Washington DC: The National Academies Press, 2010).
[2] H. F. Lynch et al., 'Of Parachutes and Participant Protection: Moving Beyond Quality to Advance Effective Research Ethics Oversight,' (2018) *Journal of Empirical Research on Human Research Ethics*, 14(3), 190–196.
[3] E. Cave and C. Nichols, 'Reforming the Ethical Review System: Balancing the Rights and Interests of Research Participants with the Duty to Facilitate Good Research', (2007) *Clinical Ethics*, 2(2), 74–79; E. S. Dove, *Regulatory Stewardship of Health Research: Navigating Participant Protection and Research Promotion* (Cheltenham: Edward Elgar, 2020).
[4] D. Moher et al., 'Increasing Value and Reducing Waste in Biomedical Research: Who's Listening?', (2016) *Lancet*, 387 (10027), 1573–1586.
[5] E. Angell et al., 'Consistency in Decision Making by Research Ethics Committees: A Controlled Comparison', (2006) *Journal of Medical Ethics*, 32(11), 662–664; E. L. Angell et al., 'Is "Inconsistency" in Research Ethics Committee Decision-Making Really a Problem? An Empirical Investigation and Reflection', (2007) *Clinical Ethics*, 2(2), 92–99; G. Silberman and K. L. Kahn, 'Burdens on Research Imposed by Institutional Review Boards: The State of the Evidence and its Implications for Regulatory Reform,' (2011) *Milbank Quarterly*, 89(4), 599–627.
[6] A. Chakladar et al., 'Paper Use in Research Ethics Applications and Study Contact', (2011) *Clinical Medicine*, 11(1), 44–47; A. G. Barnett et al., 'The High Costs of Getting Ethical and Site-Specific Approvals for Multi-Centre Research',

approval. This criticism has led to increased focus on the efficiency and effectiveness of human research regulation.[7]

In this chapter, I highlight areas that have and, I suggest, will continue to stretch health research regulation, requiring the regulatory infrastructure to adapt and evolve in order to be both effective and efficient. In doing so, I point to changes in risk assessment considerations, underlying trends toward harmonisation and streamlining of research regulation, and alternative approaches to consent. However, I also highlight countertrends that may serve to undermine these changes. Thus, like the red queen in Lewis Carroll's *Through the Looking-Glass*, I propose that the health research regulatory system runs and runs as fast as it can, only to remain in the same place.

20.2 ADAPTATION TO THE ENVIRONMENT: THE CHANGING RESEARCH LANDSCAPE

The evolution of the research regulatory landscape is shaped by new technologies and research approaches. In this next section, I consider several areas that have pushed, and I suggest will continue to push, the boundaries of human research regulation, namely: increasingly diverse and multijurisdictional research; Big Data and artificial intelligence in health research; the learning healthcare system, and; emergency/disaster research.

While multisite studies are by no means novel,[8] the scale of research has exploded, with a proliferation of data repositories, biobanks and other sources of data (see Shabani et al., Chapter 19, this volume). It is now common for research to routinely cross jurisdictional boundaries, with consequent variability in experiences with regulations that exist.[9] This has prompted discussion of ways to facilitate and improve the regulation and oversight of multijurisdictional research,[10] including the development of data sharing structures.[11]

However, the size of data is just one element. Research is now generating new types of data that stretch existing regulations. Social media, smartphones and wearable technology are being used as sources of data. Changes in hardware have been accompanied with rapid developments in analytic methods with a variety of Artificial Intelligence (AI) and other computationally heavy approaches (see Ho, Chapter 28, in this volume). These Big Data approaches have raised

---

(2016) *Research Integrity and Peer Review*, 1(1); M. D. Neuman et al., 'Time to Institutional Review Board Approval with Local versus Central Review in a Multicenter Pragmatic Trial', (2018) *Clinical Trials*, 15(1), 107–111; S. A. Page and J. Nyeboer, 'Improving the Process of Research Ethics Review', (2017) *Research Integrity and Peer Review*, 2(1).

[7] R. Ashcroft et al., 'Reforming Research Ethics Committees', (2005) *BMJ*, 331(7517), 587–588; L. Abbott and C. Grady, 'A Systematic Review of the Empirical Literature Evaluating IRBs: What We Know and What We Still Need to Learn', (2011) *Journal of Empirical Research on Human Research Ethics*, 6(1), 3–19; P. Friesen et al., 'Of Straws, Camels, Research Regulation, and IRBs', (2019) *Therapeutic Innovation & Regulatory Science*, 53(4), 526–534.

[8] C. Grady, 'Institutional Review Boards: Purpose and Challenges', (2015) *Chest*, 148(5), 1148–1155.

[9] Australian Clinical Trials Alliance (ACTA), 'Report on the 2014 National Summit of Investigator-Initiated Clinical Trials Networks', (ACTA, 2014); J. K. Alas et al., 'Regulatory Framework for Conducting Clinical Research in Canada', (2017) *Canadian Journal of Neurological Sciences*, 44(5), 469–474.

[10] E. S. Dove et al., 'Ethics Review for International Data-Intensive Research. Ad Hoc Approaches Mix and Match Existing Components', (2016) *Science*, 351(6280), 1399–1400; Grady, 'Institutional Review Boards'; S. G. Nicholls et al., 'Call for a Pan-Canadian Approach to Ethics Review in Canada', (2018) *Canadian Medical Association Journal*, 190(18), E553–E555.

[11] E. S. Dove et al., 'An Ethics Safe Harbour for International Genomics Research?', (2013) *Genome Medicine*, 5(11); E. S. Dove et al., 'Towards an Ethics Safe Harbor for Global Biomedical Research', (2014) *Journal of Law and the Biosciences*, 1(1), 3–51.

questions about access to data and its use,[12] as well the status of information generated by wearable devices[13] and whether this constitutes health data that would be protected under privacy regulations.

Finally, the contexts of learning activities have raised fundamental questions about how different activities ought to be regulated and what oversight they should be subject to. Two examples of this are the learning healthcare system (LHS), in which research is incorporated into routine clinical practice and research within humanitarian crises,[14] which has pushed regulatory and oversight processes due to the emergent and time-sensitive nature of the research (see Ganguli-Mitra and Hunt, Chapter 32 in this volume). While both of these activities blur research and practice, research conducted within disasters or humanitarian crises raise additional concerns regarding the inclusion of participants who are in a vulnerable position, the changing nature of risks in a time of crisis, and logistical issues in forming ethics review committees and conducting review in this context.[15]

20.3 SUPPORTING SOCIALLY VALUABLE RESEARCH AND THE ROLE OF RISK

In order to adapt to these changing environmental stressors, as well as the noted pressure from researchers to improve the efficiency of regulation and oversight, human health research regulations continue to evolve. Indeed, the vast increases in stored data and biological materials were an explicit driver for revisions to both the 2016 Council for International Organizations of Medical Sciences (CIOMS) International Ethical Guidelines for Health-related Research Involving Humans and the US Code of Federal Regulations, Title 45, Part 46 (45 CFR 46; herein referred to as the Common Rule),[16] while research in humanitarian crises were also an addition to CIOMS 2016, prompted by experiences with Ebola and other humanitarian crises.

One aspect to these developments has been an evolution regarding how risk is assessed and responded to. In part, this is driven by greater attention to the perceived need to balance risk against the potential social value of research with historically under-researched populations, for example, pregnant women, children, or patients with co-morbidities,[17] as discussed in the recent CIOMS revisions. The regulatory response to managing risks generated by research with traditionally under-researched or vulnerable populations – such as with disaster research – while supporting socially valuable research remains a live topic of debate.[18]

---

[12] C. Cath et al., 'Artificial Intelligence and the "Good Society": the USA , EU, and UK approach', (2018) *Science and Engineering Ethics*, 24(2), 505–528; B. Mittelstadt, 'Designing the Health-Related Internet of Things: Ethical Principles and Guidelines', (2017) *Information*, 8(3), 77.

[13] P. P. O'Rourke et al., 'Harmonization and Streamlining of Research Oversight for Pragmatic Clinical Trials', (2015) *Clinical Trials* 12(5), 449–456.

[14] M. Jansse et al., 'Advances in Multi-Agency Disaster Management: Key Elements in Disaster Research', (2009) *Information Systems Frontiers*, 12(1), 1–7; M. Hunt et al. 'The Challenge of Timely, Responsive and Rigorous Ethics Review of Disaster Research: Views of Research Ethics Committee Members', (2016) *PLoS One*, 11(6); C. M. Tansey et al., 'Familiar Ethical Issues Amplified: How Members of Research Ethics Committees Describe Ethical Distinctions between Disaster and Non-Disaster Research', (2017) *BMC Medical Ethics*, 18(1).

[15] S. Mezinska et al., 'Research in Disaster Settings: A Systematic Qualitative Review of Ethical Guidelines', (2016) *BMC Medical Ethics*, 17(1), 62.

[16] E. A. Largent, 'Recently Proposed Changes to Legal and Ethical Guidelines Governing Human Subjects Research', (2016) *Journal of Law and the Biosciences*, 3(1), 206–216; J. J. van Delden and R. van der Graaf, 'Revised CIOMS International Ethical Guidelines for Health-Related Research Involving Humans', (2017) *JAMA*, 317(2), 135–136.

[17] D. Schopper et al., 'Research Ethics Governance in Times of Ebola', (2017) *Public Health Ethics*, 10(1), 49–61.

[18] Hunt et al., 'The Challenge of Timely, Responsive and Rigorous Ethics Review'.

Further, the concept of minimal risk is increasingly raised in the regulatory context; it is a precondition in the US regulations – both Common Rule and forthcoming US Food and Drug Administration (FDA) regulations – with respect to approvals for alteration of consent, and particularly a waiver of consent, as well as featuring in EU Clinical Trials Regulation.[19] This will be particularly relevant for comparative effectiveness research within the LHS and where many interventions being assessed may be argued to be usual care and minimal risk.

Moreover, the LHS and disaster research blur the line between research and practice and several examples now exist where controversy has erupted over whether an ascribed activity would constitute research or not.[20] Indeed, the LHS, disaster research, and secondary research use of data collected for routine clinical care problematise traditional frameworks for governance given the close approximation of activities to both research and practice.[21] The blurring of the traditional distinction between research and care will continue to press regulations regarding the types of research that require regulatory approvals or ethics review, as well as the type and level of oversight they need.

20.4 RESPONDING TO CRITICISM AND SUPPORTING RESEARCH: SIMPLIFYING AND HARMONISING PROCESSES

A feature of many recent regulatory changes is the effort to reduce the burden from the research ethics review process in order to meet the efficiency demands of researchers. Changes in the USA reflect an ongoing trend of streamlining and harmonisation of ethics review structures and processes, which in some countries – such as the UK – have been ongoing over the last two decades.[22] Other regulatory reform, such as the EU Directive 2001/20/EC on clinical trials, have sought to further harmonise approaches between jurisdictions.[23]

These changes attempt to both simplify approaches by exempting certain types of research from the need for ethics review – such as happened with changes to the US Common Rule – and reduce the need for multiple review. Examples of such streamlining include changes in the Common Rule to implement a single Institutional Review Board (sIRB) as the board of record. Indeed, this is a common theme in many initiatives.[24] In Canada, one such as example is the

[19] European Union Clinical Trials Regulation 536/2014 [2014].
[20] N. E. Kass and P. J. Pronovost, 'Quality, Safety, and Institutional Review Boards: Navigating Ethics and Oversight in Applied Health Systems Research', (2011) *American Journal of Medical Quality*, 26(2), 157–159; D. A. Thompson et al., 'Variation in Local Institutional Review Board Evaluations of a Multicenter Patient Safety Study', (2012) *Journal for Healthcare Quality*, 34(4), 33–39; N. E. Kass et al., 'The Research-Treatment Distinction: A Problematic Approach for Determining Which Activities Should Have Ethical Oversight', (2013) *Hastings Center Report*, 43, S4–S15; D. Whicher and others, 'The Views of Quality Improvement Professionals and Comparative Effectiveness Researchers on Ethics, IRBs, and Oversight', (2015) *Journal of Empirical Research on Human Research Ethics*, 10(2), 132–144.
[21] Kass et al., 'The Research-Treatment Distinction'; J. Piasecki and V. Dranseika, 'Research versus Practice: The Dilemmas of Research Ethics in the Era of Learning Health-Care Systems', (2019) *Bioethics*, 33(5), 617–624; J. Piasecki and V. Dranseika, 'Learning to Regulate Learning Healthcare Systems', (2019) *Cambridge Quarterly of Healthcare Ethics*, 28(2), 369–377; Schopper et al., 'Research Ethics Governance'.
[22] A. Hedgecoe et al., 'Research Ethics Committees in Europe: Implementing the Directive, Respecting Diversity', (2006) *Journal of Medical Ethics*, 32(8), 483–486; R. Al-Shahi Salman et al., 'Increasing Value and Reducing Waste in Biomedical Research Regulation and Management', (2014) *Lancet*, 383(9912), 176–185; E. S. Dove, 'Requiring a Single IRB for Cooperative Research in the Revised Common Rule: What Lessons Can Be Learned from the UK and Elsewhere?', (2019) *Journal of Law, Medicine & Ethics*, 47(2), 264–282.
[23] Hedgecoe et al., 'Research Ethics Committees in Europe'.
[24] Nicholls et al., 'Call for a Pan-Canadian Approach to Ethics Review'.

Ontario Cancer Research Ethics Board (OCREB), which serves as the single board of record for multicentre oncology trials for twenty-six of twenty-seven cancer research sites in the province.[25]

A second trend in recent regulatory changes has been to simplify consent requirements. Indeed, it is a longstanding point of discussion that written consent forms do not necessarily facilitate informed decision-making on the part of potential research participants[26] due to their length and complexity of language.[27] Recent changes to the US Common Rule have served to try and improve these consent processes[28] and the FDA has sought to open up the possibility of waivers of consent to align with the Common Rule.[29] Moreover, alternate forms of consent were also part of the recent CIOMS revisions, which provide guidance regarding consent approaches and alternatives to prospective individual written consent.[30] Indeed, some authors have recently argued that in the context of standard-of-care comparative effectiveness trials – such as those envisioned within a LHS – consent could (or should) be waived.[31]

This trend of alternate consent models will, I propose, continue due to developments in disaster and emergency research and the LHS, where the need for timely and expedient consent approaches – or even waivers of consent – will continue the push for the simplification of, or alternative approaches to, consent processes. Meanwhile, Big Data applications and the secondary use of existing data will further press regulations regarding the types of research that require explicit consent from participants and how this is managed.

20.5 COUNTERVAILING TRENDS

Alongside these trends in the regulation of health research, there are emerging examples of regulatory or oversight processes which may run counter to those listed above and may serve to nullify the potential impact of streamlining initiatives or attempts to create a more efficient regulatory environment.

One such example has been the implementation of 'permission to contact' or 'consent to contact' policies. While such policies vary, they represent a broad consent approach to being contacted about future research and are generally asked of all patients upon arrival or intake at hospital, with the goal of expediting recruitment and easing the burden on clinical investigators by allowing researchers to contact potential patient participants. Despite the

---

[25] R. Saginur et al., 'Ontario Cancer Research Ethics Board: Lessons Learned from Developing a Multicenter Regional Institutional Review Board', (2008) *Journal of Clinical Oncology*, 26(9), 1479–1482; R. Saginur et al., 'Ethics Review of Multi-Centre Trials: Where Do We Stand?', (2009) *Health Law Review*, 17(2–3), 59–65.

[26] J. C. Brehaut et al., 'Informed Consent Documentation Necessary but Not Sufficient', (2009) *Contemporary Clinical Trials*, 30(5), 388–389; J. C. Brehaut et al., 'Informed Consent Documents Do Not Encourage Good-Quality Decision Making', (2012) *Journal of Clinical Epidemiology*, 65(7), 708–724; J. C. Brehaut et al., 'Using Decision Aids May Improve Informed Consent for Research', (2010) *Contemporary Clinical Trials*, 31(3), 218–220.

[27] C. Grady, 'Enduring and Emerging Challenges of Informed Consent', (2015) *New England Journal of Medicine*, 372(9), 855–862; C. Grady, 'The Changing Face of Informed Consent', (2017) *New England Journal of Medicine*, 376(9), 856–859.

[28] J. Menikoff et al., 'The Common Rule, Updated,' (2017) *The New England Journal of Medicine*, 376(7), 613–615.

[29] US Department of Health and Human Services, Food and Drug Administration, Office of Good Clinical Practice (OGCP), Center for Drug Evaluation and Research (CDER), Center for Biologics Evaluation and Research (CBER) and Center for Devices and Radiological Health (CDRH), 'IRB Waiver or Alteration of Informed Consent for Clinical Investigations Involving No More Than Minimal Risk to Human Subjects. Guidance for Sponsors, Investigators, and Institutional Review Boards', (2017).

[30] Largent, 'Recently Proposed Changes'.

[31] R. Dal-Ré et al., 'Low Risk Pragmatic Trials Do Not Always Require Participants' Informed Consent', (2019) *BMJ*, 364, 1756–1833.

intuitive appeal, this may, in fact, increase administrative burden and create uncertainty and inefficiencies in research. For example, studies have suggested that patients may customise what aspects of their data they could be approached about or could limit elements, meaning further review of records would be needed.[32] Others have suggested that while uptake to specific research may increase among those giving permission to contact (compared to those recruited through traditional physician-mediated contacts), there may be significant differences in the age and gender of those who agreed compared to those who declined the permission to contact form.[33]

A further consideration is whether a permission to contact agreement constitutes a valid consent. Studies indicate that there may be confusion among patients as to whether they were immediately signing up to research,[34] and this may be especially true when patients are recruited upon arrival to the hospital prior to diagnosis.[35] Consequently, permission to contact approaches not only have the potential to increase administrative burden, thus potentially negating efficiencies created elsewhere, but also raise important questions as to the status of such permission and how it should be considered within the regulatory process.

Other inefficiencies may be introduced through additional regulatory committees.[36] Friesen and colleagues, for example, note a multitude of committees that now commonly exist in the USA in addition to Institutional Review Board (IRB) review, with inconsistency in the policies and processes of many of these additional bodies.[37] A recent study indicated that the requirement to complete multiple local governance reviews had the effect that overall times to the start-up of research were no less than if multiple local IRB review – as opposed to sIRB review – had been used.[38] As such, the introduction of additional governance and oversight structures appear to negate potential benefits generated through streamlining initiatives for research ethics review procedures.

20.6 CONCLUSION: A PARTING CALL FOR THE EVALUATION OF REGULATORY CHANGES

To conclude, we appear to be moving toward an era in which research requires regulation to not only protect participants, but also to be responsive, adaptive and supportive of research; responsive in its ability to facilitate research that can address emerging and emergent topics; adaptive in so far as it is malleable and flexible enough to cope with the ever-changing technologies and data needs of health research; and supportive insofar as the regulatory responses facilitate socially valuable research with no more bureaucracy than is necessary. I have proposed several areas

---

[32] C. Papoulias et al., 'Staff and Service Users' Views on a "Consent for Contact" Research Register Within Psychosis Services: A Qualitative Study', (2014) *BMC Psychiatry*, 14(1), 377.

[33] I. Druce et al., 'Implementation of a Consent for Chart Review and Contact and Its Impact in One Clinical Centre', (2015) *Journal of Medical Ethics*, 41(5), 425–428.

[34] D. Robotham et al., 'Consenting for Contact? Linking Electronic Health Records to a Research Register Within Psychosis Services, a Mixed Method Study', (2015) *BMC Health Services Research*, 15(1), 199.

[35] A. S. Iltis, 'Timing Invitations to Participate in Clinical Research: Preliminary versus Informed Consent', (2005) *Journal of Medicine and Philosophy*, 30(1), 89–106.

[36] O'Rourke et al., 'Harmonization and Streamlining'; Friesen et al., 'Of Straws, Camels, Research Regulation, and IRBs.'

[37] Friesen et al., 'Of Straws, Camels, Research Regulation, and IRBs.'

[38] M. P. Diamond et al., 'The Efficiency of Single Institutional Review Board Review in National Institute of Child Health and Human Development Cooperative Reproductive Medicine Network-Initiated Clinical Trials', (2018) *Clinical Trials*, 16(1), 3–10.

where research regulation will continue to evolve in the near future – streamlined review processes, a focus on the management of risk, and regulatory changes to facilitate and support alternatives to traditional written consent forms.

Yet we also need to be mindful of trends that may run counter to these and which may impede progress. The noted proliferation of additional committees seems to serve only to introduce more variation and regulatory hurdles. Viewed alongside attempts to streamline ethics review, this may be seen as giving with one hand while taking away with the other. Experience in the UK with the rise of research governance following the streamlining of ethics review should serve as a historic warning.[39] Similarly, the introduction of permission to contact mechanisms may, in theory, lead to streamlined research by allowing researchers to directly contact patients – as opposed to mediated contact via physicians – who have opted into research, yet questions remain as to the status of such a consent as well as the practicalities of conducting a system which may well create the need for additional checks that could serve to increase administrative burden rather than decrease it. Based on the above, we appear doomed to repeat the failures of the past and despite running as fast as we can to make changes, we may ultimately stay in the same place.

This leads to my final parting call: the need for good evaluative research of changes to health research regulations and oversight systems. Despite all the discussion regarding changes to the regulatory landscape, this is largely an evidence-free space. Indeed, while the Notice of Proposed Rule Making issued in advance of changes to the Common Rule ran to 131 pages, there was a distinct lack of evidence regarding how the proposed changes addressed purported deficiencies in the oversight of research.[40] Moreover, there is a paucity of evidence regarding whether changes to sRB approaches have had the sought-after effect on review processes. Klitzman, for example, notes that the introduction of policies regarding the use of sIRBs has been done 'in the absence of the systematic collection of data',[41] while Rahimzadeh and colleagues note that, as yet, one cannot point to rigorous evidence that the sIRB model offers great advantages over current practice.[42]

Without empirical evaluation, we will not know whether changes in regulation or oversight are having the desired effect, or whether they increase inefficiencies and convey no benefits to research participants. This potential is borne out by the handful of studies that do exist and which show that, despite a good theoretical basis for an intervention, some may not bring about the desired improvement.[43] Thus, there is a need to develop strong theory-based

---

[39] R. Al-Shahi, 'Research Ethics Committees in the UK – The Pressure Is Now on Research and Development Department', (2005) *Journal of the Royal Society of Medicine*, 98(10), 444–447; R.A-S. Salman et al., 'Research Governance Impediments to Clinical Trials: A Retrospective Survey', (2007) *Journal of the Royal Society of Medicine*, 100(2), 101–104; A. G. H. Thompson and E. F. France, 'One Stop or Full Stop? The Continuing Challenges for Researchers Despite the New Streamlined NHS Research Governance Process', (2010) *BMC Health Services Research*, 10(1), 124.

[40] D. H. Strauss et al., 'Reform of Clinical Research Regulations', (2016) *New England Journal of Medicine*, 374(17), 1693–1694; S. G. Nicholls, 'Revisions to the Common Rule: A Proposal in Search of Evidence', (2017) *Research Ethics*, 13(2), 92–96.

[41] R. Klitzman et al., 'Single IRBs in Multisite Trials. Questions Posed by the New NIH Policy', (2017) *JAMA*, 317(20), 2061–2062.

[42] V. Rahimzadeh et al., 'The sIRB System: A Single Beacon of Progress in the Revised Common Rule?', (2017) *American Journal of Bioethics*, 17(7), 43–46.

[43] M. Dixon-Woods et al., 'Can an Ethics Officer Role Reduce Delays in Research Ethics Approval? A mixed-Method Evaluation of an Improvement Project', (2016) *BMJ Open*, 6(8), e011973; S. Sonne et al., 'Regulatory Support Improves Subsequent IRB Approval Rates in Studies Initially Deemed Not Ready for Review: A CTSA Institution's Experience', (2018) *Journal of Empirical Research on Human Research Ethics*, 13(2), 139–144.

approaches, research designs that bring rigour to the evaluation of regulatory change and a culture change that views evidence-based approaches not with scepticism but with openness.[44] Without the collection of data and open sharing of results, we cannot develop the necessary learning regulatory environment as envisaged and laid out at the outset to this collection.

---

[44] S. G. Nicholls, 'Commentary on "Regulatory Support Improves Subsequent IRB/REC Approval Rates in Studies Initially Deemed Not Ready for Review: A CTSA Institution's Experience"', (2018) *Journal of Empirical Research on Human Research Ethics*, 13(2), 145–147

Regulatory Authorities and Decision-Making in Health Research

The Institutional Dimension

Aisling M. McMahon

21.1 INTRODUCTION

Institutional theories examine the way in which policies and decisions are structurally determined by institutions. 'Institutions' traditionally included state institutions such as the legislature and executive, but can also refer to embedded systems of rules, branches of law, etc. evident within particular organisational contexts. Institutional contexts or systems of rules and practice provide the foundation upon which decision-making within that context takes place. It is also within such institutional contexts that decision-making actors – including decision-makers within regulatory entities such as the Human Tissue Authority, Health Research Authority (HRA), Medicines and Healthcare Products Regulatory Agency and NHS trusts in the United Kingdom – operate. Institutional theories broadly suggest that institutions – and specific institutional contexts within which decision-making actors operate – influence and at times constrain decision-making actors in their decisions. Yet, while there is a body of research on organisational contexts and institutions within sociological studies and political science,[1] the discussion of the effects and influences of institutional contexts on downstream decision-making outcomes and actors is limited within mainstream legal literature. Instead, such questions of institutional effects are often confined to branches of legal theory,[2] and such institutional influences consequently remain largely under-explored in discussions of decision-making within medical law, including the health research regulation (HRR) context.

This chapter seeks to fill this gap. It argues that institutional influences give rise to engrained institutional predispositions within any decision-making context, which can significantly influence decision-making actors and hence the application of, *inter alia*, legal rules and professional guidance in practice. The chapter argues that the effect of institutional frameworks is particularly acute where discretion on the application or scope of a legal provision, guidance or rule is

---

[1] Within the sociological context, see e.g. J. W. Meyer and B. Rowan, 'Institutional Organizations: Formal Structure as Myth and Ceremony', (1977) *American Journal of Sociology*, 83(2), 340–363; F. C. Wezel and A. Saka-Helmhout, 'Antecedents and Consequences of Organizational Change: "Institutionalizing" the Behavioral Theory of the Firm', (2006) *Organization Studies*, 27(2), 265–286; P. DiMaggio and W. Powell, 'The Iron Cage Revisited: Institutional Isomorphism and Collective Rationality in Organizational Fields', (1983) *American Sociological Review*, 48(2) 147–160. Within the political context, see e.g. E. Amenta and K. Ramsey, 'Institutional Theory' in T. Leicht and J. Jenkins (eds), *The Handbook of Politics: State and Civil Society in Global Perspective* (Springer, 2010) pp. 15–39; P. Hall and R. Taylor, 'Political Science and the Three Institutionalisms', (1996) *Political Studies*, 44(5), 936–957.

[2] N. MacCormick, 'Norms, Institutions and Institutional Facts', (1998) *Law and Philosophy*, 17(3) 301–345.

left to the decision-maker. Accordingly, the chapter argues that institutional factors and contexts should be very carefully scrutinised when adopting policy/legal changes, and particularly when drafting new provisions or guidance to be applied in any legal/regulatory context, including in the HRR context. Moreover, it will be argued that to achieve *effective* change within any context – 'effective' defined here as a change that fulfils the outcomes intended – it is not be sufficient merely to consider the text of a new provision to indicate how it is likely to be interpreted in practice. Instead, one must also consider the institutionalised context within which that provision or rule will operate.[3] It is only by considering a provision within its institutionalised setting that one will gain a more holistic picture of how the rule/provision is likely to be developed and applied by that decision-maker. To this one might argue that some rules leave no scope for discretion and hence must be applied in a regimented manner. This may be the case in some contexts; however, no context is static, and none should be viewed as such. This is because social or technological change often requires decision-makers to apply a rule in a situation for which the rule was not designed. Thus, discretion can emerge within contexts over time. Furthermore, the HRR context, by virtue of the constant developments within medicine and science, can be significantly affected by both social and technological change. Thus, the potential for institutional effects are vital to consider in HRR. Moreover, arguably laws/guidance must contain sufficient tolerance – in the sense of providing a space for the expansion of rules to new social/technological contexts – so that they can evolve to meet new circumstances over time.[4]

Furthermore, arguably, the effect of institutional factors cannot necessarily be accommodated or altered by training (of decision-making actors) from a top-down level. Guidance or training may help to move these decision-makers in a particular way in a particular context, but there is no guarantee that this guidance/training will be assimilated within an institutional framework to be used in other contexts. Instead, if the arguments are borne out, those designing and seeking to implement (legal/policy) change must be mindful of both the change suggested and also how this may be assimilated and interpreted within the institutional framework where decision-making bodies – responsible for the interpretation/implementation of such changes – are situated.

In making these arguments, the chapter is structured as follows: Section 21.2 sets out the nature of how institutions and institutional contexts are defined; Section 21.3 then draws on institutional theories to set out, in brief, a template of the main institutional influences in any decision-making framework – dividing these into key constraining and predictive influences. In doing so, it does not aim to provide an exhaustive list of all potential institutional influences in any context; rather, it sets out a template of key influences that are likely to form the core institutional scaffold of any context. Hence, such influences should be carefully considered when adopting decisions/changes in HRR. To demonstrate the practical significance of these arguments, examples of institutional influences within the HRR context are highlighted throughout this section. Section 21.4 concludes arguing that it is vital to take account of institutional contexts and their influence on decision-makers in HRR if we are to achieve the desired outcomes of policy and legal changes.

---

[3] This draws on the author's earlier work: A. McMahon, 'The Morality Provisions in the European Patent System: An Institutional Examination', PhD thesis, University of Edinburgh (2016).

[4] See also discussion in: T. T. Arvind and A. McMahon, 'Responsiveness and the Role of Rights in Medical Law: Lessons from Montgomery', (2020) *Medical Law Review*, 28(3), 445–477. See discussion in Part III (b) and role of responsiveness and functional suitability within law.

21.2 BACKGROUND: DEFINING INSTITUTIONS AND INSTITUTIONAL CONTEXTS

Defining what is meant by the term 'institution' in any given theory can be a complex task[5] because several 'institutions' may be identified in any process, depending on the level of decision-making and influences investigated. As Weinberger argues, institutions are so varied it is 'impossible to set down a unified class of attributes to define all of them'.[6] Indeed, '[t]here is ... no commonly accepted view of what kinds of institutions exist, or what a typology of institutions ought to look like.'[7] Despite this diversity, there is a 'general agreement on a broad conception of institutions as systems of rules that provide frameworks for social action within larger rule-governed settings'.[8] Institutions are seen as incorporating formal procedures or norms within an organisational structure, and can also include informal aspects, such as aspects of culture within broader society or even within formal organisations'.[9] As North states, institutions are:

> ... the rules of the game in society or, more formally, are the humanly devised constraints that shape human interaction. In consequence they structure incentives in human exchange, whether political, social, or economic ... Conceptually, what must be clearly differentiated are the rules from the players. The purpose of the rules is to define the way the game is played. But the objective of the team within that set of rules is to win the game ...[10]

There is disagreement among institutional theorists about whether organisations – including e.g. international organisations, regulatory entities – are institutions.[11] Nonetheless, Hodgson argues that:

> Organizations involve structures or networks, and these cannot function without rules of communication, membership, or sovereignty. The unavoidable existence of rules within organizations means that, even by North's own definition, organizations must be regarded as a type of institution ...[12]

Moreover, for the purposes of this chapter, the question of a distinction between organisations and institutions is not crucial, because although a formal organisation *per se* may not be perceived as a specific institution under a given theory, the crux of these theories is still applicable. This is because, if an institution other than a formal organisation is the main site of investigation within a given theory, it is arguably merely taking a different level of analysis or emphasis as its starting point. For example, the EU could be viewed as the overarching amalgamation of the different institutions (of the type described) which, taken together, form the overarching framework for decision-making, within which a decision-making body such as the European Medicines Agency sits. Similarly, a hospital (clinical) ethics committee will sit within a broader hospital

---

[5] J. Bengoetxea, 'Institutions, Legal Theory and EC Law', (1991) *Archiv fur Rechts-und Sozaphilosophie*, 67(2), 195–213.

[6] O. Weinberger, *Law, Institution and Legal Politics. Fundamental Problems of Legal Theory and Social Philosophy* (Dordrecht: Kluwer Academic Publishers, 1991), p. 155.

[7] Ibid., p. 158.

[8] D. Ruiter, 'A Basic Classification of Legal Institutions', (1997) *Ratio Juris*, 10(4), 357–371, 358 referring to E. Ostrum, 'An Agenda for the Study of Institutions', (1986) *Public Choice*, 48(1) 3–25 and D. Ruiter, 'Economic and Legal Institutionalism: What Can They Learn from Each Other?' (1994) *Constitutional Political Economy*, 5(1), 99–115.

[9] Amenta and Ramsey, 'Institutional Theory', p. 17.

[10] D. North, *Institutions, Institutional Change and Economic Performance* (Cambridge University Press, 1990), pp. 3–5 as cited in G. M. Hodgson, 'What Are Institutions?', (2006) *Journal of Institutional Economics*, 40(1), 1–25, 9.

[11] See Hodgson, 'What Are Institutions?', 8–9.

[12] Ibid., 9.

management system, which then sits under an overarching NHS context (in the UK). Under such conceptions, the primary argument remains the same, i.e. that the overarching organisation/institution – which may have sub-organisations or entities – comprises a framework peculiar to that entity and within which decision-making actors are situated.

Bearing in mind the foregoing definitional points, the chapter argues that such institutional frameworks offer embedded influences on decision-maker(s) within that body who apply/interpret (legal/ethical/professional guidance) provisions. Therefore, the role of institutional contexts should be carefully scrutinised in the development of effective systems of HRR.

21.3 INSTITUTIONAL FACTORS AND DECISION-MAKING: A TEMPLATE OF INFLUENCES

As Immergut states: 'institutions ... act as filters that selectively favour particular interpretations either of the goals toward which political actors strive or of the best means to achieve these ends'.[13] Institutional frameworks provide the scaffold within which decisions are taken; yet, these factors are often ignored or viewed as neutral within the decision-making context. This section challenges this view, arguing that such institutional influences can be highly significant. It argues that two main types of institutional influences can be identified, namely: (1) *prescriptive* (constraining) influences that legally constrain the scope of a decision-maker's actions, e.g. the legal competences of an adjudicative body and (2) *predictive* influences – e.g. political influences on a decision-making body; such influences, although not legally constraining, can be used to predict and/or explain the way in which decision-makers may act, particularly in relation to controversial issues.[14] Under the categories of prescriptive and predictive influences, four main institutional influences can be discerned (described below) and can be applied to any decision-making framework. Such influences, depending on the legal context applicable, can oscillate between being merely predictive to prescriptive. The template highlights two primarily *prescriptive* influences, namely: the central objectives of an institution and the path dependencies (this factor may be either constraining or predictive in nature depending on the context); and two primarily *predictive* influences, namely, the composition, decision-making structure of an institution and the inter-institutional influences (again, this factor may be constraining or predictive in nature).

21.3.1 *Central Objectives of the Decision-Making Body*

A central factor of influence for any decision-maker is the main objective of the overarching institution within which it sits, and, if relevant, the core objective of the legal provision/guidance it is responsible for applying. MacCormick's account of institutions of law highlighted the importance of having a grasp of the function or main point of an institution.[15] He stated: 'an explanation of any institution requires an account of the relevant rules set out in light of its point'.[16]

[13] E. Immergut, 'The Theoretical Core of New Institutionalism', (1998) *Politics and Society*, 26(1), 5–34, 20.

[14] See A. McMahon, 'The Morality Provisions', Chapter 2, for a detailed justification of this categorisation.

[15] N. MacCormick, *Institutions of Law* (Oxford University Press, 2007), p. 36; see also his discussions in: N. MacCormick, *Practical Reason in Law and Morality* (Oxford University Press, 2008); MacCormick, 'Norms, Institutions and Institutional Facts'.

[16] MacCormick, *Institutions of Law*, p. 36.

According to MacCormick, this does not mean that institutions cannot be used for a variety of purposes. However, if used for other purposes, then 'it is the institution that normally functions towards a given broadly-stated end – its 'final cause' – that is so adapted'.[17] As a corollary, arguably, if the institution is not adapted, tensions can arise between the new purpose and how this is carried out, and whether it aligns with the core purpose. If a new purpose moves too far from the core broadly stated aim of the institution, that new purpose unlikely will be achieved without institutional change/adaptation. Put simply, the institutional context is not fit for purpose in such contexts.

Several factors can be used to assess the final cause/objective within an institutional context, including the mission statements, self-descriptions of the overarching institution and the text/articles of founding treaties/legislation that the decision-making body is responsible for applying.

An example of this in HRR could include a scenario whereby the HRA adopted policies that went beyond its initial – albeit broadly defined – remit. These policies may be watered down in practice, as their interpretation will likely be interpreted in a way that aligns with HRA's core purpose. Similarly, if the HRA is perceived by the public/stakeholders as over-stepping its remit in adopting a new policy, the HRA's actions could be challenged/criticised and knowledge of this may also influence policy change within HRA or other bodies, particularly in areas where its remit is not entirely clear. A 'conservative' approach to policy change may be adopted to maintain its 'acceptability'.

Thus, in setting out the remit of a body in HRR, steps should be taken to ensure that its defined remit provides scope to take actions where needed or can be amended if uncertainty arises related to a body's remit. Moreover, because institutional theories suggest that the actions of decision-makers will be influenced and applied in furtherance of their central objectives, when legal change or adaptation is sought, one must consider how such change is likely to align with the overarching aims of the framework within which decision-makers are situated.

21.3.2 *Institutional Structure, Role and Composition of the Decision-Making Body and Overarching Institution*

Second, a key predictive influence is the institutional structure, role and composition of the decision-making body within which that decision-maker sits. This influence draws particularly on March and Olsen's work, who have argued that institutions 'are constitutive rules and practices prescribing appropriate behaviour for specific actors in specific situations'.[18] They emphasise how the internal rules of operation or structural elements within decision-making bodies impact upon decision-making outcomes.

The institutional structure is significant as it facilitates access to, and participation in, the decision-making process. This influences aspects such as the level of external opinion in the decision-making process and the types of actors involved or consulted by decision-making actors, thereby shaping the contours of decisions. Furthermore, the avenues in which decisions are structurally made, and shaped (and by whom), can help predict the types of issues likely to be considered by decision-makers. Key factors include (a) the decision-making structure within

---

[17] MacCormick, *Institutions of Law*, p. 37. See also MacCormick, 'Norms, Institutions and Institutional Facts'.

[18] J. March and J. Olsen, 'Elaborating the "New Institutionalism"' in R. Rhodes et al. (eds), *The Oxford Handbook of Political Institutions* (Oxford University Press, 2006), pp. 3–8, p. 3.

each institution, the composition of decision-makers, the levels of appeal (if any), and (b) the level – and avenues – of consultation, including mechanisms for public/external participation in the decision-making process.

Furthermore, the role and composition of the decision-making actors themselves may prove influential as it feeds into what Stanley Fish[19] termed an 'interpretative community'.[20] This is a community 'working with a shared set of assumptions, understandings, conventions and values that settles issues and problems of interpretation'[21] within the given system. In short, the decision-making community operates with shared understandings – depending on the composition of that decision-making body – and uses such understandings to interpret rules/provisions applicable by them in HRR and other contexts.

Support for this is gleaned by Powell and DiMaggio's work, which draws on the concept of isomorphism: a mimicking or homogenisation arising across organisations in a similar field. The authors conceptualise isomorphism as 'a constraining process ... [which] forces one unit in a population to resemble other units that face the same set of environmental constraints'.[22] One category of isomorphism is normative isomorphism whereby organisations within a field become composed of groups of similar professionals, creating:

> ... a pool of almost interchangeable individuals who occupy similar positions across a range of organisations and possess a similarity of orientation and disposition that may override variations in tradition and control that may otherwise shape organizations.[23]

Accordingly, similar ways of thinking may develop within an institutional context such that decision-making actors act in ways they are familiar with, reinforced or validated by similar groups working in parallel organisations with similar thinking styles. This can lead to similar patterns of action or outcomes across decision-making bodies in the same field – even when faced with different environmental factors such that the resulting outcome is unsuitable to their environment.

Considering the composition of decision-making actors can provide valuable lessons, as in some cases, the greater the similarity of the backgrounds of individuals on a decision-making body, the greater the likelihood for institutional pre-dispositions in favour of certain outcomes. Engrained patterns of thinking may develop and become difficult to shift.[24] Arguably, this provides a strong justification for ensuring, where possible, a multi-disciplinary approach within HRR decision-making contexts. Such an approach is crucial to help ensure conversations do not become engrained within aligned or shared understandings to the extent that broader considerations are missed by decision-makers.

---

[19] S. Fish, *Doing What Comes Naturally: Change, Rhetoric and the Practice of Theory in Literary and Legal Studies* (Durham, NC: Duke University Press, 1989).

[20] There have been criticisms of Fish's usage of this term, including: R. B. Gill, 'The Moral Implications of Interpretive Communities', (1983) *Christianity & Literature*, 33(1), 49–63; R. Scholes, 'Who Cares about the Text?', (1984) *A Forum on Fiction*, 17(2), 171–180; W. A. Davis, 'The Fisher King: Wille zur Macht in Baltimore', (1984) *Critical Inquiry*, 10(4), 668–694.

[21] P. Drahos, 'Biotechnology Patents, Markets and Morality', (1999) *European Intellectual Property Review* 21(9), 441–449, 441–442.

[22] P. DiMaggio and W. Powell, 'The Iron Cage Revisited: Institutional Isomorphism and Collective Rationality in Organizational Fields', (1983) *American Sociological Review*, 48(2), 147–160, 149.

[23] C. Perrow, 'Is Business Really Changing?', (1974) *Organizational Dynamics*, 3(1), 31–44 as cited in DiMaggio and Powell, 'The Iron Cage Revisited', 152.

[24] Also work on 'thought styles' by M. Douglas, *How Institutions Think* (Syracuse University Press, 1986).

This argument is not a criticism of the need for, or role of expertise, within the HRR context or elsewhere. Indeed, expertise is vital. However, arguably, within any regulatory context, if decision-making actors are drawn from similar backgrounds, we must be conscious that this can lead to 'stickiness',[25] or a failure to question traditional lines of thinking and engrained institutional predispositions towards certain decision-making avenues/outcomes.

Two main related challenges arise from the foregoing from an HRR perspective:

1. *Issues of accountability and risk aversion*: As each level of decision-making is accountable to the next level within the overarching institutional structure, this can lead decision-makers to adopt risk averse behaviours with positive and negative effects. On the one hand, within HRR, adopting risk-averse strategies can mean that health research bodies insist on strict compliance with research policies, such as requiring strict adherence with data protection guidance for participants. This can assist in maintaining high standards within health research. However, it can also mean that decision-makers asked to, for example, share a health dataset for research in a context where promotion of research is not explicitly part of that organisation's mission statement, may be reluctant to do so. Such decision-makers may fear reprisal if they allow data use and if there is a lack of explicit institutional policies around how data should be shared to promote health research. The 'safest' option for a decision-maker in such a context in terms of protecting itself from potential challenge/liability, is to maintain the *status quo* and refuse to share datasets. This may be entirely contrary to the public interest but could be seen by some institutions as a safer self-preservation strategy.

2. *Public v Private Duties*: Relatedly, there is an inherent tension within HRR between how to protect individual rights in health research and maintaining the public interest in conducting scientifically sound and ethically robust research which effectively promotes human health. This duty to promote both individual and public interests is enshrined within UK law,[26] but the two goals can be in tension within HRR.[27] Where tension arises, arguably the HRA will lean back on a default of protecting individual interests over public interests because institutional incentives may be stacked in this way, given that: (1) it is more likely for challenges to arise from a failure to protect named/groups of individuals, than a duty to promote the 'public interest', which often has limited teeth in practice and (2) historically, the focus within health research has been to protect individuals to avoid many of the scandals of the past. Hence, there may be engrained institutional preferences for individual interests over broader public interests. This chapter recognises the intentions behind such approaches as laudable, but we must be aware of its shortcoming in terms of promoting public interests.

In short, understanding the role and composition of the decision-making bodies within any institutional context is vital to discerning the interpretative community evident. This allows consideration of normative biases that may develop, how these may affect the provision being interpreted and ways to mitigate this where deemed necessary.

---

[25] See discussion in patent context: S. Thambisetty, 'The Learning Needs of the Patent System: Implications from Institutionalism for Emerging Technologies Like Synthetic Biology', (2013) *LSE Law, Society and Economy Working Papers* 18/2013; see also: Boettke et al., 'Institutional Stickiness and the New Development Economics', (2008) *American Journal of Economics and Sociology*, 67(2), 331–358

[26] Care Act 2014, Section 111(2).

[27] See E. Dove, *Regulatory Stewardship of Health Research: Navigating Participant Protection and Research Promotion* (Cheltenham: Edward Elgar, 2020).

21.3.3 *Path Dependencies and Historical Influences*

Path dependency is generally understood within institutional theories as the influence of historical actions on present acts.[28] At the most general level, it implies that 'what happened at an earlier point in time will affect the possible outcomes of a sequence of events occurring at a later point in time'.[29] It implies that the way a particular issue – or analogous issue – has been dealt with in the past by the institution will be influential, but not necessarily determinative, of present action(s). Within institutional theories, individual decision-makers are seen as influenced not just by past actions of the institution within which they are situated, but also by their own past actions and experiences.

In a legal/regulatory context, this influence can be either prescriptive or predictive in nature. Prescriptive or legally constraining influences can include, for instance, the principle of *stare decisis*, whereby in a common law context 'reliance upon binding precedents leads courts to begin every case with an examination of the past'.[30] If one is looking to how legal changes might be interpreted by courts, an examination of the relevant judicial/quasi-judicial context requires an investigation of how the case law on similar provisions developed in the past, whether any pattern can be discerned, analogies likely to be drawn and mapping a trajectory of influences based on what precedents are likely to bind decision-makers. Similarly, HRR decision-making bodies would likely look to their past findings/decisions for future decisions to ensure consistency in decision-making (e.g. to protect itself from review/appeal).

For example, the HRA's Confidentiality Advisory Group (CAG),[31] in developing its own approach to advising on whether to approve requests for researchers to use patient-level data for 'secondary' health research purposes within the public interest, made public its decisions in this context as precedents for its future work, and to guide future applicants.

Furthermore, past actions of a body can be relevant as suggestive of predictive influences. For instance, the past experiences and criteria for appointment (related to Section 21.3.2) of decision-makers may be instructive to the type of experiences actors in that decision-making body have. Arguably, if decision-makers are unaccustomed to taking decisions on ethical issues, they may be more reluctant to exercise discretion in such areas.[32] Furthermore, past failures in regulatory contexts may affect future decisions as decision-makers strive to ensure this does not happen again. However, as noted, this can have unwarranted effects. For example, the UK's Alder Hey organ retention scandal in the 1990s related to the storing of organs of deceased adults/children without families' knowledge or consent in many cases, and led to the adoption of strict legal reform and extensive guidance on organ donation/retention under the Human Tissue Act 2004. Such change was undoubtedly warranted, but placing issues of transplantation under this same Act arguably conflated the purposes and potentially could be viewed as having led to overly

[28] See O. Hathaway, 'Path Dependence in the Law: The Course and Pattern of Legal Change in a Common Law System', (2001) *Iowa Law Review*, 86(2), 601–665.
[29] W. H. Sewell Jr., 'Three Temporalities: Toward an Eventful Sociology' in T. McDonald, (ed) *The Historic Turn in the Human Sciences* (University of Michigan Press, 1996), pp. 245–280, pp. 262–263.
[30] Ibid.
[31] An advisory group which advises approving bodies such as HRA whether to approve requests to access confidential patient information without patient consent for research purposes. See Health Research Authority 'Confidentiality Advisory Group', (HRA), www.hra.nhs.uk/about-us/committees-and-services/confidentiality-advisory-group/.
[32] For a discussion of marginalisation of ethical issues in patent law in this context see A. McMahon, 'Gene Patents and the Marginalisation of Ethical issues', (2019) *European Intellectual Property Review*, 41(1), 608–620

restrictive policies in some contexts – such as deferring to families in practice even if the deceased consented to donation.[33]

21.3.4 *Inter-institutional Influences*

Finally, inter-institutional influences and relationships/agreements between over-arching institutions and external institutions can be highly significant on the application of rules/provisions. Sociological institutionalism highlights that institutions can affect each other, akin to a form of institutional peer pressure amounting to a potential for diffusion or homogenisation of norms/ policies across institutions operating in the same area.[34] The urge for homogenisation may be driven by types of isomorphism (discussed earlier) at play, namely:

1. *Coercive* isomorphism, whereby coercive forces exert pressure on decision-makers, such as legally or politically mandated requirements from external sources on that institution;[35]
2. *Mimetic* isomorphism, whereby in cases of uncertainty 'when the goals are ambiguous or when the environment creates symbolic uncertainty, organizations may model themselves on other organizations'.[36] This may be particularly acute within the HRR context as both societal/technological changes cause uncertainty for regulators;
3. *Normative* isomorphism whereby similar experiences/backgrounds of decision-makers lead to homogenisation given similarities in thought processes.

Understanding homogenisation across institutions is important to consider for HRR, as homogenisation may not deliver the best interests of those regulated, or public interest(s), objectively defined. Inter-institutional effects can be either constraining or predictive in nature, depending on the hierarchical legal relationships – if any – between the institutions in question. For example, a regulatory body may have legal obligations under EU law or under the European Convention on Human Rights (ECHR) system. Such influences will permeate how that body carries out its decision-making processes, particularly if it fears legal sanction for failing to abide by law. However, if the body perceives itself as having discretion in how it can apply rules to conform with EU/ECHR obligations, or if such obligations have not been strictly monitored in the past, this may limit the effect of the overarching institution's influence.

Relationships between institutions may also be highly persuasive, such as where institutions are not in a hierarchical relationship *per se*, but may seek to align functions for broader political reasons. For example, in the HRR context, section 111(4) of the Care Act 2014 imposes a legal obligation on multiple regulatory authorities to co-operate with each other to achieve a co-ordination and standardisation of UK HRR practices.

21.4 CONCLUSION

Institutional influences act as scaffolds for decision-making, and in so doing, shape and influence outcomes. Yet, these influences are often ignored and overlooked. This chapter

---

[33] See e.g. S. Harmon and A. McMahon, 'Banking (on) the Brain: From Consent to Authorisation and the Transformative Potential of Solidarity', (2014) *Medical Law Review*, 22(4), 572–605.

[34] DiMaggio and Powell, 'The Iron Cage Revisited', citing L. Coser et al., *Books: The Culture and Commerce of Publishing* (New York: Basic Books, 1982).

[35] DiMaggio and Powell, 'The Iron Cage Revisited', 151, referring to Milofsky's work on neighbourhood organisations; C. Milofsky, 'Structure and Process in Community Self-Help Organizations', (1981) *Yale Program on Non Profit Organizations, Working Paper 17*.

[36] DiMaggio and Powell, 'The Iron Cage Revisited', 151.

has argued that institutional influences require much deeper consideration within HRR and in other health contexts. Their effect is particularly acute where decision-makers have discretion on rules/provisions and can lead to engrained pre-dispositions in favour/against certain changes. Moreover, as the health context is one where social/technological change is constant, discretion and/or uncertainty on the application of provisions/rules is always evolving. Hence, we must take more seriously the effect of institutional influences on decision-making. Such influences must be actively considered and accounted for in legal/policy change in HRR and other contexts.

The Once and Future Role of Policy Advice for Health Regulation by Experts and Advisory Committees

Eric M. Meslin

There is nothing a government hates more than to be well informed, for it makes the process of arriving at decisions much more complicated and difficult.

—John Maynard Keynes

22.1 INTRODUCTION

In many countries, principally those with an established research infrastructure and a national commitment to science and technology policy, there is a loose ecosystem of advisory committees, experts, lobbyists and interested groups that are variously used to provide governments with expert advice and input on matters of policy. The government creates some of these structures for this purpose; others are formed unilaterally but at 'arm's length' from government. Some respond to requests for input from government, others volunteer it without being asked, hoping to convince government of the value and relevance of the knowledge offered. Some, as we shall see below, are more public in their work, others are more private. Common to all is that unlike the formal legislative and regulatory apparatuses of government common to civil society, this loose collection of experts and committees functions in the liminal spaces where regulation, guidelines and policies are developed, informed and debated. This helps explain why science advice to governments is more of an 'art' than a science.[1]

This chapter focuses on how governments make use of expertise to inform health regulation, where the expertise comes from sources connected to, but somewhat on the periphery of, the formal processes of policy development through legislation or judicial review. Two examples are drawn upon from direct experience (and are therefore somewhat subjective): (1) the use of expert panels supported by scholarly academies that are organised to provide input to government and (2) advisory committees established by government with a focus on the former US National Bioethics Advisory Commission. Both types play particular roles in the ecosystem of a country's policy advice regime but have different features. There is a both a rich scholarly literature[2] and a

---

[1] P. D. Gluckman, 'Policy: The Art of Science Advice to Government', (2014) *Nature*, 507(7491), 163–165.

[2] A. Fischer et al., 'Expert Involvement in Policy Development: A Systematic Review of Current Practice', (2014) *Science and Public Policy*, 41(3), 332–343; H. Douglas, *Science, Policy and the Value-Free Ideal* (University of Pittsburgh Press, 2009); S. Jasanoff, *Science and Public Reason* (New York: Routledge, 2012); R. Pielke, *The Honest Broker: Making Sense of Science in Policy and Politics* (Cambridge University Press, 2007).

grey literature on other structures and examples.[3] The main emphasis is that expert advice in its many iterations forms part of the regulatory apparatus that governments do make use of in developing regulation and other policy. However, these are often underappreciated and therefore difficult to assess with respect to impact.

22.2 THE ROLE OF EVIDENCE AND EVIDENCE GATHERING TO INFORM POLICY

In his first inauguration on 20 January 2009, President Barack Obama announced, 'We will restore science to its rightful place and wield technology's wonders to raise healthcare's quality and lower its costs'.[4] It was a statement as much about his predecessor George W. Bush's lack of support for science and evidence in decision-making, as it was about the future of the Republic. Obama's announcement was, one might say, a liminal proposition: he was looking back on almost a decade's worth of policy decisions, including restricting stem cell science, delaying the appointment of the FDA commissioner and the White science advisor, and outright ignoring science advice when making decisions about women's health led some critics to refer to Bush as 'science's worst-ever enemy'.[5] With that in the rear-view mirror, it was easy for Obama to look to a future of promise and hope. Indeed, on 21 June 2016, the White House released '100 examples of Obama's Leadership in Science, Technology, and Innovation', the first of which was that Obama 'elevated the quality and rigor of the science, technology, and innovation advice in the White House'.[6]

A similar anti-science assertion was made about Canada's prime minister, Stephen Harper, who was admonished for 'muzzling scientists' and reducing research funding during his decade in office.[7] In a manner reminiscent of Obama, and shortly after he was elected in 2015, Canadian Prime Minister Justin Trudeau asserted, 'We are a government that believes in science – and a government that believes that good scientific knowledge should inform decision-making'. Trudeau took a number of actions, including appointing a Minister of Science, prioritising the appointment of a Chief Science Advisor, and filled his first federal budget[8] with sixteen references to 'evidence', 'evidence-based decision-making', and textual support such as: the government 'understands the central role of science in a thriving, clean economy and in providing evidence for sound policy decisions'.[9] Subsequent federal budgets have made similar references to the value of evidence to inform decisions.

The USA and Canada examples were hardly unique. Indeed, it became *de rigour* for governments to embrace the value of using evidence in policy development, often referring to

[3] *Future directions for scientific advice in Whitehall*, R. Doubleday and J. Wilsdon (eds) (Cambridge: Centre for Science and Policy, 2013).
[4] The White House, 'Inaugural Address by President Barack Hussein Obama,' (*The White House President Barack Obama*), www.obamawhitehouse.archives.gov/realitycheck/the_press_office/President_Barack_Obamas_Inaugural_Address.
[5] A. McCook, 'Sizing up Bush on Science', *The Scientist* (30 September 2006).
[6] The White House Office of the Press Secretary, 'IMPACT REPORT: 100 Examples of President Obama's Leadership in Science, Technology, and Innovation', (The White House President Barack Obama, 21 June 2016) www.obamawhitehouse.archives.gov/the-press-office/2016/06/21/impact-report-100-examples-president-obamas-leadership-science.
[7] S. Zhang, 'Looking Back at Canada's Political Fight Over Science', *The Atlantic* (26 January 2017).
[8] Canada. Budget 2016. Growing the Middle Class. Ottawa (Government of Canada) 2016.
[9] Ibid.

'evidence-based' or more accurately 'evidence-informed' policy as a goal. The UK,[10] Australia[11] and New Zealand[12] are three of the most visible, and who have been emphasising the value of evidence to inform policy for decades, though the support waxes and wanes depending on who is in power. Moreover, the call to use evidence is hardly new, particularly in medicine and healthcare.[13] The justification is that data provide an objective foundation on which to develop policy, avoiding perceptions of bias, ideology or subjectivity. This approach is satisfying at many levels, though there has always been, and more recently an apparent increase, in public scepticism about the role of experts and expertise.[14]

The assertion of the value of facts alone may also unwittingly camouflage two other values of equal import: first, the value of recognising the epistemic foundation for beliefs about facts. As my colleague Alessandro Blasimme and I argue elsewhere, science has become especially challenging for policy-making, precisely because liberal democracies lack a coherent way to accommodate pluralistic views about scientific innovation.[15] This is consistent with the view that using evidence is very different in practice than in theory. As Ian Boyd, Chief Scientific Advisor, UK Department of Environment, Food and Rural Affairs once suggested:

> People don't even think *about data in* the same way: When I think of data I think of binary or hexadecimal numbers. This betrays something of my background, but it was a surprise to me when in Defra, the UK Department of State with responsibility for food and the environment, we started to talk about data and I found that other people saw data very differently. Everybody had different preconceptions about data. Some seemed to be very confused. It had become trendy to talk about data, but few people appeared to think about data.[16]

The second type of camouflage concerns the role that ethical values play, since it should be unremarkable to claim that the test of good public policy is the degree to which it is supported by good evidence, good ethics and good epistemology. Therefore, while asking for evidence may seem sensible and pragmatic, what may be delivered depends on many factors. For instance, governments might think they want *evidence* given its popularity in public discourse and as found in a typical hierarchy of evidence (systematic reviews, randomised clinical trials, etc.), when what they may *need* is something quite different. In some instances specific answers to specific questions,[17] but equally, other types of assistance including problem framing, support for a position they are intending to adopt – or oppose – or being aware of best practices by other jurisdictions. They may also ask for advice because of the perception that seeking input from elsewhere shows a degree of transparency, or provides assurance to constituents that a fair process is being undertaken to consider relevant information before a law is passed or a

[10] UK House of Commons Science and Technology Committee, 'Scientific Advice and Evidence in Emergencies: Third Report of Session 2010–11', (House of Commons, 2011). An example of a review of the use of science in government.

[11] G. Banks, 'Evidence-Based Policy Making: What Is It? How Do We Get It?', *ANU Public Lecture Series*, 4 February 2006, (Productivity Commission, Canberra).

[12] P. D. Gluckman, 'The Role of Evidence in Policy Formation and Implementation: A Report by the Prime Minister's Chief Science Advisor', (Office of the Prime Minister's Science Advisory Committee, September 2013).

[13] M. J. M. Gray, *Evidence-Based Health Care* (London: Churchill Livingstone, 1996).

[14] N. Harrison and K. Luckett, 'Experts, Knowledge and Criticality in the Age of "Alternative Facts": Re-Examining the Contribution of Higher Education', (2019) *Teaching in Higher Education*, 24(3), 259–271.

[15] E. M. Meslin and A. Blasimme, 'Towards a Theory of Science Policy for Genetics', (2013) *European Journal of Human Genetics*, 21 (Suppl 2), 360.

[16] I. Boyd, 'The Stuff and Nonsense of Open Data in Government', (2017) *Scientific Data*, 4, 170131.

[17] M. Petticrew and H. Roberts, 'Evidence, Hierarchies, and Typologies: Horses for Courses', (2003) *Journal of Epidemiol Community Health*, 57(7), 527–529.

regulation implemented (or rescinded). This accounts for the range of instruments that governments may use, reflecting an epistemic hierarchy ranging from anecdotes, cases and stories to
more organised collections of data and information, to something approaching comprehensive
knowledge. I turn now to two examples of the use of experts to advise government.

22.3 LEARNED AND ACADEMIC SOCIETIES AS EXPERTS FOR GOVERNMENT

Collections of scholars and academic experts have a long and distinguished history. Among the
oldest of these academic societies in Europe are the *Compangie du Gai Sçavoir*, founded in
1323 by seven wealthy patrons in Toulouse whose purpose was to promote the poetry of the
Occitan language; the *Academia Platonica* – also known as the Neoplatonic Florentine
Academy – in 1462–1522; and the Barber Surgeons of Edinburgh established in 1505. In their
early years, these organisations functioned as private discussion groups for their own edification
and enjoyment. Yet as my colleague Summer Johnson and I describe elsewhere,[18] it was not
until the seventeenth century when academies of science and medicine were sought by
governments for assistance in establishing public policy. Three of the most prominent were
Britain's Royal Society, established by Royal Charter in 1662; Germany's Leopoldina established
in 1652; and France's Royal Academy of Sciences, the latter beautifully depicted in Henri
Testelin's painting of Jean-Baptiste Colbert presenting the Royal Academy to King Louis XIV
at Versailles in 1667 (see Figure 22.1).

In the intervening years, and particularly in the nineteenth and twentieth centuries, learned
societies emerged across the disciplinary spectrum. The American Council of Learned
Societies, founded in 1919, lists seventy-five national or international 'member societies' in the
humanities and related social sciences.[19] The UK learned societies Wikipedia page lists more
than 230 organisations,[20] Canada's Federation for the Humanities and Social Sciences lists more
than 160 universities, colleges, and scholarly associations[21] and France counts at least thirty-six
separate organisations. As organisations that honour excellence by recognising their country's
distinguished intellectuals and practitioners – usually with the title 'Fellow' – academies
constitute a significant brain trust for any government to draw on. Increasingly, they are called
upon to contribute scholarship, testify before legislatures, and offer their expert input. It is
becoming common in the health research environment to seek out this type of expertise.[22]

In addition to the individual activities of academies in their respective countries, there are
other arrangements of these groups.[23] The first are collections of academies that are regional or
global in scope, including: the InterAcademy Partnership, the Network of African Science
Academies, Association of Academies and Societies of Sciences in Asia, InterAmerican
Network of Academies of Sciences, the European Federation of Academies of Sciences and

---

[18] E. M. Meslin and S. Johnson, 'National Bioethics Commissions and Research Ethics' in. E. J. Emanuel et al. (eds),
      The Oxford Textbook of Clinical Research Ethics (New York: Oxford University Press, 2008), pp. 187–197.
[19] 'Member Societies', (American Council of Learned Societies), www.acls.org/Member-Societies/Society-Profiles.aspx.
[20] 'Category: Learned Societies of the United Kingdom', (Wikipedia), www.en.wikipedia.org/wiki/Category:Learned_
      societies_of_the_United_Kingdom.
[21] Federation for the Humanities and Social Sciences, 'The Federation Membership: Our Community', (Federation for
      Humanities and Social Sciences), www.ideas-idees.ca/sites/default/files/sites/default/uploads/membership/member
      ship_lists_2019_web_eng.pdf.
[22] A. S. Haynes et al., 'Identifying Trustworthy Experts: How Do Policymakers Find and Assess Public Health
      Researchers Worth Consulting or Collaborating With?', (2012) *PLoS ONE*, 7(3), e32665.
[23] E. M. Meslin and C. Stachulak, 'Organizations of National Academies – A Comparison', (2017) [unpublished internal
      review undertaken at the Council of Canadian Academies, available upon request].

FIGURE 22.1. Colbert presenting the Royal Academy to King Louis XIV at Versailles in 1667 – Henri Testelin.

Reprinted with permission from the Réunion des Musées Nationaux Grand Palais

Humanities, European Academies Science Advisory Council, Council of Academies of Applied Sciences, Technology and Engineering and the Federation of European Academies of Medicine.

The second is the unique subgrouping of national academies organised within a country to provide specific expert input to governments. Only seven countries have such organisations of organisations: Australia, Belgium, Canada, Finland, Germany, Switzerland and the USA. I offer some observations about the American and Canadian versions.

The US National Academies of Sciences, Engineering, and Medicine (NASEM) is perhaps the most well-known and longest serving, with the National Academy of Sciences established in 1863, the National Academy of Engineering established in 1964 and the National Academy of Medicine – formerly Institute of Medicine – established in 1970. NASEM receives approximately US $200M annually from US Government contracts and US $100M from private or non-federal contracts,[24] undertaking about 200 different assessments, reports and projects at any given time. Numerous NASEM reports have been used to support health research regulation including early studies on primate research[25] and human subjects research regulations.[26]

Modelled after NASEM, the Government of Canada responded to a proposal developed by Canada's three main academies – the Royal Society of Canada, the Canadian Academy of Engineering, and the Canadian Academy of Health Sciences – to fund the creation of the Council of Canadian Academies (CCA) in 2005 with a mission to undertake independent

---

[24] The National Academies of Sciences, Engineering, Medicine, 'Report of the Treasurer of the National Academy of Sciences for the Year Ended December 31, 2015', (The National Academies of Sciences, Engineering, Medicine, 2016), 6.

[25] National Research Council (US) Committee on Well-Being of Nonhuman Primates, *The Psychological Well-Being of Nonhuman Primates* (Washington, DC: The National Academies Press, 1998).

[26] National Research Council, *Proposed Revisions to the Common Rule for the Protection of Human Subjects in the Behavioral and Social Sciences* (Washington, DC: The National Academies Press, 2014).

assessments of evidence to provide government decision-makers, researchers and stakeholders with high-quality information required to develop informed and innovative public policy (https://cca-reports.ca/). Each of the three Canadian academies had well deserved reputations as a result their distinguished fellowship and missions. By early 2020, the CCA had completed more than fifty assessments on diverse topics in health, environment, science and technology, energy and public safety ranging. Like the US National Academies, the CCA's assessments fill an evidence gap to support policy decision-making, including topics where regulation and legislation is ripe for review or update. The CCA's 2019 reports on *Medical Assistance in Dying* and *When Antibiotics Fail* provide concrete examples of the gap-filling expertise governments welcome in health policy. Its 2020 release of *Somatic Gene and Engineered Cell Therapies* report will inform several regulatory needs: health research, innovation and disruptive technology. Importantly, the CCA expert panels take no normative positions on the subjects they assess. Rather, the CCA undertakes assessments that answer different descriptive questions, including: *What is the state of knowledge of …? What is the socio-economic impact of …? What are Canada's strengths in …? What are the best practices that exist for …?*

The work is less about practical advice-giving to government, and more about *evaluating available evidence.* It is challenging to assess the impact of this type of work, especially on specific issues arising in health research regulation. Rarely does an evidence-focused document lead directly to a new regulation or to revision or reform of an existing one, yet there is impact. Canada's Ministry of Innovation, Science and Economic Development undertook a comprehensive evaluation and audit of the CCA in 2018, using the NASEM, the UK Royal Society, the Australian Academy of Learned Societies and Germany's Leopoldina as international comparators, and found that the CCA 'addresses a need for independent, objective, and transparent scientific knowledge to support evidence-based decision-making' and that 'the demand for CCA assessments will continue to grow given the federal government's priority for credible scientific knowledge to support evidence-based decision-making'.[27] The evaluation lists several assessments by name that have supported federal and provincial government, industry and stakeholders. Positive though these might be, such metrics can be misleading since assessment work – and evidence generally – cannot always be tracked directly to a policy outcome, which the CCA evaluators noted: 'that there is a challenge in measuring this type of impact given that the CCA does not formulate recommendations or policy advice that could be tracked and attributed directly to its assessments'.[28] This same claim can be applied to another use of experts to inform government policy development: government-based advisory committees.

22.4 THE ROLE OF BIOETHICS ADVISORY COMMITTEES

As early as the eighteenth century, specialised committees were established to report on particular topics for governments. One such panel chaired by Benjamin Franklin was convened to investigate claims made by Anton Mesmer about the healing power of animal magnetism.[29]

[27] Innovation, Science and Economic Development (ISED), 'Evaluation of the Council of Canadian Academies', (Innovation, Science and Economic Development Canada, 16 March 2018).
[28] Ibid.
[29] B. Franklin, *Animal Magnetism: Report of Dr Franklin and Other Commissioners, Charged by the King of France with the Examination of the Animal Magnetism as Practised at Paris* (London: J. Johnson, 1785).

(The claims were rejected.)[30] Today, thousands of committees, working groups, royal commissions and advisory structures have been established by governments, non-governmental organisations, philanthropic bodies and industry. Until recently, in the USA alone, there were more than 1000 federal advisory committees authorised through the Federal Advisory Committee Act – until President Trump signed an Executive Order, on 14 June 2019, requiring at least one-third of them be terminated.

Unlike the CCA, these advisory bodies are intended to *advise*, that is, to make recommendations. One sub-category of these groups is the bioethics advisory bodies that have become a regular contributor to domestic and international debate about bioethics issues, and health research in particular. The WHO maintains a database of these groups, which currently number more than 110 around the world. Among the more influential are the standing committees such as the Nuffield Council on Bioethics and France's National Consultative Committee on Ethics, while others are ad hoc groups such as WHO's Expert Advisory Committee on Developing Global Standards for Governance and Oversight of Human Genome Editing. I was fortunate to have a front row seat working for President Bill Clinton's National Bioethics Advisory Commission (NBAC) in the USA, an ad hoc advisory committee functioning between 1996 and 2001 with a focus on health research and genomics. NBAC was one in a series of such USA commissions, each of which played a key role in informing health research regulation.[31] As luck, and the advances of science, would have it, NBAC found itself occupying some of the most intriguing liminal spaces in health science regulation in a generation. Two are highlighted here.

Following the announcement of the birth of the cloned sheep Dolly, NBAC was asked by President Clinton on 24 February 1997 to 'undertake a thorough review of the legal and ethical issues associated with the use of this technology, and report back to be me within 90 days with recommendations on possible federal actions to prevent its abuse'.[32] One hundred and three days later, on 7 June 1998, NBAC delivered its report concluding, 'at this time it is morally unacceptable for anyone in the public or private sector, whether in research or clinical setting, to attempt to create a child using somatic cell nuclear transfer cloning'.[33] NBAC made six recommendations for public action, many of which Clinton accepted, including maintaining the moratorium on the use of federal funds for attempts to create a child. Perhaps the most significant impact of this work was the international conversation that began in earnest in other countries, especially the UK, Canada and Australia, and in international organisations including UNESCO, CIOMS, the World Medical Assembly and the International Convention on Harmonisation. Dolly's birth moved the debate about reproductive cloning from one about a possible future technology to one in which it was now plausible to conceive of the possibility that a variety of public and professional actors would seek ways to make use of this technology.

Less than a year later, NBAC would take up a second controversial topic: embryonic stem cell research, following from the joint scientific announcements in November 1998 that human

---

[30] K. McConkey and C. Perry, 'Franklin and Mesmerism Revisited', (2002) *International Journal of Clinical and Experiment Hypnosis*, 50(4), 320–331.

[31] H. T. Shapiro and E. M. Meslin, 'Relating to History: The Influence of the National Commission and Its Belmont Report on the National Bioethics Advisory Commission' in J. F. Childress et al. (eds), *Belmont Revisited: Ethical Principles for Research with Human Subjects*. (Washington, DC: Georgetown University Press, 2005), pp. 55–76.

[32] National Bioethics Advisory Commission, 'Cloning Human Beings: Report and Recommendations of the National Bioethics Advisory Commission,' (*National Bioethics Advisory Commission*, Rockville, MD, 1997).

[33] Ibid.

embryonic stem cells and germ cells had been cultured and derived for the first time. As with Dolly, Clinton came to NBAC to request advice, calling on the commission to 'consider the implications of such research at your meeting next week, and to report back to me as soon as possible'.[34] Yet unlike the cloning report, the Clinton White House had a different reaction to NBAC's work on stem cell science – rejecting the commission's recommendations before they were formally submitted, the context for which I have described elsewhere.[35]

NBAC's experience is not unique in the world. Bioethics-by-commission is an area of scholarly study joining the emerging literature on the use of expert commissions to advise government. Some of this literature reminds us of the risks of relying on experts only, without appealing to the public.[36] Two examples from health research are illustrative: the influential role of patient advocates in the early debates around HIV prevention and treatment trials[37] and a similar story in breast cancer research.[38] In both cases, research regulations were amended to account for patient perspectives, and helped launch a patient engagement movement in health research that thrives today.

22.5 THE FUTURE OF ADVISING ON HEALTH RESEARCH REGULATION IN A LIMINAL WORLD

It is simplistic to conceive of regulations – or policy generally – as linear: that their trajectory is arrow-straight beginning with evidence and concluding with a shiny new regulation. Policy issues emerge for governments in often-unpredictable ways, requiring different types of responses. One reason is that science and society move in unpredictable ways, often responding to a recent study or an emergent problem.[39] For example, gene therapy research was proceeding slowly but cautiously in the 1990s until the death of Jesse Gelsinger set back research for decades.[40] But another reason, alluded to above, relates the democratisation of science and its place in the public sphere with seemingly opposing consequences. Public confidence and trust in science seem to be on the rise,[41] and yet public and social media are filled with anti-science, unfounded hyped-filled assertions about medicine and health.[42]

In the midst of these developments, a new expert is also emerging: the federal science advisor. These experts are 'of' government, and therefore play a different role and are exposed to different challenges. Only five years have passed since Dr Anne Glover was removed as Chief Science

[34] National Bioethics Advisory Commission, 'Ethical Issues in Human Stem Cell Research'. (*National Bioethics Advisory Commission*, Rockville, MD, 1999).

[35] E. M. Meslin and H. T. Shapiro, 'Bioethics Inside the Beltway: Some Initial Reflections on NBAC', (2002) *Kennedy Institute of Ethics Journal*, 12(1), 95–102.

[36] See for example, M. Leinhos, 'The US National Bioethics Advisory Commission as a Boundary Organization', (2005) *Science and Public Policy*, 32(6), 423–433.

[37] M. Manganiello and M. Anderson, 'Back to Basics – HIV/AIDS Advocacy as a Model for Catalyzing Change', www .meaction.net/wp-content/uploads/2015/05/Back2Basics_HIV_AIDSAdvocacy.pdf.

[38] J. Perlmutter et al., 'Cancer Research Advocacy: Past, Present, and Future', (2013) *Cancer Research*, 73(15), 4611–4615.

[39] E. M. Meslin, 'When Policy Analysis Is Carried Out in Public: Some Lessons for Bioethics from NBAC's Experience' in James Humber and Robert Almeder (eds), *The Nature and Prospect of Bioethics: Interdisciplinary Perspectives* (Totowa, NJ : Humana Press, 2003), pp. 87–111.

[40] L. Walters, 'Gene Therapy: Overview' in T. Murray and M. J. Mehlman (eds), *Encyclopedia of Ethical, Legal and Policy Issues in Biotechnology* (New York: Wiley, 2000), pp. 336–342.

[41] See for example: Council of Canadian Academies, 'Science Culture: Where Canada Stands. Expert Panel on Canada's Science Culture', (Council of Canadian Academies, 2014); C. Funk et al., 'Trust and Mistrust in Americans' Views of Scientific Experts', (Pew Research Center, 2 August 2019).

[42] T. A. Caulfield, *Is Gwyneth Paltrow Wrong about Everything?* (Toronto, Canada: Penguin Random House, 2016).

Advisor to the European Union in October 2014.[43] Predictably – and reassuringly – the reaction from the scientific community opposing the decision was swift.[44]

Appointing and dismissing chief scientists and science advisors is itself a political act for governments and one can read both too much and too little into these decisions. It took months for George W. Bush to appoint his science advisor, and even longer for Donald Trump to appoint his. On the other hand, Justin Trudeau made appointing his Chief Science Advisor a key commitment of his Minister of Science's first mandate.[45] New administrations have the right to appoint or dismiss any un-elected position. While the worst thing one can say about appointing a science advisor, or advisory commission, is that these are optically useful moves but unlikely to improve the quality of (health) regulations, the more ominous spin about decisions to remove, not appoint – or worse – to staff them with anti-science personalities, is that they cast an odious shadow on all policy advice that emerges from their office. Examples seem to abound in the USA in the Trump era, including unqualified 'industry-captured' scientists nominated to the Environmental Protection Agency's Science Advisory Board,[46] or the muzzling of a government scientist's report on climate change by the White House.[47] A more relevant health research example has been the ping-pong policy on foetal tissue and embryo research in the USA, which has been vacillating between permissive and restrictive depending on the political philosophy of the White House and the majority party in the US Congress.[48]

Health research regulation is fraught with ethical, social and cultural challenges, particularly where the object of regulation involves fundamental matters of human health, against a backdrop of medical experimentation. Not surprisingly, legislation often takes time to craft wisely, and the regulations that follow may take even longer. A proposed revision to the main health research regulations in the USA, called the Common Rule, was first developed in 1981, revised in 1991, again in 2017, but not fully implemented until 2019 – despite substantive input from advisory commissions and expert panels, professional societies and the general public.

As the examples suggest, health research regulation is not undertaken by a single policy instrument, a single mechanism of reform or informed by single discipline or set of inputs. Health research regulation by its nature involves evidence, but also values, technical expertise, and stakeholder contributions. These are the liminal spaces in which developing, crafting and implementing these regulations exist. Fifteen years ago, President Clinton's science advisor Neal Lane recognised what was needed:

> The successful application of new knowledge and breakthrough technologies . . . will require an entirely new interdisciplinary approach to policy-making that operates in an agile problem-solving environment works effectively at the interface of science, technology business and policy, is rooted in improved understanding of people, organizations, cultures, and nations, engages the

[43] D. Butler, 'European Commission Scraps Chief Scientific Adviser Post,' (*Nature News*, 13 November 2014).

[44] Science Media Center, 'Expert Reaction to News About Abolition of Post of CSA to European Commission,' (*Science Media Center*, 13 November 2014).

[45] B. Owens, 'Canada Names New Chief Science Adviser', *Science* (26 September 2017).

[46] M. Halpern M. 'The EPA Science Advisory Board Is Being Compromised. Here's Why That Matters', (Union of Concerned Scientists, 30 October 2017).

[47] M. Bryant, 'White House "Undercutting Evidence" of Climate Crisis, Says Analyst Who Resigned', *The Guardian*, (30 July 2019).

[48] D. Wertz, 'Embryo and Stem Cell Research in the United States: History and Politics', (2002) *Gene Therapy*, 9 (11), 674–678.

nation's top social scientists, including policy experts, to work in collaboration with scientists and engineers from many fields.[49]

Lane's foresight was prescient. Health research is brimming with inter- and multidisciplinary approaches, which has led to commensurate commitment to interdisciplinary governance emphasising scientific integrity.[50] It is also evident in the encouraging commitment of young people to the future of the planet, reflected in their active engagement in climate issues,[51] and efforts in citizen science,[52] and science diplomacy.[53] The future of health research regulation will be in good hands if society is open to advice from the expertise of experts and non-experts alike.

[49] N. Lane, 'Alarm Bells Should Help Us Refocus', (2006) Science, 312(5782), 1847.

[50] A. Kretser et al., 'Scientific Integrity Principles and Best Practices: Recommendations from a Scientific Integrity Consortium', (2019) Science and Engineering Ethics, 25(2), 327–355.

[51] S. Dickson-Hoyle et al., 'Towards Meaningful Youth Participation in Science-Policy Processes: A Case Study of the Youth in Landscapes Initiative', (2018) Elementa Science of Anthropocene, 6(1).

[52] A. Irwin, 'Citizen Science Comes of Age', (2018) Nature, 562, 480–482.

[53] D. Copeland, 'Science and Diplomacy after Canada's Lost Decade: Counting the Costs, Looking Beyond', (Canadian Global Affairs Institute, 2015).

PART II

Reimagining Health Research Regulation

Private and Public Dimensions of Health Research Regulation

Introduction

Graeme Laurie

It is a common trope in discussions of human health research, particularly as to its appropriate regulation, to frame the analysis in terms of the private and public interests that are at stake. Too often, in our view, these interests are presented as being in tension with each other, sometimes irreconcilably so. In this section, the authors grapple with this (false) dichotomy, both by providing deeper insights into the nature and range of the interests in play, as well as by inviting us to rethink attendant regulatory responses and responsibilities. This is the common theme that unites the contributions.

The section opens with the chapter from Postan (Chapter 23) on the question of the return of individually relevant research findings to health research participants. Here an argument is made – adopting a narrative identity perspective – that greater attention should be paid to the informational interests of participants, beyond the possibility that findings might be of clinical utility. Set against the ever-changing nature of the researcher–participant relationship, Postan posits that there are good reasons to recognise these private identity interests, and, as a consequence, to reimagine the researcher as *interpretative partner* of research findings. At the same time, the implications of all of this for the wider research enterprise are recognised, not only in resource terms but also with respect to striking a defensible balance of responsibilities to participants while seeking to deliver the public value of research itself.

As to the concept of public interest *per se*, this has been tackled by Sorbie in Chapter 6, and various contributions in Section IB have addressed the role and importance of public engagement in the design and delivery of robust health research regulation. In this section, several authors build on these earlier chapters in multiple ways. For example, Taylor and Whitton (Chapter 24) directly challenge the putative tension between public and private interests, arguing that each is implicated in the other's protection. They offer a reconceptualisation of privacy through a public interest lens, raising important questions for existing laws of confidentiality and data protection. Their perspective requires us to recognise the common interest at stake. Most uniquely, however, they extend their analysis to show how group privacy interests currently receive short shrift in health research regulation, and they suggest that this dangerous oversight must be addressed adequately because the failure to recognise group privacy interests might ultimately jeopardise the common public interest in health research.

Starkly, Burgess (Chapter 25) uses just such an example of threats to group privacy – the *care. data* debacle – to mount a case for mobilising public expertise in the design of health research regulation. Drawing on the notion of deliberative public engagement, he demonstrates how this process cannot only counter asymmetries of power in the structural design of regulation but also how the resulting advice about what is in the public interest can bring both legitimacy and trustworthiness to resultant models of governance. This is of crucial importance, because as he states: '[i]t is inadequate to assert or assume that research and its existing and emerging regulation is in the public interest'. His contribution allows us to challenge any such assertion and to move beyond it responsibly.

The last two contributions to this section continue this theme of structural reimagining of regulatory architectures, set against the interests and values in play. Vayena and Blassime (Chapter 26) offer the example of Big Data to propose a model of adaptive governance that can adequately accommodate and respond to the diverse and dynamic interests. Following principles-based regulation as previously discussed by Sethi in Chapter 17, they outline a model involving six principles and propose key factors for their implementation and operationalisation into effective governance structures and processes. This form of adaptive governance mirrors the discussions by Kaye and Prictor in Chapter 10. Importantly, the factors identified by the current authors – of social learning, complementarity and visibility – not only lend themselves to full and transparent engagement with the range of public and private interests, they *require* it. In the final chapter of this section, Brownsword (Chapter 27) invites us to address an overarching question that is pertinent to this entire volume: 'how are the interests in pushing forward with research into potentially beneficial health technologies to be reconciled with the heterogeneous interests of the concerned who seek to push back against them?' His contribution is to push back against the common regulatory response when discussing public and private interests: namely, to seek a 'balance'. While not necessarily rejecting the balancing exercise as a helpful regulatory device at an appropriate point in the trajectory of regulatory responses to a novel technology, he implores us to place this in 'a bigger picture of lexically ordered regulatory responsibilities'. For him, morally and logically prior questions are those that ask whether any new development – such as automated healthcare – poses threats to human existence and agency. Only thereafter ought we to consider a role for the balancing exercise that is currently so prevalent in human health research regulation.

Collectively, these contributions significantly challenge the public/private trope in health research regulation, but they leave it largely intact as a framing device for engaging with the constantly changing nature of the research endeavour. This is helpful in ensuring that on-going conversations are not unduly disrupted in unproductive ways. By the same token, individually these chapters provide a plethora of reasons to rethink the nature of how we frame public and private interests, and this in turn allows us to carve out new pathways in the future regulatory landscape. Thus:

- Private interests have been expanded as to content (Postan) and extended as to their reach (Taylor and Whitton).
- Moreover, the implications of recognising these reimagined private interests have been addressed, and not necessarily in ways resulting in inevitable tension with appeals to public interest.
- The content of public interest has been aligned with deliberative engagement in ways that can increase the robustness of health research regulation as a participative exercise (Burgess).
- Systemic oversight that is adaptive to the myriad of evolving interests has been offered as proof of principle (Vayena and Blassime).
- The default of seeking balance between public and private interests has been rightly questioned, at least as to its rightful place in the stack of ethical considerations that contribute to responsible research regulation (Brownsword).

23

Changing Identities in Disclosure of Research Findings

Emily Postan

23.1 INTRODUCTION

This chapter offers a perspective on the long-running ethical debate about the nature and extent of responsibilities to return individually relevant research findings from health research to participants. It highlights the ways in which shifts in the research landscape are changing the roles of researchers and participants, the relationships between them, and what this might entail for the responsibilities owed towards those who contribute to research by taking part in it. It argues that a greater focus on the informational interests of participants is warranted and that, as a corollary to this, the potential value of findings beyond their clinical utility deserves greater attention. It proposes participants' interests in using research findings in developing their own identities as a central example of this wider value and argues that these could provide grounds for disclosure.

23.2 FEATURES OF EXISTING DISCLOSURE GUIDANCE

This chapter is concerned with the questions of whether, why, when and how individually relevant findings, which arise in the course of health research, should be offered or fed-back to the research participant to whom they directly pertain.[1] Unless otherwise specified, what will be said here applies to findings generated through observational and hands-on studies, as well as those using previously collected tissues and data.

Any discussion of ethical and legal responsibilities for disclosure of research findings must negotiate a number of category distinctions relating to the nature of the findings and the practices within which they are generated. However, as will become clear below, several lines of demarcation that have traditionally structured the debate are shifting. A distinction has historically been drawn between the intended (pertinent, or primary) findings from a study and those termed 'incidental' (ancillary, secondary, or unsolicited). 'Incidental findings' are commonly defined as individually relevant observations generated through research, but lying outwith the aims of the study.[2] Traditionally, feedback of incidental findings has been presented

---

[1] This chapter will not discuss responsibilities actively to pursue findings, or disclosures to family members in genetic research, nor is it concerned with feedback of aggregate findings. For discussion of researchers' experiences of encountering and disclosing incidental findings in neuroscience research see Pickersgill, Chapter 31 in this volume.

[2] S. M. Wolf et al., 'Managing Incidental Findings in Human Subjects Research: Analysis and Recommendations', (2008) *The Journal of Law, Medicine & Ethics*, 36(2), 219–248.

as more problematic than that of 'intended findings' (those the study set out to investigate). However, the cogency of this distinction is increasingly questioned, to the extent that many academic discussions and guidance documents have largely abandoned it.[3] There are several reasons for this, including difficulties in drawing a bright line between the categories in many kinds of studies, especially those that are open-ended rather than hypothesis-driven.[4] The relevance of researchers' intentions to the ethics of disclosure is also questioned.[5] For these reasons, this chapter will address the ethical issues raised by the return of individually relevant research results, irrespective of whether they were intended.

The foundational question of *whether* findings should be fed-back – or feedback offered as an option – is informed by the question of *why* they should. This may be approached by examining the extent of researchers' legal and ethical responsibilities to participants – as shaped by their professional identities and legal obligations – the strength of participants' legitimate interests in receiving feedback, or researchers' responsibilities towards the research endeavour. The last of these includes consideration of how disclosure efforts might impact on wider public interests in the use of research resources and generation of valuable generalisable scientific knowledge, and public trust in research. These considerations then provide parameters for addressing questions of *which* kinds of findings may be fed-back and under what circumstances. For example, which benefits to participants would justify the resources required for feedback? Finally, there are questions of *how*, including how researchers should plan and manage the pathway from anticipating the generation of such findings to decisions and practices around disclosure.

In the past two decades, a wealth of academic commentaries and consensus statements have been published, alongside guidance by research funding bodies and professional organisations, making recommendations about approaches to disclosure of research findings.[6] Some are prescriptive, specifying the characteristics of findings that ought to be disclosed, while others provide process-focused guidance on the key considerations for ethically, legally and practically robust disclosure policies. It is not possible here to give a comprehensive overview of all the permutations of responses to the four questions above. However, some prominent and common themes can be extracted.

Most strikingly, in contrast to the early days of this debate, it is rare now to encounter the bald question of *whether* research findings should ever be returned. Rather the key concerns are what should be offered and how.[7] The resource implications of identifying, validating and communicating findings are still acknowledged, but these are seen as feeding into an overall risk/benefit analysis rather than automatically implying non-disclosure. In parallel with this shift, there is less scepticism about researchers' general disclosure responsibilities. In the UK, researchers are not subject to a specific legal duty to return findings.[8] Nevertheless, there does appear to be a growing

---

[3] L. Eckstein et al., 'A Framework for Analyzing the Ethics of Disclosing Genetic Research Findings', (2014) *The Journal of Law, Medicine & Ethics*, 42(2), 190–207.

[4] B. E. Berkman et al., 'The Unintended Implications of Blurring the Line between Research and Clinical Care in a Genomic Age', (2014) *Personalized Medicine*,11(3), 285–295.

[5] E. Parens et al., 'Incidental Findings in the Era of Whole Genome Sequencing?', (2013) *Hastings Center Report*, 43(4), 16–19.

[6] For example, in addition to sources cited elsewhere in this chapter, see R. R. Fabsitz et al., 'Ethical and Practical Guidelines for Reporting Genetic Research Results to Study Participants', (2010) *Circulation: Cardiovascular Genetics*, 3(6),574–580; G. P. Jarvik et al., 'Return of Genomic Results to Research Participants: The Floor, the Ceiling, and the Choices in Between', (2014) *The American Journal of Human Genetics*, 94(6), 818–826.

[7] C. Weiner, 'Anticipate and Communicate: Ethical Management of Incidental and Secondary Findings in the Clinical, Research, and Direct-to-Consumer Contexts', (2014) *American Journal of Epidemiology*, 180(6), 562–564.

[8] Medical Research Council and Wellcome Trust, 'Framework on the Feedback of Health-Related Findings in Research', (Medical Research Council and Wellcome Trust, 2014).

consensus that researchers do have *ethical* responsibilities to offer findings – albeit limited and conditional ones.[9] The justifications offered for these responsibilities vary widely, however, and indeed are not always made explicit. This chapter will propose grounds for such responsibilities.

When it comes to determining what kinds of findings should be offered, three jointly necessary criteria are evident across much published guidance. These are captured pithily by Lisa Eckstein et al. as 'volition, validity and value'.[10] Requirements for analytic and clinical *validity* entail that the finding reliably measures and reports what it purports to. *Value* refers to usefulness or benefit to the (potential) recipient. In most guidance this is construed narrowly in terms of the information's clinical utility – construed as actionability and sometimes further circumscribed by the seriousness of the condition indicated.[11] Utility for reproductive decision-making is sometimes included.[12] Although some commentators suggest that 'value' could extend to the non-clinical, subjectively determined 'personal utility' of findings, it is generally judged that this alone would be insufficient to justify disclosure costs.[13] The third necessary condition is that the participant should have agreed *voluntarily* to receive the finding, having been advised at the time of consenting to participate about the kinds of findings that could arise and having had the opportunity to assent to or decline feedback.[14]

Accompanying this greater emphasis on the 'which' and 'how' questions is an increasing focus upon the need for researchers to establish clear policies for disclosing findings, that are explained in informed consent procedures, and an accompanying strategy for anticipating, identifying, validating, interpreting, recording, flagging-up and feeding-back findings in ways that maximise benefits and minimise harms.[15] Broad agreement among scholars and professional bodies that – in the absence of strong countervailing reasons – there is an ethical responsibility to disclose clinically actionable findings is not, however, necessarily reflected in practice, where studies may still lack disclosure policies, or have policies of non-disclosure.[16]

Below I shall advance the claim that, despite a greater emphasis upon, and normalisation of, feedback of findings, there are still gaps, which mean that feedback policies may not be as widely instituted or appropriately directed as they should be. Chief among these gaps are, first, a continued focus on researchers' inherent responsibilities considered separately from participants' interests in receiving findings and, second, a narrow conception of when these interests are engaged. These gaps become particularly apparent when we attend to the ways in which the roles of researchers and participants and relationships between them have shifted in a changing health research landscape. In the following sections, I will first highlight the nature of these changes, before proposing what these mean for participants' experiences, expectations and informational interests and, thus, for ethically robust feedback policies and practices.

23.3 THE CHANGING HEALTH RESEARCH LANDSCAPE

The landscape of health research is changing. Here I identify three facets of these changes and consider how these could – and indeed should – have an effect on the practical and ethical basis of policies and practices relating to the return of research findings.

[9] Berkman et al., 'The Unintended Implications'.
[10] Eckstein et al., 'A Framework for Analyzing'.
[11] Wolf et al., 'Managing Incidental Findings'.
[12] Ibid.
[13] Eckstein et al., 'A Framework for Analyzing'.
[14] Medical Research Council and Wellcome Trust, 'Framework on the Feedback'.
[15] Ibid.
[16] Berkman et al., 'The Unintended Implications'.

The first of these developments is a move towards 'learning healthcare' systems and translational science, in which the transitions between research and care are fluid and cyclical, and the lines between patient and participant are often blurred.[17] The second is greater technical capacities, and appetite, for data-driven research, including secondary research uses of data and tissues – sourced from patient records, prior studies, or biobanks – and linkage between different datasets. This is exemplified by the growth in large-scale and high-profile of genomic studies such as the UK's '*100,000 Genomes*' project.[18] The third development is increasing research uses of technologies and methodologies, such as functional neuroimaging, genome-wide association studies, and machine-learning, which lend themselves to open-ended, exploratory inquiries rather than hypothesis-driven ones.[19] I wish to suggest that these three developments have a bearing on disclosure responsibilities in three key respects: erosion of the distinction between research and care; generation of findings with unpredictable or ambiguous validity and value; and a decreasing proximity between researchers and participants. I will consider each of these in turn.

Much of the debate about disclosure of findings has, until recently, been premised on there being a clear distinction between research and care, and what this entails in terms of divergent professional priorities and responsibilities, and the experiences and expectations of patient and participants. Whereas it has been assumed that clinicians' professional duty of care requires disclosure of – at least – clinically actionable findings, researchers are often seen as being subject to a contrary duty to refrain from feedback if this would encourage 'therapeutic misconceptions', or divert focus and resources from the research endeavour.[20] However, as health research increasingly shades into 'learning healthcare', these distinctions become increasingly untenable.[21] It is harder to insist that responsibilities to protect information subjects' interests do not extend to those engaged in research, or that participants' expectations of receiving findings are misconceived. Furthermore, if professional norms shift towards more frequent disclosure, so the possibility that healthcare professionals may be found negligent for failing to disclose becomes greater.[22] These changes may well herald more open feedback policies in a wider range of studies. However, if these policies are premised solely on the duty of care owed in healthcare contexts to participants-as-patients, then the risk is that any expansion will fail to respond adequately to the very reasons why findings should be offered at all – to protect participants' core interests.

Another consequence of the shifting research landscape, and the growth of data-driven research in particular, lies in the nature of findings generated. For example, many results from genomic analysis or neuroimaging studies are probabilistic rather than strongly predictive, and produce information of varying quality and utility.[23] And open-ended and exploratory studies

[17] S. M. Wolf et al., 'Mapping the Ethics of Translational Genomics: Situating Return of Results and Navigating the Research-Clinical Divide', (2015) *Journal of Law, Medicine & Ethics*, 43(3), 486–501.

[18] G. Laurie and N. Sethi, 'Towards Principles–Based Approaches to Governance of Health–Related Research Using Personal Data', (2013) *European Journal of Risk Regulation*, 4(1), 43–57. Genomics England, 'The 100,000 Genomes Project', (Genomics England), www.genomicsengland.co.uk/about-genomics-england/the-100000-genomes-project/.

[19] Eckstein et al., 'A Framework for Analyzing'.

[20] A. L. Bredenoord et al., 'Disclosure of Individual Genetic Data to Research Participants: The Debate Reconsidered', (2011) *Trends in Genetics*, 27(2), 41–47.

[21] Wolf et al., 'Mapping the Ethics'.

[22] In the UK, the expected standard of duty of care is assessed to what reasonable members of the profession would do as well as what recipients want to know (see C. Johnston and J. Kaye, 'Does the UK Biobank Have a Legal Obligation to Feedback Individual Findings to Participants?', (2004) *Medical Law Review*, 12(3), 239–267.

[23] D. I. Shalowitz et al., 'Disclosing Individual Results of Clinical Research: Implications of Respect for Participants', (2005) *JAMA*, 294(6), 737–740.

pose challenges precisely because what they might find – and thus their significance to participants – are unpredictable and, especially in new fields of research, may be less readily validated. These characteristics are of ethical significance because they present obstacles to meeting the requirements (noted above) for securing validity, value and ascertaining what participants wish to receive. And where validity and value are uncertain, robust analysis of the relative risks and benefits of disclosure is not possible. Given these challenges, it is apparent that meeting participants' informational interests will require more than just instituting clear disclosure policies. Instead, more flexible and discursive disclosure practices may be needed to manage unanticipated or ambiguous findings.

Increasingly, health research is conducted using data or tissues that were collected for earlier studies, or sourced from biobanks or patient records.[24] In these contexts, in contrast to the closer relationships entailed by translational studies, researchers may be geographically, temporally and personally far-removed from the participants. This poses a different set of challenges when determining responsibilities for disclosing research findings. First, it may be harder to argue that researchers working with pre-existing data collections hold a duty of care to participants, especially one analogous to that of a healthcare professional. Second, there is the question of *who* is responsible for disclosure: is it those who originally collected materials, manage this resource or generate the findings? Third, if consent is only sought when the data or tissues are originally collected, it is implausible that a one-off procedure could address in detail all future research uses, let alone the characteristics, of all future findings.[25] And finally, in these circumstances, disclosure may be more resource-intensive where, for example, much time has elapsed or datasets have been anonymised. These observations underscore the problems of thinking of 'health research' as a homogenous category in which the respective roles and expectations of researchers and participants are uniform and easily characterised, and ethical responsibilities attach rigidly to professional identities.

Finally, it is also instructive to attend to shifts in wider cultural and legal norms surrounding our relationships to information about ourselves and the increasing emphasis on informational autonomy, particularly with respect to accessing and controlling information about our health or genetic relationships. There is increased legal protection of informational interests beyond clinical actionability, including the interest in developing one's identity, and in reproductive decision-making.[26] For example, European human rights law has recognised the right to access to one's health records and the right to know one's genetic origins as aspects of the Article 8 right to respect for private life.[27] And in the UK, the legal standard for information provision by healthcare professionals has shifted from one determined by professional judgement, to that which a reasonable patient would wish to know.[28]

When taken together, the factors considered in this section provide persuasive grounds for looking beyond professional identities, clinical utility and one-off consent and information transactions when seeking to achieve ethically defensible feedback of research findings. In the

---

[24] Laurie and Sethi, 'Towards Principles–Based Approaches'.

[25] G. Laurie and E. Postan, 'Rhetoric or Reality: What Is the Legal Status of the Consent Form in Health-Related Research?', (2013) *Medical Law Revue*, 21(3), 371–414.

[26] *Odièvre v. France* (App. no. 42326/98) [2003] 38 EHRR 871; *ABC v. St George's Healthcare NHS Trust & Others* [2017] EWCA Civ 336.

[27] J. Marshall, *Personal Freedom through Human Rights Law?: Autonomy, Identity and Integrity Under the European Convention on Human Rights* (Leiden: Brill, 2008).

[28] A. M. Farrell and M. Brazier, 'Not So New Directions in the Law of Consent? Examining *Montgomery v Lanarkshire Health Board*', (2016) *Journal of Medical Ethics*, 42(2), 85–88.

next section, I will present an argument for grounding ethical policies and practices upon the research participants' informational interests.

23.4 RE-FOCUSING ON PARTICIPANTS' INTERESTS

What emerges from the picture above is that the respective identities and expectations of researchers and participants are changing, and with them the relationships and interdependencies between them. Some of these changes render research relationships more intimate, akin to clinical care, while other makes them more remote. And the roles that each party fulfils, or are expected to fulfil, may be ambiguous. This lack of clarity presents obstacles to relying on prior distinctions and definitions and raises questions about the continued legitimacy of some existing guiding principles.[29] Specifically, it disrupts the foundations upon which disclosure of individually relevant results might be premised. In this landscape, it is no longer possible or appropriate – if indeed it ever was – simply to infer what ethical feedback practice would entail from whether not an actor is categorised as 'a researcher'. This is due not only to ambiguity about the scope of this role and associated responsibilities. It also looks increasingly unjustifiable to give only secondary attention to the nature and specificity of participants' interests: to treat these as if they are a homogenous group of narrowly health-related priorities that may be honoured, provided doing so does not get in the way of the goal of generating generalisable scientific knowledge. There is a need to revisit the nature and balance of private and public interests at stake. My proposal here is that participants' informational interests, and researchers' particular capacities to protect these interests, should comprise the heart of ethical feedback practices.

There are several reasons why it seems appropriate – particularly now – to place participants' interests at the centre of decision-making about disclosure. First, participants' roles in research are no less in flux than researchers'. While it may be true that the inherent value of any findings to participants – whether they might wish to receive them and whether the information would be beneficial or detrimental to their health, well-being, or wider interests – may not be dramatically altered by emerging research practices, their motivations, experiences and expectations of taking part may well be different. In the landscape sketched above, it is increasingly appropriate to think of participants less as passive subjects of investigation, but rather as partners in the research relationship.[30] This is a partnership grounded in the contributions that participants make to a study and in the risks and vulnerabilities incurred when they agree to take part. The role of participant-as-partner is underscored by the rise of the idea that there is an ethical 'duty to participate'.[31] This idea has escaped the confines of academic argument. Implications of such a duty are evident in in public discourse concerning biobanks and projects such as *100,000 Genomes*. For example, referring to that project, the (then) Chief Medical Officer for England has said that to achieve 'the genomic dream', we should 'agree to use of data for our own benefit and others'.[32] A further compelling reason for placing the interests of participants at the centre of

---

[29] G. Laurie, 'Liminality and the Limits of Law in Health Research Regulation: What Are We Missing in the Spaces In-Between?', (2016) *Medical Law Review*, 25 (1), 47–72.

[30] J. Kaye et al., 'From Patients to Partners: Participant-Centric Initiatives in Biomedical Research', (2012) *Nature Reviews Genetics*, 13(5), 371.

[31] J. Harris, 'Scientific Research Is a Moral Duty', (2005) *Journal of Medical Ethics*, 31(4), 242–248.

[32] S. C. Davies, 'Chief Medical Officer Annual Report 2016: Generation Genome', (Department of Health and Social Care, 2017), p. 4.

return policies is that doing so is essential to building confidence and demonstrating trustworthiness in research.[33] Without this trust there would be no participants and no research.

In light of each of these considerations, it is difficult to justify the informational benefits of research accruing solely to the project aims and the production of generalisable knowledge, without participants' own core informational interests inviting corresponding respect. That is, respect that reflects the nature of the joint research endeavour and the particular kinds of exposure and vulnerabilities participants incur.

If demonstrating respect was simply a matter of reciprocal recognition of participants' contributions to knowledge production, then it could perhaps be achieved by means other than feedback. However, research findings occupy a particular position in the vulnerabilities, dependencies and responsibilities of the researcher relationship. Franklin Miller and others argue that researchers have responsibilities to disclose findings that arise from a particular *pro tanto* ethical responsibility to help others and protect their interests within certain kinds of professional relationships.[34] These authors hold that this responsibility arises because, in their professional roles, researchers have both privileged access to private aspects of participants' lives, and particular opportunities and skills for generating information of potential significance and value to participants to which they would not otherwise have access.[35] I would add to this that being denied the opportunity to obtain otherwise inaccessible information about oneself not only fails to protect participants from avoidable harms, it also fails to respect and benefit them in ways that recognise the benefits they bring to the project and the vulnerabilities they may incur, and trust they invest, when doing so.

None of what I have said seeks to suggest that research findings should be offered without restriction, or at any cost. The criteria of 'validity, value and volition' continue to provide vital filters in ensuring that information meets recipients' interests at all. However, providing these three conditions are met, investment of research resources in identifying, validating, offering and communicating individually relevant findings, may be ethically justified, even required, when receiving them could meet non-trivial informational interests. One question that this leaves unanswered, of course, is what counts as an interest of this kind.

23.5 A WIDER CONCEPTION OF VALUE: RESEARCH FINDINGS AS NARRATIVE TOOLS

If responsibilities for feedback are premised on the value of particular information to participants, it seems arbitrary to confine this value solely to clinical actionability, unless health-related interests are invariably more critical than all others. It is not at all obvious that this is so. This section provides a rationale for recognising at least one kind of value beyond clinical utility.[36]

It is suggested here that where research findings support a participant's abilities to develop and inhabit their own sense of who they are, significant interests in receiving these findings will be engaged. The kinds of findings that could perform this kind of function might include, for example, those that provide diagnoses that explain longstanding symptoms – even where there is no effective intervention – susceptibility estimates that instigate patient activism, or indications

---

[33] Wolf et al., 'Mapping the Ethics'.
[34] F. G. Miller et al., 'Incidental Findings in Human Subjects Research: What Do Investigators Owe Research Participants?', (2008) *The Journal of Law, Medicine & Ethics*, 36(2), 271–279.
[35] Ibid.
[36] In Chapter 39 of this volume, Shawn Harmon presents a parallel argument that medical device regulations are similarly premised on a narrow conception of harm that fails to account for identity impacts.

of carrier status or genetic relatedness that allow someone to (re)assess of understand their relationships and connections to others.

The claim to value posited here goes beyond appeals to 'personal utility', as commonly characterised in terms of curiosity, or some unspecified, subjective value. It is unsurprising that, thus construed, personal utility is rarely judged to engage sufficiently significant interests to warrant the effort and resources of disclosing findings.[37] However, the claim here – which I have more fully discussed elsewhere[38] – is that information about the states, dispositions and functions of our bodies and minds, and our relationships to others (and others' bodies) – such as that conveyed by health research findings – is of value to us when, and to the extent that, it provides constitutive and interpretive tools that help us to develop our own narratives about who we are – narratives that *constitute* our identities.[39] Specifically, this value lies not in contributing to just *any* identity-narrative, but one that makes sense when confronted by our embodied and relational experiences and supports us in navigating and interpreting these experiences.[40] These experiences include those of research participation itself. A coherent, 'inhabitable' self-narrative is of ethical significance, because such a narrative is not just something we passively and inevitably acquire. Rather, it is something we develop and maintain, which provides the practical foundations for our self-understanding, interpretive perspective and values, and thus our autonomous agency, projects and relationships.[41] If we do indeed have a significant interest in developing and maintaining such a narrative, and some findings generated in health research can support us in doing so, then my claim is that these findings may be at least as valuable to us as those that are clinically actionable. As such, our critical interests in receiving them should be recognised in feedback policies and practices.

In response to concern that this proposal constitutes an unprecedented incursion of identity-related interests into the (public) values informing governance of health research, it is noted that the very act of participating in research is already intimately connected to participants' conceptions of who they are and what they value, as illustrated by choices to participate motivated by family histories of illness,[42] or objections to tissues or data being used for commercial research.[43] Participation already impacts upon the self-understandings of those who choose to contribute. Indeed, it may often be seen as contributing to the narratives that comprise their identities. Seen in this light, it is not only appropriate, but vital, that the identity-constituting nature of research participation is reflected in the responsibilities that researchers – and the wider research endeavour – owe to participants.

23.6 REVISITING ETHICAL RESPONSIBILITIES FOR FEEDING BACK FINDINGS

What would refocusing ethical feedback for research findings to encompass the kinds of identity-related interests described above mean for the responsibilities of researchers and others? I submit

[37] Eckstein et al., 'A Framework for Analyzing'.

[38] E. Postan, 'Defining Ourselves: Personal Bioinformation as a Tool of Narrative Self-Conception', (2016) *Journal of Bioethical Inquiry*, 13(1), 133–151.

[39] M. Schechtman, *The Constitution of Selves* (New York: Cornell University Press, 1996).

[40] Postan, 'Defining Ourselves'.

[41] C. Mackenzie, 'Introduction: Practical Identity and Narrative Agency' in K. Atikins and C. Mackenzie (eds), *Practical Identity and Narrative Agency* (Abingdon: Routledge, 2013), pp. 1–28.

[42] L. d'Agincourt-Canning, 'Genetic Testing for Hereditary Breast and Ovarian Cancer: Responsibility and Choice', (2006) *Qualitative Health Research*, 16(1), 97–118.

[43] P. Carter et al., 'The Social Licence for Research: Why *care.data* Ran into Trouble', (2015) *Journal of Medical Ethics*, 41(5), 404–409.

that it entails responsibilities both to look beyond clinical utility to anticipate when findings could contribute to participants' self-narratives and to act as an interpretive partner in discharging responsibilities for offering and communicating findings.

It must be granted that the question of when identity-related interests are engaged by particular findings is a more idiosyncratic matter than clinical utility. This serves to underscore the requirement that any disclosure of findings is voluntary. And while this widening of the conception of 'value' is in concert with increasing emphasis on individually determined informational value in healthcare – as noted above – it is not a defence of unfettered informational autonomy, requiring the disclosure of whatever participants might wish to see. In order for research findings to serve the wider interests described above, they must still constitute meaningful and reliable biomedical information. There is no value without validity.[44]

These two factors signal that the ethical responsibilities of researchers will not be discharged simply by disclosing findings. There is a critical interpretive role to be fulfilled at several junctures, if participants' interests are to be protected. These include: anticipating which findings could impact on participants' health, self-conceptions or capacities to navigate their lives; equipping participants to understand at the outset whether findings of these kinds might arise; and, if participants choose to receive these findings, ensuring that these are communicated in a manner that is likely to minimise distress, and enhance understanding of the capacities and limitations of the information in providing reliable explanations, knowledge or predictions about their health and their embodied states and relationships. This places the researcher in the role of 'interpretive partner', supporting participants to make sense of the findings they receive and to accommodate – or disregard – them in conducting their lives and developing their identities.

This role of interpretive partner represents a significant extension of responsibilities from an earlier era in which a requirement to report even clinically significant findings was questioned. The question then arises as to who will be best placed to fulfil this role. As noted above, dilemmas about *who* should disclose arise most often in relation to secondary research uses of data.[45] These debates err, however, when they treat this as a question focused on professional and institutional duties abstracted from participants' interests. When we attend to these interests, the answer that presents itself is that feedback should be provided by whoever is best placed to recognise and explain the potential significance of the findings to participants. And it may in some cases be that those best placed to do this are not researchers at all, but professionals performing a role analogous to genetic counsellors.

Even though the triple threshold conditions for disclosure – validity, value and volition – still apply, any widening of the definition of value implies a larger category of findings to be validated, offered and communicated. This will have resource implications. And – as with any approach to determining which findings should be fed-back and how – the benefits of doing so must still be weighed against any resultant jeopardy to the socially valuable ends of research. However, if we are not simply paying lip-service to, but taking seriously, the ideas that participants are partners in, not merely passive objects of, research, then protecting their interests – particularly those incurred through participation – is not supererogatory, but an intrinsic part of recognising their contribution to biomedical science, their vulnerability, trust and experiences of contributing. Limiting these interests to receipt of clinically actionable findings is arbitrary and

[44] E. M. Bunnik et al., 'Personal Utility in Genomic Testing: Is There Such a Thing?', (2014) *Journal of Medical Ethics*, 41(4), 322–326.
[45] S. M. Wolf et al., 'Managing Incidental Findings and Research Results in Genomic Research Involving Biobanks and Archived Data Sets', (2012) *Genetics in Medicine*, 14(4), 361–384.

out of step with wider ethico-legal developments in the health sphere. Just because these findings arise in the context of health research is not on its own sufficient reason for interpreting 'value' solely in clinical terms.

23.7 CONCLUSION

In this chapter, I have argued that there are two shortcomings in current ethical debates and guidance regarding policies and practices for feeding back individually relevant findings from health research. These are, first, a focus on the responsibilities of actors for disclosure that remains insufficiently grounded in the essential questions of when and how disclosure would meet core interests of participants; and, second, a narrow interpretation of these interests in terms of clinical actionability. Specifically, I have argued that participants have critical interests in accessing research findings where these offer valuable tools of narrative self-constitution. These shortcomings have been particularly brought to light by changes in the nature of health research, and addressing them becomes ever more important as the role participants evolves from one of an object of research, to active members of shared endeavours. I have proposed that in this new health research landscape, there are not only strong grounds for widening feedback to include potentially identity-significant findings, but also to recognise the valuable role of researchers and others as interpretive partners in the relational processes of anticipating, offering and disclosing findings.

24

Health Research and Privacy through the Lens of Public Interest

A Monocle for the Myopic?

Mark Taylor and Tess Whitton

24.1 INTRODUCTION

Privacy and public interest are reciprocal concepts, mutually implicated in each other's protection. This chapter considers how viewing the concept of privacy through a public interest lens can reveal the limitations of the narrow conception of privacy currently inherent to much health research regulation (HRR). Moreover, it reveals how the public interest test, applied in that same regulation, might mitigate risks associated with a narrow conception of privacy.

The central contention of this chapter is that viewing privacy through the lens of public interest allows the law to bring into focus more things of common interest than privacy law currently recognises. We are not the first to recognise that members of society share a common interest in both privacy and health research. Nor are we the first to suggest that public is not necessarily in opposition to private, with public interests capable of accommodating private and vice versa.[1] What is novel about our argument is the suggestion that we might invoke public interest requirements in current HRR to protect group privacy interests that might otherwise remain out of sight.

It is important that HRR takes this opportunity to correct its vision. A failure to do so will leave HRR unable to take into consideration research implications with profound consequences for future society. A failure will undermine legitimacy in HRR. It is no exaggeration to say that the value of a confidential healthcare system may come to depend on whether HRR acknowledges the significance of group data to the public interest. It is group data that shapes health policies, evaluates success, and determines the healthcare opportunities offered to members of particular groups. Individual opportunity, and entitlement, is dependent upon group classification.

The argument here is three-fold: (1) a failure to take common interests into account when making public interest decisions undermines the legitimacy of the decision-making process; (2) a common interest in privacy extends to include group interests; (3) the law's current myopia regarding group privacy interests in data protection law and the duty of confidence law can be corrected, to a varying extent, through bringing group privacy interests into view through the lens of public interest.

---

[1] The idea that both privacy and health research may be described as 'public interest causes' is also compelling developed in W. W. Lowrance, *Privacy, Confidentiality, and Health Research* (Cambridge University Press, 2012) and the relationship between privacy and the public interested in C. D. Raab, 'Privacy, Social Values and the Public Interest' in A. Busch and J. Hofmann (eds), *Politik und die Regulierung von Information [Politics and the Regulation of Information]* (Baden-Baden, Germany: Politische Vierteljahresschrift, Sonderheft 46, 2012), pp. 129–151.

24.2 COMMON INTERESTS, PUBLIC INTEREST AND LEGITIMACY

In this section, we seek to demonstrate how a failure to take the full range of common (group) interests into account when making public interest decisions will undermine the legitimacy of those decisions.

When Held described broad categories into which different theories of public interest might be understood to fall, she listed three: preponderance or aggregative theories, unitary theories and common interest theories.[2] When Sorauf earlier composed his own list, he combined common interests with values and gave the category the title 'commonly-held value'.[3] We have separately argued that a compelling conception of public interest may be formed by uniting elements of 'common interest' and 'common value' theories of public interest.[4] It is, we suggest, through combining facets of these two approaches that one can overcome the limitations inherent to each. Here we briefly recap this argument before seeking to build upon it.

Fundamental to common interest theories of the public interest is the idea that something may serve 'the ends of the whole public rather than those of some sector of the public'.[5] If one accepts the idea that there may be a common interest in privacy protection, as well as in the products of health research, then 'common interest theory' brings both privacy and health research within the scope of public interest consideration. However, it cannot explain how – in case of any conflict – they ought to be traded-off against each other – or other common interests – to determine *the* public interest in a specific scenario.

In contrast to common interest theories, commonly held *value* theories claim the 'public interest emerges as a set of fundamental values in society'.[6] If one accepts that a modern liberal democracy places a fundamental value upon all members of society being respected as free and equal citizens, then any interference with individual rights should be defensible in terms that those affected can both access and have reason to endorse[7] – with discussion subject to the principles of public reasoning.[8] Such a commitment is enough to fashion a normative yardstick, capable of driving a public interest determination. However, the object of measurement remains underspecified.

It is through combining aspects of common interest and common value approaches that a practical conception of the public interest begins to emerge: any trade-off between common interests ought to be defensible in terms of common value: for reasons that those affected by a decision can both access and have reason to endorse.[9]

---

[2] V. P. Held, *The Public Interest and Individual Interests* (New York: Basic Books, 1970).

[3] F. J. Sorauf, 'The Public Interest Reconsidered', (1957), *The Journal of Politics*, 19(4), 616–639.

[4] M. J. Taylor 'Health Research, Data Protection, and the Public Interest in Notification', (2011) *Medical Law Review*, 19(2), 267–303; M. J. Taylor and T. Whitton, 'Public Interest, Health Research and Data Protection Law: Establishing a Legitimate Trade-Off between Individual Control and Research Access to Health Data', (2020) *Laws*, 9(1), 6.

[5] M. Meyerson and E. C. Banfield, cited by Sorauf 'The Public Interest Reconsidered', 619.

[6] J. Bell, 'Public Interest: Policy or Principle?' in R. Brownsword (ed.), *Law and the Public Interest: Proceedings of the 1992 ALSP Conference* (Stuttgart: Franz Steiner, 1993) cited in M. Feintuck, *Public Interest in Regulation* (Oxford University Press, 2004), p. 186.

[7] There is a connection here with what has been described by Rawls as 'public reasons': limited to premises and modes of reasoning that are accessible to the public at large. L. B. Solum, 'Public Legal Reason', (2006) *Virginia Law Review*, 92(7), 1449–1501, 1468.

[8] 'The virtue of public reasoning is the cultivation of clear and explicit reasoning orientated towards the discovery of common grounds rather than in the service of sectional interests, and the impartial interpretation of all relevant available evidence.' Nuffield Council on Bioethics, 'Public Ethics and the Governance of Emerging Biotechnologies', (Nuffield Council on Bioethics, 2012), 69.

[9] G. Gaus, *The Order of Public Reason: A Theory of Freedom and Morality in a Diverse and Bounded World* (Cambridge University Press, 2011), p. 19. Note the distinction Gaus draws here between the Restricted and the Expansive view of Freedom and Equality.

An advantage of this hybrid conception of public interest is its connection with (social) legitimacy.[10] If a decision-maker fails to take into account the full range of interests at stake, then not only do they undermine any public interest claim, but also the legitimacy of the decision-making process underpinning it.[11] Of course, this does not imply that the legitimacy of a system depends upon everyone perceiving the 'public interest' to align with their own contingent individual or common interests. Public-interest decision-making should, however, ensure that when the interests of others displace *any* individual's interests, including those held in common, it should (ideally) be transparent why this has happened and (again, ideally) the reasons for displacement should be *acceptable* as 'good reasons' to the individual.[12] If the displaced interest is more commonly held, it is even more important for a system practically concerned with maintaining legitimacy, to transparently account for that interest within its decision-making process.

Any failure to account transparently for common interests will undermine the legitimacy of the decision-making process.

24.3 COMMON INTERESTS IN (GROUP) PRIVACY

In this section, the key claim is that a common interest in privacy extends beyond a narrow atomistic conception of privacy to include group interests.

We are aware of no 'real definition' of privacy.[13] There are, however, many stipulative or descriptive definitions, contingent upon use of the term within particular cultural contexts. Here we operate with the idea that privacy might be conceived in the legal context as representing 'norms of exclusivity' within a society: the normative expectation that *some* states of information separation are, by default, to be maintained.[14] This is a broad conception of privacy extending beyond the atomistic one that Bennet and Raab observe to be the prevailing privacy paradigm in many Western societies.[15] It is not necessary to defend a broad conception of privacy in order to

---

[10] We here associate legitimacy with 'the capacity of the system to engender and maintain the belief that the existing political institutions are the most appropriate ones for the society' S. M. Lipset, *Political Man: The Social Bases of Politics* (Baltimore, MD: John Hopkins University Press, 1981 [1959]), p. 64. This is consistent with recognition that the 'liberal principle of legitimacy states that the exercise of political power is justifiable only when it is exercised in accordance with constitutional essentials that all citizens may reasonably be expected to endorse in the light of principles and ideals acceptable to them as reasonable and rational', Solum, 'Public Legal Reason', 1472. See also D. Curtin and A. J. Meijer, 'Does Transparency Strengthen Legitimacy?', (2006) *Inform Polity II*, 11(2), 109–122, 112 and M. J. Taylor, 'Health Research, Data Protection, and the Public Interest in Notification', (2011) *Medical Law Review*, 19(2), 267–303.

[11] The argument offered is a development of one originally presented in M. J. Taylor, *Genetic Data and the Law* (Cambridge University Press, 2012), see esp. pp. 29–34.

[12] The term 'accept' is chosen over 'prefer' for good reason. M. J. Taylor and N. C. Taylor 'Health Research Access to Personal Confidential Data in England and Wales: Assessing Any Gap in Public Attitude between Preferable and Acceptable Models of Consent', (2014) *Life Sciences, Society and Policy*, 10(1), 1–24.

[13] A 'real definition' is to be contrasted with a nominal definition. A real definition may associate a word or term with elements that must necessarily be associated with the referent (a priori). A nominal definition may be discovered by investigating word usage (a posteriori). For more, see Stanford Encyclopedia of Philosophy, 'Definitions', (Stanford Encyclopedia of Philosophy, 2015), www.plato.stanford.edu/entries/definitions/.

[14] G. Laurie recognises privacy to be a state of non-access. G. Laurie, *Genetic Privacy: A Challenge to Medico-Legal Norms* (Cambridge University Press, 2002) p. 6. We prefer the term 'exclusivity' rather than 'separation' as it recognises a lack of separation in one aspect does not deny a privacy claim in another. E.g. one's normative expectations regarding use and disclosure are not necessarily weakened by sharing information with health professionals. For more see M. J. Taylor, *Genetic Data and the Law: A Critical Perspective on Privacy Protection* (Cambridge University Press, 2012), pp. 13–40.

[15] See, C. J. Bennet and C. D. Raab, *The Governance of Privacy: Policy Instruments in Global Perspective* (Ashgate, 2003), p. 13.

recognise a common interest in privacy protection. It is, however, necessary to broaden the conception in order to bring all of the *possible* common interests in privacy into view. As Bennet and Raab note, the atomistic conception of privacy

> fails to properly understand the construction, value and function of privacy within society.[16]

Our ambition here is not to demonstrate an atomistic conception to be 'wrong' in any objective or absolute sense; but, rather to recognise the possibility that a coherent conception of privacy may extend its reach and capture additional values and functions. In 1977, after a comprehensive survey of the literature available at the time, Margulis proposed the following consensus definition of privacy

> [P]rivacy, as a whole or in part, represents control over transactions between person(s) and other(s), the ultimate aim of which is to enhance autonomy and/or to minimize vulnerability.[17]

Nearly thirty years after the definition was first offered, Margulis recognised that his early attempt at a consensus definition

> failed to note that, in the privacy literature, control over transactions usually entailed limits on or regulation of access to self (Allen, 1998), sometimes to groups (e.g., Altman, 1975), and occasionally to larger collectives such as organisations (e.g., Westin, 1967).[18]

The adjustment is important. It allows for a conception of privacy to recognise that there may be relevant norms, in relation to transactions involving data, that do not relate to identifiable individuals but are nonetheless associated with normative expectation of data flows and separation. Not only is there evidence that there are already such expectations in relation to non-identifiable data,[19] but data relating to groups – rather than just individuals – will be of increasing importance.[20]

There are myriad examples of how aggregated data have led to differential treatment of individuals due to association with group characteristics.[21] Beyond the obvious examples of individual discrimination and stigmatisation due to inferences drawn from (perceived) group membership, there can be group harm(s) to collective interests including, for example, harm connected to things held to be of common cultural value and significance.[22] It is the fact that data relates to the group level that leaves cultural values vulnerable to misuse of the data.[23] This

---

[16] Ibid.

[17] S. T. Margulis 'Conceptions of Privacy: Current Steps and Next Steps', (1977) *Journal of Social Issues*, 33(3), 5–21, 10.

[18] S. T. Margulis 'Privacy as a Social Issue and a Behavioural Concept', (2003) *Journal of Social Issues*, 9(2), 243–261, 245.

[19] Department of Health, 'Summary of Responses to the Consultation on the Additional Uses of Patient Data', (Department of Health, 2008).

[20] Our argument has no application to aggregate data that does not relate to a group until or unless that association is made.

[21] A number are described for example by V. Eubanks, *Automating Inequality: How High-Tech Tools Profile, Police and Punish the Poor* (New York: St Martin's Press, 2018).

[22] See e.g. *Foster* v. *Mountford* (1976) 14 ALR 71. (Australia)

[23] An example of the kind of common purpose that privacy may serve relates to the protection of culturally significant information. A well-known example of this is the harm associated with the research conducted with the Havasupai Tribe in North America. R. Dalton, 'When Two Tribes Go to War', (2004) *Nature*, 430(6999), 500–502; A. Harmon, 'Indian Tribe Wins Fight to Limit Research of Its DNA', *The New York Times* (21 April 2010). Similar concerns had been expressed by the Nuu-chahnulth of Vancouver Island, Canada, when genetic samples provided for one purpose (to discover the cause of rheumatoid arthritis) were used for other purposes. J. L. McGregor, 'Population Genomics and Research Ethics with Socially Identifiable Groups', (2007) *Journal of Law and Medicine*, 35(3), 356–370, 362. Proposals to establish a genetic database on Tongans floundered when the ethics policy focused on the notion of individual informed consent and failed to take account of the traditional role played by the extended family in decision-making. B. Burton, 'Proposed Genetic Database on Tongans Opposed', (2002) *BMJ*, 324(7335), 443.

goes beyond a recognition that privacy may serve 'not just individual interests but also common, public, and collective purposes'.[24] It is recognition that it is not only individual privacy but group privacy norms that may serve these common purposes. In fact, group data, and the norms of exclusivity associated with it, are likely to be of increasing significance for society. As Taylor, Floridi and van der Sloot note,

> with big data analyses, the particular and the individual is no longer central. ... Data is analysed on the basis of patterns and group profiles; the results are often used for general policies and applied on a large scale.[25]

This challenges the adequacy of a narrow atomistic conception of privacy to account for what will increasingly matter to society. De-identification of an individual as a member of a group, including those groups that may be created through the research and may not otherwise exist, does not protect against any relevant harm.[26] In the next part, we suggest that not only can the concept of the public interest be used to bring the full range of privacy interests into view, but that a failure to do so will undermine the legitimacy of any public interest decision-making process.

24.4 GROUP PRIVACY INTERESTS AND THE LAW

The argument in this section is that, although HRR does not currently recognise the concept of group privacy interests, through the concept of public interest inherent to both the law of data protection and the duty of confidence, there is opportunity to bring group privacy interests into view.

24.4.1 *Data Protection Law*

The Council of Europe Convention for the Protection of Individuals with regard to Automatic Processing of Personal Data (hereafter, Treaty 108) (as amended)[27] cast the template for subsequent data protection law when it placed the individual at the centre of its object and purpose[28] and defined 'personal data' as:

> any information relating to an identified or identifiable individual ('data subject')[29]

This definition narrows the scope of data protection law even further than data relating to an individual. Data relating to unidentified or unidentifiable individuals fall outside its concern. This blinkered view is replicated through data protection instruments from the first through to the most recent: the EU General Data Protection Regulation (GDPR).

The GDPR is only concerned with personal data, defined in a substantively similar and narrow fashion to Treaty 108. In so far as its object is privacy protection, it is predicated upon a

---

[24] P. M. Regan, *Legislating Privacy: Technology, Social Values, and Public Policy* (University of North Carolina Press, 1995) p. 221.

[25] L. Taylor et al. (eds), *Group Privacy: New Challenges of Data Technologies* (New York: Springer, 2017), p. 5.

[26] Taylor et al., 'Group Privacy', p. 7.

[27] Convention for the Protection of Individuals with regard to Automatic Processing of Personal Data, Strasbourg, 28 January 1981, in force 1 October 1985, ETS No. 108, Protocol CETS No. 223.

[28] To protect every individual, whatever his or her nationality or residence, with regard to the processing of their personal data, thereby contributing to respect for his or her human rights and fundamental freedoms, and in particular the right to privacy.

[29] 'Convention for the Protection of Individuals', Article 2(a).

relatively narrow and atomistic, conception of privacy. However, if the concerns associated with group privacy are viewed through the lens of public interest, then they may be given definition and traction even within the scope of a data protection instrument like the GDPR. The term 'the public interest' appears in the GDPR no fewer than seventy times. It has a particular significance in the context of health research. This is an area, such as criminal investigation, where the public interest has always been protected.

Our argument is that it is through the application of the public interest test to health research governance in data protection law, that there is an opportunity to recognise in part common interests in group privacy. For example, any processing of personal data within material and territorial scope of the GDPR requires a lawful basis. Among the legal bases most likely to be applicable to the processing of personal data for research purposes is either that the processing is necessary for the performance of a task carried out in the public interest or in the exercise of official authority vested in the controller (Article 6(1)(e)), or, that it is necessary for the purposes of the legitimate interests pursued by the controller (Article 6(1)(f)). In the United Kingdom (UK), where universities are considered to be public authorities, universities are unable to rely upon 'legitimate interests' as a basis for lawful processing. Much health research in the UK will thus be carried out on the basis that it is necessary for the performance of a task in the public interest. Official guidance issued in the UK is that the organisations relying upon the necessity of processing to carry out a task 'in the public interest'

> should document their justification for this, by reference to their public research purpose as established by statute or University Charter.[30]

Mere assertion that a particular processing operation is consistent with an organisation's public research purpose will provide relatively scant assurance that the operation is necessary for the performance of a task in the public interest. More substantial justification would document justification relevant to particular processing operations. Where research proposals are considered by institutional review boards, such as university or NHS ethics committees, then independent consideration by such bodies of the public interest in the processing operation would provide the rationale. We suggest this provides an opportunity for group privacy concerns to be drawn into consideration. They might also form part of any privacy impact assessment carried out by the organisation. What is more, for the sake of legitimacy, any interference with group interests, or risk of harm to members of a group or to the collective interests of the group as a whole, should be subject to the test that members of the group be offered *reasons to accept* the processing as appropriate.[31] Such a requirement might support good practice in consumer engagement prior to the roll out of major data initiatives.

Admittedly, while this may provide opportunity to bring group privacy concerns into consideration where processing is carried out by a public authority (and the legal basis of processing is performance of a task carried out in the public interest), this only provides limited penetration of group privacy concerns into the regulatory framework. It would not, for example, apply where processing was in pursuit of legitimate interests or another lawful basis. There are other limited opportunities to bring group privacy concerns into the field of vision of data protection law

[30] Health Research Authority, 'Legal Basis for Processing Data', (NHS Health Research Authority, 2018). www.hra.nhs.uk/planning-and-improving-research/policies-standards-legislation/data-protection-and-information-governance/gdpr-detailed-guidance/legal-basis-processing-data/.

[31] M. J. Taylor and T. Whitton, 'Public Interest, Health Research and Data Protection Law: Establishing a Legitimate Trade-Off between Individual Control and Research Access of Health Data,' (2020) *Laws*, 9(6), 1–24, 17–19; J. Rawls, *The Law of Peoples* (Harvard University Press, 1999), pp. 129–180.

through the lens of public interest.[32] However, for as long as the gravitational orbit of the law is around the concept of 'personal data', the chances to recognise group privacy interests are likely to be limited and peripheral. By contrast, more fundamental reform may be possible in the law of confidence.

24.4.2 *Duty of Confidence*

As with data protection and privacy,[33] there is an important distinction to be made between privacy and confidentiality. However, the UK has successfully defended its ability to protect the right to respect for private and family life, as recognised by Article 8 of the European Convention on Human Rights (ECHR), by pointing to the possibility of an action for breach of confidence.[34] It has long been recognised that the law's protection of confidence is grounded in the public interest[35] but, as Lord Justice Briggs noted in R *(W,X,Y and Z)* v. *Secretary of State for Health* (2015),

> the common law right to privacy and confidentiality is not absolute. English common law recognises the need for a balancing between this right and other competing rights and interests.[36]

The argument put forward here is consistent with the idea that the protection of privacy and other competing rights and interests, such as those associated with health research, are each in the public interest. The argument here is that *when* considering the appropriate balance or trade-off between different aspects of the public interest, then a broader view of privacy protection than has hitherto been taken by English law is necessary to protect the legitimacy of decision-making. Such judicial innovation is possible.

The law of confidence has already evolved considerably over the past twenty or so years. Since the *Human Rights Act* 1998[37] came into force in 2000, the development of the common law has been in harmony with Articles 8 and 10 of the ECHR.[38] As a result, as Lord Hoffmann put it,

> What human rights law has done is to identify private information as something worth protecting as an aspect of human autonomy and dignity.[39]

Protecting private information as an aspect of *individual* human autonomy and dignity might signal a shift toward the kind of narrow and atomistic conception of privacy associated with data protection law. This would be as unnecessary as it would be unfortunate. In relation to the idea of privacy, the European Court of Human Rights has itself said that

> The Court does not consider it possible or necessary to attempt an exhaustive definition of the notion of 'private life' ... Respect for private life must also comprise to a certain degree the right to establish and develop relationships with other human beings.[40]

---

[32] E.g. The conception of public interest proposed in this chapter would allow concerns associated with processing in a third country, or an international organisation, to be taken into consideration where associated with issues of group privacy. Article 49(5) of the General Data Protection Regulation, Regulation (EU) 2016/679, OJ L 119, 4 May 2016.
[33] Although data protection law seeks to protect fundamental rights and freedoms, in particular the right to respect for a private life, without collapsing the concepts of data protection and privacy.
[34] *Earl Spencer* v. *United Kingdom* [1998] 25 EHRR CD 105.
[35] *W* v. *Egdell* [1993] 1 All ER 835. Ch. 359.
[36] *R (W, X, Y and Z)* v. *Secretary of State for Health* [2015] EWCA Civ 1034, [48].
[37] *Campbell* v. *MGN Ltd* [2004] UKHL 22, [2004] ALL ER (D) 67 (May), *per* Lord Nicholls [11].
[38] Ibid. [14].
[39] Ibid. [50].
[40] *Niemietz* v. *Germany* [1992] 13710/88 [29].

It remains open to the courts to recognise that the implications of group privacy concerns have a bearing on an individual's ability to establish and develop relations with other human beings. Respect for human autonomy and dignity may yet serve as a springboard toward a recognition by the law of confidence that data processing impacts upon the conditions under which we live social (not atomistic) lives and our ability to establish and develop relationships as members of groups. After all, human rights are due to members of a group and their protection has always been motivated by group concerns.[41]

One of us has argued elsewhere that English Law took a wrong turn when *R (Source Informatics) v. Department of Health*[42] was taken to be authority for the proposition that a duty of confidence cannot be breached through the disclosure of non-identifiable data. It is possible that the ratio in *Source Informatics* may yet be re-interpreted and recognised to be consistent with a claim that legal duties may be engaged through use and disclosure of non-identifiable data.[43] In some ways, this would simply be to return to the roots of the legal protection of privacy. In her book *The Right to Privacy*, Megan Richardson traces the origins and influence of the ideas underpinning the legal right to privacy. As she remarks, 'the right from the beginning has been drawn on to serve the rights and interests of minority groups'.[44] Richardson recognises that, even in those cases where an individual was the putative focus of any action or argument,

> Once we start to delve deeper, we often discover a subterranean network of families, friends and other associates whose interests and concerns were inexorably tied up with those of the main protagonist.[45]

As a result, it has always been the case that the right to privacy has 'broader social and cultural dimensions, serving the rights and interests of groups, communities and potentially even the public at large'.[46] It would be a shame if, at a time when we may need it most, the duty of confidence would deny its own potential to protect reasonable expectations in the use and disclosure of information simply because misuse had the potential to impact more than one identifiable individual.

24.5 CONCLUSION

The argument has been founded on the claim that a commitment to the protection of common interests in privacy and the product of health research, if placed alongside the commonly held value in individuals as free and equal persons, may establish a platform upon which one can construct a substantive idea of the public interest. If correct, then it is important to a proper calculation of the public interest to understand the breadth of privacy interests that need to be accounted for if we are to avoid subjugating the public to governance, and a trade-off between competing interests, that they have no reason to accept.

---

[41] Regan, 'Legislating Privacy', p. 8.

[42] [1999] EWCA Civ 3011.

[43] M. J. Taylor, '*R (ex p. Source Informatics) v. Department of* Health [1999]' in J. Herring and J. Wall (eds), *Landmark Cases in Medical Law* (Oxford: Hart, 2015), pp. 175–192; D. Beyleveld, 'Conceptualising Privacy in Relation to Medical Research Values' in S. A. M. MacLean (ed.), *First Do No Harm* (Farnham, UK: Ashgate, 2006), p. 151. It is interesting to consider how English Law may have something to learn in this respect from the Australian courts e.g. *Foster v. Mountford* (1976) 14 ALR 71.

[44] M. Richardson, *The Right to Privacy* (Cambridge University Press, 2017), p. 120.

[45] Ibid., p. 122.

[46] Ibid., p. 119.

Enabling access to the data necessary for health research is in the public interest. So is the protection of group privacy. Recognising this point of connection can help guide decision-making where there *is* some kind of conflict or tension. The public interest can provide a common, commensurate framing. When this framing has a normative dimension, then this grounds the claim that the full range of common interests ought to be brought into view and weighed in the balance. One must capture all interests valued by the affected public, whether individual or common in nature, to offer them a reason to accept a particular trade-off between privacy and the public interest in health research. To do otherwise is to get the balance of governance wrong and compromise its social legitimacy.

That full range of common interests must include interests in group data. An understanding of what the public interest requires in a particular situation is short-sighted if this is not brought into view. An implication is that group interests must be taken into account within an interpretation and application of public interest in data protection law. Data controllers should be accountable for addressing group privacy interests in any public interest claim. With respect to the law of confidence, there is scope for even more significant reform. If the legitimacy of the governance framework, applicable to health data, is to be assured into the future, then it needs to be able to see – so that it might protect – reasonable expectations in data relating to groups of persons and not just identifiable individuals. Anything else will be a myopic failure to protect some of the most sensitive data about people simply on the grounds that misuse does not affect a sole individual but multiple individuals simultaneously. That is not a governance model that we have any reason to accept and we have the concept of public interest at our disposal to correct our vision and bring the full range of relevant interests into view.

25

Mobilising Public Expertise in Health Research Regulation

Michael M. Burgess

25.1 INTRODUCTION

This chapter will develop the role that deliberative public engagement should have in health research regulation. The goal of public deliberation is to mobilise the expertise that members of the public have, to explore their values in relation to specific trade-offs, with the objective of recommendations that respect diverse interests. Public deliberation requires that a small group is invited to a structured event that supports informed, civic-minded consideration of diverse perspectives on public interest. Ensuring that the perspectives considered are inclusive of perspectives that might otherwise be marginalised or silenced requires explicitly designing the small group in relation to the topic. Incorporating public expertise enhances the trustworthiness of policies and governance by explicitly acknowledging and negotiating diverse public interests. Trustworthiness is distinct from trust, so the chapter begins by exploring that distinction in the context of the example of care.data and the loss of trust in the English National Health Service's (NHS) use of electronic health records for research. While better public engagement prior to the announcement might have avoided the loss of trust, subsequent deliberative public engagement may build trustworthiness into the governance of health research endeavours and contribute to re-establishing trust.

25.2 TRUSTWORTHINESS OF RESEARCH GOVERNANCE

Some activities pull at the loose threads of social trust. These events threaten to undermine the presumption of legitimacy that underlie activities directed to public interest. The English NHS care.data programme is one cautionary tale (NHS, 2013).[1] The trigger event was the distribution of a pamphlet to households. The pamphlet informed the public that a national database of patients' medical records would be used for patient care, to monitor outcomes, and that research would take place on anonymous datasets. The announcement was entirely legitimate within the existing regulatory regime. The Editor-in-Chief of the *British Medical Journal* summarises, 'But all did not go to plan. NHS England's care.data programme failed to win the public's trust and lost the battle for doctors' support. Two reports have now condemned the scheme, and last week the government decided to scrap it.'[2]

---

[1] NHS, 'News: NHS England sets out the next steps of public awareness about care.data', (NHS, 2013), www.england
.nhs.uk/2013/10/care-data/.
[2] F. Godlee, 'What Can We Salvage From care.data?', (2016) *BMJ*, 354(i3907).

The stimulation of public distrust is often characterised by a political deployment of a segment of the public, but it may lead to a wider rejection of previously non-controversial trade-offs. In the case of care.data, the first response was to ensure better education about benefits and enhanced informed consent. The Caldicott Report on the care.data programme called for better technology standards, publication of disclosure procedures, an easy opt-out procedure and a 'dynamic consent' process.[3]

There are good reasons to doubt that improved regulation and informed consent procedures alone will restore the loss, or sustain current levels, of public trust. It was unlikely that the negative reaction to care.data had to do with an assessment of the adequacy of the regulations for privacy and access to health data. Moreover, everything proposed under care.data was perfectly lawful. It is far more likely that the reaction had to do with a rejection of what was presented as a clear case of justified use of patients' medical records. The perception was that the trade-offs were not legitimate, at least to some of the public and practitioners. The destabilisation of trust that patient information was being used in appropriate ways, even with what should have been an innocuous articulation of practice, suggests a shift in how the balance between access to information and privacy are perceived. Regulatory experts on privacy and informed consent may strengthen protection or recalibrate what is protected. But such measures do not develop an understanding and response to how a wider public might assign proportionate weight to privacy and access in issues related to research regulation. Social controversy about the relative weight of important public interests demonstrates the loss of legitimacy of previous decisions and processes. It is the legitimacy of the programmes that require public input.

The literature on public understanding of science also suggests that merely providing more detailed information and technical protections is unlikely to increase public trust.[4] Although alternative models of informed consent are beyond the scope of this chapter, it seems more likely that informed consent depends on relationships of trust, and that trust, or its absence, is more of a heuristic approach that serves as a context in which people make decisions under conditions of limited time and understanding.[5] Trust is often extended without assessment of whether the conditions justify trust, or are trustworthy. It also follows that trust may not be extended even when the conditions seem to merit trust. The complicated relationship between trust and trustworthiness has been discussed in another chapter (see Chuong and O'Doherty, Chapter 12) and in the introduction to this volume, citing Onora O'Neill, who encourages us to focus on *demonstrating* trustworthiness in order to earn trust.

The care.data experience illustrates how careful regulation within the scope of law and current research ethics, and communicating those results to a wide public, was not sufficient for the plan to be perceived as legitimate and to be trusted. Regulation of health research needs to be trustworthy, yet distrust can be stimulated despite considerable efforts and on-going vigilance. If neither trust nor distrust are based on the soundness of regulation of health research, then the sources of distrust need to be explicitly addressed.

---

[3] F. Caldicott et al., 'Information: To Share Or Not to Share? The Information Governance Review', (UK Government Publishing Service, 2013).

[4] A. Irwin and B. Wynne, *Misunderstanding Science. The Public Reconstruction of Science and Technology* (Abingdon: Routledge, 1996); S. Locke, 'The Public Understanding of Science – A Rhetorical Invention', (2002) *Science Technology & Human Values*, 27(1), 87–111.

[5] K. C. O'Doherty and M. M. Burgess, 'Developing Psychologically Compelling Understanding of the Involvement of Humans in Research', (2019) *Human Arenas* 2(6), 1–18.

25.3 PATIENTS OR PUBLIC? CONCEPTUALISING WHAT INTERESTS ARE IMPORTANT

It is possible to turn to 'patients' or 'the public' to understand what may stabilise or destabilise trust and legitimacy in health research. There is considerable literature and funding opportunities related to involving patients in research projects, and related improved outcomes.[6] The distinction between public and patients is largely conceptual, but it is important to clarify what aspects of participants' lives we are drawing on to inform research and regulation, and then to structure recruitment and the events to emphasise that focus.[7] In their role as a patient, or caregivers and advocates for family and friends in healthcare, participants can draw on their experiences to inform clinical care, research and policy. In contrast, decisions that allocate across healthcare needs, or broader public interests, require consideration of a wider range of experiences, as well as the values, and practical knowledge that participants hold as members of the public. Examples of where it is important to achieve a wider 'citizen' perspective include funding decisions on drug expenditures and disinvestment, and balancing privacy concerns against benefits from access to health data or biospecimens.[8] Considerations of how to involve the public in research priorities is not adequately addressed by involving community representatives on research ethics review committees.[9]

Challenges to trust and legitimacy often arise when there are groups who hold different interpretations of what is in the public interest. Vocal participants on issues are often divide into polarised groups. But there is often also a multiplicity of public interests, so there is no single 'public interest' to be discovered or determined. Each configuration of a balance of interests also has resource implications, and the consequences are borne unevenly across the lines of inequity in society. There is democratic deficit when decisions are made without input from members of public who will be affected by the policy but have not been motivated to engage. This deficit is best addressed by 'actively seek(ing) out moral perspectives that help to identify and explore as many moral dimensions of the problem as possible'.[10] This rejects the notions that bureaucracies

[6] J. F. Caron-Flinterman et al., 'The Experiential Knowledge of Patients: A New Resource for Biomedical Research?', (2005) *Social Science and Medicine*, 60(11), 2575–2584; M. De Wit et al., 'Involving Patient Research Partners has a Significant Impact on Outcomes Research: A Responsive Evaluation of the International OMERACT Conferences', (2013) *BMJ Open*, 3(5); S. Petit-Zeman et al., 'The James Lind Alliance: Tackling Research Mismatches', (2010) *Lancet*, 376(9742), 667–669; J. A. Sacristan et al., 'Patient Involvement in Clinical Research: Why, When, and How', (2016) *Patient Preference and Adherence*, 2016(10), 631–640.

[7] C. Mitton et al., 'Health Technology Assessment as Part of a Broader Process for Priority Setting and Resource Allocation', (2019) *Applied Health Economics and Health Policy*, 17(5), 573–576.

[8] M. Aitken et al., 'Consensus Statement on Public Involvement and Engagement with Data-Intensive Health Research', (2018) *International Journal of Population Data Science*, 4(1), 1–6; C. Bentley et al., 'Trade-Offs, Fairness, and Funding for Cancer Drugs: Key Findings from a Public Deliberation Event in British Columbia, Canada', (2018) *BMC Health Services Research*, 18(1), 339–362; S. M. Dry et al., 'Community Recommendations on Biobank Governance: Results from a Deliberative Community Engagement in California', (2017) *PLoS ONE* 12(2), 1–14; R. E. McWhirter et al., 'Community Engagement for Big Epidemiology: Deliberative Democracy as a Tool', (2014) *Journal of Personalized Medicine*, 4(4), 459–474.

[9] J. Brett et al., 'Mapping the Impact of Patient and Public Involvement on Health and Social Care Research: A Systematic Review', (2012) *Health Expectations*, 17(5), 637–650; R. Gooberman-Hill et al., 'Citizens' Juries in Planning Research Priorities: Process, Engagement and Outcome', (2008) *Health Expectations*, 11(3), 272–281; S. Oliver et al., 'Public Involvement in Setting a National Research Agenda: A Mixed Methods Evaluation', (2009) *Patient*, 2(3), 179–190.

[10] S. Sherwin, 'Toward Setting an Adequate Ethical Framework for Evaluating Biotechnology Policy', (Canadian Biotechnology Advisory Committee, 2001). As cited in M. M. Burgess and J. Tansey, 'Democratic Deficit and the Politics of "Informed and Inclusive" Consultation' in E. Einsiedel (ed.), *From Hindsight to Foresight* (Vancouver: UBC Press, 2008), pp. 275–288.

and elected representatives are adequately informed by experts and stakeholders to determine what is in the interests of all who will be affected by important decisions. These decisions are, in fact, about a collective future, often funded by public funds with opportunity costs. Research regulation, like biotechnology development and policy, must explicitly consider how and who decides the relative importance of benefits and risks.

The distinction between trust and trustworthiness, between bureaucratic legitimacy and perceived social licence, gives rise to the concern that much patient and public engagement may be superficial and even manipulative.[11] Careful consideration must be given to how the group is convened, informed, facilitated and conclusions or recommendations are formulated. An earlier chapter considered the range of approaches to public and patient engagement, and how different approaches are relevant for different purposes (see Aitkin and Cunningham-Burley, Chapter 11).[12] To successfully stimulate trust and legitimacy, the process of public engagement requires working through these dimensions.

25.4 CONCEPTUALISING PUBLIC EXPERTISE: REPRESENTATION AND INCLUSIVENESS

The use of the term 'public' is normally intended to be as inclusive as possible, but it is also used to distinguish the call to public deliberation from other descriptions of members of society or stakeholders. There is a specific expertise called upon when people participate as members of the public as opposed to patients, caregivers, stakeholders or experts. Participants are sought for their broad life perspective. As perspective bearers coming from a particular structural location in society, with 'experience, history and social knowledge',[13] participants draw on their own social knowledge and capacity in a deliberative context that supports this articulation without presuming that their experiences are adequate to understand that of others', or that there is necessarily a common value or interest

'Public expertise' is what we all develop as we live in our particular situatedness, and in structured deliberative events it is blended with an understanding of other perspectives, and directed to develop collective advice related to the controversial choices that are the focus of the deliberation. Adopting Althusser's notion of hailing or 'interpellation' as ideological construction of people's role and identity, Berger and De Cleen suggest that calling people to deliberate 'offers people the opportunity to speak (thus empowering them) *and* a central aspect of how their talk is constrained and given direction (the exercise of power on people)'.[14] In deliberation, the manifestations of public expertise is interwoven with the overall framing, together co-creating the capacity to consider the issues deliberated from a collective point of view.[15] Political scientist Mark Warren suggests that '(r)epresentation can be designed to include marginalized people

[11] A. Irwin et al., 'The Good, the Bad and the Perfect: Criticizing Engagement Practice', (2013) *Social Studies of Science*, 43(1), 118–135; S. Jasanoff, *The Ethics of Invention: Technology and the Human Future* (Manhattan, NY: Norton Publishers, 2016); B. Wynne, 'Public Engagement as a Means of Restoring Public Trust in Science: Hitting the Notes, but Missing the Music?', (2006) *Community Genetics* 9(3), 211–220.

[12] J. Gastil and P. Levine, *The Deliberative Democracy Handbook: Strategies for Effective Civic Engagement in the Twenty-First Century* (Plano, TX: Jossey-Bass Publishing, 2005).

[13] I. M. Young, *Inclusion and Democracy* (Oxford University Press, 2000), p. 136.

[14] M. Berger and B. De Cleen, 'Interpellated Citizens: Suggested Subject Positions in a Deliberation Process on Health Care Reimbursement', (2018) *Comunicazioni Sociali*, 1, 91–103; L. Althusser, 'Ideology and Ideological State Apparatuses: Notes Towards an Investigation' in L. Althusser (ed.) *Lenin and Philosophy and Other Essays* (Monthly Review Press, 1971), pp. 173–174.

[15] H. L. Walmsley, 'Mad Scientists Bend the Frame of Biobank Governance in British Columbia', (2009) *Journal of Public Deliberation*, 5(1), Article 6.

and unorganized interests, as well as latent public interests'.[16] As one form of deliberation, citizen juries, captured in the name and process, the courts have long drawn on public to constitute a group of peers who must make sense and form collective judgments out of conflicting and diverse information and alternative normative weightings.[17]

Simone Chambers, in a classic review of deliberative democracy, emphasised two critiques from diversity theory, and suggested that these would be a central concern in the next generation of deliberative theorists: (1) reasonableness and reason-giving; (2) conditions of equality as participants in deliberative activities.[18] The facilitation of deliberative events is discussed below, but participants can be encouraged and given the opportunity to understand each other's perspectives in a manner that may be less restrictive than theoretical discussions suggest. For example, the use of narrative accounts to explain how participants come to hold particular beliefs or positions provide important perspectives that might not be volunteered or considered if there was a strong emphasis on justifying one's views with reasons in order for them to be considered.[19]

The definition and operationalisation of inclusiveness is important because deliberative processes are rarely large scale, focussing instead on the way that small groups can demonstrate how a wider public would respond if they were informed and civic-minded.[20] Representation or inclusiveness is often the starting place for consideration of an engagement process.[21] Steel and colleagues have described three different types of inclusiveness that provides conceptual clarity about the constitution of a group for engagement: representative, egalitarian and normic diversity.[22]

Representative diversity requires that the distribution of the relevant sub-groups in the sample reflects the same distribution as in the reference population. *Egalitarian* inclusiveness requires equal representation of people from each relevant sub-group so that each perspective is given equal representation. In contrast to representative diversity, *Egalitarian* diversity ignores the size of each sub-group in the population, and emphasises equal representation of each sub-group. *Normic* diversity requires the over-representation of sub-groups who are marginalised or over-whelmed by the larger, more influential or mainstream groups in the population. Each of these concepts aim for a symmetry, but the representative approach presumes that symmetry is the replication of the population, while egalitarian and normic concepts directly consider asymmetry of power and voice in society.

Attempts to enhance the range of perspectives considered in determining the public interest(s) is likely to draw on the normic and egalitarian concepts of diversity, and de-emphasise the representative notion. The goal of deliberative public engagement is to address a democratic deficit whereby some groups have been the dominant perspectives considered on the issues, even if none have prevailed over others. It seeks to include a wider range of input from diverse citizens about how to live together given the different perspectives on what is 'in the public

---

[16] M. E. Warren, 'Governance-Driven Democratization', (2009) *Critical Policy Studies*, 3(1), 3–13, 10.

[17] G. Smith and C. Wales, 'Citizens' Juries and Deliberative Democracy', (2000) *Political Studies*, 48(1), 51–65.

[18] S. Chambers, 'Deliberative Democratic Theory', (2003) *Annual Review of Political Science*, 6, 307–326.

[19] M. M. Burgess et al., 'Assessing Deliberative Design of Public Input on Biobanks' in S. Dodds and R. A. Ankeny (eds) *Big Picture Bioethics: Developing Democratic Policy in Contested Domains* (Switzerland: Springer, 2016), pp. 243–276.

[20] R. E. Goodin and J. S. Dryzek, 'Deliberative Impacts: The Macro-Political Uptake of Mini-Publics', (2006) *Politics & Society*, 34(2), 219–244.

[21] H. Longstaff and M. M. Burgess, 'Recruiting for Representation in Public Deliberation on the Ethics of Biobanks', (2010) *Public Understanding of Science*, 19(2), 212–24.

[22] D. Steel et al., 'Multiple Diversity Concepts and Their Ethical-Epistemic Implications', (2018) *The British Journal for the Philosophy of Science*, 8(3), 761–780.

interest. Normic diversity suggests that dominant groups are less present in the deliberating group, and egalitarian suggests that it is important to have similar representation across the anticipated diverse perspectives. The deliberation must be informed about, but not subjugated by, dominant perspectives, and one approach is to exclude dominant perspectives, including those of substance experts, from participating in the deliberation, but introduce their perspectives and related information through materials and presentations intended to inform participants. Deliberative participants must exercise their judgement and critically consider a wide range of perspectives, while stakeholders are agents for a collective identity that asserts the importance of one perspective over others.[23] It is also challenging to identify the range of relevant perspectives that give particular form to the public expertise for an issue, although demographics may be used to ensure that participants reflect a range of life experiences.[24] Specific questions may also suggest that particular public perspectives are important to include in the deliberating group. For example, in Californian deliberations on biobanks it was important to include Spanish-only speakers because, despite accounting for the majority of births, they were often excluded from research regulation issues (normic diversity), and they were an identifiable group who likely had unique perspectives compared to other demographic segments of the California population (egalitarian diversity).[25]

25.5 MOBILISING PUBLIC EXPERTISE IN DELIBERATION

As previously discussed, mobilising public expertise requires considerable support. To be credible and legitimate, a deliberative process must demonstrate that the participants are adequately informed and consider diverse perspectives. Participants must respectfully engage each other in the development of recommendations that focus on reasoned inclusiveness but fully engage the trade-offs required in policy decisions.

It seems obvious that participation in an engagement to advise research regulation must be informed about the activities to be regulated. This is far from a simple task. An engagement can easily be undermined if the information provided is incomplete or biased. It is important to provide not only complete technical details, but also ensure that social controversies and stakeholder perspectives are fairly represented. This can be managed by having an advisory of experts, stakeholders and potential knowledge users. Advisors can provide input into the questions and the range of relevant information that participants must consider to be adequately informed. It is also important to consider how best to provide information to support comprehension across participants with different backgrounds. One approach is to utilise a combination of a background booklet and a panel of four to six speakers.[26] The speakers, a combination of experts and stakeholders, are asked to be impassioned, explaining how or why they come to their particular view. This will establish that there are controversies, help draw participants into the issues and stimulate interest in the textual information.

Facilitation is another critical element of deliberative engagement. Deliberative engagement is distinguished by collective decisions supported by reasons from the participants – the recommendations and conclusions are the result of a consideration of the diverse perspectives reflected in the process and among participants. The approach to facilitation openly accepts that

[23] K. Beier et al., 'Understanding Collective Agency in Bioethics', (2016) *Medicine, Health Care and Philosophy*, 19(3), 411–422.
[24] Longstaff and Burgess, 'Recruiting for Representation'.
[25] S. M. Dry et al., 'Community Recommendations on Biobank Governance'.
[26] Burgess et al., 'Assessing Deliberative Design', pp. 270–271.

participants may re-orient the discussion and focus, and that the role of facilitation is to frame the discussion in a transparent manner.[27] Small groups of six to eight participants can be facilitated to develop fuller participation and articulation of different perspectives and interests than is possible in a larger group. Large group facilitation can be oriented to giving the participants as much control over topic and approach as they assume, while supporting exploration of issues and suggesting statements where the group may be converging. The facilitator may also draw closure to enable participants to move on to other issues by suggesting that there is a disagreement that can be captured. Identifying places of deep social disagreement identifies where setting policy will need to resolve controversy about what is genuinely in the public's interest, and where there may be a need for more nuanced decisions on a case-by-case basis. The involvement of industry and commercialisation in biobanks is a general area that has frequently defied convergence in deliberation.[28]

Even if recruitment succeeds in convening a diverse group of participants, sustaining diversity and participation across participants requires careful facilitation. The deliberative nature of the activity is dynamic. Participants increase their level of knowledge and understanding of diverse perspectives as facilitation encourages them to shift from an individual to a collective focus. Premature insistence on justifications can stifle understanding of diverse perspectives, but later in the event, justifications are crucial to produce reasons in support of conclusions. Discussion and conclusions can be inappropriately influenced by participants' personalities, as well as the tendency for some participants to position themselves as having authoritative expertise. It is well within the expertise of the public to consider whether claims to special knowledge or personalities are lacking substantive support for their positions. But self-reflective and respectful communication is not naturally occurring, and deliberation requires skilled facilitation to avoid dominance of some participants and to encourage critical reflection and participation of quieter participants. The framing of the issues and information as well as facilitation inevitably shapes the conclusions, and participants may not recognise that issues and concerns important to them have been ruled out of scope.

Assessing the quality of deliberative public engagement is fraught with challenges. Abelson and Nabatchi have provided good overviews of the state of deliberative civic engagement, assessing its impacts and assessment.[29]

There are recent considerations of whether and under what conditions deliberative public engagement is useful and effective.[30] Because deliberative engagement is expensive and resource intensive, it needs to be directed to controversies where the regulatory bodies want, and are willing, to have their decisions and policies shaped by public input. Such authorities do not thereby give up their legitimate duties and freedom to act in the public interest or to consult with experts and stakeholders. Rather, activities such as deliberative public engagement are

---

[27] A. Kadlec and W. Friedman, 'Beyond Debate: Impacts of Deliberative Issue Framing on Group Dialogue and Problem Solving', (Center for Advances in Public Engagement, 2009); H. L. Walmsley, 'Mad Scientists Bend the Frame of Biobank Governance in British Columbia', (2009) *Journal of Public Deliberation*, 5(1), Article 6.

[28] M. M. Burgess, 'Deriving Policy and Governance from Deliberative Events and Mini-Publics' in M. Howlett and D. Laycock (eds), *Regulating Next Generation Agri-Food Biotechnologies: Lessons from European, North American and Asian Experiences* (Abingdon: Routledge, 2012), pp. 220–236; D. Nicol, et al., 'Understanding Public Reactions to Commercialization of Biobanks and Use of Biobank Resources', (2016) *Social Sciences and Medicine*, 162, 79–87.

[29] J. Abelson et al., 'Bringing 'The Public' into Health Technology Assessment and Coverage Policy Decisions: From Principles to Practice', (2007) *Health Policy*, 82(1), 37–50; T. Nabatchi et al., *Democracy in Motion: Evaluating the Practice and Impact of Deliberative Civic Engagement* (Oxford University Press, 2012).

[30] D. Caluwaerts and M. Reuchamps, *The Legitimacy of Citizen-led Deliberative Democracy: The G1000 in Belgium* (Abingdon: Routledge, 2018).

supplemental to the other sources of advice, and not determinative of the outcomes. This point is important for knowledge users, sponsors and participants to understand.

How, then, might deliberative public engagement have helped avoid the negative reaction to care.data? It is first important to distinguish trust from trustworthiness. Trust, sometimes considered as social licence, is usually presumed in the first instance. As a psychological phenomenon, trust is often a heuristic form of reasoning that supports economical use of intellectual and social capital.[31] There is some evidence that trust is particularly important with regard to research participation.[32] Based on previous experiences, we develop trust – or distrust – in people and institutions. There is a good chance that many people whose records are in the NHS would approach the use of their records for other purposes with a general sense of trust. Loss of trust often flows from abrupt discovery that things are not as we presumed, which is what appears to have happened in care.data. On the other hand, trustworthiness of governance is when the governance system has the characteristics that, if scrutinised, would support that it is worthy of trust.

Given this understanding, it might have been possible to demonstrate trustworthiness of governance of the NHS data by holding deliberative public engagement and considering its recommendations for data management. Also, public trust might not have been as widely undermined if the announcement of extension of access to include commercial partners provided a basis for finding the governance trustworthy. Of course, distrust of critical stakeholders and members of public will still require direct responses to their concerns.

It is important to note that trustworthiness that can stand up to scrutiny is the goal, rather than directing efforts at increasing trust. Since trust is given in many cases without reflection, it can often be manipulated. By aiming at trustworthiness, arrived at through considerations that include deliberative public input, the authorities demonstrate that their approach is trustworthy. Articulating how controversies have been considered with input from informed and deliberating members of public would have demonstrated that the trust presumed at the outset was, in an important sense, justified. Now, after the trust has been lost and education and reinforced individual consent has not addressed the concerns, deliberation to achieve legitimate and trustworthy governance may have a more difficult time stimulating wide public trust, but it may remain the best available option.

25.6 CONCLUSION

Deliberative public engagement has an important role in the regulation of health research. Determining what trade-offs are in the public interest requires a weighing of alternatives and relative weights of different interests. Experts and stakeholders are legitimate advocates for the interests they represent, but their interests manifest an asymmetry of power. Including a well-designed process to include diverse public input can increase the legitimacy and trustworthiness of the policies. Deliberative engagement mobilises a wider public to direct their collective experience and expertise. The resulting advice about what is in the public interest explicitly builds diversity in the recruitment of the participants and in the design of the deliberation.

[31] G. Gigerenzer and P. M. Todd, 'Ecological Rationality: The Normative Study of Heuristics', in P. M. Todd and G. Gigerenzer (eds), *Ecological Rationality: Intelligence in the World* (Oxford University Press, 2012).

[32] E. Christofides et al., 'Heuristic Decision-Making About Research Participation in Children with Cystic Fibrosis', (2016) *Social Science & Medicine*, 162, 32–40; O'Doherty and Burgess, 'Developing Psychologically Compelling Understanding'; M. M. Burgess and K. C. O'Doherty, 'Moving from Understanding of Consent Conditions to Heuristics of Trust', (2019) *American Journal of Bioethics*, 19(5), 24–26.

Deliberative public engagement is helpful for issues where there is genuine controversy about what is in the public interest, but it is far from a panacea. It is an important complement to stakeholder and expert input. The deliberative approach starts with careful consideration of the issues to be deliberated and how the diversity is to be structured into the recruitment of a deliberating small group. Expert and stakeholder advisors, as well as decision-makers who are the likely recipients of the conclusions of the deliberation, can help develop the range of information necessary for deliberation on the issues to be informed. Participants need to be supported by exercises and facilitation that helps them develop a well-informed and respectful understanding of diverse perspectives. Facilitation then shifts to support the development of a collective focus and conclusions with justifications. Diversity and asymmetry of power is respected through the conceptualisation and implementation of inclusiveness, the development of information, and through facilitation and respect for different kinds of warranting. There must be a recognition that the role of event structure and facilitation means that the knowledge is co-produced with the participants, and that it is very challenging to overcome asymmetries, even in the deliberation itself. Another important feature is the ability to identify persistent disagreements and not force premature consensus on what is in the public interest. In this quality, it mirrors the need for, and nature of, regulation of health research to struggle with the issue of when research is in 'the public interest'.

It is inadequate to assert or assume that research and its existing and emerging regulation is in the public interest. It is vital to ensure wide, inclusive consideration that is not overwhelmed by economic or other strongly vested interests. This is best accomplished by developing, assessing and refining ways to better include diverse citizens in the informed reflections about what is in our collective interests, and how to best live together when those interests appear incommensurable.

Towards Adaptive Governance in Big Data Health Research

Implementing Regulatory Principles

Effy Vayena and Alessandro Blasimme

26.1 INTRODUCTION

In recent times, biomedical research has begun to tap into larger-than-ever collections of different data types. Such data include medical history, family history, genetic and epigenetic data, information about lifestyle, dietary habits, shopping habits, data about one's dwelling environment, socio-economic status, level of education, employment and so on. As a consequence, the notion of health data – data that are of relevance for health-related research or for clinical purposes – is expanding to include a variety of non-clinical data, as well as data provided by research participants themselves through commercially available products such as smartphones and fitness bands.[1] Precision medicine that pools together genomic, environmental and lifestyle data represents a prominent example of how data integration can drive both fundamental and translational research in important domains such as oncology.[2] All of this requires the collection, storage, analysis and distribution of massive amounts of personal information as well as the use of state-of-the art data analytics tools to uncover new disease-related patterns.

To date, most scholarship and policy on these issues has focused on privacy and data protection. Less attention has been paid to addressing other aspects of the wicked challenges posed by Big Data health research and even less work has been geared towards the development of novel governance frameworks.

In this chapter, we make the case for adaptive and principle-based governance of Big Data research. We outline six principles of adaptive governance for Big Data research and propose key factors for their implementation into effective governance structures and processes.

26.2 THE CASE FOR ADAPTIVE PRINCIPLES OF GOVERNANCE IN BIG DATA RESEARCH

For present purposes, the term 'governance' alludes to a democratisation of administrative decision-making and policy-making or, to use the words of sociologist Anthony Giddens, to 'a process of deepening and widening of democracy [in which] government can act in partnership with agencies in civil society to foster community renewal and development.'[3]

---

[1] fitbit Inc., 'National Institutes of Health Launches Fitbit Project as First Digital Health Technology Initiative in Landmark All of Us Research Program (Press Release)', (fitbit, 2019).

[2] D. C. Collins et al., 'Towards Precision Medicine in the Clinic: From Biomarker Discovery to Novel Therapeutics', (2017) *Trends in Pharmacological Sciences*, 38(1), 25–40.

[3] A. Giddens, *The Third Way: The Renewal of Social Democracy* (New York: John Wiley & Sons, 2013), p. 69.

Regulatory literature over the last two decades has formalised a number of approaches to governance that seem to address some of the defining characteristics of Big Data health research. In particular, adaptive governance and principles-based regulation appear well-suited to tackle three specific features of Big Data research, namely: (1) the evolving, and thus hardly predictable nature of the data ecosystem in Big Data health research – including the fast-paced development of new data analysis techniques; (2) the polycentric character of the actor network of Big Data and the absence of a single centre of regulation; and (3) the fact that most of these actors do not currently share a common regulatory culture and are driven by unaligned values and visions.[4]

Adaptive governance is based on the idea that – in the presence of uncertainty, lack of evidence and evolving, dynamic phenomena – governance should be able to adapt to the mutating conditions of the phenomenon that it seeks to govern. Key attributes of adaptive governance are the inclusion of multiple stakeholders in governance design,[5] collaboration between regulating and regulated actors,[6] the incremental and planned incorporation of evidence in governance solutions[7] and openness to cope with uncertainties through social learning.[8] This is attained by planning evidence collection and policy revision rounds in order to refine the fit between governance and public expectations; distributing regulatory tasks across a variety of actors (polycentricity); designing partially overlapping competences for different actors (redundancy); and by increasing participation in policy and management decisions by otherwise neglected social groups. Adaptive governance thus seems to adequately reflect the current state of Big Data health research as captured by the three characteristics outlined above. Moreover, social learning – a key feature of adaptive governance – can help explore areas of overlapping consensus even in a fragmented actor network like the one that constitutes Big Data research.

Principles based regulation (PBR) is a governance approach that emerged in the 1990s to cope with the expansion of the financial services industry. Just as Big Data research is driven by technological innovation, financial technologies (the so-called fintech industry) have played a disruptive role for the entire financial sector.[9] Unpredictability, accrual of new stakeholders and lack of regulatory standards and best practices characterise this phenomenon. To respond to this, regulators such as the UK Financial Services Authority (FSA), backed-up by a number of academic supporters of 'new governance' approaches,[10] have proposed principles-based regulation as a viable governance model.[11] In this model, regulation and oversight relies on broadly-stated principles that reflect regulators orientations, values and priorities. Moreover, implementation of the principles is not entirely delegated to specified rules and procedures. Rather, PBR relies on regulated actors to set up mechanism to comply with the principles.[12] Principles are

[4] E. Vayena and A. Blasimme, 'Health Research with Big Data: Time for Systemic Oversight', (2018) *The Journal of Law, Medicine & Ethics*, 46(1), 119–129.
[5] C. Folke et al., 'Adaptive Governance of Social-Ecological Systems', (2005) *Annual Review of Environment and Resources*, 30, 441–473.
[6] T. Dietz et al., 'The Struggle to Govern the Commons', (2003) *Science*, 302(5652), 1907–1912.
[7] C. Ansell and A. Gash, 'Collaborative Governance in Theory and Practice', (2008) *Journal of Public Administration Research and Theory*, 18(4), 543–571.
[8] J. J. Warmink et al., 'Coping with Uncertainty in River Management: Challenges and Ways Forward', (2017) *Water Resources Management*, 31(14), 4587–4600.
[9] R. J. McWaters et al., 'The Future of Financial Services-How Disruptive Innovations Are Reshaping the Way Financial Services Are Structured, Provisioned and Consumed', (World Economic Forum, 2015).
[10] R. A. W. Rhodes, 'The New Governance: Governing without Government', (1996) *Political Studies*, 44(4), 652–667.
[11] J. Black, 'The Rise, Fall and Fate of Principles Based Regulation', (2010) *LSE Legal Studies Working Paper*, 17.
[12] Ibid.

usually supplemented by guidance, white papers and other policies and processes to channel the compliance efforts of regulated entities. See further on PBR, Sethi, Chapter 17, this volume.

We contend that PBR is helpful to set up Big Data governance in the research space because it is explicitly focussed on the creation of some form of normative alignment between the regulator and the regulated; it creates conditions that can foster the emergence of shared values among different regulated stakeholders. Since compliance is not rooted on box-ticking nor respect for precisely-specified rules, PBR stimulates experimentation with a number of different oversight mechanisms. This bottom-up approach allows stakeholders to explore a wide range of activities and structures to align with regulatory principles, favouring the selection of more cost-efficient and proportionate mechanisms. Big data health research faces exactly this need to create stakeholders' alignment and to cope with the wide latitude of regulatory attitudes that is to be expected in an innovative domain with multiple newcomers.

The governance model that we propose below relies on both adaptive governance – as to its capacity to remain flexible to future evolutions of the field – and PBR – because of its emphasis on principles as sources of normative guidance for different stakeholders.

26.3 A FRAMEWORK TO DEVELOP SYSTEMIC OVERSIGHT

The framework we propose below provides guidance to actors that have a role in the shaping and management of research employing Big Data; it draws inspiration from the above-listed features of adaptive governance. Moreover, it aligns with PBR in that it offers guidance to stakeholders and decision-makers engaged at various levels in the governance of Big Data health research. As we have argued elsewhere, our framework will facilitate the emergence of systemic oversight functions for the governance of Big Data health research.[13] The development of systemic oversight relies on six high-order principles aimed at reducing the effects of a fragmented governance landscape and at channelling governance decisions – through both structures and processes – towards an ethically defensible common ground. These six principles do not predefine which specific governance structures and processes shall be put in place – hence the caveat that they represent *high-order* guides. Rather, they highlight governance features that shall be taken into account in the design of structures and processes for Big Data health research. Equally, our framework is not intended as a purpose-neutral approach to governance. Quite to the contrary; the six principles we advance do indeed possess a normative character in that they endorse valuable states of affairs that shall occur as a result of appropriate and effective governance. By the same token, our framework suggests that action should be taken in order to avoid certain kinds of risks that will most likely occur if left unattended. In this section, we will illustrate the six principles of systemic oversight – adaptivity, flexibility, monitoring, responsiveness, reflexivity and inclusiveness – while the following section deals with the effective interpretation and implementation of such principles in terms of both structures and processes.

Adaptivity: adaptivity is the capacity of governance structures and processes to ensure proper management of new forms of data as they are incorporated into health research practices. Adaptivity, as presented here, has also been discussed as a condition for resilience, that is, for the capacity of any given system to 'absorb disturbances and reorganize while undergoing change so as to still retain essentially the same function, structure, identity and feedbacks.'[14]

---

[13] Vayena and Blasimme, 'Health Research'.
[14] B. Walker et al., 'Resilience, Adaptability and Transformability in Social–Ecological Systems', (2004) *Ecology and Society*, 9 (2), 4.

This feature is crucial in the case of a rapidly evolving field – like Big Data research – whose future shape, as a consequence, is hard to anticipate.

Flexibility: flexibility refers to the capacity to treat different data types depending on their actual use rather than their source alone. Novel analytic capacities are jeopardising existing data taxonomies, which rapidly renders regulatory categories constructed around them obsolete. Flexibility means, therefore, recognising the impact of technical novelties and, at a minimum, giving due consideration to their potential consequences.

Monitoring: risk minimisation is a crucial aim of research ethics. With the possible exception of highly experimental procedures, the spectrum of physical and psychological harms due to participation in health research is fairly straightforward to anticipate. In the evolving health data ecosystem described so far, however, it is difficult to anticipate upfront what harms and vulnerabilities research subjects may encounter due their participation in Big Data health research. This therefore requires on-going monitoring.

Responsiveness: despite efforts in monitoring emerging vulnerabilities, risks can always materialise. In Big Data health research, privacy breaches are a case in point. Once personal data are exposed, privacy is lost. No direct remedy exists to re-establish the privacy conditions that were in place before the violation. Responsiveness therefore prescribes that measures are put in place to at least reduce the impact of such violations on the rights, interests and well-being of research participants.

Reflexivity: it is well known that certain health-related characteristics cluster in specific human groups, such as populations, ethnic groups, families and socio-economic strata. Big data are pushing the classificatory power of research to the next level, with potentially worrisome implications. The classificatory assumptions that drive the use of rapidly evolving data-mining capacities need to be put under careful scrutiny as to their plausibility, opportunity and consequences. Failing to do so will result in harms to all human groups affected by those assumptions. What is more, public support for, as well as trust in, scientific research may be jeopardised by the reputational effects that can arise if reflexivity and scrutiny are not maintained.

Inclusiveness: the last component of systemic oversight closely resonates with one of the key features of adaptive governance, that is, the need to include all relevant parties in the governance process. As more diverse data sources are aggregated, the more difficult it becomes for research participants to exert meaningful control on the expanding cloud of personal data that is implicated by their participation.[15] Experimenting with new forms of democratic engagement is therefore imperative for a field that depends on resources provided by participants (i.e. data), but that, at the same time, can no longer anticipate how such resources will be employed, how they will be analysed and with which consequences. See Burgess, Chapter 25.

These six principles can be arranged to form the acronym AFIRRM: our model framework for the governance of Big Data health research.

26.4 BIG DATA HEALTH RESEARCH: IMPLEMENTING EFFECTIVE GOVERNANCE

While there is no universal definition of the notion of effective governance, it alludes in most cases to an alignment between purposes and outcomes, reached through processes that fulfil

---

[15] E. Vayena and A. Blasimme, 'Biomedical Big Data: New Models of Control over Access, Use and Governance', (2017) *Journal of Bioethical Inquiry*, 14(4), 501–513.

constituents' expectations and which project legitimacy and trust onto the involved actors.[16] This understanding of effective governance fits well with our domain of interest: Big Data health research. In the remainder of this chapter, drawing on literature on the implementation of adaptive governance and PBR, we discuss key issues to be taken into account in trying to derive effective governance structures and oversight mechanism from the AFIRRM principles.

The AFIRRM framework endorses the use of principles as high-level articulations of what is to be expected by regulatory mechanisms for the governance of Big Data health research. Unlike the use of PBR in financial markets where a single regulator expects compliance, PBR in the Big Data context responds to the reality that governance functions are distributed among a plethora of actors, such as ethics review committees, data controllers, privacy commissioners, access committees, etc. PBR within the AFIRRM framework offers a blueprint for such a diverse array of governance actors to create new structures and processes to cope with the specific ethical and legal issues raised by the use of Big Data. Such principles have a generative function in the governance landscape, that is, in the process of being created to govern those issues.

The key advantage of principles in this respect is that they require making the reason behind regulation visible to all interested parties, including publics. This amounts to an exercise of public accountability that can bring about normative coherence among actors with different starting assumptions. The AFIRRM principles stimulate a bottom-up exploration of the values at stake and how compliance with existing legal requirements will be met. In this sense, the AFIRRM principles perform a formal, more than a substantive function, precisely because we assume the substantive ethical and legal aims of regulation that have already been developed in health research – such as the protection of research participants from the risk of harm – to hold true also for research employing Big Data. What AFIRRM principles do is to provide a starting point for deliberation and action that respects existing ethical standards and complies with pre-existing legal rules.

The AFIRRM principles do not envision actors in the space of Big Data research to self-regulate, but they do presuppose trust between regulators and regulated entities: regulators need to be confident that regulated entities will do their best to give effect to the principles in good faith. While some of the interests at stake in Big Data health research might be in tension – like the interest of researchers to access and distribute data, and the interests of data donors to control what their personal data are used for – developing efficient governance structures and processes that meet stakeholders' expectations is of advantage for all interested parties to begin with conversations based on core agreed principles. Practically, this requires all relevant stakeholders to have a say in the development and operationalisation of the principles at stake.

Adaptive governance scholarship has identified typical impediments to effective operationalisation of adaptive mechanisms. A 2012 literature review of adaptive governance, network management and institutional analysis identified three key challenges to the effective implementation of adaptive governance: ill-defined purposes and objectives, unclear governance context and lack of evidence in support of blueprint solutions.[17]

Let us briefly illustrate each of these challenges and explain how systemic oversight tries to avoid them. In the shift from centralised forms of administration and decision-making, to less

---

[16] See, for example, S. Arjoon, 'Striking a Balance between Rules and Principles-Based Approaches for Effective Governance: A Risks-Based Approach', (2006) *Journal of Business Ethics*, 68(1), 53–82; A. Kezar, 'What Is More Important to Effective Governance: Relationships, Trust, and Leadership, or Structures and Formal Processes?', (2004) *New Directions for Higher Education*, 127, 35–46.

[17] J. Rijke et al., 'Fit-for-Purpose Governance: A Framework to Make Adaptive Governance Operational', (2012) *Environmental Science & Policy*, 22, 73–84.

formalised and more distributed governance networks that occurred over the last three decades,[18] the identification of governance objectives is no longer straightforward. This difficulty may also be due to the potentially conflicting values of different actors in the governance ecosystem. In this respect, systemic oversight has the advantage of not being normatively neutral. The six principles of systemic oversight determinedly aim at fostering an ethical common ground for a variety of governance actors and activities in the space of Big Data research. What underpins the framework, therefore, is a view of what requires ethical attention in this rapidly evolving field, and how to prioritise actions accordingly. In this way, systemic oversight can provide orientation for a diverse array of governance actors (structures) and mechanisms (processes), all of which are supposed to produce an effective system of safeguards around activities in this domain. Our framework directs attention to critical features of Big Data research and promotes a distributed form of accountability that will, where possible, emerge spontaneously from the different operationalisations of its components. The six components of systemic oversight, therefore, suggest what is important to take into account when considering how to adapt the composition, mandate, operations and scope of oversight bodies in the field of Big Data research.

The second challenge to effective adaptive governance – unclear governance context – refers to the difficulty of mapping the full spectrum of rules, mechanisms, institutions and actors involved in a distributed governance system or systems. Systemic oversight requires mapping the overall governance context in order to understand how best to implement the framework in practice. This amounts to an empirical inquiry into the conditions (structures, mechanisms and rules) in which governance actors currently operate. In a recent study we showed that current governance mechanisms for research biobanks, for instance, are not aligned with the requirements of systemic oversight.[19] In particular, we showed that systemic oversight can contribute to improve accountability of research infrastructures that, like biobanks, collect and distribute an increasing amount of scientific data.

The third and last challenge to effective operationalisation of adaptive mechanisms has to do with the limits of ready-made blueprint solutions to complex governance models. Political economist and Nobel Laureate Elinor Ostrom has written extensively on this. In her work on socio-ecological systems, Ostrom has convincingly shown that policy actors have the tendency to buy into what she calls 'policy panaceas',[20] that is, ready-made solutions to very complex problems. Such policy panaceas are hardly ever supported by solid evidence regarding the effectiveness of their outcomes. One of the most commonly cited reasons for their lack of effectiveness is that complexity entails high degrees of uncertainty as to the very phenomenon that policy makers are trying to govern.

We saw that uncertainty is characteristic of Big Data research too (see Section 26.2). That is why systemic oversight refrains from prescribing any particular governance solution. While not rejecting traditional predict-and-control approaches (such as informed consent, data anonymisation and encryption), systemic oversight does not put all the regulatory weight on any particular instrument or body. The *systemic* ambition of the framework lies in its pragmatic orientation towards a plurality of tools, mechanisms and structures that could jointly stabilise the responsible

---

[18] R. A. W. Rhodes, *Understanding Governance: Policy Networks, Governance, Reflexivity, and Accountability* (Buckingham: Open University Press, 1997); R. A. W. Rhodes, 'Understanding Governance: Ten Years On', (2007) *Organization Studies*, 28(8), 1243–1264.

[19] F. Gille et al. 'Future-proofing biobanks' governance', (2020) *European Journal of Human Genetics*, 28, 989–996.

[20] E. Ostrom, 'A Diagnostic Approach for Going beyond Panaceas', (2007) *Proceedings of the National Academy of Sciences*, 104(39), 15181–15187.

use of Big Data for research purposes. In this respect, our framework acknowledges that '[a]daptation typically emerges organically among multiple centers of agency and authority in society as a relatively self-organized or autonomous process marked by innovation, social learning and political deliberation'.[21]

Still, a governance framework's capacity to avoid known bottlenecks to operationalisation is a necessary but not a sufficient condition to its successful implementation. The further question is how the principles of the systemic oversight model can be incorporated into structures and processes in Big Data research governance. With *structures* we mean actors and networks of actors involved in governance, and organised in bodies charged with oversight, organisational or policy-making responsibilities. *Processes*, instead, are the mechanisms, procedures, rules, laws and codes through which actors operate and bring about their governance objectives. Structures and processes define the polycentric, redundant and experimental system of governance that an adaptive governance model intends to promote.[22]

26.5 KEY FEATURES OF GOVERNANCE *STRUCTURES* AND *PROCESSES*

Here we follow the work of Rijke and colleagues[23] in identifying three key properties of adaptive governance structures: centrality, cohesion and density. While it is acknowledged that central-ised structures can be effective as a response to crises and emergencies, centralisation is precisely a challenge in Big Data; our normative response is to call for inclusive social learning among the broad array of stakeholders, subject to challenges of incomplete representation of relevant interests (see further below). Still, this commitment can help to promote network cohesion by fostering discussion about how to implement the principles, while also promoting the formation of links between governance actors, as required by density. In addition, this can help to ensure that governance roles are fairly distributed among a sufficiently diverse array of stakeholders and that, as a consequence, decisions are not hijacked by technical experts.

The governance space in Big Data research is already populated by numerous actors, such as IRBs, data access committees and advisory boards. These bodies are not necessarily inclusive of a sufficiently broad array of stakeholders and therefore they may not be very effective at promoting social learning. Their composition could thus be rearranged in order to be more representative of the interests at stake and to promote continued learning. New actors could also enter the governance system. For instance, data could be made available for research by data subjects themselves through data platforms.[24]

Network of actors (structures) operating in the space of health research do so through mechanisms and procedures (processes) such as informed consent and ethics review, as well as data access review, policies on reporting research findings to participants, public engagement activities and privacy impact assessment.

Processes are crucial to effective governance of health research and are a critical component of the systemic oversight approach as their features can determine the actual impact of its principles. Drawing on scholarship in adaptive governance, we present three such features

---

[21] D. A. DeCaro et al., 'Legal and Institutional Foundations of Adaptive Environmental Governance', (2017) *Ecology and Society: A Journal of Integrative Science for Resilience and Sustainability*, 22 (1), 1.

[22] B. Chaffin et al., 'A Decade of Adaptive Governance Scholarship: Synthesis and Future Directions', (2014) *Ecology and Society*, 19(3), 56.

[23] Rijke et al., 'Fit-for-Purpose Governance'.

[24] A. Blasimme et al., 'Democratizing Health Research Through Data Cooperatives', (2018) *Philosophy & Technology*, 31(3), 473–479.

(components) that are central to the appropriate interpretation of the systemic oversight principles.

Social learning: social learning refers to learning that occurs by observing others.[25] In governance settings that are open to participation by different stakeholders, social learning can occur across different levels and hierarchies of the governance structures. According to many scholars, including Ostrom,[26] social learning represents an alternative to policy blueprints (see above) – especially when it is coupled with and leading to adaptive management. Planned adaptations – that is, previously scheduled rounds of policy revision in light of new knowledge – can be occasions for governance actors to capitalise on each other's experience and learn about evolving expectations and risks. Such learning exercises can reduce uncertainty and lead to adjustments in mechanisms and rules. The premise of this approach is the realisation that in complex systems characterised by pronounced uncertainty, 'no particular epistemic community can possess all the necessary knowledge to form policy'.[27] Social learning – be it aimed at gathering new evidence, at fostering capacity building or at assessing policy outcomes – is relevant to all of the six components of systemic oversight. The French law on bioethics, for instance, prescribes periodic rounds of nationwide public consultation – the so-called Estates General on bioethics.[28] This is an example of how social learning can be fostered. Similar social learning can be triggered even at smaller scales – for instance in local oversight bodies – in order to explore new solutions and alternative designs.

Complementarity: complementarity is the capacity of governance processes to fulfil both the need for processes to be functionally compatible and to ensure procedural correspondence between processes and the phenomena they intend to regulate. *Functional* complementarity refers to the distribution of regulatory functions across a given set of processes exhibiting partial overlap (see redundancy, above). This feature is crucial for both monitoring and reflexivity. *Procedural* complementarity, on the other hand, refers to the temporal alignment between governance processes and the activities that depend on such processes. One prominent example, in this respect, is the timing of ethics review processes, or that of data access requests processing.[29] For instance, the European General Data Protection Regulation (GDPR) prescribes a maximum 72-hour delay between detection and notification of privacy breaches. This provision is an example of procedural complementarity that would be of the utmost importance for the principle of responsiveness.

Visibility: governance processes need to be visible, that is, procedures and their scope need to be as publicly available as possible to whomever is affected by them or must act accordingly to them. The notion of regulatory visibility has recently been highlighted by Laurie and colleagues, who argue for regulatory stewardship within ecosystems to help researchers clarify values and responsibilities in health research and navigate the complexities.[30] Recent work also demonstrates that currently it is difficult to access policies and standard operating procedures of prominent research institutions like biobanks. In principle, fair scientific competition may

[25] A. Bandura and R. H. Walters, *Social Learning Theory*, vol. 1 (Prentice-hall Englewood Cliffs, NJ, 1977).

[26] Ostrom, 'A Diagnostic Approach'.

[27] D. Swanson et al., 'Seven Tools for Creating Adaptive Policies', (2010) *Technological Forecasting and Social Change*, 77(6), 924–939, 925.

[28] D. Berthiau, 'Law, Bioethics and Practice in France: Forging a New Legislative Pact', (2013) *Medicine, Health Care and Philosophy*, 16(1), 105–113.

[29] G. Silberman and K. L. Kahn, 'Burdens on Research Imposed by Institutional Review Boards: The State of the Evidence and Its Implications for Regulatory Reform', (2011) *The Milbank Quarterly*, 89(4), 599–627.

[30] G. T. Laurie et al., 'Charting Regulatory Stewardship in Health Research: Making the Invisible Visible', (2018) *Cambridge Quarterly of Healthcare Ethics*, 27(2), 333–347.

militate against disclosure of technical details about data processing, but it is hard to imagine practical circumstances in which administrators of at least publicly funded datasets would not have incentives to share as much information as possible regarding the way they handle their data. Process visibility goes beyond fulfilling a pre-determined set of criteria (for instance, for auditing purposes). By disclosing governance processes and opportunities for engagement, actors actually offer reasons to be trusted by a variety of stakeholders.[31] This feature is of particular relevance for the principles of monitoring and reflexivity, as well as to improve the effectiveness of inclusive governance processes.

26.6 CONCLUSION

In this chapter, we have defended adaptive governance as a suitable regulatory approach for Big Data health research by proposing six governance principles to foster the development of appropriate structures and processes to handle critical aspects of Big Data health research. We have analysed key aspects of implementation and identified a number of important features that can make adaptive regulation operational. However, one might legitimately ask: in the absence of a central regulatory actor endowed with clearly recognised statutory prerogatives, how can it be assumed that the AFIRRM principles will be endorsed by the diverse group of stakeholders operating in the Big Data health research space? Clearly, this question does not have a straightforward answer. However, to increase likelihood of uptake, we have advanced AFIRRM as a viable and adaptable model for the creation of necessary tools that can deliver on common objectives. Our model is based on a careful analysis of regulatory scholarship vis-à-vis the key attributes of this type of research. We are currently undertaking considerable efforts to introduce AFIRRM to regulators, operators and organisations in the space of research or health policy. We are cognisant of the fact that the implementation of a model like AFIRRM needs not be temporally linear. Different actors may take initiative at different points in time. It cannot be expected that a coherent system of governance will emerge in a synchronically orchestrated manner through the uncoordinated action of multiple stakeholders. Such a path could only be imagined if a central regulator had the power and the will to make it happen. Nothing indicates, however, that regulation will assume a centralised character anytime soon. Nevertheless, polycentricity is not in itself a barrier to the emergence of a coherent governance ecosystem. Indeed, the AFIRRM principles – in line with its adaptive orientation – rely precisely on polycentric governance to cope with the uncertainty and complexity of Big Data health research.

---

[31] O. O'Neill, 'Trust with Accountability?', (2003) *Journal of Health Services Research & Policy*, 8(1), 3–4.

Regulating Automated Healthcare and Research Technologies

First Do No Harm (to the Commons)

Roger Brownsword

27.1 INTRODUCTION

New technologies, techniques, and tests in healthcare, offering better prevention, or better diagnosis and treatment, are not manna from heaven. Typically, they are the products of extensive research and development, increasingly enabled by high levels of automation and reliant on large datasets. However, while some will push for a permissive regulatory environment that is facilitative of beneficial innovation, others will push back against research that gives rise to concerns about the safety and reliability of particular technologies as well as their compatibility with respect for fundamental values. Yet, how are the interests in pushing forward with research into potentially beneficial health technologies to be reconciled with the heterogeneous interests of the concerned who seek to push back against them?

A stock answer to this question is that regulators, neither over-regulating nor under-regulating, should seek an accommodation or a balance of interests that is broadly 'acceptable'. If the issue is about risks to human health and safety, then regulators – having assessed the risk – should adopt a management strategy that confines risk to an acceptable level; and, if there is a tension between, say, the interest of researchers in accessing health data and the interest of patients in both their privacy and the fair processing of their personal data, then regulators should accommodate these interests in a way that is reasonable – or, at any rate, not manifestly unreasonable.

The central purpose of this chapter is not to argue that this balancing model is always wrong or inappropriate, but to suggest that it needs to be located within a bigger picture of lexically ordered regulatory responsibilities.[1] In that bigger picture, the paramount responsibility of regulators is to act in ways that protect and maintain the conditions that are fundamental to human social existence (the commons). After that, a secondary responsibility is to protect and respect the values that constitute a group as the particular kind of community that it is. Only after these responsibilities have been discharged do we get to a third set of responsibilities that demand that regulators seek out reasonable and acceptable balances of conflicting legitimate interests. Accordingly, before regulators make provision for a – typically permissive – framework that they judge to strike an acceptable balance of interests in relation to some particular technology, technique or test, they should check that its development, exploitation, availability and application crosses none of the community's red lines and, above all, that it poses no threat to the commons.

---

[1] See, further, R. Brownsword, *Law, Technology and Society: Re-imagining the Regulatory Environment* (Abingdon: Routledge, 2019), Ch. 4.

The chapter is in three principal parts. First, in Section 27.2, we start with two recent reports by the Nuffield Council on Bioethics – one a report on the use of Non-Invasive Prenatal Testing (NIPT),[2] and the other on genome-editing and human reproduction.[3] At first blush, the reports employ a similar approach, identifying a range of legitimate – but conflicting – interests and then taking a relatively conservative position. However, while the NIPT report exemplifies a standard balancing approach, the genome-editing report implicates a bigger picture of regulatory responsibilities. Second, in Section 27.3, I sketch my own take on that bigger picture. Third, in Section 27.4, I speak to the way in which the bigger picture might bear on our thinking about the regulation of automated healthcare and research technologies. In particular, in this part of the chapter, the focus is on those technologies that power smart machines and devices, technologies that are hungry for human data but then, in their operation, often put humans out of the loop.

27.2 NIPT, GENOME-EDITING AND THE BALANCING OF INTERESTS

In its report on the ethics of NIPT, the Nuffield Council on Bioethics identifies a range of legitimate interests that call for regulatory accommodation. On the one side, there is the interest of pregnant women and their partners in making informed reproductive choices. On the other side, there are interests – particularly of the disability community and of future children – in equality, fairness and inclusion. The question is: how are regulators to 'align the responsibilities that [they have] to support women to make informed reproductive choices about their pregnancies, with the responsibilities that [they have] … to promote equality, inclusion and fair treatment for all'?[4] In response to which, the Council, being particularly mindful of the interests of future children – in an open future – and the interest in a wider societal environment that is fair and inclusive, recommends that a relatively restrictive approach should be taken to the use of NIPT.

In support of the Council's approach and its recommendation, there is a good deal that can be said. For example, the Council consulted widely before drawing up the inventory of interests to be considered: it engaged with the arguments rationally and in good faith; where appropriate, its thinking was evidence-based; and its recommendation is not manifestly unreasonable. If we were to imagine a judicial review of the Council's recommendation, it would surely survive the challenge.

However, if the Council had given greater weight to the interest in reproductive autonomy together with the argument that women have 'a right to know' and that healthcare practitioners have an interest in doing the best that they can for their patients,[5] leading to a much less restrictive recommendation, we could say exactly the same things in its support.

In other words, so long as the Council – and, similarly, any regulatory body – consults widely and deliberates rationally, and so long as its recommendations are not manifestly unreasonable, we can treat its preferred accommodation of interests as acceptable. Yet, in such balancing deliberations, it is not clear where the onus of justification lies or what the burden of justification

[2] Nuffield Council on Bioethics, 'Non-invasive Prenatal Testing: Ethical Issues', (March 2017); for discussion, see R. Brownsword and J. Wale, 'Testing Times Ahead: Non-Invasive Prenatal Testing and the Kind of Community that We Want to Be', (2018) *Modern Law Review*, 81(4), 646–672.
[3] Nuffield Council on Bioethics, 'Genome Editing and Human Reproduction: Social and Ethical Issues', (July 2018).
[4] Nuffield Council on Bioethics, 'Non-Invasive Prenatal Testing', para 5.20.
[5] Compare N. J. Wald et al., 'Response to Walker', (2018) *Genetics in Medicine*, 20(10), 1295; and in Canada, see the second phase of the Pegasus project, Pegasus, 'About the Project', www.pegasus-pegase.ca/pegasus/about-the-project/.

is; and, in the final analysis, we cannot say why the particular restrictive position that the Council takes is more or less acceptable than a less restrictive position.

Turning to the Council's second report, it hardly needs to be said that the development of precision gene-editing techniques, notably CRISPR-Cas9, has given rise to considerable debate.[6] Addressing the ethics of gene editing and human reproduction, the Council adopted a similar approach to that in its report on NIPT. Following extensive consultation – and, in this case, an earlier, more general, report[7] – there is a careful consideration of a range of legitimate interests, following which a relatively conservative position is taken. Once again, although the position taken is not manifestly unreasonable, it is not entirely clear why this particular position is taken.

Yet, in this second report, there is a sense that something more than balancing might be at stake.[8] For example, the Council contemplates the possibility that genome editing might inadvertently lead to the extinction of the human species – or, conversely, that genome editing might be the salvation of humans who have catastrophically compromised the conditions for their existence. In these short reflections about the interests of 'humanity', we can detect a bigger picture of regulatory responsibilities.

27.3 THE BIGGER PICTURE OF REGULATORY RESPONSIBILITIES

In this part of the chapter, I sketch what I see as the bigger – three-tier – picture of regulatory responsibilities and then speak briefly to the first two tiers.

27.3.1 *The Bigger Picture*

My claim is that regulators have a first-tier 'stewardship' responsibility for maintaining the pre-conditions for any kind of human social community ('the commons'). At the second tier, regulators have a responsibility to respect the fundamental values of a particular human community, that is to say, the values that give that community its particular identity. At the third tier, regulators have a responsibility to seek out an acceptable balance of legitimate interests. The responsibilities at the first tier are cosmopolitan and non-negotiable. The responsibilities at the second and third tiers are contingent, depending on the fundamental values and the interests recognised in each particular community. Conflicts between commons-related interests, community values and individual or group interests are to be resolved by reference to the lexical ordering of the tiers: responsibilities in a higher tier always outrank those in a lower tier. Granted, this does not resolve all issues about trade-offs and compromises because we still have to handle horizontal conflicts *within* a particular tier. But, by identifying the tiers of responsibility, we take an important step towards giving some structure to the bigger picture.

27.3.2 *First-Tier Responsibilities*

Regulatory responsibilities start with the existence conditions that support the particular biological needs of humans. Beyond this, however, as *agents*, humans characteristically have the capacity to pursue various projects and plans whether as individuals, in partnerships, in groups, or in whole communities. Sometimes, the various projects and plans that they pursue will be

[6] See, e.g., J. Harris and D. R. Lawrence, 'New Technologies, Old Attitudes, and Legislative Rigidity' in R. Brownsword et al. (eds) *Oxford Handbook of Law, Regulation and Technology* (Oxford University Press, 2017), pp. 915–928.
[7] Nuffield Council on Bioethics, 'Genome Editing: An Ethical Review', (September 2016).
[8] Nuffield Council on Bioethics, 'Genome Editing and Human Reproduction', paras 3.72–3.78.

harmonious; but often – as when the acceptability of the automation of healthcare and research is at issue – human agents will find themselves in conflict with one another. Accordingly, regulators also have a responsibility to maintain the conditions – conditions that are entirely neutral between the particular plans and projects that agents individually favour – that constitute the context for agency itself.

Building on this analysis, the claim is that the paramount responsibility for regulators is to protect, preserve, and promote:

- the essential conditions for *human* existence (given *human* biological needs);
- the generic conditions for human *agency* and self-development; and,
- the essential conditions for the development and practice of moral *agency*.

These, it bears repeating, are imperatives in all regulatory spaces, whether international or national, public or private. Of course, determining the nature of these conditions will not be a mechanical process. Nevertheless, let me indicate how the distinctive contribution of each segment of the commons might be elaborated.

In the first instance, regulators should take steps to maintain the natural ecosystem for human life.[9] At minimum, this entails that the physical well-being of humans must be secured: humans need oxygen, they need food and water, they need shelter, they need protection against contagious diseases, if they are sick they need whatever treatment is available, and they need to be protected against assaults by other humans or non-human beings. When the Nuffield Council on Bioethics discusses catastrophic modifications to the human genome or to the ecosystem, it is this segment of the commons that is at issue.

Second, the conditions for meaningful self-development and agency need to be constructed: there needs to be sufficient trust and confidence in one's fellow agents, together with sufficient predictability to plan, so as to operate in a way that is interactive and purposeful rather than merely defensive. Let me suggest that the distinctive capacities of prospective agents include being able: to form a sense of what is in one's own *self*-interest; to choose one's own ends, goals, purposes and so on ('to do one's own thing'); and to form a sense of one's own identity ('to be one's own person').

Third, the commons must secure the conditions for an aspirant moral community, whether the particular community is guided by teleological or deontological standards, by rights or by duties, by communitarian or liberal or libertarian values, by virtue ethics, and so on. The generic context for moral community is impartial between competing moral visions, values, and ideals; but it must be conducive to 'moral' development and 'moral' agency in the sense of forming a view about what is the 'right thing' to do relative to the interests of both oneself and others.

On this analysis, each human agent is a stakeholder in the commons where this represents the essential conditions for human existence together with the generic conditions of both self-regarding and other-regarding agency. While respect for the commons' conditions is binding on all human agents, it should be emphasised that these conditions do not rule out the possibility of prudential or moral pluralism. Rather, the commons represents the pre-conditions for both individual self-development and community debate, giving each agent the opportunity to develop his or her own view of what is prudent, as well as what should be morally prohibited, permitted or required.

[9] Compare, J. Rockström et al., 'Planetary Boundaries: Exploring the Safe Operating Space for Humanity' (2009) *Ecology and Society*, 14(2); K. Raworth, *Doughnut Economics* (Random House Business Books, 2017), pp. 43–53.

27.3.3 *Second-Tier Responsibilities*

Beyond the stewardship responsibilities, regulators are also responsible for ensuring that the fundamental values of their particular community are respected. Just as each individual human agent has the capacity to develop their own distinctive identity, the same is true if we scale this up to communities of human agents. There are common needs and interests but also distinctive identities.

In the particular case of the United Kingdom: although there is not a general commitment to the value of social solidarity, arguably this is actually the value that underpins the NHS. Accordingly, if it were proposed that access to NHS patient data – data, as Philip Aldrick has put it, that is 'a treasure trove ... for developers of next-generation medical devices'[10] – should be part of a transatlantic trade deal, there would surely be an uproar because this would be seen as betraying the kind of healthcare community that we think we are.

More generally, many nation states have expressed their fundamental (constitutional) values in terms of respect for human rights and human dignity.[11] These values clearly intersect with the commons' conditions and there is much to debate about the nature of this relationship and the extent of any overlap – for example, if we understand the root idea of human dignity in terms of humans having the capacity freely to do the right thing for the right reason,[12] then human dignity reaches directly to the commons' conditions for moral agency.[13] However, those nation states that articulate their particular identities by reference to their commitment to respect for human dignity are far from homogeneous. Whereas in some communities, the emphasis of human dignity is on individual empowerment and autonomy, in others it is on constraints relating to the sanctity, non-commercialisation, non-commodification and non-instrumentalisation of human life.[14] These differences in emphasis mean that communities articulate in very different ways on a range of beginning-of-life and end-of-life questions as well as on questions of acceptable health-related research, and so on.

Given the conspicuous interest of today's regulators in exploring technological solutions, an increasingly important question will be whether, and if so, how far, a community sees itself as distinguished by its commitment to regulation by rule and by human agents. In some smaller-scale communities or self-regulating groups, there might be resistance to a technocratic approach because automated compliance compromises the context for trust and for responsibility. Or, again, a community might prefer to stick with regulation by rules and by human agents because it is worried that with a more technocratic approach, there might be both reduced public participation in the regulatory enterprise and a loss of flexibility in the application of technological measures.

If a community decides that it is generally happy with an approach that relies on technological measures rather than rules, it then has to decide whether it is also happy for humans to be

---

[10] P. Aldrick, 'Make No Mistake, One Way or Another NHS Data Is on the Table in America Trade Talks', *The Times*, (8 June 2019), 51.
[11] See R. Brownsword, 'Human Dignity from a Legal Perspective' in M. Duwell et al. (eds), *Cambridge Handbook of Human Dignity* (Cambridge University Press, 2014), pp. 1–22.
[12] For such a view, see R. Brownsword, 'Human Dignity, Human Rights, and Simply Trying to Do the Right Thing' in C. McCrudden (ed), *Understanding Human Dignity – Proceedings of the British Academy 192* (The British Academy and Oxford University Press, 2013), pp. 345–358.
[13] See R. Brownsword, 'From Erewhon to Alpha Go: For the Sake of Human Dignity Should We Destroy the Machines?', (2017) *Law, Innovation and Technology*, 9(1), 117–153.
[14] See D. Beyleveld and R. Brownsword, *Human Dignity in Bioethics and Biolaw* (Oxford University Press, 2001); R. Brownsword, *Rights, Regulation and the Technological Revolution* (Oxford University Press, 2008).

out of the loop. Furthermore, once a community is asking itself such questions, it will need to clarify its understanding of the relationship between humans and robots – in particular, whether it treats robots as having moral status, or legal personality, and the like.

These are questions that each community must answer in its own way. The answers given speak to the kind of community that a group aspires to be. That said, it is, of course, essential that the fundamental values to which a particular community commits itself are consistent with (or cohere with) the commons' conditions.

27.4 AUTOMATED HEALTHCARE AND THE BIGGER PICTURE OF REGULATORY RESPONSIBILITY

One of the features of the NHS *Long Term Plan*[15] – in which the NHS is described as 'a hotbed of innovation and technological revolution in clinical practice'[16] – is the anticipated role to be played by technology in 'helping clinicians use the full range of their skills, reducing bureaucracy, stimulating research and enabling service transformation'.[17] Moreover, speaking about the newly created unit, NHSX (a new joint organisation for digital, data and technology), the Health Secretary, Matt Hancock, said that this was 'just the beginning of the tech revolution, building on our Long Term Plan to create a predictive, preventative and unrivalled NHS'.[18]

In this context, what should we make of the regulatory challenge presented by smart machines and devices that incorporate the latest AI and machine learning algorithms for healthcare and research purposes? Typically, these technologies need data on which to train and to improve their performance. While the consensus is that the collection and use of personal data needs governance and that big datasets (interrogated by state of the art algorithmic tools) need it a fortiori, there is no agreement as to what might be the appropriate terms and conditions for the collection, processing and use of personal data or how to govern these matters.[19]

In its recent final report on Ethics Guidelines for Trustworthy AI,[20] the European Commission (EC) independent high-level expert group on artificial intelligence takes it as axiomatic that the development and use of AI should be 'human-centric'. To this end, the group highlights four key principles for the governance of AI, namely: respect for human autonomy, prevention of harm, fairness and explicability. Where tensions arise between these principles, then they should be dealt with by 'methods of accountable deliberation' involving 'reasoned, evidence-based reflection rather than intuition or random discretion'.[21] Nevertheless, it is emphasised that there might be cases where 'no ethically acceptable trade-offs can be identified. Certain fundamental rights and correlated principles are absolute and *cannot be subject to a balancing exercise* (e.g. human dignity)'.[22]

---

[15] NHS, 'NHS Long Term Plan', (January 2019), www.longtermplan.nhs.uk.
[16] Ibid., 91.
[17] Ibid.
[18] Department of Health and Social Care, 'NHSX: New Joint Organisation for Digital, Data and Technology', (19 February 2019), www.gov.uk/government/news/nhsx-new-joint-organisation-for-digital-data-and-technology.
[19] Generally, see R. Brownsword, 'Law, Technology and Society', Ch. 12; D. Schönberger, 'Artificial Intelligence in Healthcare: A Critical Analysis of the Legal and Ethical Implications', (2019) *International Journal of Law and Information Technology*, 27(2), 171–203.
  For the much-debated collaboration between the Royal Free London NHS Foundation Trust and Google DeepMind, see, J. Powles, 'Google DeepMind and healthcare in an age of algorithms', (2017) *Health and Technology*, 7(4), 351–367.
[20] European Commission, 'Ethics Guidelines for Trustworthy AI', (8 April 2019).
[21] Ibid.,13.
[22] Ibid., emphasis added.

In line with this analysis, my position is that while there might be many cases where simple balancing is appropriate, there are some considerations that should never be put into a simple balance. The group mentions human rights and human dignity. I agree. Where a community treats human rights and human dignity as its constitutive principles or values, they act – in Ronald Dworkin's evocative terms – as 'trumps'.[23] Beyond that, the interest of humanity in the commons should be treated as even more foundational (so to speak, as a super-trump).

It follows that the first question for regulators is whether new AI technologies for healthcare and research present any threat to the existence conditions for humans, to the generic conditions for self-development, and to the context for moral development. It is only once this question has been answered that we get to the question of compatibility with the community's particular constitutive values, and, then, after that, to a balancing judgment. If governance is to be 'human-centric', it is not enough that no individual human is exposed to an unacceptable risk or is not actually harmed. To be fully human-centric, technologies must be designed to respect both the commons and the constitutive values of particular human communities.

Guided by these regulatory imperatives, we can offer some short reflections on the three elements of the commons and how they might be compromised by the automation of research and healthcare.

27.4.1 *The Existence Conditions*

Famously, Stephen Hawking remarked that 'the advent of super-intelligent AI would be either the best or the worst thing ever to happen to humanity'.[24] As the best thing, AI would contribute to '[the eradication of] disease and poverty'[25] as well as '[helping to] reverse paralysis in people with spinal-cord injuries'.[26] However, on the downside, some might fear that in our quest for greater safety and well-being, we will develop and embed ever more intelligent devices to the point that there is a risk of the extinction of humans – or, if not that, then a risk of humanity surviving 'in some highly suboptimal state or in which a large portion of our potential for desirable development is irreversibly squandered'.[27] If this concern is well-founded, then communities will need to be extremely careful about how far and how fast they go with intelligent devices.

Of course, this is not specifically a concern about the use of smart machines in the hospital or in the research facility: the concern about the existential threat posed to humans by smart machines arises across the board; and, indeed, concerns about existential threats are provoked by a range of emerging technologies.[28] In such circumstances, a regulatory policy of precaution and zero risk is indicated; and while stewardship might mean that the development and application of some technologies that we value has to be restricted, this is better than finding that they have compromised the very conditions on which the enjoyment of such technologies is predicated.

---

[23] R. Dworkin, *Taking Rights Seriously*, revised edition (London: Duckworth, 1978).

[24] S. Hawking, *Brief Answers to the Big Questions* (London: John Murray, 2018) p. 188.

[25] Ibid., p. 189.

[26] Ibid., p. 194.

[27] See, N. Bostrom, *Superintelligence* (Oxford University Press, 2014), p. 281 (note 1); M. Ford, *The Rise of the Robots* (London: Oneworld, 2015), Ch. 9.

[28] For an indication of the range and breadth of this concern, see e.g. 'Resources on Existential Risk', (2015), www .futureoflife.org/data/documents/Existential%20Risk%20Resources%20(2015-08-24).pdf.

27.4.2 *The Conditions for Self-Development and Agency*

The developers of smart devices are hungry for data: data from patients, data from research participants, data from the general public. This raises concerns about privacy and data protection. While it is widely accepted that our privacy interests – in a broad sense – are 'contextual',[29] it is important to understand not just that 'there are contexts and contexts' but that there is a Context in which we all have a common interest. What most urgently needs to be clarified is whether any interests that we have in privacy and data protection touch and concern the essential conditions (the Context).

If, on analysis, we judge that privacy reaches through to the interests that agents necessarily have in the commons' conditions – particularly in the conditions for self-development and agency – it is neither rational nor reasonable for agents, individually or collectively, to authorise acts that compromise these conditions (unless they do so in order to protect some more important condition of the commons). As Bert-Jaap Koops has so clearly expressed it, privacy has an 'infrastructural character', 'having privacy spaces is an important presupposition for autonomy [and] self-development'.[30] Without such spaces, there is no opportunity to be oneself.[31] On this reading, privacy is not so much a matter of protecting goods – informational or spatial – in which one has a personal interest, but protecting infrastructural goods in which there is either a common interest (engaging first-tier responsibilities) or a distinctive community interest (engaging second-tier responsibilities).

By contrast, if privacy – and, likewise, data protection – is simply a legitimate informational interest that has to be weighed in an all things considered balance of interests, then we should recognise that what each community will recognise as a privacy interest and as an acceptable balance of interests might well change over time. To this extent, our reasonable expectations of privacy might be both 'contextual' and contingent on social practices.

27.4.3 *The Conditions for Moral Development and Moral Agency*

As I have indicated, I take it that the fundamental aspiration of *any* moral community is that regulators and regulatees alike should try to do the right thing. However, this presupposes a process of moral reflection and then action that accords with one's moral judgment. In this way, agents exercise judgment in trying to do the right thing and they do what they do for the right reason in the sense that they act in accordance with their moral judgment. Accordingly, if automated research and healthcare relieves researchers and clinicians from their moral responsibilities, even though well intended, this might result in a significant compromising of their dignity, qua the conditions for moral agency.[32]

Equally, if robots or other smart machines are used for healthcare and research purposes, some patients and participants might feel that this compromises their 'dignity' – robots might not

---

[29] See, for example, D. J. Solove, *Understanding Privacy* (Cambridge, MA: Harvard University Press, 2008); H. Nissenbaum, *Privacy in Context* (Palo Alto, CA: Stanford University Press, 2010).

[30] B. Koops, 'Privacy Spaces', (2018) *West Virginia Law Review*, 121(2), 611–665, 621.

[31] Compare, too, M. Brincker, 'Privacy in Public and the Contextual Conditions of Agency' in T. Timan, et al. (eds), *Privacy in Public Space* (Cheltenham: Edward Elgar, 2017), pp. 64–90; M. Hu, 'Orwell's *1984* and a Fourth Amendment Cybersurveillance Nonintrusion Test', (2017) *Washington Law Review*, 92(4), 1819–1904, 1903–1904.

[32] Compare K. Yeung and M. Dixon-Woods, 'Design-Based Regulation and Patient Safety: A Regulatory Studies Perspective', (2010) *Social Science and Medicine*, 71(3), 502–509.

physically harm humans, but even caring machines, so to speak, 'do not really care'.[33] The question then is whether regulators should treat the interests of such persons as a matter of individual interest to be balanced against the legitimate interests of others, or as concerns about dignity that speak to matters of either (first-tier) common or (second-tier) community interest.

In this regard, consider the case of Ernest Quintana whose family were shocked to find that, at a particular Californian hospital, a 'robot' displaying a doctor on a screen was used to tell Ernest that the medical team could do no more for him and that he would soon die.[34] What should we make of this? Should we read the family's shock as simply expressing a preference for the human touch or as going deeper to the community's constitutive values or even to the commons' conditions? Depending on how this question is answered, regulators will know whether a simple balance of interests is appropriate.

27.5 CONCLUSION

In this chapter, I have argued that it is not always appropriate to respond to new technologies for healthcare and research simply by enjoining regulators to seek out an acceptable balance of interests. My point is not that we should eschew either the balancing approach or the idea of 'acceptability' but that regulators should respond in a way that is sensitised to the full range of their responsibilities.

To the simple balancing approach, with its broad margin for 'acceptable' accommodation, we must add the regulatory responsibility to be responsive to the red lines and basic values that are distinctive of the particular community. Any claimed interest or proposed accommodation of interests that crosses these red lines or that is incompatible with the community's basic values is 'unacceptable' – but this is for a different reason to that which applies where a simple balancing calculation is undertaken.

Most fundamentally, however, regulators have a stewardship responsibility in relation to the anterior conditions for *humans* to exist and for them to function as a *community of agents*. We should certainly say that any claimed interest or proposed accommodation of interests that is incompatible with the maintenance of these conditions is totally 'unacceptable' – but it is more than that. Unlike the red lines or basic values to which a particular community commits itself – red lines and basic values that may legitimately vary from one community to another – the commons' conditions are not contingent or negotiable. For human agents to compromise the conditions upon which human existence and agency is itself predicated is simply unthinkable.

Finally, it should be said that my sketch of the regulatory responsibilities is incomplete – in particular, concepts such as the 'public interest' and the 'public good' need to be located within this bigger picture; and, there is more to be said about the handling of horizontal conflicts and tensions *within* a particular tier. Nevertheless, the 'take home message' is clear. Quite simply: while automated healthcare and research might be efficient and productive, new technologies should not present unacceptable risks to the legitimate interests of humans; beyond mere balancing, new technologies should be compatible with the fundamental values of particular communities; and, above all, these technologies should do no harm to the commons' conditions – supporting human existence and agency – on which we all rely and which we undervalue at our peril.

---

[33] Compare R. Brownsword, 'Regulating Patient Safety: Is It Time for a Technological Response?', (2014) *Law, Innovation and Technology*, 6(1), 1–29.

[34] See M. Cook, 'Bedside Manner 101: How to Deliver Very Bad News', *Bioedge* (17 March 2019), www.bioedge.org/bioethics/bedside-manner-101-how-to-deliver-very-bad-news/12998.

Widening the Lens

Introduction

Agomoni Ganguli-Mitra

The sheer diversity of topics in health research makes for a daunting task in the development, establishment, and application of oversight mechanisms and various methods of governance. The authors of this section illustrate how this task is made even more complex by emerging technologies, applications and context, as well as the presence of a variety of actors both in the research and the governance landscape. Nevertheless, key themes emerge, and these sometimes trouble existing paradigms and parameters, and shift and widen our regulatory lenses. A key anchor is the relationship between governance and time: be it the urgent nature of research conducted in global health emergencies; the appropriate weight given to historical data in establishing evidence, anticipating future risk, benefit or harm; or the historical and current forces that have shaped regulatory structures as we meet them today. The perspectives explored in this section can be seen to illustrate different kinds of liminality, which result in regulatory complexity but also offer potential for new kinds of imaginaries, norms and processes.

A first kind of shift in lens is created by the nature of research contexts: for example, whether research is carried out in labs, in clinical settings, traditional healing encounters or, indeed, in a pandemic. These spaces might be the site where values, interests or rules conflict, or they might be characterised by the absence of regulation. Additional tension might be brought about in the interaction of *what* is being regulated, with *how* it is being regulated: emerging interventions in already established processes, traditional interventions in more recently developed but strongly established paradigms, or marginal interventions precipitated to the centre by outside forces (crises, economic profit, unexpected findings, imminent or certain injury or death). These shifts give rise to considerations of flexibility and resilience in regulation, of the legitimacy and authority of different actors, and the epistemic soundness in the development and deployment of innovative, experimental, or less established practices.

In Chapter 28, Ho addresses the key concept of risk, and its role within the governance of artificial intelligence (AI) and machine learning (ML) as medical devices. Using the illustration of AI/ML as clinical decision support in the diagnosis of diabetic retinopathy, the author situates their position in qualified opposition to those who perceive governance as an impediment to development and economic gain and those who favour more oversight of AI/ML. In managing such algorithms as risk objects in governance, Ho advocates a governance structure that re-characterises risk as a form of iterative learning process, rather than a rule-based one-time evaluation and regulatory approval based on the quantification of future risk.

The theme of regulation as obstacle is also explored in the following chapter (Chapter 29) by Lipworth et al., in the context of autologous mesenchymal stem cell-based interventions. Here, too, the perspective of the authors is set against those who see traditional governance and translational pathways as an impediment to addressing life-threatening and debilitating illnesses. They also resist the reimagination of healthcare as a marketplace (complete with aggressive marketing and dubious claims) where the patient is seen as a consumer, and the decision to access emerging and novel (unproven and potentially risky) interventions merely as a matter of shared decision-making between patient and clinician. The authors recommend the strengthening a multipronged governance framework, which includes professional regulation, marketplace regulation, regulation of therapeutic products, and research oversight.

In Chapter 30, Haas and Cloatre also explore the difficult task of aligning interventions and products within established regulatory and translational pathways. Here, however, the challenge is not novel or emerging interventions, but traditional or non-conventional medicine, which challenges establishes governance frameworks based on the biomedical paradigm, and yet which millions of patients worldwide rely on as their primary form of healthcare. Here, uncertainty relates to the epistemic legitimacy of non-conventional forms of knowledge gathering. Actors in conflict with established epistemic processes are informed by historical and contextual evidence and practices that far predate the establishment of current frameworks. Traditional and non-conventional interventions are, nevertheless, pushed towards hegemonic governance pathways, often in the 'scientised and commercial' forms, in order to gain recognition and legitimacy.

When considering pathways to legitimacy, a key role is played by ethics, in its multiple forms. In Chapter 31, Pickersgill explores ethics in its multiple forms through the eyes of neuroscience researchers, who in their daily practice experience the ethical dimensions of neuroscience and negotiate ethics as a regulatory tool. Ethics can be seen as obstacle to good science, and the (institutional) ethics of human research is often seen as prone to obfuscation and in lack of clear guidance. This results in novel practices and norms within the community, which are informed by a commitment to doing the right thing and by institutional requirements. In order to minimise potential subversion (even well-meant) of ethics in research, Pickersgill advocates the development of governance that arises not only from collaborations between scientists and regulators but also those who can act as critical friends to both of these groups of actors.

Ethics guidance and ethical practices are also explored by Ganguli-Mitra and Hunt (Chapter 32), this time in the context of research carried out in global health emergencies (GHEs). These contexts are characterised by various factors that complicate ethical norms and practices, as well as trouble existing frameworks and paradigms. GHEs are sites of multiple kinds of practices (humanitarian, medical, public health, development) and of multiple actors, whose goals and norms of conduct might be in conflict in a context that is characterised by urgency and high risk of injury and death. Using the examples of recent emergencies, the authors explore the changing nature of ethics and ethical practices in extraordinary circumstances.

In the final chapter of this section (Chapter 33), Arzuaga offers an illustration of regulatory development, touching upon the many actors, values, interests, and forces explored in the earlier chapters. Arzuaga reports on the governance of advanced therapeutic medicinal products (ATMPs) in Argentina, moving from a situation of non-intervention on the part of the state, to the establishment of a governance framework. Here, the role of hard and soft law as adding both resilience and flexibility to regulation is explored, fostering innovation without abdicating ethical concerns. Arzuaga describes early, unsuccessful attempts at regulating stem cell-based interventions, echoing the concerns presented by Lipworth et al., before exploring a more promising exercise in legal foresighting, which included a variety of actors and collaboration, as well a combination of top-down models and bottom-up, iterative processes.

28

When Learning Is Continuous

Bridging the Research–Therapy Divide in the Regulatory Governance of Artificial Intelligence as Medical Devices

Calvin W. L. Ho

28.1 INTRODUCTION

The regulatory governance of Artificial Intelligence and Machine Learning (AI/ML) technologies as medical devices in healthcare challenges the regulatory divide between research and clinical care, which is typically of pharmaceutical products. This chapter considers the regulatory governance of an AI/ML clinical decision support (CDS) software for the diagnosis of diabetic retinopathy as a 'risk object' by the Food and Drug Administration (FDA) in the United States (US). The FDA's regulatory principles and approach may play an influential role in how other countries govern this and other software as a medical device (SaMD). The disruptions that AI/ML technologies can cause are well publicised in the lay and academic media alike, although the more serious 'risks' of harm are still essentially anticipatory. In some quarters, there is a prevailing sense that a 'light-touch' approach to regulatory governance should be adopted to ensure that the advancement of AI – particularly in ways that are expected to generate economic gain – should not be unduly burdened. Hence, in response to the question of whether regulation of AI is needed now, scholars like Chris Reed have responded with a qualified 'No'. As Reed explains, the use of the technology in medicine is already regulated by the profession, and regulation will be adapted piecemeal as new AI technologies come into use anyway.[1] A 'wait and see' approach is likely to produce better long-term results than hurried regulation based on a very partial understanding of what needs to be regulated. It is also perhaps consistent with this mind-set that the commercial development and application of AI and AI-based technologies remain largely unregulated.

This chapter takes a different view on the issue, and argues that the response should be a qualified 'Yes' instead, partly because there is already an existing regulatory framework in place that may be adapted to meet anticipated challenges. As a 'risk object', the regulation of AI/ML medical devices cannot be understood and managed separately from a broader 'risk culture' within which it is embedded. Contrary to what an approach in 'command-and-control' suggests, regulatory governance of AI/ML medical devices should not be understood merely as the application of external forces to contain ills that must somehow be managed in order to derive the desired effects. Arguably, it is this limited conception of 'risks' and its relationship with

---

[1] C. Reed, 'How Should We Regulate Artificial Intelligence?', (2018) *Philosophical Transactions of the Royal Society*, Series A 376(2128), 20170360.

regulation that give rise to liminality. As Laurie and others clearly explains,[2] a liminal space is created contemporaneously with the uncertainties generated by new and emerging technologies. Drawing on the works of Arnold van Gennep and Victor Turner, 'liminality' is presented as an analytic to engage with the processual and experiential dynamics of transitional and transformational inter-structural boundary or marginal spaces. It is itself an intermediary process in a three-part pattern of experience, that begins with separation from an existing order, and concludes with re-integration into a new world.[3] Mapping liminal spaces and the changing boundaries entailed can help to highlight gaps in regulatory regimes.[4]

Risk-based evaluation is often a feature of such liminal spaces, and when they become sites for battles of power and values, ethical issues arise. Whereas liminality has been applied to account for human experiences within regulated spaces, this chapter considers the epistemic quality of 'risks' and its situatedness within regulatory governance as a discursive practice and as a matter of social reality. In this respect, regulation is not necessarily extrinsic to its regulatory object, but constitutive of it. Concerns about 'risks' from technological innovations and the need to tame them have been central to regulatory governance.[5] Whereas governance has been a longstanding cultural phenomenon that relates to 'the system of shared beliefs, values, customs, behaviours and artifacts that members of society use to cope with their world and with one another, and that are transmitted from generation to generation through learning',[6] it is the regulatory turn that is especially instructive. Here, regulatory response is taken to reduce the uncertainty and instability of mitigating potential risks and harms and by directing or influencing actors' behaviour to accord with socially accepted norms and/or to promote desirable social outcomes, and regulation encompasses any instrument (legal or non-legal in character) that is designed to channel group behaviour.[7] The high connectivity of AL/ML SaMDs that are capable of adapting to their digital environment in order to optimise performance suggests that the research agenda persists beyond what may be currently limited to the pilot or feasibility stages of medical device trials. If continuous risk-monitoring is required to support the use of SaMDs in a learning healthcare system, more robust and responsive regulatory mechanisms are needed, not less.[8]

28.2 AI/ML SOFTWARE AS CLINICAL DECISION SUPPORT

In April 2018, the FDA granted approval for IDx-DR (DEN180001) to be marketed as the first AI diagnostic system that does not require clinician interpretation to detect greater than a mild level of diabetic retinopathy in adults diagnosed with diabetes.[9] In essence, this SaMD applies an AI

---

[2] G. Laurie, 'Liminality and the Limits of Law in Health Research Regulation: What Are We Missing in the Spaces In-Between?', (2016) *Medical Law Review*, 25(1), 47–72; G. Laurie, 'What Does It Mean to Take an Ethics+ Approach to Global Biobank Governance?', (2017) *Asian Bioethics Review*, 9(4), 285–300; S. Taylor-Alexander et al., 'Beyond Regulatory Compression: Confronting the Liminal Spaces of Health Research Regulation', (2016) *Law Innovation and Technology*, 8(2), 149–176.

[3] Laurie, 'Liminality and the Limits of Law', 69.

[4] Taylor-Alexander et al., 'Beyond Regulatory Compression', 172.

[5] R. Brownsword et al., 'Law, Regulation and Technology: The Field, Frame, and Focal Questions' in R. Brownsword et al. (eds), *The Oxford Handbook of Law, Regulation and Technology* (Oxford University Press, 2017), pp. 1–36.

[6] D. G. Bates and F. Plog, *Cultural anthropology*. (New York: McGraw-Hill, 1990), p. 7.

[7] R. Brownsword, *Law, Technology and Society: Re-Imaging the Regulatory Environment* (Abingdon: Routledge, 2019), p. 45.

[8] B. Babic et al., 'Algorithms on Regulatory Lockdown in Medicine: Prioritizing Risk Monitoring to Address the 'Update Problem", (2019) *Science*, 366(6470), 1202–1204.

[9] Food and Drug Administration, 'FDA Permits Marketing of Artificial Intelligence-Based Device to Detect Certain Diabetes-related Eye Problems', (FDA New Release, 11 April 2018).

algorithm to analyse images of the eye taken with a retinal camera that are uploaded to a cloud server. A screening decision is made by the device as to whether the individual concerned is detected with 'more than mild diabetic retinopathy' and, if so, is referred to an eye care professional for medical attention. Where the screening result is negative, the individual will be rescreened in twelve months. IDx-DR was reviewed under the FDA's De Novo premarket review pathway and was granted Breakthrough Device designation,[10] as the SaMD is novel and of low to moderate risk. On the whole, the regulatory process did not detract substantially from the existing regulatory framework for medical devices in the USA. A medical device is defined broadly to include low-risk adhesive bandages to sophisticated implanted devices. In the USA, a similar approach is adopted in the definition of the term 'device' in Section 201(h) of the Federal Food, Drug and Cosmetic Act.[11]

For regulatory purposes, medical devices are classified based on their intended use and indications for use, degree of invasiveness, duration of use, and the risks and potential harms associated with their use. At the classification stage, a manufacturer is not expected to have gathered sufficient data to demonstrate that its proposed product meets the applicable marketing authorisation standard (e.g. data demonstrating effectiveness). Therefore, the focus of the FDA's classification analysis is on how the product is expected to achieve its primary intended purposes.[12] The FDA has established classifications for approximately 1700 different generic types of devices and grouped them into sixteen medical specialties referred to as 'panels'. Each of these generic types of devices is assigned to one of three regulatory classes based on the level of control necessary to assure the safety and effectiveness of the device. The class to which the device is assigned determines, among other things, the type of premarketing submission/application required for FDA clearance to market. All classes of devices are subject to General Controls,[13] which are the baseline requirements of the FD&C Act that apply to all medical devices. Special Controls are regulatory requirements for Class II devices, and are usually device-specific and include performance standards, postmarket surveillance, patient registries, special labelling requirements, premarket data requirements and operational guidelines. For Class III devices, active regulatory review in the form of premarket approval is required (see Table 28.1).

Clinical trials of medical devices, where required, are often non-randomised, non-blinded, do not have active control groups, and lack hard endpoints, since randomisation and blinding of patients or physicians for implantable devices will in many instances be technically challenging and ethically unacceptable.[14] Table 28.2 shows key differences between clinical trials of pharmaceuticals in contrast to medical devices.[15] Class I and some Class II devices may be introduced

---

[10] Food and Drug Administration, 'Classification of Products as Drugs and Devices & Additional Product Classification Issues: Guidance for Industry and FDA Staff', (FDA, 2017).

[11] Federal Food, Drug, and Cosmetic Act (25 June 1938), 21 USC §321(h).

[12] The regulatory approaches adopted in the European Union and the United Kingdom are broadly similar to that of the FDA. See: J. Ordish et al., *Algorithms as Medical Devices* (Cambridge: PHG Foundation, 2019).

[13] Food and Drug Administration, 'Regulatory Controls', (FDA, 27 March 2018), www.fda.gov/medical-devices/overview-device-regulation/regulatory-controls.

[14] G. A. Van Norman, 'Drugs and Devices: Comparison of European and US Approval Processes', (2016) *JACC Basic to Translational Science*, 1(5), 399–412.

[15] Details in Table 28.2 are adapted from the following sources: International Organization of Standards, 'Clinical Investigation of Medical Devices for Human Subjects – Good Clinical Practice', (ISO, 2019), ISO/FDIS 14155 (3rd edition); Genesis Research Services, 'Clinical Trials – Medical Device Trials', (Genesis Research Services, 5 September 2018), www.genesisresearchservices.com/clinical-trials-medical-device-trials/; B. Chittester, 'Medical Device Clinical Trials – How Do They Compare with Drug Trials?', (Master Control, 7 May 2020), www.mastercontrol.com/gxp-lifeline/medical-device-clinical-trials-how-do-they-compare-with-drug-trials-/.

TABLE 28.1. *FDA classification of medical devices by risks*

Class	Risk	Level of regulatory controls	Whether clinical trials required	Examples
I	Low	General	No	Gauze, adhesive bandages, toothbrush
II	Moderate	General and special	Maybe	Suture, diagnostic X-rays
III	High	General and premarket approval	Yes	Pacemakers, implantable defibrillators, spinal cord stimulators

TABLE 28.2. *Comparing pharmaceutical trial phases and medical device trial stages*

Pharmaceuticals			Medical devices		
Phase	Participants	Purpose	Stage	Participants	Purpose
0 (Pilot/ exploratory ; not all drugs undergo this phase)	10–15 participants with disease or condition	Test very small (subtherapeutic) dosage to study effects and mechanisms	Pilot/early feasibility/ first-in-human	10–15 participants with disease or condition	Collect preliminary safety and performance data to guide development
I (Safety and toxicity)	10–100 healthy participants	Test safety and tolerance Determine dosing and major adverse effects	Feasibility	20–30 participants with disease or condition	Assess safety and efficacy of near-final or final device design Guides design of pivotal study
II (Safety and effectiveness)	50–200 participants with disease or condition	Test safety and effectiveness Confirm dosing and major adverse effects			
III (Clinical effectiveness)	>100–1000 participants with disease or condition	Test safety and effectiveness Determine drug–drug interaction and minor adverse effects	Pivotal	>100–300 participants with disease or condition	Establish clinical efficacy, safety and risks
IV (Post-approval study)	>1000	Collect long-term data and adverse effects	Post-approval study	>1000	Collect long-term data and adverse effects

into the US market without having been tested in humans through an approval process that is based on predicates. Through what is known as the 510(k) pathway, a manufacturer needs to show that its 'new' device is at least as safe and effective as (or substantially equivalent to) a legally marketed predicate device (as was the case for IDx-DR).[16]

The nature of regulatory control is changing; regulatory control does not arise solely through the exertion of regulatory power over a regulated entity but also acts intrinsically from within the entity itself. It is argued that risk-based regulation draws on different knowledge domains to constitute the AI/ML algorithm as a 'risk object', and not merely to subjugate it. Risk objectification renders the regulated entity calculable. Control does not thereby arise because the regulated entity behaves strictly in adherence to specific commands but rather because of the predictability of its actions. Where risk cannot be precisely calculated however, liminal spaces

[16] J. P. Jarow and J. H. Baxley, 'Medical Devices: US Medical Device Regulation', (2015) *Urologic Oncology: Seminars and Original Investigations*, 33(3), 128–132.

may help to articulate various 'scenarios' with different degrees of plausibility. These liminal spaces are thereby themselves a means by which uncertainty is managed. Typically, owing to conditions that operate outside of direct regulatory control, liminal spaces can either help to maintain a broader regulatory space to which they are peripheral, or contribute to its re-configuration through a 'domaining effect'. This aspect will be considered in the penultimate section of this chapter.

28.3 RE-EMBEDDING RISK AND A RETURN TO SOCIALITY

The regulatory construction of IDx-DR as a 'risk object' is accomplished by linking the causal attributes of economic and social risks, and risks to human safety and agency, to its constitutive algorithms reified as a medical device.[17] This 'risk object' is made epistemically 'real' when integrated through a risk discourse, by which risk attributions and relations have come to define identities, responsibilities, and socialities. While risk objectification has been effective in paving a way forward to market approval for IDx-DR, this technological capability is pushed further into liminality. The study that supported the FDA's approval was conducted under highly controlled conditions where a relatively small group of carefully selected patients had been recruited to test a diagnostic system that had a narrow usage criteria.[18] It is questionable whether the AI/ML feature was itself tested, since the auto-didactic aspect of the algorithm was locked prior to the clinical trial, which greatly constrained the variability of the range of outputs.[19] At this stage, IDx-DR is not capable of evaluating the most severe forms of diabetic retinopathy that requires urgent ophthalmic intervention. However, IDx-DR is capable of ML, which is a subset of AI and refers to a set of methods that have the ability to automatically detect patterns in data in order to predict future data trends or for decision-making under uncertain conditions.[20] Deep learning (DL) is in turn a subtype of ML (and a subfield of representation learning) that is capable of delivering a higher level of performance, and does not require a human to identify and compute the discriminatory features for it. From the 1980s onwards, DL software has been applied in computer-aided detection systems, and the field of radiomics (a process that extracts large number of quantitative features from medical images) is broadly concerned with computer-aided diagnosis systems, where DL has enabled the use of computer-learned tumour signatures.[21] It has the potential to detect abnormalities, make differential diagnoses and generate preliminary radiology reports in the future, but only a few methods are able to manage the wide range of radiological presentations of subtle disease states. In the foreseeable future, unsuper-vised AI/ML will test the limits of conventional means of regulation of medical devices.[22] The challenges to risk assessment, management and mitigation will be amplified as AI/ML medical devices change rapidly and become less predictable.[23]

[17] A. Bowser et al., *Artificial Intelligence: A Policy-Oriented Introduction* (Washington, DC: Woodrow Wilson International Center for Scholars, 2017).

[18] M. D. Abràmoff et al., 'Pivotal Trial of an Autonomous AI-Based Diagnostic System for Detection of Diabetic Retinopathy in Primary Care Offices', (2018) *NPJ Digital Medicine*, 39(1), 1.

[19] P. A. Keane and E. J. Topol, 'With an Eye to AI and Autonomous Diagnosis', (2018) *NPJ Digital Medicine*, 1, 40.

[20] K. P. Murphy, *Machine Learning: A Probabilistic Perspective* (Cambridge, MA: MIT Press, 2012).

[21] M. L. Giger, 'Machine Learning in Medical Imaging', (2018) *Journal of the American College of Radiology*, 15(3), 512–520.

[22] E. Topol, *Deep Medicine: How Artificial Intelligence Can Make Healthcare Human Again* (New York: Basic Books, 2019); A. Tang et al., 'Canadian Association of Radiologists White Paper on Artificial Intelligence in Radiology', (2018) *Canadian Association of Radiologists Journal*, 69(2), 120–135.

[23] Babic et al., 'Algorithms'; M. U. Scherer, 'Regulating Artificial Intelligence Systems: Risks, Challenges, Competencies and Strategies', (2016) *Harvard Journal of Law & Technology*, 29(2), 354–400.

Regulatory conservatism reflects a particular positionality and related interests that are at stake. For many high-level policy documents on AI, competitive advantage for economic gain is a key interest.[24] This position appears to support a 'light touch' approach to regulatory governance of AI in order to sustain technological development and advance national economic interests. If policy-makers, as a matter of socio-political construction, consider regulation as impeding technological development, then regulatory governance is unlikely to see meaningful progression. Not surprisingly, the private sector has had a dominant presence in defining the agenda and shape of AI and related technologies. While this is not in and of itself problematic, the narrow regulatory focus and absence of broader participation could be. For instance, it is not entirely clear to what extent the development of AI/ML algorithms is determined primarily by sectorial interests.[25]

Initial risk assessment is essentially consequentialist in its focus on intended use of the SaMD to achieve particular clinical outcomes. Risk characterisation is abstracted to two factors:[26] (1) significance of the information provided by the SaMD to the healthcare decision; and (2) state of the healthcare situation or condition. Risk is thereby derived from 'objective' information that is provided by the manufacturer on intended use of the information provided by the SaMD in clinical management. Such use may be significant in one of three ways: (1) to treat or to diagnose, (2) to drive clinical management or (3) to inform clinical management. The significance of an intended use is then associated with a healthcare situation or condition (i.e. critical, serious or non-serious). Schematically, Table 28.3 presents the risk characterisation framework based on four different levels of impact on the health of patients or target populations. Level IV of the framework (e.g. SaMD that performs diagnostic image analysis for making treatment decisions in patients with acute stroke, or screens for mutable pandemic outbreak that can be highly communicable through direct contact or other means) relates to the highest impact while Level I (e.g. SaMD that analyses optical images to guide next diagnostic action of astigmatism) relates to the lowest.[27]

To counter the possible deepening of regulatory impoverishment, regulatory governance as concept and process will need to re-characterise risk management as a form of learning and experimentation rather than rule-based processes, thus placing stronger reliance on human capabilities to imagine alternative futures instead of quantitative ambitions to predict the future. Additionally, a regulatory approach that is based on total project lifecycle needs to be taken up. This better accounts for modifications that will be made to the device through real-world learning and adaptation. Such adaptation enables a device to change its behaviour over time based on new data and optimise its performance in real time with the goal of improving health outcomes. As the FDA's conventional review procedures for medical devices discussed above are not adequately responsive to assess adaptive AI/ML technologies, the FDA has proposed for a premarket review mechanism to be developed.[28] This mechanism seeks to introduce a

[24] Executive Office of the President, *Artificial Intelligence, Automation and the Economy* (Washington, DC: US Government, 2016); House of Commons, Science and Technology Committee (2016), *Robotics and Artificial Intelligence: Fifth Report of Session 2016–17*, HC 145 (London, 12 October 2016).
[25] C. W. L. Ho et al., 'Governance of Automated Image Analysis and Artificial Intelligence Analytics in Healthcare', (2019) *Clinical Radiology*, 74(5), 329–337.
[26] IMDRF Software as a Medical Device (SaMD) Working Group, 'Software as a Medical Device: Possible Framework for Risk Categorization and Corresponding Considerations', (International Medical Device Regulators Forum, 2014), para. 4.
[27] Ibid., p. 14, para. 7.2.
[28] Food and Drug Administration, 'Proposed Regulatory Framework for Modifications to Artificial Intelligence/Machine Learning (AI/ML)-Based Software as a Medical Device (SaMD): Discussion Paper and Request for Feedback', (US Department of Health and Human Services, 2019).

TABLE 28.3. *Risk characterisation framework for software as a medical device*

State of healthcare situation or condition	Significance of information provided by SaMD to healthcare decision		
	Treat or diagnose	Drive clinical management	Inform clinical management
Critical	IV	III	II
Serious	III	II	I
Non-serious	II	I	I

predetermined change control plan in the premarket submission, in order to give effect to the risk categorisation and risk management principles, as well as the total product lifecycle approach, of the IMDRF. The plan will include the types of anticipated modifications (or pre-specifications) and associated methodology that is used to implement the changes in a controlled manner while allowing risks to patients to be managed (referred to as Algorithm Change Protocol). In essence, the proposed changes will place on manufacturers a greater responsibility of monitoring the real-world performance of their medical devices and to make available the performance data through periodic updates on what changes were made as part of the approved pre-specifications and the Algorithm Change Protocol. In totality, these proposed changes will enable the FDA to evaluate and monitor, collaboratively with manufacturers, an AI/ML software as a medical device from its premarket development to postmarket performance. The nature of the FDA's regulatory oversight will also become more iterative and responsive in assessing the impact of device optimisation on patient safety.

As the IMDRF also explains, every SaMD will have its own risk category according to its definition statement even when it is interfaced with other SaMD, other hardware medical devices or used as a module in a larger system. Importantly, manufacturers are expected to have an appropriate level of control to manage changes during the lifecycle of the SaMD. The IMDRF labels any modifications made throughout the lifecycle of the SaMD, including its maintenance phase, as 'SaMD Changes'.[29] Software maintenance is in turn defined in terms of post-marketing modifications that could occur in the software lifecycle processes identified by the International Organization for Standardization.[30] It is generally recognised that testing of software is not sufficient to ensure safety in its operation. Safety features need to be built into the software at the design and development stages, and supported by quality management and post marketing surveillance after the SaMD has been installed. Post market surveillance includes monitoring, measurement and analysis of quality data through logging and tracking of complaints, clearing technical issues, determining problem causes and actions to address, identify, collect, analyse and report on critical quality characteristics of products developed. However, monitoring software quality alone does not guarantee that the objectives for a process are being achieved.[31]

As a concern of Quality Management System (QMS), the IMDRF requires that maintenance activities preserve the integrity of the SaMD without introducing new safety, effectiveness, performance and security hazards. It recommends that a risk assessment, including

[29] IMDRF SaMD Working Group, 'Software as a Medical Device (SaMD): Key Definitions', (International Medical Decive Regulators Forum, 2013).

[30] International Organization of Standards, 'ISO/IEC 14764:2006 Software Engineering – Software Life Cycle Processes – Maintenance (2nd Edition)', (International Organization of Standards, 2006).

[31] IMDRF SaMD, 'Software as a Medical Device (SaMD): Application of Quality Management System. IMDRF/SaMD WG/N23 FINAL 2015', (International Medical Device Regulators Forum, 2015), para. 7.5.

considerations in relation to patient safety and clinical environment and technology and systems environment, should be performed to determine if the changes affect the SaMD categorisation and the core functionality of SaMD as set out in its definition statement. The proposed QMS complements the risk categorisation framework through its goal of incorporating good software quality and engineering practices into the device. Principles underscoring QMS are set out in terms of organisational support structure, lifecycle support processes, and a set of realisation and use processes for assuring safety, effectiveness and performance. These principles have been endorsed by the FDA in its final guidance to describe an internally agreed upon understanding (among regulators) of clinical evaluation and principles for demonstrating the safety, effectiveness and performance of the device, and activities that manufacturers can take to clinically evaluate their device.[32]

28.4 REGULATORY GOVERNANCE AS PARTICIPATORY LEARNING SYSTEM

In this penultimate section of this chapter, it is argued that the regulatory approach considered in the preceding sections is intended to support a participatory learning system comprising at least two key features: (1) a platform and/or mechanisms that enable constructive engagement with, and participation of, members of society; and (2) the means by which a common fund of knowledges (to be explained below) may be pooled to generate an anticipatory knowledge that could guide collective action. In some instances, institutionalisation could advance this agenda, but it is beyond the scope of this manuscript to examine this possibility to a satisfactory degree.

There is a diverse range of modalities through which constituents of a society engage in collaborative learning. As Annelise Riles's PAWORNET illustrates, each modality has its own goals, character, strengths and limitations. In her study, Riles observes that networkers did not understand themselves to share a set of values, interests or culture.[33] Instead, they understood themselves to be sharing their involvement in a certain network that was a form of institutionalised association devoted to information sharing. What defined networkers most of all was the fact that they were personally and institutionally connected or knowledgeable about the world of specific institutions and networks. In particular, it was the work of creating documents, organising conferences or producing funding proposals that generated a set of personal relations that drew people together and also created divisions of its own. In the author's own study,[34] ethnographic findings illustrate how the 'publics' of human stem cell research and oocyte donation were co-produced with an institutionalised 'bioethics-as-public-policy' entity known as the Bioethics Advisory Body. In that context, the 'publics' comprised institutions and a number of individuals – often institutionally connected – that represented a diverse set of values, interests and perhaps cultures (construed in terms of their day-to-day practices in the least). These 'publics' resemble a network in a number of ways. They were brought into a particular set of relationship within a deliberative space created mainly by the consultation papers and reinforced through a variety of means that included public meetings, conferences, and feedback sessions. Arguably, even individual feedback from a public outreach platform known as 'REACH' encompassed a certain kind of pre-existing (sub-) network that has been formed with a view to soliciting relatively more spontaneous and independent, uninvited forms of civil

---

[32] Food and Drug Administration, 'Software as a Medical Device (SAMD): Clinical Evaluation', (US Department of Health and Human Services, 2017).

[33] A. Riles, *The Network Inside Out* (Ann Arbor, MI: University of Michigan Press, 2001), pp. 58–59 and p. 68.

[34] C. W. L. Ho, *Juridification in Bioethics* (London: Imperial College Press, 2016).

participatory action. But this 'network' is not a static one. It varied with, but was also shaped by, the broader phenomenon of science and expectations as to how science ought to be engaged. In this connection, Riles's observation is instructive: 'It is not that networks "reflect" a form of society, therefore, nor that society creates its artifacts . . . Rather, it is all within the recursivity of a form that literally speaks about itself'.[35]

A 'risk culture' that supports learning and experimentation rather than rule-based processes must embed the operation of AI and related technologies as 'risk objects' within a common fund of knowledges. Legal processes are inherent to understanding the risk, such as that of a repeat sexual offence under 'Megan's Law', which encompasses the US community notification statutes relating to sexual offenders.[36] Comprising three tiers, this risk assessment process determines the scope of community notification. In examining the constitutional basis of Megan's Law, Mariana Valverde et al. observe that 'the courts have emphasised the scientific expertise that is said to be behind the registrant risk assessment scale (RRAS) in order to argue that Megan's Law is not a tool of punishment but rather an objective measure to regulate a social problem'.[37] However, reliance on Megan's Law as grounded in objective scientific knowledge has given rise to an 'intermediary knowledge in which legal actors – prosecutors and judges – are said not only to be more fair but even more reliable and accurate in determining a registrant's risk of re-offence'.[38] In this, the study also illustrates a translation from scientific knowledge and processes to legal ones, and how the 'law' may be cognitively and normatively open.

Finally, the articulation of possible harms and dangers as 'risks' involves the generation of 'anticipatory knowledge', which is defined as 'social mechanisms and institutional capacities involved in producing, disseminating, and using such forms [as] . . . forecasts, models, scenarios, foresight exercises, threat assessments, and narratives about possible technological and societal futures'.[39] Like Ian Hacking's 'looping effect', anticipatory knowledge is about knowledge-making about the future, and could operate as a means to gap-filling. The study by Hugh Gusterson of the Reliable Replacement Warhead (RRW) program is illustrative of this point, where US weapons laboratories could design new and highly reliable nuclear weapons that are safe to manufacture and maintain.[40] Gusterson shows that struggle over the RRW Program, initiated by the US Congress in 2004, occurred across four intersecting 'plateaus of nuclear calculations' – geopolitical, strategic, enviropolitical, and technoscientific – each with its own contending narratives of the future. He indicates that 'advocates must stabilise and align anticipatory knowledge from each plateau of calculation into a coherent-enough narrative of the future in the face of opponents seeking to generate and secure alternative anticipatory knowledges'.[41] Hence the *interconnectedness* of the four plateaus of calculation, including the trade-offs entailed, was evident in the production of anticipatory knowledge vis-à-vis the RRW program. In addition, the issues of performativity and 'social construction of ambiguity' were also evident. Gusterson observes that being craft items, no two nuclear weapons are exactly alike. However, the proscription of testing through detonation meant that both performativity and

[35] Riles, *The Network*, p. 69.
[36] M. Valverde et al., *Legal Knowledges of Risk. In Law Commission of Canada, Law and Risk* (Vancouver, BC: University of British Columbia Press, 2005), pp. 86–120, p. 103 and p. 106.
[37] Ibid., p. 106.
[38] Ibid.
[39] N. Nelson et al., 'Introduction: The Anticipatory State: Making Policy-relevant Knowledge About the Future', (2008) *Science and Public Policy*, 35(8), 546–550.
[40] H. Gusterson, 'Nuclear Futures: Anticipator Knowledge, Expert Judgment, and the Lack that Cannot Be Filled', (2008) *Science and Public Policy*, 35(8), 551–560.
[41] Ibid., 553.

ambiguity over reliability became matters of speculation, determined through extrapolation from the past to fill knowledge 'gaps' in the present and future. This attempt at anticipatory knowledge creation also prescribed a form that the future was to take. Applying a similar analysis from a legal standpoint, Graeme Laurie and others explain that foresighting as a means of devising anticipatory knowledge is neither simple opinion surveying nor mere public participation.[42] It must instead be directed at the discovery of shared values, the development of shared lexicons, the forging of a common vision of the future and the taking of steps to realise the vision with the understanding that this is being done from a position of partial knowledge about the future. As we have considered earlier on in this chapter, this visionary account captures the approach that has been adopted by the IMDRF impressively well.

28.5 CONCLUSION

Liminality highlights the need for a processual-oriented mode of regulation in order to recognise the flexibility and fluidity of the regulatory context (inclusive of its objects and subjects) and the need for iterative interactions, as well as to possess the capacity to provide non-directive guidance.[43] If one considers law as representing nothing more than certainty, structure and directed agency, then we should rightly be concerned as to whether the law can envision and support the creation of genuinely liminal regulatory spaces, which is typified by uncertainty, anti-structure and an absence of agency.[44] The crucial contribution of regulatory governance however, is its conceptualisation of law as an epistemically open enterprise, and in respect of which learning and experimentation are possible.

---

[42] G. Laurie et al., 'Foresighting Futures: Law, New Technologies, and the Challenges of Regulating for Uncertainty', (2012) *Law, Innovation and Technology*, 4(1), 1–33.
[43] Laurie, 'Liminality and the Limits of Law', 68–69; Taylor-Alexander et al., 'Beyond Regulatory Compression', 158.
[44] Laurie, 'Liminality and the Limits of Law', 71.

29

The Oversight of Clinical Innovation in a Medical Marketplace

*Wendy Lipworth, Miriam Wiersma, Narcyz Ghinea, Tereza Hendly, Ian Kerridge,
Tamra Lysaght, Megan Munsie, Chris Rudge, Cameron Stewart
and Catherine Waldby*

29.1 INTRODUCTION

Clinical innovation is ubiquitous in medical practice and is generally viewed as both necessary and desirable. While innovation has been the source of considerable benefit, many clinical innovations have failed to demonstrate evidence of clinical benefit and/or caused harm. Given uncertainty regarding the consequences of innovation, it is broadly accepted that it needs some form of oversight. But there is also pushback against what is perceived to be obstruction of access to innovative interventions. In this chapter, we argue that this pushback is misguided and dangerous – particularly because of the myriad competing and conflicting interests that drive and shape clinical innovation.

29.2 CLINICAL INNOVATION AND ITS OVERSIGHT

While the therapeutics lifecycle is usually thought of as one in which research precedes clinical application, it is common for health professionals to offer interventions that differ from standard practice, and that have either not (yet) been shown to be safe or effective or have been shown to be safe but not yet subjected to large phase 3 trials. This practice is often referred to as 'clinical innovation'.[1] The scope of clinical innovation is broad, ranging from minor alterations to established practice – for example using a novel suturing technique – to more significant departures from standard practice – for example using an invasive device that has not been formally tested in *any* population.

For the most part, clinical innovation is viewed as necessary and desirable. Medicine has always involved the translation of ideas into treatment and it is recognised that ideas originate in the clinic as well as in the research setting, and that research and practice inform each other in an iterative manner.[2] It is also recognised that the standard trajectory of research followed by health technology assessment, registration and subsidisation may be too slow for patients with life-limiting or debilitating diseases and that clinical innovation can provide an important

---

[1] W. Lipworth et al., 'The Need for Beneficence and Prudence in Clinical Innovation with Autologous Stem Cells', (2018) *Perspectives in Biology and Medicine*, 61(1), 90–105.

[2] P. L. Taylor, 'Overseeing Innovative Therapy without Mistaking It for Research: A Function-Based Model Based on Old Truths, New Capacities, and Lessons from Stem Cells', (2010) *The Journal of Law, Medicine & Ethics*, 38(2), 286–302.

avenue for access to novel treatments.[3] There are also limitations to the systems that are used to determine what counts as 'standard' practice because it is up to – usually commercial – sponsors to seek formal registration for particular indications.[4]

While many clinical innovations have positively transformed medicine, others have failed to demonstrate evidence of clinical benefit,[5] or exposed patients to considerable harm – for example, the use of transvaginal mesh for the treatment of pelvic organ prolapse.[6] Many innovative interventions are also substantially more expensive than traditional treatments,[7] imposing costs on both patients and health systems. It is therefore broadly accepted that innovation requires some form of oversight. In most jurisdictions, oversight of innovation consists of a combination of legally based regulations and less formal governance mechanisms. These, in turn, can be focused on:

1. the oversight of clinical practice by professional organisations, medical boards, healthcare complaints bodies and legal regimes;
2. the registration of therapeutic products by agencies such as the US Food and Drug Administration, the European Medicines Agency and Australia's Therapeutic Goods Administration;
3. consumer protection, such as laws aimed at identifying and punishing misleading advertising; and
4. the oversight of research when innovation takes place in parallel with clinical trials or is accompanied by the generation of 'real world evidence' through, for example, clinical registries.

The need for some degree of oversight is relatively uncontroversial. But there is also pushback against what is perceived to be obstruction of access to innovative interventions.[8] There are two main arguments underpinning this position. First, it is argued that existing forms of oversight create barriers to clinical innovation. Salter and colleagues, for example, view efforts to assert external control over clinical innovation as manifestations of conservative biomedical hegemony that deliberately hinders clinical innovation in favour of more traditional translational

---

[3] B. Salter et al., 'Hegemony in the Marketplace of Biomedical Innovation: Consumer Demand and Stem Cell Science', (2015) *Social Science & Medicine*, 131, 156–163.

[4] N. Ghinea et al., 'Ethics & Evidence in Medical Debates: The Case of Recombinant Activated Factor VII', (2014) *Hastings Center Report*, 44(2), 38–45.

[5] C. Davis, 'Drugs, Cancer and End-of-Life Care: A Case Study of Pharmaceuticalization?', (2015) *Social Science & Medicine*, 131, 207–214; D. W. Light and J. Lexchin, 'Pharmaceutical Research and Development: What Do We Get for All That Money?', (2012) *BMJ*, 345, e4348; C. Y. Roh and S. H. Kim, 'Medical Innovation and Social Externality', (2017) *Journal of Open Innovation: Technology, Market, and Complexity*, 3(1), 3; S. Salas-Vega et al., 'Assessment of Overall Survival, Quality of Life, and Safety Benefits Associated with New Cancer Medicines', (2017) *JAMA Oncology*, 3(3), 382–390.

[6] K. Hutchinson and W. Rogers, 'Hips, Knees, and Hernia Mesh: When Does Gender Matter in Surgery?', (2017) *International Journal of Feminist Approaches to Bioethics*, 10(1), 26.

[7] Davis, 'Drugs, Cancer'; T. Fojo et al., 'Unintended Consequences of Expensive Cancer Therapeutics – The Pursuit of Marginal Indications and a Me-Too Mentality that Stifles Innovation and Creativity: The John Conley Lecture', (2014) *JAMA Otolaryngology – Head and Neck Surgery*, 140(12), 1225–1236; S. C. Overley et al., 'Navigation and Robotics in Spinal Surgery: Where Are We Now?', (2017) *Neurosurgery*, 80(3S), S86.

[8] D. Cohen, 'Devices and Desires: Industry Fights Toughening of Medical Device Regulation in Europe', (2013) *BMJ*, 347, f6204; C. Di Mario et al., 'Commentary: The Risk of Over-regulation', (2011) *BMJ*, 342, d3021; O. Dyer, 'Trump Signs Bill to Give Patients Right to Try Drugs', (2018) *BMJ*, 361, k2429; S. F. Halabi, 'Off-label Marketing's Audiences: The 21st Century Cures Act and the Relaxation of Standards for Evidence-based Therapeutic and Cost-comparative Claims', (2018) *American Journal of Law & Medicine*, 44(2–3), 181–196; M. D. Rawlins, 'The "Saatchi Bill" will Allow Responsible Innovation in Treatment', (2014) *BMJ*, 348, g2771; Salter et al., 'Hegemony in the Marketplace'.

pathways.[9] It has also been argued that medical negligence law deters clinical innovation[10] and that health technology regulation is excessively slow and conservative, denying patients the 'right to try' interventions that have not received formal regulatory approval.[11]

Second, it is argued that barriers are philosophically and politically inappropriate on the grounds that patients are not actually 'patients', but rather 'consumers'. According to these arguments, consumers should be free to decide for themselves what goods and services they wish to purchase without having their choices restricted by regulation and governance systems – including those typically referred to as 'consumer' (rather than 'patient') protections. Following this line of reasoning, Salter and colleagues[12] argue that decisions about access to innovative interventions should respect and support 'the informed health consumer' who:

> assumes she/he has the right to make their own choices to buy treatment in a health care market which is another form of mass consumption. . .'[13]

and who is able to draw on:

> a wide range of [information] sources which include not only the formally approved outlets of science and state but also the burgeoning information banks of the internet.'[14]

There are, however, several problems with these arguments. First, there is little evidence to support the claim that there is, in fact, an anti-innovative biomedical hegemony that is creating serious barriers to clinical innovation. While medical boards can censure doctors for misconduct, and the legal system can find them liable for trespass or negligence, these wrongs are no easier to prevent or prove in the context of innovation than in any other clinical context. Product regulation is similarly facilitative of innovation, with doctors being free to offer interventions 'off-label' and patients being allowed to apply for case-by-case access to experimental therapies. The notion that current oversight systems are anti-innovative is therefore not well founded.

Second, it is highly contestable that patients are 'simply' consumers – and doctors are 'simply' providers of goods and services – in a free market. For several reasons, healthcare functions as a very imperfect market: there is often little or no information available to guide purchases; there are major information asymmetries – exacerbated by misinformation on the internet; and patients may be pressured into accepting interventions when they have few, if any, other therapeutic options.[15] Furthermore, even if patients *were* consumers acting in a marketplace, it would not follow that the marketplace should be completely unregulated, for even the most libertarian societies have regulatory structures in place to prevent bad actors misleading people or exploiting them financially (e.g. through false advertising, price fixing or offering services that they are unqualified to provide).

This leaves one other possible objection to the oversight of clinical innovation – that patients are under the care of professionals who are able to collaborate with them in making decisions through shared decision-making. Here, the argument is that innovation (1) *should not* be overseen because it is an issue that arises between a doctor and a patient, and (2) *does not need* to be overseen

---

[9] Salter et al., 'Hegemony in the Marketplace'.
[10] Rawlins, 'The "Saatchi Bill"'.
[11] Dyer, 'Trump Signs Bill'.
[12] Salter et al., 'Hegemony in the Marketplace'.
[13] Ibid., 159.
[14] Ibid.
[15] T. Cockburn and M. Fay, 'Consent to Innovative Treatment', (2019) *Law, Innovation and Technology*, 11(1), 34–54; T. Hendl, 'Vulnerabilities and the Use of Autologous Stem Cells for Medical Conditions in Australia', (2018) *Perspectives in Biology and Medicine*, 61(1), 76–89.

because doctors are professionals who have their patients' interests at heart. These are compelling arguments because they are consistent with both the emphasis on autonomy in liberal democracies and with commonly accepted ideas about professionals and their obligations.

Two objections can, however, be raised. First, these arguments ignore the fact that professionalism is concerned not only with patient well-being but also with commitments to the just distribution of finite resources, furthering scientific knowledge and maintaining public trust.[16] The second problem with these arguments is that they are premised on the assumption that all innovating clinicians are consistently alert to their professional obligations and willing to fulfil them. Unfortunately, this assumption is open to doubt. To illustrate this point, we turn to the case of autologous mesenchymal stem cell-based interventions.

29.3 THE CASE OF AUTOLOGOUS MESENCHYMAL STEM CELL INTERVENTIONS

Stem cell-based interventions are procedures in which stem cells – cells that have the potential to self-replicate and to differentiate into a range of different cell types – or cells derived from stem cells are administered to patients for therapeutic purposes. Autologous stem cell-based interventions involve administering cells to the same person from whom they were obtained. The two most common sources of such stem cells are blood and bone marrow (haematopoietic) cells and connective tissue (mesenchymal) cells.

Autologous haematopoietic stem cells are extracted from blood or bone marrow and used to reconstitute the bone marrow and immune system following high dose chemotherapy. Autologous mesenchymal cells are extracted most commonly from fat and then injected – either directly from the tissue extracts or after expansion in the laboratory – into joints, skin, muscle, blood stream, spinal fluid, brain, eyes, heart and so on, in order to 'treat' degenerative or inflammatory conditions. The hope is that because mesenchymal stem cells may have immunomodulatory properties they may support tissue regeneration.

The use of autologous haematopoietic stem cells is an established standard of care therapy for treating certain blood and solid malignancies and there is emerging evidence that they may also be beneficial in the treatment of immunological disorders, such as multiple sclerosis and scleroderma. In contrast, evidence to support the use of autologous mesenchymal stem cell interventions is weak and limited to only a small number of conditions (e.g. knee osteoarthritis).[17] And even in these cases, it is unclear what the precise biological mechanism is and whether the cells involved should even be referred to as 'stem cells'[18] (we use this phrase in what follows for convenience).

Despite this, autologous mesenchymal stem cell interventions (henceforth AMSCIs) are offered for a wide range conditions for which there is *no* evidence of effectiveness, including spinal cord injury, motor neuron disease, dementia, cerebral palsy and autism.[19] Clinics offering these and other claimed 'stem cell therapies' have proliferated globally, primarily in the private

[16] Medical Professionalism Project, 'Medical Professionalism in the New Millennium: A Physicians' Charter', (2002) *Lancet*, 359(9305), 520–522.

[17] H. Iijima et al., 'Effectiveness of Mesenchymal Stem Cells for Treating Patients with Knee Osteoarthritis: A Meta-analysis Toward the Establishment of Effective Regenerative Rehabilitation', (2018) *NPJ Regenerative Medicine*, 3(1), 15.

[18] D. Sipp et al., 'Clear Up this Stem-cell Mess', (2018) *Nature*, 561, 455–457.

[19] M. Munsie et al., 'Open for Business: A Comparative Study of Websites Selling Autologous Stem Cells in Australia and Japan', (2017) *Regenerative Medicine*, 12(7); L. Turner and P. Knoepfler, 'Selling Stem Cells in the USA: Assessing the Direct-to-Consumer Industry', (2016) *Cell Stem Cell*, 19(2), 154–157.

healthcare sector – including in jurisdictions with well-developed regulatory systems – and there are now both domestic markets and international markets based on stem cell tourism.[20]

While AMSCIs are relatively safe, they are far from risk-free, with harm potentially arising from the surgical procedures used to extract cells (e.g. bleeding from liposuction), the manipulation of cells outside of the body (e.g. infection) and the injection of cells into the bloodstream (e.g. immunological reactions, fever, emboli) or other tissues (e.g. cyst formation, microcalcifications).[21] Despite these risks, many of the practitioners offering AMSCIs have exploited loopholes in product regulation to offer these interventions to large numbers of patients.[22] To make matters worse, these interventions are offered without obvious concern for professional obligations, as evident in aggressive and misleading marketing, financial exploitation and poor-quality evidence-generation practices.

First, despite limited efficacy and safety, AMSCIs are marketed aggressively through clinic websites, advertisements and appearances in popular media.[23] This is inappropriate both because the interventions being promoted are experimental and should therefore be offered to the minimum number of patients outside the context of clinical trials, and because marketing is often highly misleading. In some cases, this takes the form of blatant misinformation – for example, claims that AMSCIs are effective for autism, dementia and motor neuron disease. In other cases, consumers are misled by what have been referred to as 'tokens of legitimacy'. These include patient testimonials, references to incomplete or poor-quality research studies, links to scientifically dubious articles and conference presentations, displays of certification and accreditation from unrecognised organisations, use of meaningless titles such as 'stem cell physician' and questionable claims of ethical oversight. Advertising of AMSCIs is also rife with accounts of biological processes that give the impression that autologous stem cells are entirely safe – because they come from the patient's own body – and possess almost magical healing qualities.[24]

Second, AMSCIs are expensive, with patients paying thousands of dollars (not including follow-up care or the costs associated with travel).[25] In many cases, patients take drastic measures to finance access to stem cells, including mortgaging their houses and crowd-sourcing funding from their communities. Clinicians offering AMSCIs claim that such costs are justified given the complexities of the procedures and the lack of insurance subsidies to pay for them.[26] However, the costs of AMSCIs seem to be determined by the business model of the industry and by a determination of 'what the market will bear' – which in the circumstances of illness, is substantial. Furthermore, clinicians offering AMSCIs also conduct 'pay-to-participate' clinical trials and ask patients to pay for their information to be included in clinical registries. Such

---

[20] I. Berger et al., 'Global Distribution of Businesses Marketing Stem Cell-based Interventions',(2016) *Cell Stem Cell*, 19(2), 158–162; D. Sipp et al., 'Marketing of Unproven Stem Cell–Based Interventions: A Call to Action', (2017) *Science Translational Medicine*, 9(397); M. Sleeboom-Faulkner and P. K. Patra, 'Experimental Stem Cell Therapy: Biohierarchies and Bionetworking in Japan and India', (2011) *Social Studies of Science*, 41(5), 645–666.

[21] G. Bauer, et al., 'Concise Review: A Comprehensive Analysis of Reported Adverse Events in Patients Receiving Unproven Stem Cell-based Interventions', (2018) *Stem Cells Translational Medicine*, 7(9), 676–685; T. Lysaght et al., 'The Deadly Business of an Unregulated Global Stem Cell Industry', (2017) *Journal of Medical Ethics*, 43, 744–746.

[22] Sipp et al., 'Clear Up'.

[23] T. Caulfield et al., 'Confronting Stem Cell Hype', (2016) *Science*, 352(6287), 776–777; A. K. McLean et al., 'The Emergence and Popularisation of Autologous Somatic Cellular Therapies in Australia: Therapeutic Innovation or Regulatory Failure?', (2014) *Journal of Law and Medicine*, 22(1), 65–89; Sipp et al., 'Clear Up'.

[24] Munsie et al., 'Open for Business'; Sipp et al., 'Marketing'.

[25] A. Petersen et al., 'Therapeutic Journeys: The Hopeful Travails of Stem Cell Tourists', (2014) *Sociology of Health and Illness*, 36(5), 670–685.

[26] Worldhealth.net, 'Why Is Stem Cell Therapy So Expensive?', (WorldHealth.Net, 2018), www.worldhealth.net/news/why-stem-cell-therapy-so-expensive/.

practices are generally frowned upon as they exacerbate the therapeutic misconception and remove any incentive to complete and report results in a timely manner.[27]

Finally, contrary to the expectation that innovating clinicians should actively contribute to generating generalisable knowledge through research, clinics offering AMSCIs have proliferated in the absence of robust clinical trials.[28] Furthermore, providers of AMSCIs tend to overstate what is known about efficacy[29] and to misrepresent what trials are for, arguing that they simply 'measure and validate the effect of (a) new treatment'.[30] Registries that have been established to generate observational evidence about innovative AMSCIs are similarly problematic because participation is voluntary, outcome measures are subjective and results are not made public. There are also problems with the overall framing of the registries, which are presented as alternatives – rather than supplements – to robust clinical trials.[31] And because many AMSCIs are prepared and offered in private practice, there is lack of oversight and independent evaluation of what is actually administered to the patient, making it impossible to compare outcomes in a meaningful way.[32]

While it is possible that doctors offering autologous stem cell interventions simply lack awareness of the norms relating to clinical innovation, this seems highly unlikely, as many of these clinicians are active participants in policy debates about innovation and are routinely censured for behaviour that conflicts with accepted professional obligations. A more likely explanation, therefore, is that the clinicians offering autologous stem cell interventions are motivated not (only) by concern for their patients' well-being, but also by other interests such as the desire to make money, achieve fame and satisfy their intellectual curiosity. In other words, they have competing and conflicting interests that override their concerns for patient well-being and the generation of valid evidence.

29.4 IMPLICATIONS FOR OVERSIGHT OF CLINICAL INNOVATION

Unfortunately, the case of AMSCIs is far from unique. Other situations in which clinicians appear to be abusing the privilege of using their judgement to offer non-evidence-based therapies include orthopaedic surgeons over-using arthroscopies for degenerative joint disease,[33] assisted reproductive technology specialists who offer unproven 'add-ons' to traditional in-vitro fertilisation[34] and health professionals engaging in irresponsible off-label prescribing of psychotropic medicines.[35]

[27] D. Sipp, 'Pay-to-Participate Funding Schemes in Human Cell and Tissue Clinical Studies', (2012) *Regenerative Medicine*, 7(6s), 105–111.
[28] Sipp et al., 'Clear Up'.
[29] Sipp et al., 'Marketing'.
[30] R. T. Bright, 'Submission to the TGA Public Consultation: Regulation of Autologous Stem Cell Therapies: Discussion Paper for Consultation', (Macquarie Stem Cell Centres of Excellence, 2015), 4, www.tga.gov.au/sites/default/files/submissions-received-regulation-autologous-stem-cell-therapies-msc.pdf.
[31] Adult Stem Cell Foundation, 'Adult Stem Cell Foundation', www.adultstemcellfoundation.org; M. Berman and E. Lander, 'A Prospective Safety Study of Autologous Adipose-Derived Stromal Vascular Fraction Using a Specialized Surgical Processing System', (2017) *The American Journal of Cosmetic Surgery*, 34(3), 129–142; International Cellular Medicine Society, 'Open Treatment Registry', (ICMS, 2010), www.cellmedicinesociety.org/attachments/184_ICMS%20Open%20Treatment%20Registry%20-%20Overview.pdf.
[32] Sipp et al., 'Marketing'.
[33] P. F. Stahel, 'Why Do Surgeons Continue to Perform Unnecessary Surgery?', (2017) *Patient Safety in Surgery*, 11(1), 1.
[34] J. Wise, 'Show Patients Evidence for Treatment "Add-ons", Fertility Clinics are Told', (2019) *BMJ*, 364, l226.
[35] P. Sugarman et al., 'Off-Licence Prescribing and Regulation in Psychiatry: Current Challenges Require a New Model of Governance', (2013) *Therapeutic Advances in Psychopharmacology*, 3(4), 233–243.

Clinicians in all of these contexts are embedded in a complex web of financial and non-financial interests such as the desire to earn money, create product opportunities, pursue intellectual projects, achieve professional recognition and career advancement, and develop knowledge for the good of future patients[36] – all of which motivate their actions. Clinicians are also susceptible to biases such as the 'optimism bias', which might lead them to over-value innovative technologies and they are impacted upon by external pressures, such as industry marketing[37] and pressure from patients desperate for a 'miracle cure'.[38]

With these realities in mind, arguments against the oversight of innovation – or, more precisely, a reliance on consumer choice – become less compelling. Indeed, it could be argued that the oversight of innovation needs to be strengthened in order to protect patients from exploitation by those with competing and conflicting interests. That said, it is important that the oversight of clinical innovation does not assume that all innovating clinicians are motivated primarily by personal gain and, correspondingly, that it does not stifle responsible clinical innovation.

In order to strike the right balance, it is useful – following Lysaght and colleagues[39] – for oversight efforts to be framed in terms of, and account for, three separate functions: a *negative function* (focused on protecting consumers and sanctioning unacceptable practices, such as through tort and criminal law); a *permissive function* (concerned with frameworks that license health professionals and enable product development, such as through regulation of therapeutic products); and a *positive function* (dedicated to improving professional ethical behaviour, such as through professional registration and disciplinary systems). With that in mind, we now present some examples of oversight mechanisms that could be employed.

Those with responsibility for overseeing clinical practice need to enable clinicians to offer innovative treatments to selected patients outside the context of clinical trials, while at the same time preventing clinicians from exploiting patients for personal or socio-political reasons. Some steps that could be taken to both encourage responsible clinical innovation and discourage clinicians from acting on conflicts of interest might include:

- requiring that all clinicians have appropriate qualifications, specialisation, training and competency;
- mandating disclosure of competing and conflicting interests on clinic websites and as part of patient consent;
- requiring that consent be obtained by an independent health professional who is an expert in the patient's disease (if necessary at a distance for patients in rural and remote regions);
- ensuring that all innovating clinicians participate in clinical quality registries that are independently managed, scientifically rigorous and publicly accessible;

---

[36] T. E. Chan, 'Legal and Regulatory Responses to Innovative Treatment', (2012) *Medical Law Review*, 21(1), 92–130; T. Keren-Paz and A. J. El Haj, 'Liability versus Innovation: The Legal Case for Regenerative Medicine', (2014) *Tissue Engineering Part A*, 20(19–20), 2555–2560; J. Montgomery, 'The "Tragedy" of Charlie Gard: A Case Study for Regulation of Innovation?', (2019) *Law, Innovation and Technology*, 11(1), 155–174; K. Raus, 'An Analysis of Common Ethical Justifications for Compassionate Use Programs for Experimental Drugs', (2016) *BMC Medical Ethics*, 17(1), 60; P. L. Taylor, 'Innovation Incentives or Corrupt Conflicts of Interest? Moving Beyond Jekyll and Hyde in Regulating Biomedical Academic-Industry Relationships', (2013) *Yale Journal of Health Policy, Law, and Ethics*, 13(1), 135–197.

[37] Chan, 'Legal and Regulatory Responses'; Taylor, 'Innovation Incentives'.

[38] Chan, 'Legal and Regulatory Responses'.

[39] T. Lysaght et al., 'A Roundtable on Responsible Innovation with Autologous Stem Cells in Australia, Japan and Singapore', (2018) *Cytotherapy*, 20(9), 1103–1109.

- requiring independent oversight to ensure that appropriate product manufacturing standards are met;
- ensuring adequate pre-operative assessment, peri-operative care and post-operative monitoring and follow-up;
- ensuring that patients are not charged excessive amounts for experimental treatments, primarily by limiting expenses to cost-recovery; and
- determining that some innovative interventions should be offered only in a limited number of specialist facilities.

Professional bodies (such as specialist colleges), professional regulatory agencies, clinical ethics committees, drugs and therapeutics committees and other institutional clinical governance bodies would have an important role to play in ensuring that such processes are adhered to.

There may also be a need to extend current disciplinary and legal regimes regarding conflicts of interest (or at least ensure better enforcement of existing regimes). Many professional codes of practice already require physicians to be transparent about, and refrain from acting on, conflicts of interest. And laws in some jurisdictions already recognise that financial interests should be disclosed to patients, that patients should be referred for independent advice and that innovating clinicians need to demonstrate concern for patient well-being and professional consensus.[40]

With respect to advertising, there is a need to prevent aggressive and misleading direct-to-consumer advertising while still ensuring that all patients who might benefit from an innovative intervention are aware that such interventions are being offered. With this in mind, it would seem reasonable to strengthen existing advertising oversight (which, in many jurisdictions, is weak and *ad hoc*). It may also be reasonable to prohibit innovating clinicians from advertising interventions directly to patients – including indirectly through 'educational' campaigns and media appearances – and instead develop systems that alert referring doctors to the existence of doctors offering innovative interventions.

Those regulating access to therapeutic products need to strike a balance between facilitating timely access to the products that patients want, and ensuring that those with competing interests are not granted licence to market products that are unsafe or ineffective. In this regard, it is important to note that product regulation is generally lenient when it comes to clinical innovation and it is arguable that there is a need to push back against current efforts to accelerate access to health technologies – efforts that are rapidly eroding regulatory processes and creating a situation in which patients are being exposed to an increasing number of ineffective and unsafe interventions.[41] In addition, loopholes in therapeutic product regulation that can be exploited by clinicians with conflicts of interest should be predicted and closed wherever possible.

Although clinical innovation is not under the direct control of research ethics and governance committees, such committees have an important role to play in ensuring that those clinical trials and registries established to support innovation are not distorted by commercial and other imperatives. The task for such committees is to strike a balance between assuming that all researcher/innovators are committed to the generation of valid evidence and placing excessive burdens on responsible innovators who wish to conduct high-quality research. In this regard, research ethics committees could:

---

[40] Cockburn and Fay, 'Consent'; Keren-Paz and El Haj, 'Liability versus Innovation'.
[41] J. Pace et al., 'Demands for Access to New Therapies: Are There Alternatives to Accelerated Access?', (2017) *BMJ*, 359, j4494.

- ensure that participants in trials and registries are informed about conflicts of interest;
- ensure that independent consent processes are in place so that patients are not pressured into participating in research or registries; and
- consider whether it is ever acceptable to ask patients to 'pay to participate' in trials or in registries.

Research ethics committees also have an important role in minimising biases in the design, conduct and dissemination of innovation-supporting research. This can be achieved by ensuring that:

- trials and registries have undergone rigorous, independent scientific peer review;
- data are collected and analysed by independent third parties (e.g. Departments of Health);
- data are freely available to any researcher who wants to analyse it; and
- results – including negative results – are widely disseminated in peer-reviewed journals.

While this chapter has focused on traditional 'top-down' approaches to regulation and professional governance, it might also be possible to make use of what Devaney has referred to as 'reputation-affecting' regulatory approaches.[42] Such approaches would reward those who maintain their independence or manage their conflicts effectively with reputation-enhancing measures such as access to funding and publication in esteemed journals. In this regard, other parties not traditionally thought of as regulators – such as employing institutions, research funders, journal reviewers and editors and the media – might have an important role to play in the oversight of clinical innovation.

Importantly, none of the oversight mechanisms we have suggested here would discourage responsible clinical innovation. Indeed, an approach to the oversight of clinical innovation that explicitly accounts for the realities of competing and conflicting interests could make it easier for well-motivated clinicians to obtain the trust of both individual patients and broader social licence to innovate.

29.5 CONCLUSION

Clinical innovation has an important and established role in biomedicine and in the development and diffusion of new technologies. But it is also the case that claims about patients' – or consumers' – rights and about the sanctity of the doctor–patient relationship, can be used to obscure both the risks of innovation and the vested interests that drive some clinicians' decision to offer innovative interventions. In this context, adequate oversight of clinical innovation is crucial. After all, attempts to exploit the language and concept of innovation not only harms patients, but also threatens legitimate clinical innovation and undermines public trust. Efforts to push back against the robust oversight of clinical innovation need, therefore, to be viewed with caution.

[42] S. Devaney, 'Enhancing the International Regulation of Science Innovators: Reputation to the Rescue?', (2019) *Law, Innovation and Technology*, 11(1), 134–154.

30

The Challenge of 'Evidence'

Research and Regulation of Traditional and Non-Conventional Medicines

Nayeli Urquiza Haas and Emilie Cloatre

30.1 INTRODUCTION

Governments and stakeholders have struggled to find a common ground on how to regulate research for different ('proven' or 'unproven') practices. Research on traditional, alternative and complementary medicines is often characterised as following weak research protocols and as producing evidence too poor to stand the test of systematic reviews, thus rendering individual case studies results insignificant. Although millions of people rely on traditional and alternative medicine for their primary care needs, the *regulation* of research into, and practice of, these therapies is governed by biomedical parameters. This chapter asks how, despite efforts to accommodate other forms of evidence, regulation of research concerning traditional and alternative medicines is ambiguous as to what sort of evidence – and therefore what sort of research – can be used by regulators when deciding how to deal with practices that are not based on biomedical epistemologies. Building on ideas from science and technology studies (STS), in this chapter we analyse different approaches to the regulation of traditional and non-conventional medicines adopted by national, regional and global governmental bodies and authorities, and we identify challenges to the inclusion of other modes of 'evidence' based on traditional and hybrid epistemologies.

30.2 BACKGROUND

Non-conventional medicines are treatments that are not integrated to conventional medicine and are not necessarily delivered by a person with a degree in medical science. This may include complementary, alternative and traditional healers who may derive their knowledge from local or foreign knowledges, skills or practices.[1] For the World Health Organization (WHO), traditional medicine may be based on explicable or non-explicable theories, beliefs and experiences of different indigenous cultures.[2] That being said, traditional medicine is often included within the umbrella term of 'non-conventional medicine' in countries where biomedicine is the norm. However, this is often considered a misnomer insofar as traditional medicine may be the main source of healthcare in many countries, independent of its legitimate or illegitimate status. Given the high demand for traditional and non-conventional therapies, governments have

---

[1] P. Lannoye, 'Report on the Status of Non-Conventional Medicine', (Committee on the Environment, Public Health and Consumer Protection, 6 March 1997).

[2] WHO, 'WHO Global Report on Traditional and Complementary Medicine 2019', (WHO, 2019).

sought to bring these therapies into the fold of regulation, yet, the processes involved to accomplish this task have been complicated by the tendency to rely on biomedicine's standards of practice as a baseline. For example, the absence of and/or limited data produced by traditional and non-conventional medicine research and the unsatisfactory methodologies that do not stand the test of internationally recognised norms and standards for research involving human subjects have been cited as common barriers to the development of legislation and regulation of traditional and non-conventional medicine.[3] In 2019, the WHO reported that 99 out of 133 countries considered the absence of research as one of the main challenges to regulating these fields.[4] At the same time, governments have been reluctant to integrate traditional and non-conventional medicines as legitimate healthcare providers because their research is not based on the 'gold standard', namely multi-phase clinical trials.[5] Without evidence produced through conventional research methodologies, it is argued that people are at risk of falling prey to charlatans who peddle magical cures – namely placebos without any concrete therapeutic value – or that money is wasted on therapies and products based on outdated or disparate bodies of knowledge rather than systematic clinical research.[6] While governments have recognised to some extent the need to accommodate traditional and non-conventional medicines for a variety of reasons[7] – including the protection of cultural rights, consumer rights, health rights, intellectual property and biodiversity[8] – critics suggest that there is no reason why these modalities of medicine should be exempted from providing quality evidence.[9]

Picking up on some of these debates, this chapter charts the challenges arising from attempts to regulate issues relevant to research in the context of traditional and alternative medicine. From the outset, it explores what kinds of evidence and what kinds of research are accepted in the contemporary regulatory environment. It outlines some of the sticky points arising out of debates about research of traditional and non-conventional medicines, in particular, the role of placebo effects and evidence. Section 30.4 explores two examples of research regulation: WHO's Guidelines for Methodologies on Research and Evaluation of Traditional Medicine and the European Directive on Traditional Herbal Medicine Products (THMPD). Both incorporate mixed methodologies into research protocols and allow the use of historical data as evidence of efficacy, thus recognising the specificity of traditional medicine and non-conventional medicine. However, we argue that these strategies may themselves become subordinated to the biomedical logics, calling into question the extent to which other epistemologies or processes are allowed to shape what is considered as acceptable evidence. Section 30.5 focuses on the UK, as an example of how other processes and rationalities, namely economic governmentalities, shape the spaces that non-conventional medicine can inhabit. Section 30.6 untangles and

[3] E. Ernst, 'Commentary on: Close et al. (2014) A Systematic Review Investigating the Effectiveness of Complementary and Alternative Medicine (CAM) for the Management of Low Back and/or Pelvic Pain (LBPP) in Pregnancy', (2014) *Journal of Advanced Nursing*, 70(8), 1702–1716; WHO, 'General Guidelines for Methodologies on Research and Evaluation of Traditional Medicine', (WHO, 2000).

[4] WHO, 'Global Report on Traditional and Complementary Medicine 2019', (WHO, 2019).

[5] House of Lords, Select Committee on Science and Technology: Sixth Report (2000, HL).

[6] M. K. Sheppard, 'The Paradox of Non-evidence Based, Publicly Funded Complementary Alternative Medicine in the English National Health Service: An Explanation', (2015) *Health Policy*, 119(10), 1375–1381.

[7] The International Bioethics Committee (IBC) of the United Nations Educational, Scientific and Cultural Organization (UNESCO), the World Intellectual Property Organisation (WIPO), the World Trade Organisation (WTO) and WHO have stated support for the protection of traditional knowledges, including traditional medicines.

[8] Such as the European Red List of Medicinal Plants, which documents species endangered by human economic activities and loss of biodiversity.

[9] K. Hansen and K. Kappel, 'Complementary/Alternative Medicine and the Evidence Requirement' in M. Solomon et al. (eds), *The Routledge Companion to Philosophy of Medicine* (New York and Abingdon: Routledge, 2016).

critically analyses the assumptions and effects arising out of the process of deciding what counts as evidence in healthcare research regulations. It suggests that despite attempts to include different modalities, ambiguities persist due to acknowledged and unacknowledged hierarchies of knowledge-production explored in this chapter. The last section opens up a conversation about what is at stake when the logic underpinning the regulation of research creates a space for difference, including different medical traditions and what counts as evidence.[10]

<h2>30.3 EVIDENCE-BASED MEDICINE AND PLACEBO CONTROLS</h2>

Evidence-based medicine (EBM) stands for the movement which suggests that the scientific method allows researchers to find the best evidence available in order to make informed decisions about patient care. To find the best evidence possible, which essentially means that the many is more significant than the particular, EBM relies on multiple randomised controlled trials (RCTs) and evidence from these is eventually aggregated and compared.[11] Evidence is hierarchically organised, whereby meta-reviews and systematic reviews based on RCTs stand at the top, followed by non-randomised controlled trials, observational studies with comparison groups, case series and reports, single case studies, expert opinion, community evidence and individual testimonies at the bottom. In addition to reliance on quantity, the quality of the research matters. Overall, it means that the best evidence is based on data from blinded trials, which show a causal relation between therapeutic interventions and the effect, and isolates results from placebo-effects.

From a historical perspective, the turn to blinded tests represented a significant shift in medical practice insofar as it diminished the relevance of expert opinion, which was itself based on a hierarchy of knowledge that tended to value authority and theory over empirical evidence. Physicians used to prescribe substances, such as mercury, that although believed to be effective for many ailments, were later found to be highly toxic.[12] Thus, the notion of evidence arising out of blinded trials closed the gap between science and practice, and also partially displaced physicians' authority. Blinded trials and placebo controls had other effects: they became a tool to demarcate 'real' medicine from 'fake' medicine, proper doctors from 'quacks' and 'snake-oil' peddlers. By exposing the absence of a causal relationships between the therapy and the physical effect, some therapies and knowledges associated with them were rebranded as fraudulent or as superstitions. While the placebo effect might retrospectively explain why some of these dis-carded therapies were seen as effective, in practice, EBM's hierarchy of evidence dismisses patients' subjective accounts.[13] While explanations about the placebo effect side-lined the role of autosuggestion in therapeutic interventions, they did not clarify either the source or the benefits of self-suggestion.

---

[10] M. Zhan, *Other Wordly: Making Chinese Medicine through Transnational Frames* (London: Duke University Press, 2009); C. Schurr and K. Abdo, 'Rethinking the Place of Emotions in the Field through Social Laboratories', (2016) *Gender, Place and Culture*, 23(1), 120–133.
[11] D. L. Sackett et al., 'Evidence Based Medicine: What It Is and What It Isn't', (1996) *British Medical Journal*, 312(7023), 71–72.
[12] R. Porter, *The Greatest Benefit to Mankind: A Medical History of Humanity from Antiquity to the Present* (New York: Fontana Press, 1999).
[13] A. Wahlberg, 'Above and Beyond Superstition – Western Herbal Medicine and the Decriminalizing of Placebo', (2008) *History of the Human Sciences*, 21(1), 77–101; A. Harrington, 'The Many Meanings of the Placebo Effect: Where They Came From, Why They Matter', (2006) *BioSocieties*, 1(2), 181–193; P. Friesen, 'Mesmer, the Placebo Effect, and the Efficacy Paradox: Lessons for Evidence Based Medicine and Complementary and Alternative Medicine Medicine', (2019) *Critical Public Health*, 29(4), 435–447.

Social studies suggest that the role of imagination has been overlooked as a key element mediating therapeutic interactions. Phoebe Friesen argues that, rather than being an 'obstacle' that modern medicine needed to overcome, imagination 'is a powerful instrument of healing that can, and ought to be, subjected to experimental investigations.'[14] At the same time, when the positive role of the placebo effect and self-suggestion has been raised, scholarship research has pointed out dilemmas that remain unsolved, for example: Is it ethical to give a person a placebo in the conduct of research on non-orthodox therapies, and when is it justifiable, and for which conditions? Or, could public authorities justify the use of tax-payers money for so-called 'sham' treatments when people themselves, empowered by consumer choice rhetoric and patient autonomy, demand it? As elaborated in this chapter, some governments have been challenged for using public money to fund therapies deemed to be 'unscientific', while others have tightened control, fearing that self-help gurus, regarded as 'cultish' sect-leaders, are exploiting vulnerable patients.

To the extent that physiological mechanisms of both placebo and nocebo effects are still unclear, there does not seem to be a place in mainstream public healthcare for therapies that do not fit the EBM model because it is difficult to justify them politically and judicially, especially as healthcare regulations rely heavily on science to demonstrate public accountability.[15] And yet, while the importance of safety, quality and efficacy of therapeutic practices cannot be easily dismissed, the reliance on EBM as a method to demarcate effective from non-effective therapies dismisses too quickly the reasons why people are attracted to these therapies. When it comes to non-conventional medicines, biomedicine and the scientific method do not factor in issues such as patient choice or the social dimension of medical practice.[16] In that respect, questions as to how non-conventional medicine knowledges can demonstrate whether they are effective or not signal broader concerns. First, is it possible to disentangle the reliance of public accountability from science in order to solve the ethical, political, social and cultural dilemmas embedded in the practice of traditional and alternative medicine? Second, if we are to broaden the scope of how evidence is assessed, are there other processes or actors that shape what is considered effective from the perspective of healthcare regulation, for example, patient choice or consumer rights? And, finally, if science is not to be considered as the sole arbiter of healing, what are the spaces afforded for other epistemologies of healing? Without necessarily answering all of these questions, the aim of this chapter is to signpost a few sticky points in these debates. The next section explores three examples, at different jurisdictional levels – national, regional and international – of how healthcare regulators have sought to provide guidelines on how to incorporate other types of evidence into research dealing with traditional and non-conventional medicine.

30.4 INTEGRATION AS SUBORDINATION: GUIDELINES AND REGULATIONS ON EVIDENCE AND RESEARCH METHODOLOGIES

Traditional medicine has been part of the WHO's political declarations and strategies born in the lead up to the 1978 Declaration of Alma Ata.[17] Since then, the WHO has been at the

---

[14] Friesen, 'Mesmer', 436.

[15] B. Goldacre, 'The Benefits and Risks of Homeopathy', (2007) *Lancet*, 370(9600), 1672–1673.

[16] E. Cloatre, 'Regulating Alternative Healing in France, and the Problem of "Non-Medicine"', (2018) *Medical Law Review*, 27(2), 189–214.

[17] WHO, 'Declaration of Alma-Ata, International Conference on Primary Health Care, Alma-Ata, USSR, 6–12', (WHO, September 1978).

forefront of developing regulations aimed at carving out spaces for traditional medicines. However, the organisation has moved away from its original understanding of health, which was more holistic and focused on social practices of healing. Regional political mobilisations underpinned by postcolonial critiques of scientific universalism were gradually replaced again by biomedical logics of health from the 1980s onwards.[18] This approach, favouring biomedical standards of practice, can be appreciated to some extent in the 'General Guidelines for the Research of Non-Conventional Medicines', which is prefaced by the need to improve research data and methodologies with a view of furthering the regulation and integration of traditional herbal medicines and procedure-based therapies.[19] The guidelines state that conventional methodologies should not hamper people's access to traditional therapies; and instead, reaffirms the plurality of non-orthodox practices.[20] Noting the great diversity of practices and epistemologies framing traditional medicine, the guidelines re-organised them around two broad classifications – medicines and procedure-based therapies.

Based on these categories, the guidelines suggest that efficacy can be demonstrated through different research methodologies and types of evidence, including historical evidence of traditional use. To ensure safety and efficacy standards are met, herbal medicines ought to be first differentiated through botanical identification based on scientific Latin plant names. Meanwhile, the guidelines leave some room for the use of historical records of traditional evidence of efficacy and safety, which should be demonstrated through a variety of sources including literature reviews, theories and concepts of system of traditional medicine, as well as clinical trials. It also affirms that post-marketing surveillance systems used for conventional medicines are relevant in monitoring, reporting and evaluating adverse effects of traditional medicine.

More importantly, the guidelines contemplate the use of mixed methodologies, whereby EBM can make up for the gaps of evidence of efficacy in traditional medicine. And, where claims are based on different traditions, for example, Traditional Chinese Medicine (TCM) and Western Herbalism, the guidelines require evidence linking them together; and where there is none, scientific evidence should be the basis. If there are any contradictions between them, 'the claim used must reflect the truth, on balance of the evidence available'.[21] Although these research methodologies give the impression of integrating traditional medicine into the mainstream, the guidelines reflect policy transformations since the late 1980s, when plants appeared more clearly as medical objects in the Declaration of Chiang Mai.[22] Drawing on good manufacturing practice guidelines as tools to assess the safety and quality of medicines, WHO's guidelines and declarations between 1990 and 2000 increasingly framed herbal medicines as an object of both pharmacological research and healthcare governance.[23]

WHO's approach resonates with contemporary European Union legislation, namely the Directive 2004/24/EC on the registration of traditional herbal medicines.[24] This Directive also

---

[18] S. Langwick, 'From Non-aligned Medicines to Market-Based Herbals: China's Relationship to the Shifting Politics of Traditional Medicine in Tanzania', (2010) *Medical Anthropology*, 29(1), 15–43.

[19] WHO, 'General Guidelines for Methodologies on Research and Evaluation of Traditional Medicine', (WHO, 2000).

[20] Ibid.

[21] Ibid., 42.

[22] O. Akerele et al. (eds) *Conservation of Medicinal Plants* (Cambridge University Press, 1991).

[23] M. Saxer, *Manufacturing Tibetan Medicine: The Creating of an Industry and the Moral Economy of Tibetanness* (New York: Berghan Books, 2013).

[24] Directive 2004/24/EC of the European Parliament and of the Council of 31 March 2004 amending, as regards traditional herbal medicinal products, Directive 2001/83/EC on the Community code relating to medicinal products for human use, OJ 2004 No. L136, 30 April 2004.

appears to be more open to qualitative evidence based on historical sources, but ultimately subordinates evidence to the biomedical mantra of safety and quality that characterises the regulation of conventional medicines. Traditional herbal medicine applications should demonstrate thirty years of traditional use of the herbal substances or combination thereof, of which fifteen years should be in the European Union (EU). In comparison with conventional medicines requiring multiphase clinical trials in humans, the Directive simplifies the authorisation procedure by admitting bibliographic evidence of efficacy. However, applications must be supplemented with non-clinical studies – namely, toxicology studies – especially if the herbal substance or preparation is not listed in the Community Pharmacopeia.[25] In the end, these regulations subordinate traditional knowledges to the research concepts and methodologies of conventional medicine. Research centres of non-conventional medicines in the EU also align mission statements to integration-based approaches, whereby inclusion of traditional and non-conventional medicine is premised on their modernisation through science.[26] However, as we argue in the next section, science is not the sole arbiter of what comes to be excluded or not in the pursuit of evidence. Indeed, drawing on the UK as a case study, we argue that economic rationalities are part of the regulatory environment shaping what is or is not included as evidence in healthcare research.

30.5 BEYOND EVIDENCE: THE ECONOMIC REASONING OF CLINICAL GUIDELINES

Despite there being no specific restrictions preventing the use of non-conventional treatments within the National Health Service (NHS), authorities involved in the procurement of health or social care work have been under increasing pressure to define the hierarchy of scientific evidence in public affairs. For example, under pressure of being judicially reviewed, the Charities Commission opened up a consultation that produced new guidance for legal caseworkers assessing applications from charities promoting the use of complementary and alternative medicine. Charities have to define their purpose and how this benefits publics. For example, if the declared purpose is to cure cancer through yoga, it will have to demonstrate evidence of public benefit, based on accepted sources of evidence and EBM's 'recognised scales of evidence'. Although observations, personal testimonies or expert opinion are not excluded per se, they cannot substitute scientific medical explanation.[27] For the Commission, claims that fail the scientifically-based standard are meant to be regarded as cultural or religious beliefs.

There have also been more conspicuous ways in which evidence, as understood through a 'scientific-bureaucratic-medicine' model, has been used to limit the space for non-conventional medicines.[28] Clinical guidelines are a key feature of this regulatory model – increasingly institutionalised in the UK since the 1980s. The main body charged with this task is the National Institute for Health and Care Excellence (NICE), a non-departmental public body

[25] T. P. Fan et al., 'Future Development of Global Regulations of Chinese Herbal Products', (2012) *Journal of Ethnopharmacology*, 140(3), 568–586.

[26] V. Fønnebø et al., 'Legal Status and Regulation of CAM in Europe Part II – Herbal and Homeopathic Medicinal Products', (CAMbrella, 2012).

[27] Charity Commission for England and Wales, 'Operational Guidance (OG) 304 Complementary and Alternative Medicine', (Charity Commission for England and Wales, 2018).

[28] S. Harrison and K. Checkland, 'Evidence-Based Practice in UK Health Policy' in J. Gabe and M. Calnan (eds), *The New Sociology of Health Service* (Abingdon: Routledge, 2009).

with statutory footing through the Health and Social Care Act 2012. The purpose of NICE clinical guidelines is to reduce variability in both quality and availability in the delivery of treatments and care and to confirm an intervention's effectiveness. Although not compulsory, compliance with the guidelines is the norm and exceptions are 'both rare and carefully documented'[29] because institutional performance is tied to their implementation and non-adherence may have a financial impact.[30] Following a campaign by 'The Good Thinking Society', an anti-pseudoscience charity, NHS bodies across London, Wales and the North of England have stopped funding homeopathic services.[31] Meanwhile, an NHS England consultation also led to the ban of the prescription of products considered to be of 'low clinical value', such as homeopathic and herbal products. Responding to critics, the Department of Health defended its decision to defund non-conventional medicine products stating they were neither clinically nor cost effective.[32] However, it is also worth noting that outside of the remit of publicly funded institutions, traditional and non-conventional medicines have been tolerated, or even encouraged, as a solution to relieve the pressure from austerity healthcare policies. For example, the Professional Standards Authority (PSA) has noted that accredited registered health and social care practitioners – which include acupuncturists, sports therapists, aromatherapy practitioners, etc. – could help relieve critical demand for NHS services.[33] This raises questions about what counts as evidence and how different regulators respond to specific practices that are not based on biomedical epistemologies, particularly what sort of research is acceptable in healthcare policy-making. What we have sought to demonstrate in this section is the extent to which, under the current regulatory landscape, the production of knowledge has become increasingly enmeshed with various layers of laws and regulations drafted by state and non-state actors.[34] Although the discourse has focused on problems with the kind of evidence and research methodologies used by advocates of non-conventional medicine, a bureaucratic application of EBM in the UK has limited access to traditional and non-conventional medicines in the public healthcare sector. In addition to policing the boundaries between 'fake' and 'real' medicines, clinical guidelines also delimit which therapies should be funded or not by the state. Thus, this chapter has sketched the links between evidence-based medicine and law, and the processes that influence what kind of research and what kind of evidence are appropriate for the purpose of delivering healthcare. Regulation, whether through laws implementing the EU Directives on the registration of traditional herbal medicines, or clinical guidelines produced by NICE, can be seen as operating as normative forces shaping healthcare knowledge production. The final section analyses the social and cultural dimensions of knowledge production and it argues that contemporary regulatory approaches discussed in the preceding sections assume non-conventional knowledges follow a linear development. Premised upon notions of scientific

---

[29] Ibid., p. 126.

[30] R. McDonald and S. Harrison, 'The Micropolitics of Clinical Guidelines: An Empirical Study', (2004) *Policy and Politics*, 32(2), 223–239.

[31] The Good Thinking Society, 'NHS Homeopathy Spending', (The Good Thinking Society, 2018), www .goodthinkingsociety.org/projects/nhs-homeopathy-legal-challenge/nhs-homeopathy-spending/.

[32] UK Government and Parliament, 'Stop NHS England from Removing Herbal and Homeopathic Medicines', (UK Government and Parliament, 2017), www.petition.parliament.uk/petitions/200154.

[33] Professional Standards Authority, 'Untapped Resources: Accredited Registers in the Wider Workforce', (Professional Standards Authority, 2017).

[34] M. Jacob, 'The Relationship between the Advancement of CAM Knowledge and the Regulation of Biomedical Research' in J. McHale and N. Gale (eds), *The Routledge Handbook on Complementary and Alternative Medicine: Perspectives from Social Science and Law* (Abingdon: Routledge, 2015), p. 359.

progress and modernity, this view ultimately fails to grasp the complexity of knowledge-production and the hybrid nature of healing practices.

30.6 REGULATING FOR UNCERTAINTY: MESSY KNOWLEDGES AND PRACTICES

Hope for a cure, dissatisfaction with medical authority, highly bureaucratised healthcare systems or limited access to primary healthcare, are among some of the many reasons that drive people to try untested as well as the unregulated pills and practice-based therapies from traditional and non-conventional medicines. While EBM encourages a regulatory environment averse to the miracle medicines or testimonies of overnight cures and home-made remedies, Lucas Richert argues 'unknown unknowns fail to dissuade the sick, dying or curious from experimenting with drugs'.[35] The problem, however, is the assumption that medicines, and also law, progress in a linear trajectory. In other words, that unregulated drugs became regulated through standardised testing and licensing regulations that carefully assess medicines quality, safety and efficacy before and after they are approved into the market.[36]

Instead, medicines' legal status may not always follow this linear evolution. We have argued so far that the regulatory environment of biomedicine demarcates boundaries between legitimate knowledge-makers/objects and illegitimate ones, such as street/home laboratories and self-experimenting patients.[37] But 'evidence' also acts as a signpost for a myriad of battles to secure some kind of authority over what is legitimate or not between different stakeholders (patient groups, doctors, regulators, industry, etc.).[38] Thus, by looking beyond laboratories and clinical settings, and expanding the scope of research to the social history of drugs, STS scholarship suggests that the legal regulation of research and medicines is based on more fragmented and dislocated encounters between different social spaces where experimentation happens.[39] For example, Mei Zhan argues that knowledge is 'always already impure, tenuously modern, and permanently entangled in the networks of people, institutions, histories, and discourses within which they are produced'.[40] This means neither 'Western' biomedical science or 'traditional' medicines have ever been static and hermeneutically sealed spaces. Instead, therapeutic interventions and encounters are often 'uneven' and messy, linking dissimilar traditions and bringing together local and global healing practices, to the point that they constantly disturb assumptions about 'the Great Divides' in medicine. For example, acupuncture's commodification and marketisation in Western countries reflects how Traditional Chinese Medicine has been transformed through circulation across time and space, enlisting various types of actors from different professional healthcare backgrounds – such as legitimate physicians, physiotherapists, nurses, etc. – as well as lay people who have not received formal training in a biomedical profession. New actors with different backgrounds take part in the negotiations for medical legitimacy and authority that are central to the

---

[35] L. Richert, *Strange Trips: Science, Culture, and the Regulation of Drugs* (Montreal: McGill University Press, 2018), p. 174.

[36] J. Barnes, 'Pharmacovigilance of Herbal Medicines: A UK Perspective', (2003) *Drug Safety*, 26(12), 829–851.

[37] E. Cloatre, 'Law and Biomedicine and the Making of "Genuine" Traditional Medicines in Global Health', (2019) *Critical Public Health*, 29(4), 424–434.

[38] Richert, *Strange Trips*, pp. 56–76.

[39] J. Kim, 'Alternative Medicine's Encounter with Laboratory Science: The Scientific Construction of Korean Medicine in a Global Age', (2007) *Social Studies of Science*, 37(6), 855–880.

[40] Zhan, *Other Wordly*, p. 72.

reinvention of traditional and non-conventional medicine. These are processes of 'transloca-tion' – understood as the circulation of knowledges across different circuits of exchange value – which reconfigure healing communities worldwide.[41]

So, in the process of making guidelines, decisions and norms about research on traditional and non-conventional medicines, the notion of 'evidence' could also signify a somewhat impermanent conclusion to a struggle between different actors. As a social and political space, the integration of traditional medicine and non-conventional medicine is not merely a procedural matter dictated by the logic of medical sciences. Instead, what is accepted or not as legitimate is constantly 'remodelled' by political, economic and social circumstances.[42] In that sense, Stacey Langwick argues that evidence stands at the centre of ontological struggles rather than simply being contestations of authority insofar it is a 'highly politicized and deeply intimate battle over who and what has the right to exist'.[43] For her, determination of what counts as evidence is at the heart of struggles of postcoloniality. When regulations based on EBM discard indigenous epistemologies of healing or the hybrid practices of individuals and communities who pick up knowledge in fragmented fashion, they also categorise their experi-ences, histories and effects as non-events. This denial compounds the political and economic vulnerability of traditional and non-conventional healers insofar as *their survival* depends on their ability to adapt their practice to conventional medicine, by mimicking biomedical practices and norms.[44] Hence, as Marie Andree Jacobs argues, the challenge for traditional and non-conventional medicines lies in translating 'the alternativeness of its knowledge into genuinely alternative research practices' and contributes to reimagining alternative models of regulation.[45]

30.7 CONCLUSION

This chapter analysed how regulators respond to questions of evidence of traditional and non-conventional medicines. It argued that these strategies tend to subordinate data that is not based on EBM's hierarchies of evidence, allowing regulators to demarcate the boundaries of legitimate research as well as situating the 'oddities' of non-conventional medicines outside of science (e.g. as 'cultural' or 'religious' issues in the UK's case). In order to gain legitimacy and authority, as exemplified through the analysis of specific guidelines and regulations of research of traditional and non-conventional medicines, the regulatory environment favours the translation and trans-formation of traditional and non-conventional medicines into scientised and commercial versions of themselves. Drawing on STS scholarship, we suggested understanding these debates as political and social struggles reflecting changes about how people heal themselves and others in social communities that are in constant flux. More importantly, they reflect struggles of healing communities seeking to establish their own viability and right to exist within the dominant scientific-bureaucratic model of biomedicine. This chapter teased out limits of research regulation on non-conventional medicines, insofar practices and knowledges are already immersed in constantly shifting processes, transformed by the very efforts to pin them

---

[41] Ibid., p. 18
[42] Richert, *Strange Trips*, p. 172.
[43] S. A. Langwick, *Bodies, Politics and African Healing: The Matter of Maladies in Tanzania* (Indiana University Press, 2011), p. 233.
[44] Ibid., p. 223.
[45] Jacob, 'CAM Knowledge', p. 358.

down into coherent and artificially closed-off systems. By pointing out the messy configurations of social healing spaces, we hope to open up a space of discussion with the chapters in this section. Indeed, how can we widen the lens of research regulation, and accommodate non-conventional medicines, without compromising the safety and quality of healthcare interventions? At the very minimum, research on regulation could engage with the social and political context of medicine-taking, and further the understanding of how and why patients seek one therapy over another.

31

Experiences of Ethics, Governance and Scientific Practice in Neuroscience Research

Martyn Pickersgill

31.1 INTRODUCTION[1]

Over the last decade or so, sociologists and other social scientists concerned with the development and application of biomedical research have come to explore the lived realities of regulation and governance in science. In particular, the instantiation of ethics as a form of governance within scientific practice – via, for instance, research ethics committees (RECs) – has been extensively interrogated.[2] Social scientists have demonstrated the reciprocally constitutive nature of science and ethics, which renders problematic any assumption that ethics simply follows (or stifles) science in any straightforward way.[3]

This chapter draws on and contributes to such discussion through analysing the relationship between neuroscience (as one case study of scientific work) and research ethics. I draw on data from six focus groups with scientists in the UK (most of whom worked with human subjects) to reflect on how ethical questions and the requirements of RECs as a form of regulation are experienced within (neuro)science. The focus groups were conducted in light of a conceptual concern with how 'issues and identities interweave'; i.e. how personal and professional identities relate to how particular matters of concern are comprehended and engaged with, and how those engagements themselves participate in the building of identities.[4] The specific analysis

---

[1] This chapter revisits and reworks a paper previous published as: M. Pickersgill, 'The Co-production of Science, Ethics and Emotion', (2012) *Science, Technology & Human Values*, 37(6), 579–603. Data are reproduced by kind permission of the journal and content used by permission of the publisher, SAGE Publications, Inc.

[2] M. M. Easter et al., 'The Many Meanings of Care in Clinical Research', (2006) *Sociology of Health & Illness*, 28(6), 695–712; U. Felt et al., 'Unruly Ethics: On the Difficulties of a Bottom-up Approach to Ethics in the Field of Genomics', (2009) *Public Understanding of Science*, 18(3), 354–371; A. Hedgecoe, 'Context, Ethics and Pharmacogenetics', (2006) *Studies in History and Philosophy of Biological and Biomedical Sciences*, 37(3), 566–582; A. Hedgecoe and P. Martin, 'The Drugs Don't Work: Expectations and the Shaping of Pharmacogenetics', (2003) *Social Studies of Science*, 33(3), 327–364; B. Salter 'Bioethics, Politics and the Moral Economy of Human Embryonic Stem Cell Science: The Case of the European Union's Sixth Framework Programme', (2007) *New Genetics & Society*, 26(3), 269–288; S. Sperling, 'Managing Potential Selves: Stem Cells, Immigrants, and German Identity', (2004) *Science & Public Policy*, 31(2), 139–149; M. N. Svendsen and L. Koch, 'Between Neutrality and Engagement: A Case Study of Recruitment to Pharmacogenomic Research in Denmark', (2008) *BioSocieties*, 3(4), 399–418; S. P. Wainwright et al., 'Ethical Boundary-Work in the Embryonic Stem Cell Laboratory', (2006) *Sociology of Health & Illness*, 28(6), 732–748.

[3] M. Pickersgill, 'From "Implications" to "Dimensions": Science, Medicine and Ethics in Society', (2013) *Health Care Analysis*, 21(1), 31–42.

[4] C. Waterton and B. Wynne, 'Can Focus Groups Access Community Views?' in R. S. Barbour and J. Kitzinger (eds), *Developing Focus Group Research: Politics, Theory and Practice* (London: Sage, 1999), pp. 127–143, 142. The methodology of these focus groups is more fully described in the following: M. Pickersgill et al., 'Constituting Neurologic Subjects: Neuroscience, Subjectivity and the Mundane Significance of the Brain', (2011) *Subjectivity*, 4(3), 346–365;

presented is informed by the work of science and technology studies (STS) scholar Sheila Jasanoff and other social scientists who have highlighted the intertwinement of knowledge with social order and practices.[5] In what follows, I explore issues that the neuroscientists I spoke with deem to be raised by their work, and characterise how both informal ideas about ethics and formal ethical governance (e.g. RECs) are experienced and linked to their research. In doing so, I demonstrate some of the lived realities of scientists who must necessarily grapple with the heterogenous forms of health-related research regulation the editors of this volume highlight in their Introduction, while seeking to conduct research with epistemic and social value.[6]

31.2 NEGOTIATING THE ETHICAL DIMENSIONS OF NEUROSCIENCE

It is well known that scientists are not lovers of the bureaucracies of research management, which are commonly taken to include the completion of ethical review forms. This was a topic of discussion in the focus groups: one scientist, for instance, spoke of the 'dread' (M3, Group 5) felt at the prospect of applying for ethical approvals. Such an idiom will no doubt be familiar to many lawyers, ethicists and regulatory studies scholars who have engaged with life scientists about the normative dimensions of their work.

Research governance – specifically, ethical approvals – could, in fact, be seen as having the potential of hampering science, without necessarily making it more ethical. In one focus group (Group 1), three postdoctoral neuroscientists discussed the different terms ethics committees had asked them to use in recruitment materials. One scientist (F3) expressed irritation that another (F2) was required to alter a recruitment poster, in order that it clearly stated that participants would receive an 'inconvenience allowance' rather than be 'paid'. The scientists did not think that this would facilitate recruitment into a study, nor enable it to be undertaken any more ethically. F3 described how 'it's just so hard to get subjects. Also if you need to get subjects from the general public, you know, you *need* these tricks'. It was considered that changing recruitment posters would not make the research more ethical – but it might prevent it happening in the first place.

All that being said, scientists also feel motivated to ensure their research is conducted 'ethically'. As the power of neuroimaging techniques increases, it is often said that it becomes all the more crucial for neuroscientists to engage with ethical questions.[7] The scientists in my focus groups shared this sentiment, commonly expressed by senior scientists and ethicists. As one participant reflected, 'the ethics and management of brain imaging is really becoming a very key feature of [...] everyday imaging' (F2, Group 4). Another scientist (F1, Group 2) summarised the perspectives expressed by all those who participated in the focus groups:

I think the scope of what we can do is broadening all the time and every time you find out something new, you have to consider the implications on your [research] population.

M. Pickersgill et al., 'The Changing Brain: Neuroscience and the Enduring Import of Everyday Experience', (2015), *Public Understanding of Science*, 24(7), 878–892; Pickersgill, 'The Co-production of Science'.

[5] S. Jasanoff, S. (ed.) *States of Knowledge: The Co-Production of Science and Social Order*, Oxford (Routledge, 2004), pp. 1–12; P. Brodwin, 'The Coproduction of Moral Discourse in US Community Psychiatry', (2008) *Medical Anthropology Quarterly*, 22(2), 127–147.

[6] See Introduction of this volume; A. Ganguli-Mitra, et al., 'Reconfiguring Social Value in Health Research through the Lens of Liminality', (2017) *Bioethics*, 31(2), 87–96.

[7] M. J. Farah, 'Emerging Ethical Issues in Neuroscience', (2002) *Nature Neuroscience*, 5(11), 1123–1129; T. Fuchs, 'Ethical Issues in Neuroscience', (2006) *Current Opinion in Psychiatry*, 19(6), 600–607; J. Illes and É. Racine, 'Imaging or Imagining? A Neuroethics Challenge Informed by Genetics', (2005) *American Journal of Bioethics*, 5(2), 5–18.

What scientists consider to be sited within the territory of the 'ethical' is wide-ranging, underscoring the scope of neuroscientific research, and the diverse institutional and personal norms through which it is shaped and governed. One researcher (F1, Group 2) reflected that ethical research was not merely that which had been formally warranted as such:

> I think when I say you know 'ethical research', I don't mean research passed by an ethics committee I mean ethical to what I would consider ethical and I couldn't bring myself to do anything that I didn't consider ethical in my job even if it's been passed by an ethics committee. I guess researchers should hold themselves to that standard.

Conflicts about what was formally deemed ethical and what scientists felt was ethical were not altogether rare. In particular, instances of unease and ambivalence around international collaboration were reflected upon in some of the focus group discussions. Specifically, these were in relation to collaboration with nations that the scientists perceived as having relatively lax ethical governance as compared to the UK. This could leave scientists with a 'slight uneasy feeling in your stomach' (F2, Group 4). Despite my participants constructing some countries as being more or less 'ethical', no focus group participant described any collaborations having collapsed as a consequence of diverging perspectives on ethical research. However, the *possibility* that differences between nations exist, and that these difference could create problems in collaboration, was important to the scientists I spoke with. There was unease attached to collaborating with a 'country that doesn't have the same ethics' (F2, Group 4). To an extent, then, an assumption of a shared normative agenda seemed to have significance as an underpinning for cross-national team science.

The need to ensure confidentiality while also sharing data with colleagues and collaborators was another source of friction. This was deemed to be a particularly acute issue for neuroscience, since neuroimaging techniques were seen as being able to generate and collect particularly sensitive information about a person (given both the biological salience of the brain and the role of knowledge about it in crafting identities).[8] The need to separate data from anything that could contribute to identifying the human subject it was obtained from impacted scientists' relationships with their research. In one focus group (Group 3), M3 pointed out that no longer were scientists owners of data, but rather, they were responsible chaperones for it.

Fears were expressed in the focus groups that neuroscientific data might inadvertently impact upon research participants, for instance, affecting their hopes for later life, legal credibility and insurance premiums. Echoing concerns raised in both ethics and social scientific literatures, my participants described a wariness about any attempt to predict 'pathological' behaviours, since this could result in the 'labelling' (F1, Group 4) or 'compartmentalising' (F2, Group 4) of people.[9] As such, these scientists avoided involving themselves in research that necessarily entailed children, prisoners, or 'vulnerable people' (F2, group 4). Intra-institutional tensions could emerge when colleagues were carrying out studies that the scientists I spoke with did not regard as ethically acceptable.

Some focus group participants highlighted the hyping of neuroscience, and argued that it was important to resist this.[10] These scientists nevertheless granted the possibility that some of the

[8] E. Postan, 'Defining Ourselves: Personal Bioinformation as a Tool of Narrative Self-conception', *Journal of Bioethical Inquiry*, 13(1), 133–151. See also Postan, Chapter 23 in this volume.
[9] Farah, 'Emerging Ethical Issues'; Illes and Racine, 'Imaging or Imagining?'; M. Gazzaniga, *The Ethical Brain* (Chicago: Dana Press, 2005).
[10] Hedgecoe and Martin, 'The Drugs Don't Work', 8.

wilder promises made about neuroscience (e.g. 'mind reading') *could* one day be realised – generating ethical problems in the process:

> there's definitely a lot of ethical implications on that in terms of what the average person thinks that these methods can do and can't do, and what they actually can do. And if the methods should get to the point where they *could* do things like that, to what extent is it going to get used in what way. (F1, group 1)

Scientists expressed anxiety about 'develop[ing] your imaging techniques' but then being unable to 'control' the application of these (F2, Group 4). Yet, not one of my participants stated that limits should be placed on 'dangerous' research. Developments in neuroscience were seen neither as intrinsically good nor as essentially bad, with nuclear power sometimes invoked as a similar example of how, to their mind, normativity adheres to deployments of scientific knowledge rather than its generation. More plainly: the rightness or wrongness of new research findings were believed to 'come down to the people who use it' (F1, Group 1), not to the findings per se. Procedures almost universally mandated by RECs were invoked as a way of giving licence to research: 'a good experiment is a good experiment as long as you've got full informed consent, actually!' (F1, Group 3). Another said:

> I think you can research any question you want. The question is how you design your research, how ethical is the design in order to answer the question you're looking at. (F2, Group 2)

Despite refraining from some areas of work themselves, due to the associated social and ethical implications my participants either found it difficult to think of anything that should not be researched at all, or asserted that science should not treat anything as 'off-limits'. One scientist laughed in mock horror when asked if there were any branches of research that should not be progressed: 'Absolutely not!' (F1 Group 3). This participant described how 'you just can't stop research', and prohibitions in the UK would simply mean scientists in another country would conduct those studies instead. In this specific respect, ethical issues seemed to be somewhat secondary to the socially produced sense of competition that appears to drive forward much biomedical research.

31.3 INCIDENTAL FINDINGS WITHIN NEUROIMAGING RESEARCH

The challenge of what to do with incidental findings is a significant one for neuroscientists, and a matter that has exercised ethicists and lawyers (see Postan, Chapter 23 in this volume).[11] They pose a particular problem for scientists undertaking brain imaging. Incidental findings have been defined as 'observations of potential clinical significance unexpectedly discovered in healthy subjects or in patients recruited to brain imaging research studies and unrelated to the purpose or variables of the study'.[12] The possibilities and management of incidental findings were key issues in the focus group discussions I convened, with a participant in one group terming them 'a whole can of worms'

---

[11] T. C. Booth et al., 'Incidental Findings in "Healthy" Volunteers during Imaging Performed for Research: Current Legal and Ethical Implications', (2010) *British Journal of Radiology*, 83(990), 456–465; N. A. Scott et al., 'Incidental Findings in Neuroimaging Research: A Framework for Anticipating the Next Frontier', (2012) *Journal of Empirical Research on Human Research Ethics*, 7(1), 53–57; S. A. Tovino, 'Incidental Findings; A Common Law Approach', (2008) *Accountability in Research*, 15(4), 242–261.

[12] J. Illes et al., 'Incidental Findings in Brain Imaging Research', *Science*, 311(5762), 783–784, 783.

(F1, Group 3). Another scientist reflected on the issue, and their talk underscores the affective dimensions of ethically challenging situations:

> I remember the first time [I discovered an incidental finding] 'cos we were in the scanner room we were scanning the child and we see it online basically, that there might be something. It's a horrible feeling because you then, you obviously at this point you know the child from a few hours, since a few hours already, you've been working with the child and it's … you have a personal investment, emotional investment in that already but the important thing is then once the child comes out of the scanner, you can't say anything, you can't let them feel anything, you know realise anything, so you have to be just really back to normal and pretend there's nothing wrong. Same with the parents, you can't give any kind of indication to them at all until you've got feedback from an expert, which obviously takes so many days, so on the day you can't let anything go and no, yeah it was, not a nice experience. (F2, Group 2)

Part of the difficulties inherent in this ethically (and emotionally) fraught area lies in the relationality between scientist and research subject. Brief yet close relationships between scientists and those they research are necessary to ensure the smooth running of studies.[13] This intimacy, though, makes the management of incidental findings even more challenging. Further, the impacts of ethically significant issues on teamwork and collaboration are complex; for instance, what happens if incidental findings are located in the scans of co-workers, rather than previously unknown research subjects? One respondent described how these would be 'even more difficult to deal with' (F1, Group 1). Others reflected that they would refrain from 'helping out' by participating in a colleague's scan when, for instance, refining a protocol. This was due to the potential of neuroimaging to inadvertently reveal bodily or psychological information that they would not want their colleagues to know.

The challenge of incidental findings is one that involves a correspondence between a particular technical apparatus (i.e. imaging methods that could detect tumours) and an assemblage of normative imperatives (which perhaps most notably includes a duty of care towards research participants). This correspondence is reciprocally impactful: as is well known, technoscientific advances shift the terrain of ethical concern – but so too does the normative shape the scientific. In the case of incidental findings, for example, scientists increasingly felt obliged to cost in an (expensive) radiologist into their grants, to inspect each participant's scan; a scientist might 'feel uncomfortable showing anybody their research scan without having had a radiologist look at it to reassure you it was normal' (F1, Group 3). Hence, 'to be truly ethical puts the cost up' (F2, Group 4). Not every scientist is able to command such sums from funders, who might also demand more epistemic bang for the buck when faced with increasingly costly research proposals. What we can know is intimately linked to what we can, and are willing to, spend. And if being 'truly ethical' indeed 'puts the cost up', then what science is sponsored, and who undertakes this, will be affected.

31.4 NORMATIVE UNCERTAINTIES IN NEUROSCIENCE

Scientific research using human and animal subjects in the UK is widely felt to be an amply regulated domain of work. We might, then, predict that issues like incidental findings can be

---

[13] S. Cohn, 'Making Objective Facts from Intimate Relations: The Case of Neuroscience and Its Entanglements with Volunteers', (2008) *History of the Human Sciences*, 21(4), 86–103; S. Shostak and M. Waggoner, 'Narration and Neuroscience: Encountering the Social on the "Last Frontier of Medicine"', in M. D. Pickersgill and I. van Keulen, (eds), *Sociological Reflections on the Neurosciences* (Bingley: Emerald, 2011), pp. 51–74.

rendered less challenging to deal with through recourse to governance frameworks. Those neuroscientists who exclusively researched animals indeed regarded the parameters and procedures defining what was acceptable and legal in their work to be reasonable and clear. In fact, strict regulation was described as enjoining self-reflection about whether the science they were undertaking was 'worth doing' (F1, Group 6). This was not, however, the case for my participants working with humans. Rather, they regarded regulation in general as complicated, as well as vague: in the words of two respondents, 'too broad' and 'open to interpretation' (F1, Group 2), and 'a bit woolly' and 'ambiguous' (F2, group 2). Take, for instance, the Data Protection Act: in one focus group (Group 3) a participant (F1) noted that a given university would 'take their own view' about what was required by the Act, with different departments and laboratories in turn developing further – potentially diverging – interpretations.

Within the (neuro)sciences, procedural ambiguity can exist in relation to what scientists, practically, should do – and how ethically valorous it is to do so. Normative uncertainty can be complicated further by regulatory multiplicity. The participants of one focus group, for example, told me about three distinct yet ostensibly nested ethical jurisdictions they inhabited: their home department of psychology, their university medical school and their local National Health Service Research Ethics Committee (NHS REC). The scientists I spoke with understood these to have different purviews, with different procedural requirements for research, and different perspectives on the proper way enactment of ethical practices, such as obtaining informed consent in human subjects research.

Given such normative uncertainty, scientists often developed what we might term 'ethical workarounds'. By this, I mean that they sought to navigate situations where they were unsure of what, technically, was the 'right' thing to do by establishing their own individual and community norms for the ethical conduct of research, which might only be loosely connected to formal requirements. In sum, they worked around uncertainty by developing their own default practices that gave them some sense of surety. One participant (F1, Group 2) described this in relation to drawing blood from people who took part in her research. To her mind, this should be attempted only twice before being abandoned. She asserted that this was not formally required by any research regulation, but instead was an informal standard to which she and colleagues nevertheless adhered.

In the same focus group discussion, another scientist articulated a version of regulatory underdetermination to describe the limits of governance:

> not every little detail can be written down in the ethics and a lot of it is in terms of if you're a researcher you have to you know make your mind up in terms of the ethical procedures you have to adhere to yourself and what would you want to be done to yourself or not to be done ... (F2, Group 2)

Incidental findings were a key example of normative uncertainty and the ethical workarounds that resulted from this. Although 'not every little detail can be written down', specificity in guidelines can be regarded as a virtue in research that is seen to have considerable ethical significance, and where notable variations in practice were known to exist. The scientist quoted above also discussed how practical and ethical decisions must be made as a result of the detection of clinically relevant incidental findings, but that their precise nature was uncertain: scientists were 'struggling' due to being 'unsure' what the correct course of action should be. Hence, 'proper guidelines' were invoked as potentially useful, but these were seemingly considered to be hard to come by.

The irritations stimulated by a perceived lack of clarity on the ethically and/or legally right way to proceed are similarly apparent in the response of this scientist to a question about her feelings upon discovering, for the first time, a clinically relevant incidental finding in the course of her neuroimaging work:

> It was unnerving! And also because it was the first time I wasn't really sure how to deal with it all, so I had to go back in the, see my supervisor and talk to them about it and, try to find out how exactly we're dealing now with this issue because I wasn't aware of the exact clear guidelines. (F2, Group 2)

Different scientists and different institutions were reported to have 'all got a different way of handling' (F2, Group 4) the challenge of incidental findings. Institutional diversity was foregrounded, such as in the comments of F1 (Group 1). She described how when working at one US university 'there was always a doctor that had to read the scans so it was just required'. She emphasised how there was no decision-making around this on behalf of the scientist or the research participant: it was simply a requirement. On the other hand, at a different university this was not the case – no doctor was on call to assess neuroimages for incidental findings.

An exchange between two researchers (F1 and F2, Group 2) also illustrates the problems of procedural diversity. Based in the same university but in different departments, they discussed how the complexities of managing incidental findings was related, in part, to practices of informed consent. Too lengthy a dialogue to fully reproduce here, two key features stood out. First, differences existed in whether the study team would, in practice, inform a research subject's physician in the event of an individual finding: in F2's case, it was routine for the physician to be contacted, but F1's participants could opt out of this. However, obtaining physician contact details was itself a tricky undertaking:

> we don't have the details of the GP so if we found something we would have to contact them [the participant] and we'd have to ask them for the GP contact and in that case they could say no, we don't want to, so it's up to them to decide really, but we can't actually say anything directly to them what we've found or what we think there might be because we don't know, 'cos the GP then will have to send them to proper scans to determine the exact problem, 'cos our scans are obviously not designed for any kind of medical diagnosis are they? So I suppose they've still got the option to say no. (F2, Group 2)

It is also worth noting at this point the lack of certitude of the scientists I spoke with about where directives around ethical practice came from, and what regulatory force these had. F1 (Group 1) and F2 (Group 2) above, for instance, spoke about how certain processes were 'just required' or how they 'have to' do particular things to be 'ethical'. This underscores the proliferation and heterogeneity of regulation the editors of this volume note in their Introduction, and the challenges of comprehending and negotiating it in practice by busy and already stretched professionals.

31.5 DISCUSSION

The ethical aspects of science often require discursive and institutional work to become recognised as such, and managed thereafter. In other words, for an issue to be regarded as specifically ethical, scientists and universities need to, in some sense, *agree* that it is; matters that ethicists, for instance, might take almost for granted as being intrinsically normative can often escape the attention of scientists themselves. After an issue has been characterised by researchers *as* ethical, addressing it can necessitate bureaucratic innovation, and the reorganisation of work practices (including new roles and changing responsibilities). Scientists are not always satisfied with the extent to which they are able, and enabled, to make these changes. The ethics of neuroscience, and the everyday conversations and practices that come into play to deal with them, can also have epistemic effects: ethical issues can and do shape scientists relationships with the work, research participants, and processes of knowledge-production itself.

The scientists I spoke with listed a range of issues as having ethical significance, to varying degrees. Key among these were incidental findings. The scientists also engaged in what sociologist Steven Wainwright and colleagues call 'ethical boundary work'; i.e. they some-times erected boundaries between scientific matters and normative concerns, but also collapsed these when equating good science with ethical science.[14] This has the effect of enabling scientists to present research they hold in high regard as being normatively valorous, while also bracketing off ethical questions they consider too administratively or philosophical challenging to deal with as being insufficiently salient to science itself to necessitate sustained engagement.

Still, though, ethics is part and parcel of scientific work and of being a scientist. Normative reflection is, to varying degrees, embedded within the practices of researchers, and can surface not only in focus group discussions but also in corridor talk and coffee room chats. This is, in part, a consequence of the considerable health-related research regulation to which scientists are subject. It is also a consequence of the fact that scientists are moral agents: people who live and act in a world with other persons, and who have an everyday sense of right and wrong. This sense is inevitably and essentially context-dependent, and it inflects their scientific practice and will be contoured in turn by this. It is these interpretations of regulation in conjunction with the mundane normativity of daily life that intertwine to constitute scientists' ethical actions within the laboratory and beyond, and in particular that cultivate their ethical workarounds in conditions of uncertainty.

31.6 CONCLUSION

In this chapter I have summarised and discussed data regarding how neuroscientists construct and regard the ethical dimensions of their work, and reflected on how they negotiate health-related research regulation in practice. Where does this leave regulators? For a start, we need more sustained, empirical studies of how scientists comprehend and negotiate the ethical dimensions of their research in actual scientific work, in order to ground the development and enforcement of regulation.[15] What is already apparent, however, is that any regulation that demands actions that require sharp changes in practice, to no clear benefit to research partici-pants, scientists, or wider society, is unlikely to invite adherence. Nor are frameworks that place demands on scientists to act in ways they consider unethical, or which place unrealistic burdens (e.g. liaising with GPs without the knowledge of research participants) on scientists that leave them anxious and afraid that they are, for instance, 'breaking the law' when failing to act in a practically unfeasible way.

It is important to recognise that scientists bring to bear their everyday ethical expertise to their research, and it is vital that this is worked with rather than ridden over. At the same time, it takes

---

[14] Wainwright et al., 'Ethical Boundary-Work'.

[15] M. Pickersgill et al., 'Biomedicine, Self and Society: An Agenda for Collaboration and Engagement', (2019) *Wellcome Open Research*, 4(9).

a particular kind of scientist to call into question the ethical basis of their research or that of close colleagues, not least given an impulse to conflate good science with ethical science. Consequently, developing regulation in close collaboration with scientists also needs the considered input of critical friends to both regulators and to life scientists (including but not limited to social scientific observers of the life sciences). This would help mitigate the possibility of the inadvertent reworking or even subverting of regulation designed to protect human subjects by well-meaning scientists who inevitably want to do good (in every sense of the word) research.

32

Humanitarian Research

Ethical Considerations in Conducting Research during Global Health Emergencies

Agomoni Ganguli-Mitra and Matthew Hunt

32.1 INTRODUCTION

Global health emergencies (GHEs) are situations of heightened and widespread health crisis that usually require the attention and mobilisation of actors and institutions beyond national borders. Conducting research in such contexts is both ethically imperative and requires particular ethical and regulatory scrutiny. While global health emergency research (GHER) serves a crucial function of learning how to improve care and services for individuals and communities affected by war, natural disasters or epidemics, conducting research in such settings is also challenging at various levels. Logistics are difficult, funding is elusive, risks are elevated and likely to fluctuate, social and institutional structures are particularly strained, infrastructure destroyed. GHER is diverse. It includes biomedical research, such as studies on novel vaccines and treatments, or on appropriate humanitarian and medical responses. Research might also include the development of novel public health interventions, or measures to strength public health infrastructure and capacity building. Social science and humanities research might also be warranted, in order to develop future GHE responses that better support affected individuals and populations. Standard methodologies, including those related to ethical procedures, might be particularly difficult to implement in such contexts.

The ethics of GHER relates to a variety of considerations. First are the ethical and justice-based justifications to conduct research at all in conditions of emergency. Second, the ethics of GHER considers whether research is designed and implemented in an ethically robust manner. Finally, ethical issues also relate to questions arising in the course of carrying out research studies. GHER is characterised by a heterogeneity (of risk, nature, contexts, urgency, scope) which itself gives rise to various kinds of ethical implications:[1] why research is done, who conducts research, where and when it is conducted, what kind of research is done and how. It is therefore difficult to fully capture the range of ethical considerations that arise, let alone provide a one-size-fits-all solution to such questions. Using illustrations drawn from research projects conducted during GHEs, we discuss key ethical and governance concerns arising in GHER – beyond those traditionally associated with biomedical research – and explore the future direction of oversight for GHER. After setting out the complex context of GHER, we illustrate the various ethical issues associated with justifying research, as well as considerations related to

---

[1] M. Hunt et al., 'Ethical Implications of Diversity in Disaster Research', (2012) *American Journal of Disaster Medicine*, 7(3), 211–221.

context, social value and engagement with the affected communities. Finally, we explore some of the new orientations and lenses in the governance of GHER through recent guidelines and emerging practices.

32.2 THE CONTEXT OF GLOBAL HEALTH EMERGENCY RESEARCH

GHEs are large-scale crises that affect health and that are of global concern (epidemics, pandemics, as well as health-related crises arising from conflicts, natural disasters or forced displacement). They are characterised by various kinds of urgency, driven by the need to rapidly and appropriately respond to the needs of affected populations. However, effective responses, treatments or preventative measures require solid evidence bases, and the establishment of such knowledge is heavily dependent on findings from research (including biomedical research) carried out in contexts of crises.[2] As the Council for International Organizations of Medical Sciences (CIOMS) guidelines point out: 'disasters can be difficult to prevent and the evidence base for effectively preventing or mitigating their public health impact is limited'.[3] Generating relevant knowledge in emergencies is therefore necessary to enhance the care of individuals and communities, for example through treatments, vaccines or improved palliative care. Research can also consolidate preparedness for public health and humanitarian responses (including triage protocols) and contribute to capacity building (for example, by training healthcare professionals) in order to strengthen health systems in the long run. Ethical consideration and regulation must therefore adapt to both immediate and urgent issues, as well as contribute to developing sustainable and long-term processes and practices.

Adding to this is the fact that responses to GHEs involve a variety of actors: humanitarian responders, health professionals, public health officials, researchers, media, state officials, armed forces, national governments and international organisations. Actors conducting both humanitarian work and research can encounter particular ethical challenges, given the very different motivations behind response and research. Such dual roles might, at times, pull in different directions and therefore warrant added ethical scrutiny and awareness, even where such actors might be best placed to deliver both aims, given their presence and knowledge of the context, and especially if they have existing relationships with affected communities.[4] Medical and humanitarian responses to GHEs are difficult contexts for ethical deliberation – for ethics review and those involved in research governance – where various kinds of motivations and values collide, potentially giving rise to conflicting values and aims, or to incompatible lines of accountability[5] (for example, towards humanitarian versus research organisations or towards national authorities versus international organisations).

Given the high level of contextual and temporal complexity, and the heightened vulnerability to harm of those affected by GHEs, there is a broad consensus within the ethics literature that research carried out in such contexts requires both a higher level of justification and careful ongoing ethical scrutiny. Attention to vulnerability is, of course, not new to research ethics.

---

[2] N. M. Thielman et al., 'Ebola Clinical Trials: Five Lessons Learned and a Way Forward', (2016) *Clinical Trials*, 13(1), 86–86.

[3] Council for International Organizations of Medical Sciences, 'International Ethical Guidelines for Health-related Research Involving Humans', (CIOMS, 2016), Guideline 20.

[4] A. Levine, 'Academics Are from Mars, Humanitarians Are from Venus: Finding Common Ground to Improve Research during Humanitarian Emergencies,' (2016) *Clinical Trials*, 13(1), 79–82.

[5] Nuffield Council on Bioethics, 'Research in Global Health Emergencies: Ethical Issues,' (*Nuffield Council on Bioethics*, 2020).

It has catalysed many developments in the field, such as the establishment of frameworks, principles, and rules aiming to ensure that participants are not at risk of additional harm, and that their interests are not sacrificed to the needs and goals of research. It has also been a struggle in research governance, however, to find appropriate regulatory measures and measures of oversight that are not overly protectionist; ones that do not stereotype and silence individuals and groups but ensure that their interests and well-being are protected. The relationship between research and vulnerability becomes particularly knotty in contexts of emergency. How should we best attend to vulnerability when it is pervasive?[6] On the one hand, all participants are in a heightened context of vulnerability when compared to populations under ordinary circumstances. On the other hand, those individuals who suffer from systematic and structural inequality, disadvantage and marginalisation, will also see their potential vulnerabilities exacerbated in conditions of public health and humanitarian emergencies. The presence of these multiple sources and forms of vulnerability adds to the difficulty in determining whether research and its design are ethically justified.

32.3 JUSTIFYING RESEARCH: WHY, WHERE AND WHEN?

While research is rightly considered an integral part of humanitarian and public health responses to GHEs,[7] and while there may indeed, as the WHO suggests, be an 'ethical obligation to learn as much as possible, as quickly as possible',[8] research must be ethically justified on various fronts. At a minimum, GHER must not impede current humanitarian and public health responses, even as it is deployed with the aim of improving future responses. Nor should it drain existing resource and skills. Additionally, the social value of such research derives from its relevance to the particular context and the crisis at hand.[9] Decisions regarding location, recruitment of participants, as well as study design (including risk–benefit calculations) must ensure that scientific and social value are not compromised[10] in the process. The Working Group on Disaster Research Ethics (WGDRE), formed in response to the 2004 Indian Ocean tsunami, has argued that while ethical research can be conducted in contexts of emergencies, such research must respond to local needs and priorities, in order avoid being opportunistic.[11] Similar considerations were reiterated during the 2014–2016 Ebola outbreak. Concern was expressed that 'some clinical sites could be perversely incentivized to establish research collaborations based on resources promised, political pressure or simply the powers of persuasion of prospective researchers – rather than a careful evaluation of the merits of the science or the potential benefit for patients. Some decision-makers at clinical sites may not have the expertise to evaluate the scientific merits of the research being proposed'.[12] Such observation reflects considerations that have been identified in a range of GHE settings.

The question of social value is not only related to the ultimate or broad aims of research. Specific research questions can only be justified if these cannot be investigated in non-crisis

---

[6] C. Tansey et al., 'Familiar Ethical Issues Amplified', (2017) *BMC Medical Ethics*, 1891, 1–12.
[7] Thielman et al., 'Ebola Clinical Trials'.
[8] WHO, 'Guidance For Managing Ethical Issues In Infectious Disease Outbreaks', (WHO, 2016), 30.
[9] Ibid.
[10] CIOMS, 'International Ethical Guidelines', Commentary to Guideline 20.
[11] A. Sumathipala et al., 'Ethical Issues in Post-disaster Clinical Interventions and Research: A Developing World Perspective. Key Findings from a Drafting and Consensus Generating Meeting of the Working Group on Disaster Research Ethics (WGDRE) 2007', (2010) *Asian Bioethics Review*, 2(2), 124–142.
[12] Thielman et al. 'Ebola Clinical Trials', 85.

conditions,[13] and as specified above, where the answers to such questions is expected to be of benefit to the community in question – or to relevantly similar communities, be it during the current emergency, or in the future. Relatedly, research should be conducted within settings that are most likely to benefit from the generation of such knowledge, perhaps because they are the site of cyclical disasters or endemic outbreaks that frequently disrupt social structures. Given the heightened precarity of GHE contexts, the risk of exposing study participants to additional harm is particularly salient, and such potential risk must therefore be systematically justified. If considerations of social value are key, these need to extend to priority-setting in GHER. Yet, the funding and development of GHER is not immune to how research priority is set globally. Consequently, this divergence (between the kind of research that is currently being funded and developed, and the research that might be required in specific contexts of crisis) will present particular governance challenges at the local, national, and global levels. Stakeholders from contexts of scarce resources have warned that priority-setting in GHE might mirror broader global research agendas, where the health concerns and needs of low- and middle-income countries (LMICs) are systematically given lower priority.[14] The global research agenda is not generally directed by the specific needs arising from crises (especially crisis in resource-poor contexts), and yet the less well-funded and less resilient health systems of LMICs frequently bear the brunt and severity of crises. The ethical challenges associated with conducting research in contexts of crisis therefore are consistently present at all levels, from the broader global research agenda, to the choice of context and participants, from how research is designed and conducted, to how research data and findings are used and shared.

32.4 JUSTIFYING RESEARCH: WHAT AND HOW?

GHER includes a wide range of activities, from minimally invasive collection of data[15] and systems research aimed at strengthening health infrastructure,[16] to more controversial procedures including testing of experimental therapeutics and vaccines.[17] A common issue of GHER, one that has arisen prominently during recent epidemics and infectious disease outbreaks, is the challenge to long-established standards and trials designs, in particular to what is known as the 'gold standard': randomised, double-blind clinical trials as the standard developmental pathway for new drugs and interventions. The ethical intuitions and debates often pull in different direction. As discussed earlier in the chapter, the justification for conducting research in crises must be ethically robust, as must research design and deployment. Equally, in the context of the COVID-19 pandemic, a strong argument has been made for the need to ensure methodologically rigorous research design and not to accept lower scientific standards as a form of 'pandemic research exceptionalism'.[18] At the time of writing, human challenge trials – the intentional infection of research participants with the virus – proposed as a way to accelerate the development of a vaccine for the novel coronavirus, remain ethically and scientifically controversial. While some commentators have suggested that this may be a rapid and necessary route to

[13] Tansey et al., 'Familiar Ethical Issues'.
[14] Sumathipala et al., 'Ethical Issues'.
[15] Nuffield Council on Bioethics, 'Briefing Note: Zika – Ethical Considerations', (Nuffield Council on Bioethics, 2016).
[16] S. Qari et al., 'Preparedness and Emergency Response Research Centers: Early Returns on Investment in Evidence-based Public Health Systems Research', (2014) *Public Health Reports*, 129(4), 1–4.
[17] A. Rid and F. Miller, 'Ethical Rationale for the Ebola "Ring Vaccination" Trial Design', (2016) *American Journal of Public Health*, 106(3), 432–435.
[18] A. J. London and J. Kimmelman, 'Against Pandemic Research Exceptionalism', (2020) *Science*, 368(6490), 476–477.

vaccine development,[19] others have argued that the criteria for ethical justification of human challenge studies, including social value and fair participation selection, are not likely to be met.[20]

Such tensions are particularly heightened in contexts of high infectious rates, morbidity and mortality. During the 2014–2016 Ebola outbreak in West Africa, several unregistered interventions were approved for use as investigational therapeutics. Importantly, while these were approved for emergency use, they were to be deployed under the MEURI scheme: 'monitored emergency use of unregistered and experimental interventions (MEURI)',[21] that is, through a process where results of an intervention's use are shared with the scientific and medical community, and not under the medical label of 'compassionate use'. This approach allows clinical data to be compiled and thus contributes to the process of generating generalisable evidence. The deployment of experimental drugs was once again considered – alongside the deployment of experimental vaccines – early during the 2018 Ebola outbreak in the Democratic Republic of the Congo.[22] This time, regulatory and ethical frameworks were in place to approve access to five investigational therapeutics under the MEURI scheme,[23] two of which have since shown promise during the clinical trials conducted in 2018. The first Ebola vaccine, approved in 2019, was tested through ring vaccine trials first conducted during the 2014–2016 West African outbreak. Methods and study designs need to be aligned with the needs of the humanitarian response, and yet it is not an easy task to translate the values of humanitarian responses onto research design. How experimental interventions should be deployed under the MEURI scheme was heavily debated and contested by local communities, who saw these interventions as their only and last resort against the epidemic.

While success stories in GHER heavily depend on global cooperation, suitable infrastructure, and often collaboration between the public and private sector, such interventions are unlikely to succeed without the collaboration and engagement of local researchers and communities, and without establishing a relationship based on trust. Engaging with communities and establishing relationships of trust and respect are key to successful research endeavours in all contexts, but are particularly crucial where social structures have broken down and where individuals and communities are at heightened risk of harm. Community engagement, especially for endeavours not directly related to response and medical care, is also particularly challenging to implement. These challenges are most significant in sudden-onset GHE such as earthquakes,[24] if prior relationships do not exist between researchers and the communities. During the 2014–2016 Ebola outbreak, the infection and its spread caused 'panic in the communities by the lack of credible information and playing to people's deepest fears'.[25] Similarly, distrust arose during the subsequent outbreak in eastern DRC, a region already affected by conflict, where low trust in institutions and officials resulted in low acceptance of vaccines and a spread of

[19] E. Zamrozik and M. J. Selgelid, 'Covid-19 Human Challenge Studies: Ethical Issues', (2020) *Lancet Infectious Disease*, www.thelancet.com/journals/laninf/article/PIIS1473-3099(20)30438-2/fulltext.

[20] S. Holm, 'Controlled Human Infection with SARS-CoV-2 to Study COVID-19 Vaccine and Treatments: Bioethics in Utopia,' (2020) *Journal of Medical Ethics*, 0, 1–5.

[21] WHO, 'Ethical Issues Related to Study Design for Trials on Therapeutics for Ebola Virus Disease', (WHO, 2014), 2.

[22] E. C. Hayden, 'Experimental Drugs Poised for Use in Ebola Outbreak,' *Nature* (18 May 2018), www.nature.com/articles/d41586-018-05205-x.

[23] WHO, 'Ebola Virus Disease – Democratic Republic of Congo', *WHO* (31 August 2018), www.who.int/csr/don/31-august-2018-ebola-drc/en/.

[24] Tansey et al., 'Familiar Ethical Issues', 24.

[25] A. Saxena and M. Gomes, 'Ethical Challenges to Responding to the Ebola Epidemic: The World Health Organization Experience', (2016) *Clinical Trials*, 13(1), 96–100.

the virus.[26] Similarly, in the aftermath of Hurricane Katrina there was widespread frustration and distrust of the US federal response by those engaged in civil society and community-led responses.[27] However, such contexts have also given rise to new forms solidarity and cooperation. The recent Ebola outbreaks, the aftermath of Katrina, the 2004 Indian Ocean tsunami and the Fukushima disaster have also given rise to unprecedent levels engagement and leadership by members of the affected communities.[28] Given that successful responses to GHEs are heavily dependent on trust as well as on the engagement and ownership of response activities by local communities, there is little doubt that successful endeavours in GHER will also depend on establishing close, trustworthy and respectful collaborations between researchers, responders, local NGOs, civil society and members of the affected population.

32.5 GOVERNANCE AND OVERSIGHT: GUIDELINES AND PRACTICES

The difficulty of conducting GHER is compounded by much complexity at the level of regulation, governance and oversight. Those involved in research in these contexts are working in and around various ethical frameworks including humanitarian ethics, medical ethics, public health ethics and research ethics. Each framework has traditionally been developed with very different actors, values and interests in mind. Navigating these might result in various kinds of conflicts or dissonance, and at the very least make GHER a particularly challenging endeavour. Such concerns are then compounded by regulatory complexity, including existing national laws and guidelines, international regulations and guidance produced by different international bodies (for example, the International Health Regulations 2005 by the WHO and Good Clinical Practice by the National Institute for Health Research in the United Kingdom), all of which are engaged in a context of urgency, shifting timelines and rapidly evolving background conditions. Two recent pieces of guidance are worth highlighting in this context. The first are the revised CIOMS guidelines, published in 2016, which have a newly added entry (Guideline 20) specifically addressing GHER. The CIOMS guidelines recognise that '[d]isasters unfold quickly and study designs need to be chosen so that studies will yield meaningful data in a rapidly evolving situation. Study designs must be feasible in a disaster situation but still appropriate to ensure the study's scientific validity'.[29] While reaffirming the cornerstones of research ethics, Guideline 20 also refers to the need to ensure equitable distribution of risks and benefits; the importance of community engagement; the need for flexibility and expediency in oversight while providing due scrutiny; and the need to ensure the validity of informed consent obtained under conditions of duress. CIOMS also responds to the need for flexible and alternative study designs and suggests that GHER studies should ideally be planned ahead and that generic versions of protocols could be pre-reviewed prior to a disaster occurring.

Although acting at a different governance level to CIOMS, the Nuffield Council on Bioethics has also recently published a report on GHER,[30] engaging with emerging ethical issues and echoing the central questions and values reflected in current discussions and regulatory frame-works. Reflecting on the lessons learned from various GHEs over the last couple of decades, the report encourages the development of an ethical compass for GHER that focuses on respect,

---

[26] P. Vince et al., 'Institutional Trust and Misinformation in the Response to the 2018–2019 Ebola Outbreak in North Kivu, DR Congo: A Population-based Survey', (2019) *Lancet*, 19(5), 529–356.

[27] Nuffield Council on Bioethics, 'Research in Global Health Emergencies', 41.

[28] Ibid., 32–36.

[29] CIOMS, 'International Ethical Guidelines', Guideline 20.

[30] Nuffield Council on Bioethics, 'Research in Global Health Emergencies'.

reducing suffering, and fairness.[31] The report is notable for recommending that GHER endeavours attend not just to whose needs are being met (that is, questions of social value and responsiveness) but also to who has been involved in defining those needs. In other words, the report reminds us that beyond principles and values guiding study design and implementation, ethical GHER requires attention to a wider ethics ecosystem that includes all stakeholders, and that upholding fairness is not only a feature of recruitment or access to the benefits of research, but must also exist in collaborative practices with local researchers, authorities and communities.

All guidelines and regulations need interpretation on the ground,[32] at various levels of governance, as well as by researcher themselves. The last couple of decades have seen a variety of innovative and adaptive practices being developed for GHER, including the establishment of research ethics committees specifically associated with humanitarian organisations. Similarly, many research ethics committees that are tasked with reviewing GHER protocols have adapted their standard procedures in line with the urgency and developing context of GHEs.[33] Such strategies include convening ad-hoc meetings, prioritising these protocols in the queue for review, waiving deadlines, having advisors pre-review protocols and conducting reviews by teleconference.[34] Another approach can be found in the development of pre-approved, or pre-reviewed protocol templates, which allow research ethics committees to conduct an initial review of proposed research ahead of a crisis occurring, or to review generic policies for the transfer of samples and data. Following their experience in reviewing GHER protocols during the 2014–2016 Ebola outbreak, members of the World Health Organization Ethics Review Committee recommended the formation of a joint research ethics committee for future GHEs.[35] A need for greater involvement and interaction between ethics committees and researchers has been indicated by various commentators, pointing to the need for ethical review to be an ongoing and iterative process. One such model for critical and ongoing engagement, entitled 'real-time responsiveness',[36] proposes a more dynamic approach to ethics oversight for GHER, including more engagement between researchers, research ethics committees, and advisors once the research is underway. An iterative review process has been proposed for research in ordinary contexts[37] but is particularly relevant to GHER, given the urgency and rapidly changing context.

It is important to also consider how to promote and sustain the ethical capacities of researchers in humanitarian settings. Such capacities include the following, which have been linked to ethical humanitarian action:[38] foresighting (the ability to anticipate potential for harms), attentiveness (especially for the social and relational dynamics of particular GHE contexts), and responsiveness to the often-shifting features of a crisis, and their implications for the conduct of the research. These capacities point to the role of virtues, in addition to guidelines and

[31] Ibid., xvi–xvii.

[32] Ibid., 29.

[33] M. Hunt et al., 'The Challenge of Timely, Responsive and Rigorous Ethics Review of Disaster Research: Views of Research Ethics Committee Members', (2016) *PLoS ONE*, 11(6), e0157142.

[34] Ibid.

[35] E. Alirol et al., 'Ethics Review of Studies during Public Health Emergencies – The Experience of the WHO Ethics Review Committee During the Ebola Virus Disease Epidemic', (2017) *BMC Medical Ethics*, 18(1), 8.

[36] L. Eckenwiler et al., 'Real-Time Responsiveness for Ethics Oversight During Disaster Research', (2015) *Bioethics*, 29(9), 653–661.

[37] A. Ganguli-Mitra et al., 'Reconfiguring Social Value in Health Research Through the Lens of Liminality', (2017) *Bioethics*, 31(2), 87–96.

[38] N. Pal et al., 'Ethical Considerations for Closing Humanitarian Projects: A Scoping Review', (2019) *Journal of International Humanitarian Action*, 4(1), 1–9.

principles, in the context of GHER. As highlighted by O'Mathuna, humanitarian research calls for virtuous action on the part of researchers in crisis settings 'to ensure that researchers do what they believe is ethically right and resist what is unethical'.[39] Ethics therefore is not merely a feature of approval or bureaucratic procedure. It must be actively engaged with at various levels and also by all involved, including by researchers themselves.

32.6 NEW ORIENTATIONS AND LENSES

As outlined above, GHEs present a distinctive context for the conduct of research. Tailored ethics guidance for GHER has been developed by various bodies, and it has been acknowledged that GHER can be a challenging fit for standard models to ethics oversight and review. As a result, greater flexibility in review procedures has been endorsed, while emphasising the importance of upholding rigorous appraisal of protocol. Particular attention has been given to the proportionality of ethical scrutiny to the ethical concerns (risk of harm, issues of equity, situations of vulnerability) associated with particular studies. Novel approaches, such as the preparation and pre-review of generic protocols, have also been incorporated into more recent guidance documents (e.g. CIOMS) and implemented by research ethics committees associated with humanitarian organisations.[40] These innovations reflect the importance of temporal sensitivity in GHER and in its review. As well as promoting timely review processes for urgent protocols, scrutiny is also needed to identify research that does not need to be conducted in an acute crisis and whose initiation ought to be delayed.

Discussions about GHER, and on disaster risk reduction more broadly, also point to the importance of preparedness and anticipation. Sudden onset events and crises often require quick response and reaction. Nonetheless, there are many opportunities to lay advance groundwork for research and also for research ethics oversight. In this sense, pre-review of protocols, careful preparation of standard procedures, and even research ethics committees undertaking their own planning procedures for reviewing GHER, are all warranted. It also suggests that while methodological innovation and adaptive designs may be required, methodological standards should be respected in crisis research and can be promoted with more planning and preparation.

32.7 CONCLUSION

Research conducted in GHEs present a particularly difficult context in terms of governance. While each kind of emergency presents its own particular challenge, there are recurring patterns, characterised by urgency in terms of injury and death, extreme temporal constraints, and uncertainty in terms of development and outcome. Research endeavours have to be ushered through a plethora of regulation at various levels, not all of which have been developed with GHER in mind. Several sectors are necessarily involved: humanitarian, medical, public health, and political to name just a few. Conducting research in these contexts is necessary, however, in order to contribute to a robust evidence-base for future emergencies. Ethical considerations are crucial in the implementation and interpretation of guidance, and in rigorously evaluating justification for research. Governance must find a balance between the protection of research

---

[39] D. O'Mathúna, 'Research Ethics in the Context of Humanitarian Emergencies', (2015) *Journal of Evidence-Based Medicine*, 8(1), 31–35, 31.

[40] D. Schopper et al., 'Innovations in Research Ethics Governance in Humanitarian Settings', (2015) *BMC Medical Ethics*, 16(1), 7–8.

participants, who find themselves in particular circumstances of precarity, and the need for flexibility, preparedness, and responsiveness as emergencies unfold. Novel ethical orientations suggest the need, at times, to rethink established procedures, such as one-off ethics approval, or gold standard clinical trials, as well as to establish novel ethical procedures and practice, such as specially trained ethics committees, and pre-approval of research protocols. However, the ethics of such research also suggest that time, risk and uncertainty should not work against key ethical considerations relating to social value, fairness in recruitment or against meaningful and ongoing engagement with the community in all phases of response and research. A dynamic approach to the governance of GHER will also require supporting the ability of researchers, ethics committees and those governing research to engage with and act according to the ethical values at stake.

33

A Governance Framework for Advanced Therapies in Argentina

Regenerative Medicine, Advanced Therapies, Foresight, Regulation and Governance

Fabiana Arzuaga

33.1 INTRODUCTION

Research in the field of regenerative medicine, especially that which uses cells and tissues as therapeutic agents, has given rise to new products called 'advanced therapies' or advanced therapeutic medicinal products (ATMPs). These cutting-edge advances in biomedical research have generated new areas for research at both an academic and industrial level and have posed new challenges for existing regulatory regimes applicable to therapeutic products. The leading domestic health regulatory agencies in the world, such as the US Food and Drug Administration (FDA) and the European Medicines Agency (EMA), have regulated therapeutic tissues and cells as biological medicines and are currently making efforts to establish a harmonised regulatory system that facilitates the process of approval and implementation of clinical trials.

In the mid-2000s, the Argentine Republic did not have any regulations governing ATMPs, and governance approaches to them were weak and diverse. Although the process of designing a governance framework posed significant challenges, Argentina started to develop a regulatory framework in 2007. After more than ten years of work, this objective was achieved thanks to local efforts and the support of academic institutions and regulatory agencies from countries with more mature regulatory frameworks. In 2019, however, Argentina was leading in the creation of harmonised regulatory frameworks in Latin America.

In this chapter I will show how the framework was developed from a position of state non-intervention to the implementation of a governance framework that includes hard and soft law. I will identify the main objectives that drove this process, the role of international academic and regulatory collaborations, milestones and critical aspects of the construction of normative standards and the ultimate governance framework, and the lessons learned, in order to be able to transfer them to other jurisdictions.

33.2 THE EVOLUTION OF REGULATION OF BIOTECHNOLOGY IN ARGENTINA: AGRICULTURAL STRENGTH AND HUMAN HEALTH FRAGMENTATION

Since its advent in the middle of the 1990s, modern biotechnology has represented an opportunity for emerging economies to build capacity alongside high-income countries, thereby blurring the developed/developing divide in some areas (i.e. it represents a 'leapfrog' technology similar to mobile phones). For this to occur, and for maximum benefit to be realised, an innovation-friendly environment had to be fostered. Such an environment does not abdicate moral limits or public oversight but is characterised by regulatory clarity and flexibility.[1] The development of biotechnology in the agricultural sector in Argentina is an example of this. Although it had not been a technology-producing country, Argentina faced a series of favourable conditions that allowed the rapid adoption of genetically modified crops.[2] At the same time, significant institutional decisions were made, especially with regard to biosecurity regulations, with the creation of the National Commission for Agricultural Biotechnology (CONABIA) in 1991.[3] These elements, together with the fact that Argentina has 26 million hectares of arable land, made the potential application of these technologies in Argentina – and outside the countries of origin of the technology, especially the USA – possible. This transformed Argentina into an exceptional 'landing platform' for the rapid adoption of these biotechnological developments. The massive incorporation of Roundup Ready (RR) soybean is explained by the reduction of its production costs and by the expansion of arable land. This positioned Argentina as the world's leading exporter of genetically modified soybean and its derivatives.[4]

The development of biotechologies directed at human health was more complex and uncertain, and unfolded in a more contested and dynamic setting, which resulted in it evolving at a much slower pace, with regulation also developing more slowly, involving a greater number of stakeholders. This context, as will be demonstrated below, offered opportunities for developing new processual mechanisms aimed at soliciting and developing the views and concerns of diverse stakeholders.[5]

33.3 FIRST STEPS IN THE CREATION OF A GOVERNANCE FRAMEWORK FOR CELL THERAPIES

The direct antecedent of stem cells for therapeutic purposes is the hematopoietic progenitor cell (HPC), which has been extracted from bone marrow to treat blood diseases for more than fifty years and is considered an 'established practice'.[6] HPC transplantation is regulated by the Transplant Act 1993, and its regulatory authority is the National Institute for Transplantation (INCUCAI), which adopted regulations governing certain technical and procedural aspects of this practice in 1993 and 2007.[7] This explains the rationale by which many countries – including Argentina – started regulating cell therapies under the a transplantation legal framework. However, Argentina's active pursuit of regenerative medicine research aimed at developing stem

[1] E. Da Silva, 'Biotechnology: Developing Countries and Globalization', (1998) *World Journal of Microbiology and Biotechnology*, 14(3), 463–486.
[2] There was a seed industry in the country in which national firms and subsidiaries of multinational companies actively participated as well as public institutions and had a long tradition of germplasm renewal.
[3] In 2014, the Food and Agriculture Organization (FAO) recognised CONABIA as a centre of reference for biosecurity of genetically modified organisms worldwide.
[4] E. Trigo et al., 'Los transgénicos en la agricultura argentina', (2002) *Libros del Zorzal*, I, 165–178.
[5] G. Laurie et al., 'Law, New Technologies, and the Challenges of Regulating for Uncertainty', (2012) *Law, Innovation & Technology*, 4(1), 1–33.
[6] F. Arzuaga, 'Stem Cell Research and Therapies in Argentina: The Legal and Regulatory Approach', (2013) *Stem Cells and Development*, 22(S1), 4–43.
[7] Organs and Anatomic Human Material Transplantation, Act No. 24.193, of 24 March 1993 and amendments. INCUCAI Resolution No. 307/2007 establishes the classification of medical indications for autologous, allogeneic and unrelated transplantation of HPC. It also regulates procedures for tissue banking, including the banking of stem cells from umbilical cord blood (UCB), which is an alternative source of HPC used in transplants in replacement of bone marrow.

cell solutions to health problems required something more, and despite its efforts to promote this research, there were no regulations or studies related to ethics and the law in this field.[8]

In 2007, the Advisory Commission on Regenerative Medicine and Cellular Therapies (Commission) was created under the National Agency of Promotion of Science and Technology (ANPCYT) and the Office of the Secretary of Science and Technology was transformed into the Ministry of Science, Technology and Productive Innovation (MOST) in 2008.[9] The Commission comprised Argentinian experts in policy, regulation, science and ethics, and was set up initially with the objective of advising the ANPCYT in granting funds for research projects in regenerative medicine.[10] However, faced with a legal gap and the increasing offer of unproven stem cells treatments to patients, this new body became the primary conduit for identifying policy needs around stem cell research and its regulation, including how existing regulatory institutions in Argentina such INCUCAI and the National Administration of Drugs, Food and Medical Technology (ANMAT), would be implicated.

The Commission promoted interactions with a wide range of stakeholders from the public and private sectors, the aim being to raise awareness and interest regarding the necessity of forging a governance framework for research and products approval in the field of regenerative medicine. In pursuing this ambitious objective, the Commission wanted to benefit from lessons from other regions or countries.[11] In 2007, it signed a Collaborative Agreement between the Argentine Secretary of Science and Technology and the University of Edinburgh's AHRC SCRIPT Centre (the Arts and Humanities Research Council Research Centre for Studies in Intellectual Property & Technology Law).[12] This collaboration, addressed in greater detail below, extended to 2019 and was a key factor in the construction of the regulatory framework for ATMPs in Argentina.

33.4 FROM TRANSPLANTS TO MEDICINES

In 2007, in an attempt to halt the delivery of untested stem cell–based treatments that were not captured by the current regulatory regime applicable to HPCs, the Ministry of Health issued Resolution MS 610/2007, under the Transplant Act 1993. The 610/2007 Resolution states 'activities related to the use of human cells for subsequent implantation in humans fall within the purview of the Transplant Authority (INCUCAI)'.[13] This Resolution formally recognises INCUCAI's competence to deal with activities involving the implantation of cellular material

---

[8] S. Harmon, 'Emerging Technologies and Developing Countries: Stem Cell Research (and Cloning) Regulation and Argentina', (2008) *Developing World Bioethics*, 8(2), 138–150.

[9] National Agency of Promotion of Science and Technology, which in 2008 became the Ministry of Science, Technology and Productive Innovation (MOST).

[10] Resolution ANPCYT N° 214/06 creates the Advisory Commission in Cellular Therapies and Regenerative Medicine with the objective to advise the National Agency of Promotion of Science and Technology in the evaluation of research projects in regenerative medicine (RM) that request funding for research as well as to study regulatory frameworks on RM in other jurisdictions.

[11] S. Harmon and G. Laurie, *The Regulation of Human Tissue and Regenerative Medicine in Argentina: Making Experience Work. SCRIPT Opinions*, No. 4 (AHRC Research Centre for Studies in Intellectual Property and Technology Law, 2008).

[12] AHRC/SCRIPT was directed by Professor Graeme Laurie.

[13] The direct antecedent of the use of stem cells for therapeutic purposes is the hematopoietic progenitor cells (HPC) transplantation from bone marrow to treat blood diseases. This practice has been performed for more than fifty years and is considered an 'established practice'. HPC transplantation is regulated by the Transplant Act 1993, and its regulatory authority is INCUCAI, which has issued regulations governing certain technical and procedural aspects of this practice. INCUCAI Resolution 307/2007 establishes the classification of medical indications for autologous, allogeneic and unrelated transplantation of HPC. It also covers procedures for tissue banking, including the banking

into humans. However, it is very brief and does not specify which type of cell it applies to, nor any specific procedures (kind of manipulation) to which cells it can be subject, an issue that is, in any event, beyond the scope of the Act.[14] This Resolution is supplemented by Regulatory Decree 512/95, which, in Article 2, states that 'any practice that involves implanting of human cells that does not fall within HPC transplantation is radically new and therefore is considered as experimental practice until it is demonstrated that it is safe and effective'.

To start a new experimental practice, researchers or medical practitioners must seek prior authorisation from INCUCAI by submitting a research protocol signed by the medical professional or team leader who will conduct the investigation, complying with all requirements of the regulations, including the provision of written informed consent signed by the research subjects, who must not be charged any monies to participate in the procedure. In May 2012, INCUCAI issued Resolution 119/2012, a technical standard to establish requirements and procedures for the preparation of cellular products. Substantively, it is in harmony with international standards of good laboratory and manufacturing practices governing this matter. However, very few protocols have been filed with INCUCAI since 2007, and the delivery of unproven stem cell treatments continued to grow, a situation that exposed INCUCAI's difficulties in policing the field and reversing the growth of health scams.[15]

Another attempt to regulate was the imposition of obligations to register some cellular-based products as biological medicaments. The ANMAT issued two regulations under the Medicines Act 1964:[16] Dispositions 7075/2011 and 7729/2011. These define 'biological medicinal products' as 'products derived from living organisms like cells or tissues', a definition that captures stem cell preparations, and they are categorised in both Dispositions as ATMPs. Cellular-based or biological medicaments must be registered with the National Drugs Registry (REM), and approval for marketing, use and application in humans falls within the scope of the Medicines Act and its implementing regulations. Cellular medicine manufacturers must register at the ANMAT as manufacturing establishments, and they must request product registration before marketing or commercialising their products.

Importantly, the ANMAT regulations do not apply in cases where ATMPs are manufactured entirely by an authorised medical centre, to be used exclusively in that centre. In that case, the local health authority maintains the right for approval. Like all regulations issued by the national Ministry of Health under the Medicines Act, the provisions of Dispositions 7075/2011 and 7729/2011 apply only in areas of national jurisdiction, in cases where interprovincial transit is implicated, or where ATMPs are imported or exported. In short, the Medicines Act is not applicable so long as the product does not leave the geographic jurisdiction of the province. And within the provinces, regulatory solutions were inconsistent; for example, in one they were regulated as transplants and in another as medicines.

As alluded to above, while imperfect regulatory attempts were pursued, the offer of unproven treatments with cells continued to grow. As in many countries, it was usual to find publications in the media reporting the –'almost magical – healing power of stem cells, with little or no

---

of stem cells from umbilical cord blood (UCB), which is an alternative source of HPC used in transplants in replacement of bone marrow.

[14] Arzuaga, 'Stem Cells Research in Argentina'.

[15] In eleven years, Incucai has approved four research protocols using outologous cells. Details of protocols can be accessed on: 'Tratamientos existentes', (Ministerio de Ciencia, Tecnología e Innovación Productiva, Presidencia de la Nación), www.celulasmadre.mincyt.gob.ar/tratamientos.php.

[16] Commercialization Regime of Medicinal Products Act, Act 16.463, of 8 August 1964, and Decree 9763/1964 and amendments.

supporting evidence, and such claims have great impact on public opinion and on the decisions of individual patients. Moreover, the professionals offering these 'treatments' took refuge in the independence of medical practice and the autonomy that it offers, but it seems clear that some of the practices reported were directly contrary to established professional ethics, and they threatened the safety of patients receiving the treatments.[17] In addition to the safety issues, given that these were experimental therapies (that have not been proven to be safe and effective), health insurers have stated their refusal to cover them (and one can anticipate the same antipathy to indemnifying patients who chose to accept them and are injured by them). Indeed, patients filed judicial actions demanding payment of such treatments by both health insurance institutions and the national and provincial state (as guarantors of public health).[18]

The regulatory regime – by virtue of its silence, its imperfect application to regenerative medicine and concomitant practices, and its shared authority between national and provincial bodies – permitted unethical practices to continue, and decisions of some courts have mandated the transfer of funds from the state (i.e. the social welfare system) to the medical centres offering these experimental cellular therapies. In short, the regime established a poorly coordinated regulatory patchwork that was proving to be insufficient to uniformly regulate regenerative medicine – and stem cell – research and its subsequent translation into clinical practice and treatments as ATMPs. Moreover, attempts by regulatory authorities to stop these practices, though valiant, also proved ineffective.

33.5 KEY DRIVERS FOR THE CONSTRUCTION OF THE GOVERNANCE FRAMEWORK

The landscape described above endured until 2017, when the Interministerial Commission for Research and Medicaments of Advanced Therapies (Interminsterial Commission) was created. This new body, jointly founded by the Ministry of Science and Technology (MOST) and the Ministry of Health (MOH), which also oversaw INCUCAI and ANMAT, was set up to:

1. Advise the MOST and MOH in the subjects of their competence.
2. Review current regulations on research, development and approval of products in order to propose and raise for the approval of the competent authority, a comprehensive and updated regulatory framework for advanced therapies.
3. Promote dissemination within the scientific community and the population more broadly on the state of the art relating to ATMPs.

Led by a coordinator appointed by the MOST, the Interministerial Commission focused its efforts first on adopting a new regulatory framework that was harmonised with the EMA and FDA, and that recognised the strengths of local institutions in fulfilling its objectives. The

---

[17] C. Krmpotic, 'Creer en la cura. Eficacia simbólica y control social en las prácticas del Dr. M.,' (2011) *Scripta Ethnológica*, (XXXIII), 97–116.

[18] The National Constitution of Argentina establishes a right to health, and stipulates that the private or public health system of the provinces or federal authorities is guarantor of the right. The following are examples of judicial cases that were reported by the Legal Department of OSDE (Social Security Organization for Company Managers): 'Jasminoy, María Cristina c / Osde Binario s /Sumarísimo' (Expte. 4008 / 03), Court of First Instance in Civil and Commercial Matters No. 11, Secretariat No. 22. The treatment was covered by OSDE. Diagnosis: Multiple Sclerosis; 'Silenzi de Stagni de Orfila Estela c / Osde Binario S. A s / Amparo' (Expte. 4475 / 05), National Civil Court No. 11. The treatment was covered by OSDE. Diagnosis: Multiple Sclerosis; 'Ferrreira Mariana c / Osde Binario y otros / Sumarísimo' (Expte. 8342 / 06), Civil and Commercial Federal Court No. 9, Secretariat No. 17. The court decision ordered the coverage of the treatment but it could not be implemented because the plaintiff died. Diagnosis: Leukaemia.

strategy to create the governance framework was centred in three levels of norms: federal law, regulation and soft law. The proposal was accepted by both Ministries and efforts were made to put in force, first, the regulatory framework and soft law in order to stop the delivery of unproven treatments. These elements would then be in force while a bill of law was sent to the National Parliament.

On September 2018, the new regulatory framework was issued through ANMAT Disposition 179/2018 and an amendment to the Transplant Law giving competence to INCUCAI to deal with hematopoietic progenitor cells (CPH) in their different collection modalities, the cells, tissues and/or starting materials that originate, compose or form part of devices, medical products and medicines, as well as cells of human origin of autologous use used in the same therapeutic procedure with minimal manipulation and to perform the same function of origin.

The Interministerial Commission benefitted immensely from the work of the original Commission, which was formed in 2007 and which collaborated across technical fields and jurisdictional borders for a decade, moving Argentina from a position of no regulation for ATMPs, to one of imperfect regulation (limited by the conditions of the time). The original Commission undertook the following:

1. Undertaking studies on the legislation of Argentina and other countries to better understand how these technical developments might be shaped by law (i.e. through transplant, medicines or a *sui generis* regime).
2. Proposing a governance framework adapted to the Argentine legal and cultural context, harmonised with European and US normative frameworks.
3. Communicating this initiative to all interested sectors and managing complex relationships to promote debate in society, and then translate learnings from that debate into a normative/governance plan.[19]

The work of the Commission was advanced through key collaborations; first and foremost with the University of Edinburgh (2007–2019). This collaboration had several strands and an active institutional relationship.[20]

Other collaborations involved the Spanish Agency for Medicaments, the Argentine judiciary[21] and the creation of the Patient Network for Advanced Therapies (APTA Network) to provide patients with accurate information about advances in science and their translation into healthcare applications. All this was accompanied by interactions with a range of medical societies in order to establish a scientific position in different areas of medicine against the offer of unproven treatments.[22]

---

[19] V. Mendizabal et al., 'Between Caution and Hope: The Role of Argentine Scientists and Experts in Communicating the Risks Associated with Stem Cell Tourism',(2013) *Perspectivas Bioéticas*, 35–36, 145–155.

[20] An ESRC-funded research project, *Governing Emerging Technologies: Social Values in Stem Cell Research Regulation in Argentina*, explored various stakeholders' regulatory values, ambitions and tolerances. The institutional relationship resulted in the training of researchers and members of the Commission, the hosting of eight international seminars at which experts from various countries – mainly the UK – shared their experiences, and the holding of fellowships which facilitated research visits to academic and regulatory institutions in the UK.

[21] Which resulted in engagement activities with judicial associations so as to raise awareness among judges about the problem of experimental treatments, and the need to avoid ordering the transfer of resources from the health system to unscrupulous medical doctors

[22] See more at: 'Red argentina de pacientes', (Argentina.gob.ar), www.argentina.gob.ar/ciencia/celulasmadre/red-argen tina-de-pacientes.

33.6 CURRENT LEGAL/REGULATORY FRAMEWORK

The current legal framework in force and proposed by the Interministerial Commission is the result of a collaboration work focused on identifying the different processes involved in research and approval of ATMPs and set up an effective articulation between its parts. It consists of laws and regulations and establishes a coordinated intervention of both authorities, ANMAT and INCUCAI, in the process of approval of research and products. The system operates as follows:

1. Medicaments Law establishes ANMAT with competence to regulate the scientific and technical requirements at national level applicable to clinical pharmacology studies, the authorisation of manufacturing establishments, production, registration and authorisation of commercialisation, and surveillance of Advanced Therapy Medicaments.[23]
2. Transplants Law establishes INCUCAI with competencies to regulate the stages of donation, obtaining, and control of cells and/or tissues from human beings when they are used as starting material in the production of an ATMP.[24]
3. Manufacturing establishments that produce ATMP must be authorised by ANMAT.
4. When an ATMP is developed and used within the same facility, the donation, procurement, production and control stages are ruled under the INCUCAI regulations. INCUCAI must request the intervention of ANMAT for the evaluation and technical assistance in the stages of the manufacturing process, in order to guarantee that they meet the same standards as the rest of the Advanced Therapy Medications.
5. Cell preparations containing cells of human origin with minimal manipulation are not considered medications and will be under the INCUCAI regulations.

Finally, the newly amended Argentine Civil Commercial Code 2015 establishes the ethico-legal requirements for clinical trials. Specifically, Article 58 states that investigations in human beings through interventions, such as treatments, preventative methods, and diagnostic or predictive tests, whose efficacy or safety are not scientifically proven, can only be carried out if specific requirements are met relating to consent, privacy, and a protocol that has received ethical approval, etc.

Laws and regulations above described combine to form a reasonably comprehensive normative system applicable to research, market access approval and pharmacovigilance for ATMPs, harmonised with international standards.

Importantly, and interestingly, though many stakeholders in the period 2011–2017 reported a preference for command-and-control models of regulation (i.e. state-led, top-down approaches)[25] and many elements of the prevailing regime do now reflect this, the framework itself emerged through a bottom-up, iterative process, which sought to connect abstract concepts and models of governance with actual experience and the national social and legal normative culture. While the Commission, together with a key circle of actors, shaped the process, a wide variety of stakeholders from academia, regulatory bodies, medical societies, researchers, patients

[23] ANMAT Disposition 179/2018.
[24] Law No. 27.447/2018 y su Decreto Reglamentario No. 16/2019.
[25] S. Harmon, 'Argentina Unbound: Governing Emerging Technologies: Social Values in Stem Cell Regulation in Argentina', (2008) Presented at *European Association of Health Law, 'The Future of Health Law in Europe'* (*Conference, 10–11 April 2008, Edinburgh*).

and social media cooperated to advance the field. Their efforts were very much an example, imperfectly realised, of legal foresighting.[26]

To complete the normative framework currently in force, it would be advisable to maintain a soft law design to provide support to regulatory bodies to maintain updated proceedings as well as the flexibility to accompany the advances of science. Finally, it would be prudent to count on a federal law that regulates clinical research, and fundamentally to provide the regulatory authority a robust policy power to stop the advance of eventual unproven treatments across the country as a legal warranty for the protection of patients and research human subjects.

33.7 CONCLUSION

The design and adoption of a governance framework for regenerative medicine research and ATMPs in the Argentine Republic has been a decade-long undertaking that has relied on the strengths and commitment of key institutions like MOST, MOH, ANMAT and INCUCAI and on the ongoing engagement with a range of stakeholders.

To achieve the current normative framework, it was necessary to amend existing legal instruments and issue new laws and regulations.

The new framework exemplifies a more joined-up regime that is harmonised with other important regulatory agencies like EMA and the FDA. This is important because the development of ATMPs is increasingly global in nature, and it is expected that Argentine regulators will work closely with international partners in multiple ways to support safe and effective innovation that will benefit a wider segment of the population, including, importantly, traditionally marginalised groups.

---

[26] G. Laurie et al., 'Foresighting Futures: Law, New Technologies, and the Challenges of Regulating for Uncertainty', (2012) *Law, Innovation and Technology*, 4(1), 1–33.

Towards Responsive Regulation

Introduction

Catriona McMillan

This section of the volume offers a contemporary selection of examples of where existing models of law and regulation are pushed to their limits. Novel challenges are arising that require reflection on appropriate and adaptive regulatory responses, especially where ethical concerns raise questions about the acceptability of the research itself. The focus in this section is on how these examples create disturbances within regulatory approaches and paradigms, and how these remain a stubborn problem if extant approaches are left untouched. The reference to 'responsive' here highlights the temporally limited nature of law and regulation, and the reflexivity and adaptability that is required by these novel challenges to health research regulation. The choice of examples is illustrative of existing and novel research contexts where the concepts, tools, and mechanisms discussed in Part I come into play.

The first part of this section speaks to nascent challenges in the field of reproductive technologies, an area of health research often characterised by its disruptiveness to particular legal and social norms. The first two contributions to this theme focus on human gene editing, a field of research that erupted in global public ethical and policy debates when the live birth of twin girls, Lulu and Nana, whose genes had been edited *in vitro*, was announced by biophysician He Jiankui in 2018. For Isasi (Chapter 34), recent crises such as this provide opportunity to transform not only global policy on human germline gene editing, but collective behaviours in this field. In this chapter, she analyses the commonalities and divergences in international normative systems that regulate gene editing. For Isasi, a policy system that meaningfully engages global stakeholders can only be completely effective if we achieve both societal consensus and governance at local and global levels. For Chan (Chapter 35), the existence of multiple parallel discourses highlighted by Isasi can be used to facilitate broader representation of views within any policy solution. Chan considers the wider lessons that the regulatory challenges of human germline gene editing pose for the future of health research regulations. She posits that human germline gene editing is a 'contemporary global regulatory experiment-in progress', which we can use to revisit current regulatory frameworks governing contentious science and innovation.

For the authors of the next two chapters, the order upon which existing regulatory approaches were built is being upended by new, dynamic sociotechnical developments that call into question the boundaries that law and regulation has traditionally relied upon. First, Hinterberger and Bea (Chapter 36) challenge us to consider how we might reconsider normative regulatory boundaries in their chapter on human animal chimeras – an area of biomedical research where our normative distinctions between human and animal are becoming more blurred as research advances. Here, the authors highlight the potential of interspecies research to perturb lasting, traditional regulatory models in the field of biomedical research. Next, McMillan (Chapter 37) examines the fourteen-day limit on embryo research as a current example of an existing regulatory tool – here a legal 'line in the sand' – that is being pushed to its scientific limits. She argues that recent advancements in *in vitro* embryo research challenge us to disrupt our existing legal framework governing the processual entity that is the embryo *in vitro*. For McMillan, disrupting our existing regulatory paradigms in embryo research enables essential policy discussion surrounding how we can, and whether we should, implement enduring regulatory frameworks in such a rapidly changing field.

For the second part of this section, the final two chapters examine the downstream effects of health research regulation in two distinct contexts. For these authors, it is clear that innovation in research practice and its applications requires us not only to disrupt our normative regulatory frameworks and systems, but to do so in a way that meaningfully engages stakeholders (see Laurie, Introduction). Jackson (Chapter 38) challenges the sufficiency of giving patients information about the limited evidence-base behind 'add-on' treatments that are available in fertility clinics, as a mechanism for safely controlling their use. For Jackson, regulation of these add-ons needs to go further; she argues that these treatments should be deemed by the Human Fertilisation and Embryology Authority – the regulator of fertility clinics and research centres in the UK – as 'unsuitable practices'. She highlights the combination of a poor evidence base for the success of these 'add-on' treatments and patients' understandable enthusiasm that these might improve fertility treatment outcomes. Her contribution confronts this 'perfect storm' of the uncertain yet potentially harmful nature of these add-ons, which are routinely 'oversold' in these clinics, yet under-researched. Jackson's offering gives us an example of an ongoing and increasing practice and process that requires us to disrupt prevailing regulatory norms. In the final chapter of this section, Harmon (Chapter 39) offers human enhancement as an example of how a regulatory regime, catalysed by disruptive research and innovation, has failed to capture key concepts. For Harmon, greater integration of humans and technology requires our regulatory frameworks to engage with 'identity' and 'integrity' more deeply, yet the current regulatory regime's failure to do so provides a lack of support and protection for human wellbeing.

Together, these chapters provide detailed analyses of carefully chosen examples and/or contexts that instantiate the necessity for reflexivity in a field where paradigms are (and should be) disturbed by health research and innovation. It is clear that particular regulatory feedback loops within and across particular regulatory spaces need to be closed in order to deliver authentic learning back to the system and to its users (see Laurie, Introduction and Afterword). A key theme in this section is the call to approach health research regulation as a dynamic endeavour, continually constituted by scientific processes and engaged with stakeholders and beneficiaries. In doing so, this section provides grounded assessments of HRR, showing the positive potential of responsive regulation as new approaches to health research attempt to meet the demands of an ever-changing world.

34

Human Gene Editing

Traversing Normative Systems

Rosario Isasi

34.1 INTRODUCTION

Gene editing technologies consist of a set of engineering tools, such as CRISPR/Cas9, that seek to deliberately target and modify specific DNA sequences of living cells.[1] They can enable both *ex vivo* and *in vivo* deletions and additions to DNA sequences at both somatic and germline cell levels. While technical and safety challenges prevail, particularly regarding germline applications, these technologies are touted as transformational for the promotion and improvement of health and well-being. Furthermore, their enhanced simplicity, efficiency, precision, and affordability had spurred their development. This in turn, has brought to the fore scientific and socio-political debates concerning their wide range of actual and potential applications together with their inexorable ethical implications.

The term 'inevitable' refers to the certainty or the unavoidability of an occurrence. Such was the worldwide response after the 2018 announcement – and later confirmation[2]– of the live birth of twin girls whose genomes were edited during *in vitro* fertilisation procedures. While foreseeable, shock followed and ignited intense national and international debates. China was placed at the epicentre of controversy, as the ubiquitous example of inadequate governance and moral failure. Yet, as the facts of the case unfolded, it became clear that the global community shared a critical level of responsibility.[3] Crisis can provoke substantial changes in governance and fundamentally alter the direction of a given policy system. While the impact of the shock is still being felt, the subsequent phase of readjustment has yet to take place. A 'window of opportunity' is thereby present for collective assessment of its impact, for ascertaining accountability, and for enacting resulting responses. Reactionary approaches can be predicted, as demonstrated by the wave of policies in the 1990s and 2000s following the derivation of the first human embryonic stem cell line or the birth of 'Dolly' the cloned mammal. Indeed, the 'embryo-centric' approach that characterised these past debates is still present.[4] Additionally, the globalisation phenomenon has permeated the genomics field, reshuffling the domain of debate and action from the

---

[1] K. E. Ormond et al., 'Human Germline Genome Editing', (2017) *The American Journal of Human Genetics*, 101(2), 167–176.

[2] D. Normile, 'Government Report Blasts Creator of CRISPR Twins', (2019) *Science*, 363(6425), 328.

[3] J. Qiu, 'American Scientist Played More Active Role in "CRISPR Babies" Project than Previously Known', (Stat News, 31 January 2019), www.statnews.com/2019/01/31/crispr-babies-michael-deem-rice-he-jiankui/.

[4] B. M. Knoppers et al., 'Genetics and Stem Cell Research: Models of International Policy-Making' in J. M. Elliot et al. (eds), *Bioethics in Singapore: The Ethical Microcosm* (Singapore: World Scientific Publishing, 2010), pp. 133–163.

national to the international. A case in point are the past International Gene Editing Summits aimed at fostering global dialogue.[5]

So far, human gene editing (HGE) has stimulated a new wave of policy by an extensive range of national and international actors (e.g. governments, professional organisations, funding agencies, etc.). This chapter outlines some of the socio-ethical issues raised by HGE technologies, with focus on human germline interventions (HGI), and addresses a variety of policy frameworks. It further analyses commonalities as well as divergences in approaches traversing a continuum of normative models.

34.2 NAVIGATING NORMATIVE SYSTEMS FOR HGE

Across jurisdictions, the regulation of genomics research has generally followed a linear path combining 'soft' and 'hard' approaches that widely consider governance as a 'domestic matter'.[6] Driven by scientific advances and changes in societal attitudes that resulted in greater technological uptake, genomics has increasingly become streamlined. This is reflected in the departure from the exceptionalist regulation of somatic gene therapy, now ruled by the general biomedical research framework, or in the increasing acceptance of reproductive technologies, where pre-implantation genetic diagnosis is no longer considered as an experimental treatment.

Normative systems cluster a broad range of rules or principles governing and evaluating human behaviour, thereby establishing boundaries between what should be considered acceptable or indefensible actions. They are influenced by local historical, socio-cultural, political and economic factors. Yet, international factors are not without effect. These systems are enacted by a recognised legitimate authority and unified by their purpose, such as the protection of a common good. Often, they encompass set criteria for imposing punitive consequences in the form of civil and criminal sanctions, or by moral ones, in the form of social condemnation for deviations. The boundaries normative systems impose are sometimes set arbitrarily, while in others, these divisions are systematically designed. Thus, they either create invisible or discernible ethical thresholds by making explicit the principles and values underpinning them.

At the same time, normative systems are often classified by their coercive or binding nature, as exemplified in the binary distinction between 'soft' and 'hard' law. While this categorisation is somewhat useful, it is important to note that 'hard' and 'soft' laws are not necessarily binary; rather, they often act as mutually reinforcing or complementary instruments. The term 'soft law' refers to policies that are not legally binding or are of voluntary compliance, such as those emanating from self-regulatory bodies (e.g. professional guidelines, codes of conduct) or by international agencies (e.g. declarations) without formal empowered mechanisms to enforce compliance, including sanctions. In turn, 'hard law' denotes policies that encompass legally enforceable obligations, such regulations. They are of binding nature to the parties involved and can be coercively enforced by an appropriate authority (e.g. courts).

In the context of HGI, normative systems have opted for either a public ordering model consisting of state-led, top-down legislative approaches, or a private ordering one, which adopts a bottom-up, self-regulatory approach. In between them, there is also a mix of complex public–private models. Normative systems are present in a continuum from permissive, to intermediate,

---

[5] Human Genome Editing Initiative, 'New International Commission on Clinical Use of Heritable Human Genome Editing', (National Academies of Science Engineering Medicine, 2019), www.nationalacademies.org/gene-editing/index.htm.

[6] R. Isasi et al., 'Genetic Technology Regulation: Editing Policy to Fit the Genome?', (2016) *Science*, 351(6271), 337–339.

and to restrictive, reflecting attitudes towards scientific innovation, risk tolerance and consider-ations for proportional protections to cherished societal values (e.g. dignity, identity, integrity, equality and other fundamental freedoms). The application of HGE technologies in general, and HGI in particular, are regulated in over forty countries by a complex set of legislation, professional guidelines, international declarations, funding policies and other instruments.[7] Given their diverse nature, these norms vary in their binding capacity (e.g. legislation vs self-regulation), their breadth and their scope (e.g. biomedical research vs clinical applications vs medical innovation). Notwithstanding all the previously stated heterogeneity in normative models, harmonised core elements are still present between them.

Resistance towards applying HGE in the early stages of development commonly rest on beliefs regarding the moral – and fortiori legal – status of the embryo, social justice and welfare concerns. Their inheritable capacity, in turn, brings to the fora issues such as intergenerational responsibility and the best interests of the future child, together with concerns regarding their population (e.g. genetic diversity), societal (e.g. discrimination, disability) and political impacts (e.g. public engagement, democracy).[8] Remaining safety and efficacy challenges are also of chief importance and often cited to invoke the application of the 'precautionary principle'. Lastly, fears over 'slippery slopes' leading to problematic (e.g. non-medical or enhancements) uses and eugenic applications are at the centre of calls for restrictive normative responses.[9] However, across these systems the foundational principles underpinning a given norm and reflecting a society's or an institution's common vision and moral values are not always sufficiently substantiated, if at all articulated. As such, calls for caution to protect life, dignity and integrity, or against eugenic scenarios, appear as mere blanket or rhetorically arguments used for political expediency. As a consequence, the thresholds separating what is deemed as an acceptable or indefensible practice remain obscure and leave an ambiguous pathway to resolve the grey areas, mostly present in the transition towards clinical applications.

An unprecedented level of policy activity followed the rapid development of HGE. National and international scientific organisations, funding and regulatory agencies, as well advocacy groups have responded to these advances by enacting 'soft laws' appealing for caution, while others have opted for assessing the effectiveness of extant 'hard' and 'soft' policies.

34.2.1 *National Policy Frameworks*

Normative systems are often conceptualised using a hierarchy that differentiates between restrictive, intermediate and permissive approaches. Under this model, restrictive policies set up ethical and political boundaries by employing upstream limits – blank bans or moratoria – to interventions irrespective of their purpose. Pertaining to the application of HGI, restrictive approaches essentially outlaw or tightly regulate most embryo and gamete research. Supported by concerns over degrading dignity and fostering commodification of potential life, these approaches are based on attributing a moral – personhood or special – status to embryos, and thus advocating for robust governmental controls. Stipulations forbidding 'genetic engineering

---

[7] Isasi et al. 'Genetic Technology Regulation'.
[8] Ormond et al., 'Human Germline Genome Editing'.
[9] D. Baltimore et al., 'Biotechnology: A Prudent Path Forward for Genomic Engineering and Germline Gene Modification', (2015) *Science*, 348(6230), 36–38.

on human germ cells, human zygotes or human embryos"[10] or stating that no 'gene therapy shall be applied to an embryo, ovum or fetus'"[11] exemplify this model.

While apparently wide-ranging, restrictive policies contain several potential loopholes. Among their major shortcomings are their reliance on research exceptions for therapeutic interventions that are deemed beneficial or life preserving to the embryo, or which are necessary in order to achieve a pregnancy. Terminological imprecisions will render as inapplicable a norm once a particular intervention could be considered as medical innovation or standard medical practice. Similar gaps are present in norms referencing specific technologies and in legal definitions of what constitute a embryo or a gamete, as all of these could later be outpaced by scientific advances, such as those brought by developments in the understanding of embryogenesis, organoids, and pluripotent stem cells. Indeed, the growth of HGE technologies has brought back to centre stage reflections over what is a reproductive cell. Evocative of the debates that took place during the peak of the stem cell era, the scientific, legal, and moral status of these entities continue to be tested, while at the same time remaining as the most prevalent policy benchmark. Whether silent or overtly present in distinct conceptualisations (e.g. developmental capacity or precise time period), criteria defining these early stages of human development are at the core of policies directing the permissibility of certain interventions.

The most favoured policy position is, however, an intermediate one, in which restrictions are applied downstream by banning research with reproductive purposes. Yet, this position considers permissible the practices that are directed at fundamental scientific research activities, such as investigating basic biology or aspects of the methodology itself. Policies adopted in countries such the Netherlands,[12] reflect this moderate perspective by outlawing any intervention directed at initiating – including attempts to initiate – a pregnancy with an embryo – or a reproductive cell – that has been subject to research or whose germline has been intentionally altered. Balancing social and scientific concerns, this approach calls for modest governance structures, yet close oversight. Nevertheless, it is at the risk of internal inconsistencies and ambiguities, given that norms are often the result of political compromises, which seem necessary in order to achieve policy adoption. A case in point are those research policies that confer moral and legal status to the human embryo while – at the same time – mandating their destruction after a certain period of time, or in ambiguous norms regarding the permissibility of clinical translation.

Largely misinterpreted, liberal models do not necessarily postulate a laissez-faire or a blanket unregulated approach. Rather, they provide significant scientific freedom predicated on the strength of their governance frameworks. They seek to promote scientific advances as a tool for social progress. In the context of HGI, liberal policies[13] allow for basic and reproductive research while banning clinical implementation. Given that these approaches depend on the effectiveness of their governance structures (e.g. licensing, oversight) with decisions often on a case-by-case or a de-facto basis, they are at the risk of arbitrary applications and system failure. Moreover, when the model rests on self-regulatory approaches devoid of effective enforcement mechanisms, they risk being – or being perceived to be – self-serving and following a market consumer model.

---

[10] Biosafety Law, Law No. 11, 2005 (Brazil).

[11] Bioethics and Safety Act 2013 (South Korea).

[12] Act Containing Rules Relating to the Use of Gametes and Embryos [The Embryos Act] 2002 (The Netherlands).

[13] Isasi et al. 'Genetic Technology Regulation'; S. Lingqiao and R. Isasi, *The Regulation of Human Germline Genome Modification in China. Human Germline Genome Modification and the Right to Science: A Comparative Study of National Laws and Policies* (Cambridge: Cambridge University Press; 2019).

Throughout policy models, the progression from research to clinical purposes is at times blurred in the peculiarities of such approaches. In fact, uncertainty regarding the scope of requirements is particularly present when there are permissible exceptions to norms forbidding HGE in reproductive cells. This is the case of Israel, which outlaws 'using reproductive cells that have undergone a permanent intentional genetic modification (germline gene therapy) in order to cause the creation of a person',[14] yet permits to apply to a research licence 'for certain types of genetic intervention' provided that 'human dignity will not be prejudiced'.[15] Similarly in France where 'eugenic practice aimed at organizing the selection of persons' and alteration(s) 'made to genetic characteristics in order to modify the offspring of a person' are banned,[16] yet at the same time the law exempts interventions aiming 'for the prevention and treatment of genetic diseases'[17] without providing further guidance.

Notwithstanding heterogeneous normative approaches, these models share a common objective: fostering scientific innovation and freedoms while protecting their vision of a common good, mostly expressed in safeguarding human dignity. In order to do so, sanctions and other coercive mechanisms are often adopted as deterrents. Indeed, the global HGE policy landscape is frequently accompanied by some form of sanctions, ranging from criminal to pecuniary and other social penalties. In particular, when such systems are based on legislative models, criminal penalties – substantial imprisonment and fines – are the standard. Upholding criminal law in biomedical research is an exceptional approach, and societies around the world use this tool to send the strongest condemnatory message. Here, as in other fields, criminal law serves as a tool for moral education and for achieving retribution, denunciation, and/or deterrence. But other type of penalties, such as moral sanctions, could be equally powerful. A radical example of the latter is China's 'social credit system'[18] where research misconduct is sanctioned by a wide umbrella of actors, which can impose an equally wide set of penalties and can even reach far beyond the traditional academic setting – from employment to funding, insurance, and banking eligibility. However, employing criminal law can be problematic because it often requires intentionality (mens rea). In the context of HGI, criminal law could create loopholes for downstream interventions when restrictions are limited to certain applications. For instance, German law bans the 'artificial' alteration of 'the genetic information of a human germ line cell'[19] and the use of such cell for fertilisation. Yet, such prohibition would not be applicable 'if any use of it for fertilisation has not been ruled out.'[20] While under Canadian legislation, it is an offense to 'knowingly' 'alter the genome of a cell of a human being or in vitro embryo such that the alteration is capable of being transmitted to descendants.'[21]

Comparably, an issue of shared concern across normative systems are references to the eugenic potential of HGI. Fears over the ability to alter the germline infringing dignity and integrity have been widely articulated in policies. These concerns are best illustrated in France, where a new crime against the integrity of the human species has been typified and which

---

[14] Prohibition of Genetic Intervention (Human Cloning and Genetic Manipulation of Reproductive Cells) Law 1999 last renewed, 2009 (Israel).

[15] Prohibition of Genetic Intervention.

[16] Bioethics Law/Loi No. 2004-800 du aout 6 2004 relative à la bioethique and Code Civil (1804) 2004 last amendment 2015 (France).

[17] Bioethics Law.

[18] D. Cyranoski, 'China Introduces 'Social' Punishments for Scientific Misconduct', (*Nature*, 14 December 2018).

[19] Embryo Protection Act. 1990 (Germany).

[20] Embryo Protection Act.

[21] An Act respecting human assisted reproduction and related research (Assisted Human Reproduction Act) 2004 (Canada).

forbids 'carrying out a eugenic practice aimed at organizing the selection of persons.'[22] Similarly, Indian guidelines restrict 'eugenic genetic engineering for selection against personality, character, formation of body organs, fertility, intelligence and physical, mental and emotional characteristics.'[23] In the same vein, Belgium outlaws carrying out 'research or treatments of eugenic nature that is to say, focused on the selection or amplification of non-pathological genetic characteristics of the human species.'[24] However, these policies provide little guidance for interpretation: when should interventions seeking to repair deleterious gene mutations or confer disease immunity – at the individual or population level – be considered eugenic interventions? Or a non-medical or enhancement practice? Selecting or de-selecting traits, while not an ethically neutral intervention, is not *per se* eugenics. Therefore, contextualising thresholds and defining the paraments for scientific and ethical acceptability of such interventions are required not only to provide much needed legal clarity, but also to avoid being perceived as simply rhetorical calls for political expediency.

34.2.2 *International Policy Frameworks*

Significant policy activity followed the refinement of HGE. A wide range of professional organisations, funding and regulatory agencies, quickly reacted to these developments with statements reflecting an equally varied range of positions.[25]A common theme among them is a circumspect attitude with appeals for the protection of dignity and integrity. While these positions endorse different normative approaches, they all pay particular attention to intergenerational responsibilities in their calls for principled restrictions to reproductive HGI.

Among the earliest international instruments addressing HGI are several non-binding Declarations adopted under the United Nations' framework. First, are the UNESCO's Universal Declaration on the Human Genome and Human Rights and the ensuing report on HGE by their International Bioethics Committee, which conceptualise the genome as the 'heritage of humanity' and in that vein, they plea for a moratorium on HGI that is based on prevailing 'concerns about the safety of the procedure and its ethical implications.'[26] Succeeding UNESCO's efforts, and after a failed attempt to adopt legally binding policy, the United Nations passed the UN Declaration on Human Cloning, calling on states 'to adopt the measures necessary to prohibit the application of genetic engineering techniques that may become contrary to human dignity.'[27] The pleas raised by these UN bodies remain a contemporary mandate, appealing for concrete measures to implement moral commitments into national legislation with the necessary enforcement measures.

Following the human rights approach enshrined in the abovementioned instruments, two important regional policies were enacted: the Council of Europe's Oviedo Convention[28] and

[22] Bioethics Law.
[23] Indian Council of Medical Research, 'Ethical Guidelines for Biomedical Research on Human Participants', (Indian Council of Medical Research, 2000 last amendment 2006).
[24] Act on Research on Embryos In Vitro – Loi relative à la recherché sur les embryons in vitro 2003 (Belgium).
[25] National Academies of Sciences, Engineering, and Medicine, *Human Genome Editing: Science, Ethics, and Governance*, (The National Academies Press, 2017); HUGO Ethics Committee, 'Statement on Gene Therapy Research', (Human Genome Organisation, 2001).
[26] UNESCO Constitution, 'Universal Declaration on the Human Genome and Human Rights', (United Nations Educational, Scientific, and Cultural Organization, 1997)
[27] United Nations, 'United Nations Declaration on Human Cloning', (United Nations, 2005).
[28] Council of Europe, 'Convention for the Protection of Human Rights and Dignity of the Human Being with regard to the Application of Biology and Medicine: Convention on Human Rights and Biomedicine', (Council of Europe, 1997).

the European Union Clinical Trials Regulation.[29] These remain to date as the only international legally binding instruments governing HGI. The Oviedo Convention – as a general rule – explicitly forbids research and clinical interventions seeking to modify the genome. Yet, it exempts interventions that are 'undertaken for preventive, diagnostic or therapeutic purposes' when the aim is 'not to introduce any modification in the genome of any descendants'.[30] In turn, the cited EU Regulation focuses on gene therapy, banning clinical trials resulting 'in modifications to the subject's germ line genetic identity'.[31] Yet, no guidance has been provided to define or interpret the notion of 'genetic identity' in order to fully grasp the scope and breadth of such provisions.

Actors from different fields and parts of the world[32] have been quite prolific in articulating their positions with regards to HGE and in conveying how they envisage – or not – a path forward to reproductive HGE.[33] Even in China after the birth of the HGE twins, funding and professional organisations have swiftly publicised their positions,[34] aligning to mainstream ones. Indeed, all of these statements share several common threads. First, they all endorse a guarded approach to HGI, calling for temporary halts or moratoria, rather than advocating for permanent bans. The scope and breadth of such restrictions vary, from positions that seek to prevent clinical applications but allow reproductive research, to those that condemn any use. Second, a prospective approach also characterises them. While recent developments might render prevention a futile goal, precautionary measures fostering scientific integrity are still relevant. Third, they are by far based on scientific concerns, given the current inability to fully assess HGE's safety and efficacy. Notably, societal considerations focusing on protecting human rights are also prevalent. Lastly, appeals for public engagement are widespread, including calls for participatory, inclusive and transparent dialogue in order to empower stakeholders, inform policy-making efforts, and foster trustworthiness.[35]

34.3 THE ROAD TO HARMONISATION

Reactionary responses often follow the advent of scientific developments deemed to be disruptive to notions of integrity and dignity, such as with HGI. A concomitant result of the debates over genetic engineering techniques that started decades ago, is an overall fraught policy landscape that generally seeks to condemn such interventions but is void of global governance. However, they steered a level of policy convergence.

---

[29] European Union Clinical Trials Regulation 536/2014, OJ No. L 158/1, 2014.
[30] Council of Europe, 'Convention for the Protection of Human Rights'.
[31] European Union Clinical Trials Regulation.
[32] Ormond et al., 'Human Germline Genome Editing'; National Academies, 'Human Genome Editing'; Genetic Alliance Germline Gene Editing, 'A Call for Moratorium on Germline Gene Editing, Commentary by Genetic Alliance', (Genetic Alliance, 2019), www.geneticalliance.org/advocacy/policyissues/germline_gene_editing; National Academies of Sciences, Engineering, and Medicine, *International Summit on Human Gene Editing: A Global Discussion*, (The National Academies Press, 2015); National Academies of Sciences, Engineering, and Medicine, *Second International Summit on Human Genome Editing: Continuing the Global Discussion Proceedings of a Workshop—in Brief*, (The National Academies Press, 2019); International Society for Stem Cell Research, 'The ISSCR Statement on Human Germline Genome Modification', (ISSCR: International Society for Stem Cell Research, 2015).
[33] C. Brokowski, 'Do CRISPR Germline Ethics Statements Cut It?', (2018) *The CRISPR Journal*, 1(2), 115–125.
[34] Enforcement of Scientific Ethics Committee, Academic Division of the Chinese Academy of Sciences (CASAD), 'Statement About CCR5 Gene-edited Babies', (CASAD, 2018) www.english.casad.cas.cn/bb/201811/t20181130_201704 .html; Chinese Society for Stem Cell Research & Genetics Society of China, 'Condemning the Reproductive Application of Gene Editing on Human Germline', (Chinese Society for Cell Biology, 2018), www.cscb.org.cn/news/20181127/2988.html.
[35] M. Allyse et al., 'What Do We Do Now?: Responding to Claims of Germline Gene Editing in Humans', (2019) *Genetics in Medicine*, 21(10), 2181–2183.

The plethora of social debates and policies emanating in the context of HGE demonstrate that across the globe, policy harmonisation remains a laudable objective. These efforts seek convergence in fundamental ethical safeguards for research participants – and future patients – coupled with criteria for regulating the application of these technologies. Throughout the world, and with diverse levels of success, governance mechanisms have been established empowering authorities with granting licences, conducting ethical oversight and enforcing compliance. However, for these requirements to be effective, consistent implementation is needed in a manner that respects scientific integrity and freedoms.

Harmonisation is therefore apparent in convergent criteria that bar or condemn HGI. Yet, in some cases, these positions are only transitory by virtue of established moratoria or other precautionary temporary measures. Thus, they remain effective only while extant safety and other technical concerns remain. In fact, some responses seemed to be solely based on our current state of knowledge, as exemplified below:

> Although our report identifies circumstances in which genome interventions of this sort should not be permitted, we do not believe that there are absolute ethical objections that would rule them out in all circumstances, for all time. If this is the case, there are moral reasons to continue with the present lines of research and to secure the conditions under which heritable genome editing interventions would be permissible.[36]

Additional examples of the latter are found in Singapore policy forbidding HGI due to 'insufficient knowledge of potential long-term consequences'[37] and pending 'scientific evidence that techniques to prevent or eliminate serious genetic disorders have been proven effective'.[38] The same rationale underpins Indian policy restricting 'gene therapy for enhancement of genetic characteristics (so called designer babies)' based on 'insufficient information at present to understand the effects of attempts to alter/enhance the genetic machinery of humans'.[39]

Despite diverse normative systems and societal contexts, the world seems to be disposed towards harmonisation.[40] Which factors help explain this phenomenon? Policy transfer and emulation[41] might be factors supporting policy growth and the emergence of global convergence. However, such consensus is still quite precarious as best exemplified by the level of international involvement and the strength of the response to recent developments.[42] Scepticism over the stability of an emerging or actual consensus is based on the fact that policy responses thus far are grounded in distinct rationale. While they all call for 'action' and 'caution', they legitimately differ in their significance and understanding of such terms. As we have seen, in some instances a cautious approach has been translated in voluntary moratoria. This is the

[36] Nuffield Council on Bioethics. 'Genome Editing and Human Reproduction: Social and Ethical Issues', (Nuffield Council on Bioethics, 2018), 154.
[37] Bioethics Advisory Committee Singapore, 'Ethics Guidelines for Human Biomedical Research', (Bioethics Advisory Committee Singapore, 2015), 50.
[38] Ibid.
[39] Indian Council of Medical Research, 'Ethical Guidelines for Biomedical Research'.
[40] Nuffield Council on Bioethics, 'Genome Editing and Human Reproduction'; The Hinxton Group, 'Statement on Genome Editing Technologies and Human Germline Genetic Modification', (The Hinxton Group: An International Consortium on Stem Cells, Ethics, & Law, 2015).
[41] European Academies' Science Advisory Council, 'Genome Editing: Scientific Opportunities, Public Interests and Policy Options in the European Union', (EASAC: European Academies' Science Advisory Council, 2017).
[42] R. Isasi, 'Human Genome Editing: Reflections on Policy Convergence and Global Governance' in ZfMER (eds), *Genomeditierung – Ethische, rechtliche und kommunikations – wissenschaftliche Aspekte im Bereich der molekularen Medizin un Nutzplanzenzüchtung, Zeitschrift für Medizin-Ethik-Recht*, (Nomos, 2017), pp. 287–298.

temporarily halting of certain types of clinical interventions or in promoting public engagement[43] so as to allow for policy to reflect changes in scientific knowledge or societal values. In other instances, precautionary responses – under vigilant oversight – purposely do not deter or outlaw research given the need for evidence in quantifying risks and benefits. Finally, in other circumstances, caution has signified enacting blank legal prohibitions.[44]

Conceptual misunderstandings between the notion of harmonisation[45]and standardisation are often present.[46] As such, appeals for standardisation frequently do not realise that they entail the creation of uniform legal and ethical standards, which are not only highly unachievable, but also undesirable particularly with respect to HGE. In the latter, sovereignty and moral diversity must be respected. Harmonisation[47] processes do not seek uniformity as the end result, they rather entail substantial correspondence between fundamental ethical principles present across the continuum of normative responses. They aim to foster cross-jurisdictional collaboration and thus governance. Still, harmonisation is not without challenges, particularly in regards to criteria for evaluating policy convergence and assessing variations in the regulation of fundamental ethical requirements, where thresholds for determining the significance of a given policy can vary. The latter is of great importance as variations could potentially undermine the integrity of ethical safeguards or societal values.

34.4 CONCLUSION

For the sceptics, attempts to meaningfully engage a global community of stakeholders to adopt binding policy and governance will inevitably end in 'pyrrhic' victories[48] – as in the past. History seems to be full of examples to support this position.[49] Indeed, thus far the inability to form a representative community to reconcile conflicting interests – economic and otherwise – and to prevent egregious actions, has taught us that sole condemnation of a particular intervention is futile for preventing abuses absent morally binding obligations and 'actionable' regulatory frameworks. For the optimists, the level of societal engagement, emergent policy convergence and swift condemnatory responses following the most contemporaneous and appalling gross violations of human rights and scientific standards[50] are grounds to believe that a level of policy harmonisation remain a realistic endeavour. Crisis provides the opportunity to significant alter the direction and strength of a given policy system, including reshaping governance mechanisms and reconfiguring the power of stakeholders. It therefore has the ability to transform more than policy; it can stir real change in collective behaviour. In the aftermath of this crisis, the central lesson must be that without defining and achieving societal consensus and governance at both the local and global level, no policy system would ever be completely effective.

[43] E. S. Lander et al., 'Adopt a Moratorium on Heritable Genome Editing', (2019) *Nature*, 567(7747), 165–168; Allyse et al., 'What Do We Do Now?'.

[44] Allyse et al. 'What Do We Do Now?'.

[45] M. Boodman, 'The Myth of Harmonization of Laws', (1991) *The American Journal of Comperative Law*, 39(4), 699–724.

[46] R. Isasi, 'Policy Interoperability in Stem Cell Research: Demystifying Harmonization', (2009) *Stem Cell Reviews and Reports*, 5(2), 108–115.

[47] Oxford English Dictionary, 'Harmonization', (2019) *Lexico*, https://en.oxforddictionaries.com/definition/harmonization.

[48] R. Isasi and G. J. Annas, 'To Clone Alone: The United Nations Human Cloning Declaration', (2006) *Revista de Derecho y Genoma Humano*, 49(24), 13–26.

[49] United Nations, 'United Nations Declaration on Human Cloning'; D. Lodi et al., 'Stem Cells in Clinical Practice: Applications and Warnings', (2011) *Journal of Experimental & Clinical Cancer Research*, 30(1), 9.

[50] Normile, 'Government Report', 328.

Towards a Global Germline Ethics?

Human Heritable Genetic Modification and the Future of Health Research Regulation

Sarah Chan

35.1 INTRODUCTION

Human germline genetic modification (HGGM) has been the subject of bioethical attention for over four decades. Recently, however, two areas of biomedical technology have revived debates over HGGM. First, the development of 'mitochondrial replacement therapy' (MRT) represents, some have argued, a form of HGGM, since it affects the genetic makeup of the resulting children in a way that may be passed on to future generations. Second, the advent of genome editing[1] technologies has made heritable genetic modification of humans for the first time a genuinely practicable possibility: one that was dramatically and prematurely realised when, in November 2019, it was announced that two genome-edited babies had already been born.[2] Amid renewed scrutiny of human genome editing, emerging clinical uses of MRTs, and the increasing globalisation of science and of health technology markets, the question of how HGGM can and should be regulated has gained new salience. Moreover, having been so long contested and in relation to such fundamental concepts as 'human dignity' and 'human nature', the issue of germline modification has assumed a significance beyond its likely direct consequences for human health.

The current 'regulatory moment' with respect to HGGM thus perhaps represents something of a watershed for the global governance of science more generally. Further, both the potential impacts of the technology, and the moral and political power of the human genome as a metaphor through which to negotiate competing visions of human nature and society, require us to consider these issues at a global scale. This also creates an opportunity for critical exploration of novel approaches to regulation.

Following on from the previous chapter's analysis, this chapter considers broader lessons we might learn from examining the challenges of HGGM for the future of health research regulation. HGGM, I suggest, is a contemporary global regulatory experiment-in-progress through which we can re-imagine the regulation of (in particular, ethically contentious) science and innovation: what it should address, what its purposes might be, and how, therefore, we should go about shaping global scientific regulation. Through examining this, I argue that such regulation should focus on processes and practices, rather than objects; and that its utility lies

---

[1] Early discussions of these technologies often referred to 'gene editing'; in this chapter I employ the term 'genome editing', a usage that has since become more standard.

[2] D. Cyranoski and H. Ledford, 'Genome-Edited Baby Claim Provokes International Outcry', (2018) *Nature*, 563(7733), 607–608.

more in mediating these processes than in establishing absolute prohibitions or bright lines. Especially in the case of emerging and controversial technologies, regulation plays an important role in negotiating ideas of responsibility within the science–society discourse. In so doing, it also affects, and should be shaped by attention to, the global dynamics of science and the consequences for global scientific justice.

35.2 GERMLINE TECHNOLOGIES: A BRIEF OVERVIEW

Earlier techniques used for genetic modification were inefficient and simply impractical to allow the creation of genetically engineered humans.[3] The 'game-changing' aspect of genome editing technologies[4] is their ability to achieve more precise gene targeting, with much higher efficiency, in a wide range of cell types including human embryos. The best-known genome editing technology is the CRISPR/Cas9 system, the use of which was described in 2012.[5] Following the publication of the CRISPR/Cas9 method, it rapidly became clear that HGGM needed urgently to be reconsidered as a real possibility. In 2015, reports of the first genome editing of human embryos[6] spurred scientists to call for restrictions on the technology,[7] and prompted further investigations of the associated ethical and policy issues by various national and international groups.[8]

Notably, many reports, including those of the US National Academies[9] and the UK's Nuffield Council on Bioethics,[10] concluded that heritable human genome editing could be acceptable, providing certain conditions were met. These conditions included further research to ensure safety before proceeding to clinical application, and sufficient time for broad and inclusive engagement on governance. Neither of these conditions were fulfilled, however, when He Jiankui announced to the supposedly unsuspecting world[11] that he had already attempted the procedure.

Somewhat before genome editing technologies came onto the scene, MRTs were already being developed as a treatment for certain forms of mitochondrial disease.[12] According to the

---

[3] The methods developed in the 1980s for producing transgenic mice, for example (see B. H. Koller and O. Smithies, 'Altering Genes in Animals by Gene Targeting', [1992] *Annual Review of Immunology*, 10, 705–730), required extensive manipulation of embryonic stem cells (ESC) *in vitro*, followed by injecting these cells to form chimeric embryos, genetically screening a large number of progeny, and then selectively cross-breeding them to produce the desired genetic makeup – all steps ethically unthinkable to perform in humans.

[4] H. Ledford, 'CRISPR, the Disruptor', (2015) *Nature*, 522(7554), 20–24.

[5] M. Jinek et al., 'A Programmable Dual-RNA-Guided DNA Endonuclease in Adaptive Bacterial Immunity', (2012) *Science*, 337(6096), 816–821.

[6] P. Liang et al., 'CRISPR/Cas9-Mediated Gene Editing in Human Tripronuclear Zygotes', (2015) *Protein Cell*, 6(5), 363–372.

[7] D. Baltimore et al., 'Biotechnology. A Prudent Path Forward for Genomic Engineering and Germline Gene Modification', (2015) *Science*, 348(6230), 36–38; E. Lanphier et al., 'Don't Edit the Human Germ Line', (2015) *Nature*, 519(7544), 410–441.

[8] Reviewed in C. Brokowski, 'Do CRISPR Germline Ethics Statements Cut It?', (2018) *The CRISPR Journal*, 1(2), 115.

[9] Committee on Human Gene Editing, *Human Genome Editing: Science, Ethics and Governance* (Washington, DC: N. A. Press, 2017).

[10] Nuffield Council on Bioethics, 'Genome Editing and Human Reproduction', (Nuffield Council on Bioethics, 2018).

[11] Although, it would later transpire, more than a few international academics knew of He's work prior to the announcement, provoking questions as to why the work was not flagged earlier (N. Kofler, 'Why Were Scientists Silent over Gene-Edited Babies?', (2019) *Nature*, 566(7745), 427).

[12] Mitochondria are numerous organelles within each cell that produce energy via chemical reactions and carry their own genome, separate to nuclear DNA. Some of the genes required for mitochondrial function are encoded within the nuclear DNA, while others are in the mitochondrial genome (mtDNA) itself. Since most of the cytoplasm of a developing embryo comes from the egg, mitochondria are transmitted almost exclusively from the oocyte to offspring,

possibility foreseen in the 2008 amendments to the Human Fertilisation and Authority Act, and following an extensive consultation process, in 2015 it became legal for MRT to be licensed in the UK.[13] In the USA, the Institute of Medicine likewise concluded that MRT could be acceptable,[14] though it is not currently legal in the USA. Pre-empting the regulatory process, however, in September 2016 John Zhang, an American scientist, announced that he had already performed the first successful use of MRT in the clinic,[15] using embryos created in the USA and shipped to Mexico for intra-uterine transfer.

While reams have been written on the ethics of HGGM, the most pressing questions with respect to HGGM no longer concern whether we *ought* to do it at all, but how; where; by and for whom; and with (or without) what authority it *will* be done. This is not a claim about the inadequacy of regulation in the face of technological inevitability but a statement of where things currently stand ethically and legally, as well as scientifically. MRT is legal and being carried out in a number of countries; heritable genome editing, while not yet legalised, has been deemed ethically acceptable in principle. One way or another, HGGM is becoming a reality; regulation can guide this process. To do so effectively will require careful consideration of what is regulated and how, with what justifications, and with whose participation.

35.3 WHAT ARE WE REGULATING? WHAT *SHOULD* WE REGULATE?

As pointed out in the previous chapter (see Isasi, Chapter 34 in this volume) regulation can serve to articulate normative concerns but does not always do so coherently or consistently. In setting out to regulate 'germline modification' or 'the human genome', what concerns might be entangled?

The term 'germline modification' is itself subject to interpretation. Technically speaking, 'the germline' can encompass any cell that is part of the germ lineage, including gametes and embryos; thus a prohibition on modifying the germline might be taken to preclude *any* use of genome editing in human embryos, including for basic research. Early calls for a moratorium favoured this highly restrictive approach: some argued that because 'genome editing in human embryos ... could have unpredictable effects on future generations ... scientists should agree not to modify the DNA of human reproductive cells'.[16] This, however, ignores that genome editing of human embryos is only *likely* to have direct effects on future generations if those embryos become people! Context, in other words, is key.

Moreover, novel technologies may potentially render 'the germline' an impossibly broad category. It is now possible to reprogramme somatic cells to pluripotent cells,[17] and to turn pluripotent cells into gametes.[18] *Any* cell could therefore in theory become part of the germline, meaning any genetic modification of a somatic cell could potentially be a 'germline'

with little if any contribution from the sperm. Diseases caused by mtDNA mutations are thus 'maternally inherited', that is, passed on from mother to child.

[13] The Human Fertilisation and Embryology (Mitochondrial Donation) Regulations 2015.

[14] National Academies of Sciences, Engineering, and Medicine, *Mitochondrial Replacement Techniques: Ethical, Social, and Policy Considerations*, (Washington, DC: T. N. A. Press, 2016).

[15] J. Hamzelou, 'Exclusive: World's First Baby Born with New "3 Parent" Technique', (*New Scientist*, 27 September 2016).

[16] Lanphier et al., 'Don't Edit the Human Germ Line', 410.

[17] K. Takahashi et al., 'Induction of Pluripotent Stem Cells from Adult Human Fibroblasts by Defined Factors', (2007) *Cell*, 131(5), 861–872; K. Takahashi and S. Yamanaka, 'Induction of Pluripotent Stem Cells from Mouse Embryonic and Adult Fibroblast Cultures by Defined Factors', (2006) *Cell*, 126(4), 663–676.

[18] S. Hendriks et al., 'Artificial Gametes: A Systematic Review of Biological Progress towards Clinical Application', (2015) *Human Reproduction Update*, 21(3), 285–296.

modification. It is not, however, 'the germline' in the abstract, but the continuity or otherwise of *particular*, modified or unmodified germlines, that should be our concern.

The 'human genome' is likewise a nebulous concept: does it refer to an individual's genome, or the combined gene pool of humanity? References to the human genome as the basis of 'the fundamental unity of all members of the human family' and 'the heritage of humanity'[19] seem to suggest a collective account, but it is hard to see how the 'collective genome' could be regulated. Indeed, one might argue that the human genome, in the sense of the collective gene pool of humanity, would not be altered were genome editing to be used to introduce a sequence variant into an *individual* human genome that already exists within the gene pool.[20]

Even the term 'modification' raises questions. MRT involves no change to DNA sequence, only a new combination of nuclear and mitochondrial DNA; the Institute of Medicine report, however, recommended that its use be limited to having male children only, to avoid this 'modification' being transmitted. Yet this combination of nuclear and mitochondrial DNA might also have arisen by chance rather than design, naturally rather than via MRT. The same can be said about genome editing to introduce existing genetic sequence variants. In regulating technology, we should consider whether the focus should be on outcomes, or the actions (or inactions) leading to them – and why.

The difficulty of regulating the 'germline' or the 'human genome' highlights the problem of regulating static objects rather than the dynamic relations and practices through which these objects move. In regulating a 'thing' in itself, the law tends to fix and define it, thereby rendering it inflexible and unable to evolve to match developments in technology (see McMillan, Chapter 37 in this volume). Especially in the area of biomedicine, both the pace of research and the propensity of science to discover new and often unexpected means to its ends can result in overly specific legislative provisions becoming rapidly obsolete or inapplicable.

Examples of this can be seen in previous legislative attempts to define 'the human embryo' and 'cloning'. The original Human Fertilisation and Embryology Act (1990) s1(1)(a) defined an embryo as 'a live human embryo where fertilisation is complete'. The development of somatic cell nuclear transfer technology, the process by which Dolly the sheep was cloned, immediately rendered this definition problematic, since embryos produced via this technique do not undergo fertilisation at all. Addressing this legal lacuna necessitated the hurried passage of the Human Reproductive Cloning Act 2001,[21] before the eventual decision of the House of Lords brought 'Dolly'-style embryos back within the Act's purview.

A similar situation occurred in Australia, before the passage of uniform federal laws: in the state of Victoria, the embryo was defined as 'any stage of human embryonic development at and from syngamy',[22] leaving unclear the status of embryos produced via nuclear transfer, in which syngamy never takes place. In South Australia, meanwhile, the law prohibited cloning, but defined 'cloning' specifically as referring to embryo splitting, again leaving nuclear transfer embryos unregulated.

These examples illustrate the pitfalls of over-determining the objects of regulation. Attempts to regulate HGGM, though, may suffer not only from being too specific in defining their objects, but also from being too vague. For example, references to 'eugenic practices' in national and

[19] UNESCO, 'Universal Declaration on the Human Genome and Human Rights', (1998).
[20] See I. de Miguel Beriain, 'Should Human Germ Line Editing Be Allowed? Some Suggestions on the Basis of the Existing Regulatory Framework', (2019) *Bioethics*, 33(1), 105–111.
[21] D. Morgan and M. Ford, 'Cell Phoney: Human Cloning after Quintavalle', (2004) *Journal of Medical Ethics*, 30(6), 524–526.
[22] Infertility Treatment Act 1995 (Vic), s3(1).

international legal and policy instruments leave open the question of what actually constitutes a eugenic practice (see Isasi, Chapter 34 in this volume). Without any *processes* in place to determine how such terms should be interpreted, their inclusion tends to obfuscate rather than clarify the scope of regulation.

Similar examples abound: the EU Clinical Trials Directive declared: 'No gene therapy trials may be carried out which result in modifications to the subject's germ line genetic identity',[23] a position further affirmed by the replacement Clinical Trials Regulation.[24] But what exactly is 'germ line genetic identity'? UNESCO's Declaration opposes 'practices that could be contrary to human dignity, such as germ-line interventions',[25] but does not indicate how or why germline interventions 'could be contrary to dignity' – making it difficult to distinguish whether they actually are.

The requirements of being neither too specific nor too vague may seem a Goldilocks-style demand with respect to defining the appropriate targets of regulation. What this illustrates, however, is that regulation is important for the *processes* and *practices* it establishes, as much as the definitions of objects to which these pertain.

35.4 RESEARCH OR REPRODUCTION? THE IMPORTANCE OF CONTEXT

In regulating HGGM, our concern should be, not whether a modification is in principle heritable but whether it is in fact inherited. The context, both social and scientific, in which the modification procedure is carried out therefore matters a great deal. Attempting to regulate this solely in terms of permitting or prohibiting particular technologies would be extremely limiting.

Instead, we should consider how our concerns can be addressed by regulating practices with respect to assisted reproduction; and relations between healthcare practitioners, healthcare systems, patients, research participants and the market. Such practices and relations are key to the processes by which future generations, and our relationships with them, are created. Regulation of this sort can be effective at transnational as well as intra-national level: cross-border surrogacy is another situation where particular practices and relations, not just the technology itself, create ethical concerns – India's regulatory response represents an example of correlative attempts to address them.

Focusing our regulatory attention on processes and relations also allows us better to distinguish desirable versus undesirable *contexts* for the application of technology. Basic research on embryos never destined for implantation is very different to the creation of genetically modified human beings; regulation ought accordingly to enable us clearly to separate these possibilities. This might be done in various ways, as can be seen by considering UK and USA examples.

The UK's Human Fertilisation and Embryology Act to some extent regulates embryos relationally and in terms of practices: what may be done to or with an embryo depends on the relationships among actors connected to the embryo, their relationships with the embryo itself, and the embryo's own relational context, in terms of its origin, ontology, and history. By creating the category of 'permitted embryos' as the only embryos that may be implanted, the Act effectively separates reproductive use from other applications.[26]

[23] Clinical Trials Directive, 2001/20/EC, Article 9(6).
[24] Clinical Trials Regulation, 536/2014.
[25] UNESCO, 'Universal Declaration on the Human Genome and Human Rights', (1998).
[26] The definition of 'permitted embryo' requires that 'no nuclear or mitochondrial DNA of any cell of the embryo has been altered' (see Human Fertilisation and Embryology Act 2008, S. 3ZA(4)(b)), which *prima facie* prevents

In comparison, US federal regulation affecting HGGM incorporates aspects of both object-focused and contextual regulation. Laws prohibiting federal funding of human embryo research apply to research with any and all embryos, regardless of origin, context, or intended destination.[27] When it comes to genetically modifying those embryos, however, context matters: via the FDA's jurisdiction, current laws effectively prevent any clinical applications involving modified embryos,[28] while basic research falls outside this domain.

Looking ahead to the possible futures of HGGM, what sorts of purposes and processes might we be concerned to regulate? Many of the worries that have been expressed over HGGM can be addressed via regulation (in the broad sense) of processes across different contexts. For example, the dystopian vision of a society in which parents visit a 'baby supermarket' to choose their perfect designer child is quite different to one in which the healthcare system permits parents to access reproductive interventions that have been accepted as safe (enough) for particular purposes within defined contexts. This being the case, it is far from evident that our response to these possible futures should be to forgo exploring the potential benefits of gene therapy for fear of 'designer babies': the two possibilities may be mediated via the same technologies but involve very different contexts, relationships, roles and practices. These can be differently regulated; and regulation in turn can shape which practices emerge and how they evolve.

One possible regulatory position, often motivated by a 'slippery slope' argument, is that we should prohibit all embryo genome editing research, in order to avoid the extreme dystopian futures it might one day enable. As argued above, however, context is key and focusing on technology alone fails to take account of this. Restricting research today in order to prevent one of the distant possible futures it might enable also forecloses any beneficial outcomes it might produce. To prohibit something that is *prima facie* acceptable merely because it may make possible the unacceptable is drawing the line in the wrong place.

Although concerns over technological development often invoke the 'slippery slope', this metaphor ignores the fact that science is not a single, uni-directional process with a defined endpoint. Instead, research and the applications it might enable are more like a 'garden of forking paths',[29] a labyrinth of infinite possibilities. Regulatory slippery-slopeism, for fear of one of those possibilities, would foreclose the remainder.

That said, it might be true that what seems unacceptable from our present perspective will, from halfway down the slope, be less so. Studies of public opinions show greater acceptance of novel genetic technologies among younger demographics who have grown up in the age of IVF and genetic screening; and it was suggested with respect to MRT that this might be a slippery slope to other forms of HGGM such as genome editing[30] – though this would be difficult to prove with certainty.

---

implantation of genetically modified embryos. MRT is rendered legal via specific provision for regulations to include within the 'permitted category' embryos that have undergone 'a prescribed process designed to prevent the transmission of serious mitochondrial disease' (see S. 3ZA(5)). This provision was implemented in the Human Fertilisation and Embryology (Mitochondrial Donation) Regulations 2015.

[27] Note, however, that this does not constitute a ban on embryo research across the board, only on federal funding.

[28] I. G. Cohen and E. Y. Adashi, 'The FDA Is Prohibited from Going Germline', (2016) *Science*, 353(6299), 545–546.

[29] S. Chan and M.-d-J. Medina Arellano, 'Genome Editing and International Regulatory Challenges: Lessons from Mexico', (2016) *Ethics, Medicine and Public Health*, 2(3), 426–434; S. Chan, 'Embryo Gene Editing: Ethics and Regulation', in K. Appasani (eds), *Genome Editing and Engineering: From TALENs, ZFNs and CRISPRs to Molecular Surgery* (Cambridge: Cambridge University Press, 2018), pp. 454–463.

[30] T. Ishii, 'Potential Impact of Human Mitochondrial Replacement on Global Policy Regarding Germline Gene Modification', (2014) *Reprod Biomed Online*, 29(2), 150–155; F. Baylis, 'Human Nuclear Genome Transfer (So-Called Mitochondrial Replacement): Clearing the Underbrush', (2017) *Bioethics*, 31(1), 7–19.

The response to slippery-slope fears is often to try to draw a 'bright line'. Any lines we might draw, however, are liable to suffer from the above-mentioned problem of either over-determination or vagueness. Some distinctions themselves may be less clear. For example, a commonly held position in relation to genome editing is that it should be used only for therapy, not for enhancement; but as much bioethical scholarship has revealed, the line between therapy and enhancement is not so easily defined. In a similar way, however, the definition of 'serious' disability, illness, or medical condition, for which the HFEA permits pre-implantation embryo testing, is subject to interpretation; yet its provisions have, nonetheless, been effective because there is a regulatory process for legitimate decision-making in the case of ambiguity.

Moreover, as the scientific and ethical landscape shifts, bright lines may eventually become grey areas: for example, the fourteen-day rule on embryo research was a regulatory (if not ethical) bright line for decades yet is now once again the subject of discussion(see McMillan, Chapter 37 in this volume). We should not assume that we are currently at the pinnacle of ethical understanding such that the only way is down: what we now perceive as 'slipping' might in future generations be understood as moral progress. Regulation on the slippery slope might sometimes involve drawing lines, but these should be seen as pragmatic necessities, not moral absolutes.

One supposed 'bright line' in HGGM that may not prove so clear is the somatic / germline distinction: is this really as legible or significant as it has been made out to be? Publics might not think so: recent engagement initiatives have shown widespread acceptance for therapeutically oriented genome editing, including heritable HGGM,[31] while in the wake of He's attempt at creating genetically modified babies, crossing this supposed ethical Rubicon, the projected public backlash does not seem to have manifested. Moreover, in considering the possible consequences and balance of risks involved in somatic versus germline modification, we might argue that the two are not as dissimilar as might be assumed.[32] Neither then in regulation should the germline be assumed to be a bright line in perpetuity: as for the fourteen-day rule, its importance as a line lies in the processes invoked when considering whether to cross it.[33]

35.5 REGULATION, RESPONSIBILITY AND COOPERATIVE PRACTICE

Given the above, what is the justification for regulating HGGM? Clearly, it is not absolute protection of the germline or genome itself: nothing stops someone visiting a plutonium refinery and exposing their germ cells to ionising radiation, or wearing too-tight underwear, and then subsequently engaging in reproduction via natural means, even though both of these processes are likely to result in heritable genetic changes. Nor would we consider it appropriate to attempt to regulate such activities.[34] What, then, is regulation doing here?

[31] A. Van Mil et al., 'Potential Uses for Genetic Technologies: Dialogue and Engagement Research Conducted on Behalf of the Royal Society', (Hopkins van Mil, 2017).

[32] S. Chan, 'Playing It Safe? Precaution, Risk, and Responsibility in Human Genome Editing', (2020) *Perspectives in Biology and Medicine*, 63(1), 111–125.

[33] McMillan, Chapter 37 in this volume ; G. Cavaliere, 'A 14-Day Limit for Bioethics: The Debate over Human Embryo Research', (2017) *BMC Medical Ethics*, 18(1), 38; S. Chan, 'How to Rethink the Fourteen-Day Rule', (2017) *Hastings Center Report*, 47(3), 5–6; S. Chan, 'How and Why to Replace the 14-Day Rule', (2018) *Current Stem Cell Reports*, 4 (3), 228–234.

[34] The principle of procreative autonomy, or reproductive liberty, is well-established ethically: see for example J. A. Robertson, *Children of Choice: Freedom and the New Reproductive Technologies* (Princeton University Press, 1994); R. Dworkin, *Life's Dominion: An Argument about Abortion, Euthanasia and Individual Freedom* (New York: Vintage, 1993).

It may seem peculiar to allow reckless random genetic modification by individuals while the much more controlled deliberate use of directed technology should be prohibited. But the aim of regulation is not simply, or not only, to prevent certain factual outcomes. A shift in the language points us to what is at stake here: before the era of genome editing, HGGM was considered 'too risky'; now, instead, 'it would be irresponsible'.

The question then becomes what 'responsibility' requires and how should it be enacted. This highlights an important role of regulation in relation to risk, specifically in determining how we understand risk and responsibility when something goes wrong. Regulatory responsibility is not about assigning blame to individual actors, be they scientists or clinicians, but instead deciding as a society how much and what kind of risk we are collectively willing to take responsibility for. That is, in regulating to permit something, we are implicitly accepting a certain degree of accountability for the practice and for its consequences. Even as scientific responsibility has been theorised in ways that go beyond the individual scientist to the collective community,[35] wider social responsibility for science requires a consideration of the interplay between social norms, regulation, and scientific practice.

Regulation therefore can also, and should, facilitate cooperative practices among different actors. At the Second International Summit on Human Genome Editing, David Baltimore described He Jiankui's attempts to create genome-edited children despite all scientific and ethical advice to the contrary as representing 'a failure of self-regulation'.[36] But this is only necessarily true if we understand the primary purpose of regulation as being the absolute prevention of particular outcomes. Moreover, for 'the scientific community' to assume all of the blame for failing adequately to police its members ignores the function of states and the existence of state regulation, while arrogating what is arguably a disproportionate level of self-governance to scientists.

In fact, as Chinese bioethicists and legal scholars were quick to assert,[37] there were various existing regulatory instruments that were breached by He's work. Genome-edited babies may have been created, but the real test of regulation is what happens next: how regulators and policy-makers (broadly categorised) respond, to this case specifically and in terms of regulating HGGM more generally, will determine whether regulation can be judged to have succeeded or failed. Notably, the imposition of a prison sentence for He[38] signals that, although scientific convention and existing oversight mechanisms may not have been sufficient deterrent beforehand, the criminal law was, nonetheless, capable of administering appropriate *post hoc* judgment. While the criminal law in regulating science may serve a partly symbolic function in assuring social licence for morally contentious research,[39] its value in this role must be backed by a willingness to invoke its 'teeth' when needed: He's punishment aptly demonstrates this.

[35] H.-J. Ehni, 'Dual Use and the Ethical Responsibility of Scientists', (2008) *Archivum Immunologiae et Therapiae Experimentalis*, 56(3), 147–152; H. Jonas, *The Imperative of Responsibility* (Chicago University Press, 1984). Further analysis is warranted of which collective responsibilities, particularly with respect to complicity, might have been at stake in the He case.

[36] National Academies of Sciences, Engineering, and Medicine, 'Second International Summit on Human Genome Editing: Continuing the Global Discussion: Proceedings of a Workshop – in Brief', (Washington, DC: N. A. P. (US), 2019).

[37] X. Zhai et al., 'Chinese Bioethicists Respond to the Case of He Jiankui', (Hastings Bioethics Forum, 7 February 2019), www.thehastingscenter.org/chinese-bioethicists-respond-case-jiankui/.

[38] D. Cyranoski, 'What CRISPR-Baby Prison Sentences Mean for Research', (2020) *Nature*, 577(7789), 154–155.

[39] M. Brazier, 'Regulating the Reproduction Business?', (1999) *Medical Law Review*, 7(2), 166–193; A. Alghrani and S. Chan, 'Scientists in the Dock: Criminal Law and the Regulation of Science,' in A. Alghrani et al. (eds), *The Criminal Law and Bioethical Conflict: Walking the Tightrope* Cambridge: Cambridge University Press, 2013), pp. 121–139.

As this case illustrates, regulation is not just about absolute prevention, but involves mediating a complex set of relationships. Rather than viewing scientific self-regulation as a law unto itself and a separate domain, we should consider how scientists can effectively contribute to the broad project of regulation, understood as a combination of law and policy, process and practice, at multiple levels from individual to community, local to global.

35.6 GLOBAL REGULATION AND SCIENTIFIC JUSTICE

A common theme in discussions of the regulation of HGGM is that decisions about these technologies need to involve global participation. In order to determine what 'global participation' ought to consist of, it is worth considering *why* a global approach is appropriate.

Some have suggested a global approach is required because in affecting the human genome, HGGM affects all of us. George Annas and colleagues, for example, write that 'a decision to alter a fundamental characteristic in the definition of human should not be made ... without wide discussion among all members of the affected population ... Altering the human species is an issue that directly concerns all of us'.[40] Yet the sum of the reproductive choices being made by millions of individual humans in relation to 'natural reproduction' is vastly greater than the potential effect of what will be, in the short term at least, a tiny proportion of parents seeking to use genome editing or MRTs.

In fact, many areas of science and policy will have far-reaching consequences for humanity, some probably much more so and more immediately than HGGM. Environmental policy, for example, and the development of renewable energy sources are likely to have far greater impact on the survival and future of our species, and affect far more people now and in the future, than HGGM at the scale it is likely initially to be introduced.

Doom-laden predictions overstating the possible consequences of HGGM for 'the human race' or 'our species as a whole' tend to demand precaution, in the sense of a presumption against action, as a global approach. Such calls to action have rhetorical force and appeal to emotion, but rest on shaky premises. Overblown claims that in altering the genome we are somehow interfering with the fundamental nature of humanity are a form of genetic essentialism in themselves, implying that what makes us worthy of respect as persons and what should unite us as a moral community is nothing but base (literally!) biology.

Nevertheless, the political history of genetics has reified the moral and metaphorical power of the 'germline' concept, as something quintessential and common to all humans. This history has seen heredity, 'the germline', and 'the genome' used as a tool both for division and for unification, from eugenics to the Human Genome Project,[41] imbuing genetics with a significance well beyond the mere scientific. While the biological genome as the 'heritage of humanity' and the basis of human dignity[42] does not stand up well to analysis, the *political* genome, as object of multiple successive sociotechnical imaginaries, has acquired tremendous power as a regulatory fulcrum.

It is therefore genome alteration as a social and political practice, not its direct biological consequences, that we should be concerned to regulate. Beyond just requiring a global approach, this creates an *opportunity* to develop one. HGGM represents a socio-techno-

---

[40] G. J. Annas et al., 'Protecting the Endangered Human: Toward an International Treaty Prohibiting Cloning and Inheritable Alterations', (2002) *American Journal of Law and Medicine*, 28(2–3), 151–178.

[41] M. Meloni, *Political Biology* (London: Palgrave Macmillan, 2016).

[42] UNESCO, 'Universal Declaration on the Human Genome and Human Rights', (1998), Art. 1.

regulatory 'event horizon', the significance of which has been contributed to by the long historical association of genetics with politics, and which aligns with a broader trend towards engaging publics in discourse over science with the aim of democratising its governance. The immediate consequences of HGGM for human reproduction are likely to be fairly small-scale, and while opening up the possibility of clinical applications of genome editing will no doubt influence the direction of the field, as MRTs are already doing for related technologies, human genome editing is just one area of the vast landscape of scientific endeavour. Yet, in providing both opportunity and momentum to produce new approaches to global regulation, HGGM may have much wider implications for the broader enterprise of science as a whole.

An important feature of any attempt to develop a global approach to regulation is that it should account for and be responsive to transnational dynamics. This requires attention to equity in terms of scientific and regulatory capacity, as well as the ability to participate in and develop ethical and social discourses over science. When it comes to emerging and contested technologies such as embryo research, cell and gene therapies, and now genome editing, countries with more advanced scientific capacity have tended also to lead in developing regulation, and to dominate ethical discussion. The resulting global regulatory 'patchwork' creates the possibility of scientific tourism, which in turn combines with uneven power over regulatory and ethical discourse to reproduce and increase global scientific inequities. This can be seen, for example, in the consequences of Zhang's Mexican MRT tourism and its effects on global scientific justice.[43]

Another feature of the variegated regulatory landscape for controversial technologies is concern, among countries with high scientific capacity but restrictive regulation, about remaining internationally competitive, when researchers in other countries may take advantage of lower regulatory thresholds to forge ahead. This was a prominent factor in the embryo research debates of the early 2000s. Examining the expressions of concern with respect to human embryo genome editing research in China, about 'the science … going forward before there's been the general consensus after deliberation that such an approach is medically warranted',[44] versus in the UK being described 'an important first … a strong precedent for allowing this type of research',[45] it seems that international dynamics and 'keeping pace' may also be a consideration here. Dominant actors may seek to control this pace to their advantage by re-asserting ethical and regulatory superiority, in the process reinforcing existing hegemonies.

The significance of the present 'regulatory moment' with respect to HGGM is that it offers opportunities to disrupt and re-evaluate these hegemonies, across geographic, cultural, disciplinary, political, and epistemic boundaries. This should include critical attention to the internalised narratives of science: in particular, the problem of characterising science as a competitive activity. The race to be first, the scientific 'cult of personality' and narratives of scientific heroism (or in the case of He Jiankui, anti-heroism) may serve to valorise and promote scientific achievement, but also drive secrecy and create incentives for 'rogue science'. What alternative narratives might we develop, to chart a better course?

---

[43] S. Chan et al., 'Mitochondrial Replacement Techniques, Scientific Tourism, and the Global Politics of Science', (2017) *Hastings Center Report*, 47(5), 7–9.

[44] E. Callaway, 'Second Chinese Team Reports Gene Editing in Human Embryos', (2016) *Nature*, doi:10.1038/nature.2016.19718.

[45] E. Callaway, 'Embryo Editing Gets Green Light', (2016) *Nature*, 530(7588), 18.

35.7 CONCLUSION: WHERE NEXT FOR GLOBAL GERMLINE REGULATION?

Seeking a new paradigm for global health research regulation will require a conscious effort to be more inclusive. We need to examine what constitutes effective engagement in a global context and how to achieve this, across a plurality of cultural and political backgrounds, varying levels of scientific capacity and science capital, and different existing regulations.

Consider, for example, the contrasts that might emerge between the UK and China, where expectations over discourse, governance and participation differ from those in which UK public engagement has been theorised and developed.[46] Distinct challenges will arise for engagement in Latin America, where the politics of gender and reproduction overtly drive regulation, and where embryo research and reproductive technologies are heavily contested. The discourse is not necessarily uniform among all countries: discussing embryo genome editing is more challenging where IVF itself is still controversial. In approaching these issues, we need also to be aware of potential negative impacts the discussion may have, for example on women's access to reproductive health services.

Furthermore, we need to engage not just with publics and not just 'the scientific community' but with scientific *communities*. As we recognise in the field of engagement that there is not just a single unitary Public but a wide range of publics with different perspectives, values and beliefs, we need also to acknowledge pluralism of values, practices and motivations among scientists. In thinking about the governance of science, we must consider what factors might influence these and how, as an indirect form of regulating research. Some attention has already been given to the potential of actors such as journal publishers and funders to shape scientific culture and influence behaviour; further research might more clearly delineate these evolving regulatory roles, their limitations and how they work in tandem with 'harder' forms of regulation such as criminal law.

At the time of writing, the various proposed approaches to global governance of human genome editing have yet to coalesce into a single solution. The He incident triggered renewed calls for a moratorium.[47] While the scientific academies behind the International Summits were already considering aspects of regulation, the process was probably likewise hastened by these events, resulting in the formation of an International Commission; the WHO have launched their own inquiry;[48]; and numerous statements have been published over the past five years,[49] with many initiatives ongoing and proposals issued.

It seems clear that a moratorium is unlikely to emerge as the answer, despite reactions to He's transgression. In the first place, it is far from clear that a moratorium would have prevented He's experiment: the consensus of scientific communities publicly expressed was already that it should not be done. A moratorium without enforcement mechanisms would have been no more effective than the existing guidelines; and any symbolic value a moratorium would have would be rapidly eroded if it were not respected. Other proposals include, as per the WHO, a registry to promote greater information and transparency and facilitate the involvement of wider scientific players including funders and publishers; a global observatory[50] is another proposed

[46] J. Zhang, 'Comment: Transparency Is a Growth Industry', (2017) *Nature*, 545(7655), S65.

[47] E. S. Lander et al., 'Adopt a Moratorium on Heritable Genome Editing', (2019) *Nature*, 567(7747), 165–168.

[48] World Health Organisation, 'Global Health Ethics: Human Genome Editing', www.who.int/ethics/topics/human-genome-editing/en/.

[49] Brokowski, 'Do CRISPR Germline Ethics Statements Cut It?'.

[50] J. Benjamin Hurlbut et al., 'Building Capacity for a Global Genome Editing Observatory: Conceptual Challenges', (2018) *Trends in Biotechnology*, 36(7), 639–641; S. Jasanoff et al., 'Democratic Governance of Human Germline Genome Editing', (2019) *The CRISPR Journal*, 2(5), 266–271; K. Saha et al., 'Building Capacity for a Global Genome Editing Observatory: Institutional Design', (2018) *Trends in Biotechnology*, 36(8), 741–743.

mechanism to enable governance. With any of these, we will still need to attend to the dynamics of discourse, which interests are represented and how.

With that in mind, perhaps a single solution is not what we should be seeking. The proliferation of initiatives aimed at determining principles and frameworks for acceptable governance of HGGM may lead some to wonder whether we really need so many cooks for this broth, and to raise objections regarding potential inconsistencies when multiple bodies are charged with a similar task. Yet, even among the number of bodies that currently exist, the full range of diverse views has not been represented. The existence of multiple parallel discourses is not necessarily a bad thing: more can be better if it allows for broader representation. The meta-solution of integrating these is the challenge; approaching regulation as dynamic, constituted by practices and concerned with processes and relations, may be a way to meet it.

36

Cells, Animals and Human Subjects

Regulating Interspecies Biomedical Research

Amy Hinterberger and Sara Bea

36.1 INTRODUCTION

The availability of new cellular technologies, such as human induced pluripotent stem cells (iPSCs), has opened possibilities to significantly 'humanise' the biology of experimental and model organisms in laboratory settings. With greater quantities of genetic sequences being manipulated and advances in embryo and stem cell technologies, it is increasingly possible to replace animal tissues and cells with human tissues and cells. The resulting chimeric embryos and organisms are used to support basic research into human biology. According to some researchers, such chimeras might be used to grow functional human organs for transplant inside an animal like a pig. These types of interspecies biomedical research confound long-established regulatory and legal orders that have traditionally structured biomedicine. In contexts where human cells are inserted into animal embryos, or in the very early stages of animal development, regulators face a conundrum: they need to continue to uphold the differences in treatment and protections between humans and animals, but they also want to support research that is producing ever-more intimate entanglements between human and animal species.

In research terms, human beings fall into the regulatory order of human subjects protection, a field of law and regulation that combines elements of professional care with efforts to preserve individual autonomy.[1] Animals, however, belong to a much different regulatory order and set of provisions relating to animal welfare.[2] To this end, animals have been used, and continue to be used, for understanding and researching human physiology and disease where such experiments would be unethical in humans. Researchers can do things to animals, and use animal cells, tissues and embryos in ways that are very different from human cells, tissues and embryos. Traditionally, ethical concerns and political protection have focused on the human subject in biomedical research, with ensuing allowances to address animal welfare and human embryos. However, such divisions are now under immense strain and are undergoing substantial revision. This chapter investigates these transformations in the area of interspecies mammalian chimera. We ask: what forms of regulation and law are drawn on to maintain boundaries between human

---

[1] For a current discussion of human subject regulation see: I. G. Cohen and H. F. Lynch (eds), *Human Subjects Research Regulation: Perspectives on the Future* (Cambridge, MA: MIT Press, 2014).

[2] For a current discussion of animal welfare regulation see: G. Davies et al., 'Science, Culture, and Care in Laboratory Animal Research: Interdisciplinary Perspectives on the History and Future of the 3Rs', (2018) *Science, Technology, & Human Values*, 43(4), 603–621.

research subjects and experimental animals in interspecies research? What kinds of reasoning are explicitly and implicitly used? What kinds of expertise are invoked and legitimised?

We will begin with a brief overview of the chimeric organisms in the context of new cellular technologies. We will then explore significant national moments and debates in the UK and USA that highlight the tacit presumptions of regulatory institutions to explore where disagreement and contestation have arisen and how resolutions were reached to accommodate interspecies chimeras within the existing regulatory landscape. Through these national snapshots, the chapter will explore how human–animal chimeras become objects of regulatory controversy and agreement depending on the concepts, tools and materials used to make them. The final sections of the chapter provide some reflections on the future of chimera-based research for human health that, as we argue, calls forth a reassessment of regulatory boundaries between human subjects and experimental animals. We argue that interspecies research poses pressing questions for the regulatory structures of biomedicine, especially health research regulation systems' capacity to simultaneously care for and realign the human and animal vulnerabilities at stake within interspecies chimera research and therapeutic applications.

36.2 FROM DISH TO ANIMAL HOST

Chimeric organisms, containing both human and non-human animal cells, sit at the interface between different regulatory orders. The 'ethical choreography' that characterises health research regulation on interspecies mixtures is densely populated with human and animal embryos, pluripotent stem cells, human subjects and experimental animals.[3] Much depends on the types of human cells being used, the species of the host animal that will receive the cells, along with age of the animal and the region where human cells are being delivered. Regulation thus includes institutional review board approval for using human cells from living human subjects. There also needs to be approval from animal care and use committees that assess animal welfare issues. Depending on the country, there might also be review from a stem cell oversight committee, which must deliberate on whether the insertion of human cells into an animal may give it 'human contributions'.[4] There are significant national differences in regulatory regimes, making for diverse legal and regulatory environments at both national and international levels because countries regulate human and animal embryos, stem cells and animal welfare very differently.[5]

In the biological sciences, the term chimera is a technical term, but it does not necessarily refer to one specific entity or process. Generally speaking, chimeras are formed by mixing together whole cells originating from different organisms.[6] It is a polyvalent term and can refer to entities resulting from both natural and engineered processes.[7] Historians of science have explored how species divides, especially between humans and other animals, are culturally

---

[3] C. Thompson, *Good Science: The Ethical Choreography of Stem Cell Research* (Cambridge, MA: MIT Press, 2013).

[4] As discussed below, the term 'human contributions' is used in the NAS Guidelines for stem cell research oversight.

[5] See, for example, I. Geesink et al., 'Stem Cell Stories 1998–2008', (2008) *Science as Culture* 17(1), 1–11; L. F. Hogle. 'Characterizing Human Embryonic Stem Cells: Biological and Social Markers of Identity', (2010) *Medical Anthropology Quarterly*, 24(4), 433–450.

[6] For a history of the use of the term chimera in developmental biology and stem cell science see: A. Hinterberger, 'Marked 'H' for Human: Chimeric Life and the Politics of the Human', (2018) *BioSocieties*, 13(2), 453–469.

[7] On natural chimerism see: A. Martin, 'Ray Owen and the History of Naturally Acquired Chimerism', (2015) *Chimerism*, 6(1–2), 2–7.

produced and historically situated both inside and outside the laboratory.[8] The regulatory practices we explore in this chapter are not separate, but rather embedded in these larger structures of cultural norms about differences – and similarities – between humans and animals. As the life sciences continue to create new types of organisms, there are currently many groups and regulatory actors in different countries involved in producing definitions and forms of regulation for new human-animal mixtures. As we will see below, it is precisely the debates over the naming and classification of these new entities where the regulatory boundary work between the human and animal categories is illuminated.

36.3 ANIMALS CONTAINING HUMAN MATERIAL

In the following two sections, we will explore national snapshots from advisory and regulatory bodies in the UK and USA. We will examine how they are confronting issues of responsibility and jurisdiction for boundary crossing entities that cannot easily be siphoned into the traditional legal and regulatory orders of either human or animal. We will show that while each country's response via report or guidelines focuses on the human and animal division as primary to maintain in research practices, they each provide different solutions to the problems raised by interspecies mammalian chimera. These two sections of the chapter thus illuminate how interspecies chimera confound long-standing regulatory divisions in health research that challenge the law's capacity to simply encompass new entities.

In 2011, the UK's Academy of Medical Sciences released what is regarded as the first comprehensive recommendations to regulate the creation and use of chimeric organisms, called *Animals Containing Human Material*. The central conclusion of the Academy's report is that research that uses animals containing human material is likely to advance basic biology and medicine without compromising ethical boundaries. The report itself was part of a much longer history of deliberation around the status of the human embryo in the UK where specific forms of human–animal mixtures have been proposed, debated and, in the end, legislated. The UK regulatory landscape is significant in this respect as no other nation has written into law human–animal mixtures – which in UK law are called 'human admixed embryos'. The term human admixed embryo was introduced in 2008 amendments to the Human Fertilisation and Embryology Act 1990 (HFE Act). While it was the 'cybrid embryo' debate that became the most controversial and well-known related to these new legal entities, the legislation outlines a number of different kinds of human and animal mixtures that fall under its remit, including chimeric embryos. According to the Act, a human admixed embryo is any embryo that 'contains both nuclear or mitochondrial DNA of a human and nuclear or mitochondrial DNA of an animal but in which the animal DNA *does not predominate*'.[9]

A 2008 debate in the House of Lords over the revised HFE Act and the term 'human admixed' highlights the classification conundrums of how boundaries between human and animal are drawn. The Parliamentary Under-Secretary of State for the Department of Health explained, regarding the term 'human admixed embryo': 'It was felt that the word "human" should be used to indicate that these entities are at the human end of the spectrum of this research'.[10] Responding to this notion of the spectrum, the Archbishop of Canterbury responded that:

[8] For a recent account see: N. C. Nelson, 'Modeling Mouse, Human, and Discipline: Epistemic Scaffolds in Animal Behavior Genetics', (2013) *Social Studies of Science*, 43(1), 3–29.

[9] Human Fertilisation and Embryology Act 2008 (emphasis added).

[10] UK House of Lords debate, 15 January 2008, Column 1183.

'the human end of the spectrum' seems to introduce a very unhelpful element of uncertainty. Given that some of the major moral reservations around this Bill ... pivot upon the concern that this legislation is gradually but inexorably moving towards a more instrumental view of how we may treat human organisms, any lack of clarity in this area seems fatally compromising and ambiguous.[11]

This lack of clarity referred to by the Archbishop, which may 'be fatally compromising', sought to be addressed by the *Animals Containing Human Material* report.[12] Clarity, in this case, is provided by carefully considered boundaries and robust regulation, to remove elements of uncertainty. In the UK, the Human Fertilisation and Embryology Authority (HFEA) is the central body responsible for addressing proposals for embryo research in the UK. It is the body that licenses human embryonic stem cell research, oversees IVF treatment and the use of human embryos.

Violations of the licensing requirements of the HFEA are punishable under criminal law, which is both a literal and symbolic marker of respect for the conflicting and contested views on embryo research in the UK.[13] However, the HFEA only has jurisdiction over human embryos (not animal embryos). Research on animal embryos is governed and regulated by an entirely different body, The Home Office, which regulates the use of animals in scientific procedures through the Animals (Scientific Procedures) Act 1986 (ASPA).

Assessing whether the human or animal DNA is most predominant may be harder with chimeric research embryos since their cellular make-up may change over time. Thus, it can become unclear whether their regulation should fall within the remit of the HFEA or the Home Office. Any mixed embryo judged to be 'predominantly human' is regulated by HFEA and cannot be kept beyond the 14-day stage, whereas currently in the UK an animal embryo, or one judged to be predominantly animal, is unregulated until the mid-point of gestation and can in principle be kept indefinitely. Whether or not an admixed embryo is predominantly 'human' is, according to the Academy's report, an expert judgement. However, it recommended that the Home Office and HFEA, two government bodies that had not previously been connected, needed to work together to create an operational interface at the boundaries of their new areas of responsibility.

The Academy report purifies, both through language and regulatory approach, ambiguities raised by chimeric organisms by trying, as best as possible, to compartmentalise research into human or animal regulatory orders. The term 'animals containing human material' itself highlights this goal. According to the report, animals containing human material are animals first and foremost. In this respect, the report places the regulatory responsibility for these new chimeric entities squarely in the already regulated domain of animal research. To this end, the UK remains a highly regulated but permissive research environment for different types of chimera-based research, and is the only country to formerly write into law the protection of biological chimeras containing human and animal cells.

36.4 ASSESSING 'HUMAN CONTRIBUTIONS' TO EXPERIMENTAL ANIMALS

Unlike the UK, in the USA there is no formal legal regulation of interspecies chimera research. The 2005 National Academy of Science (NAS) Guidelines continue to be the cornerstone of

---

[11] Ibid.
[12] The Academy of Medical Sciences, 'Animals Containing Human Material', (2011).
[13] S. Franklin, 'Drawing the Line at Not-Fully-Human: What We Already Know', (2003) *The American Journal of Bioethics*, 3(3), 25–27.

scientific research involving embryos, stem cell biology and mammalian development. The Academy is not a governmental agency, nor does it have enforcement power but the guidelines are viewed to be binding by governmental and institutional authorities. The NAS guidance acts as the principal reference on the recommendations applicable to research using interspecies chimera involving human embryonic stem cells and other stem cell types.

Stem Cell Research Oversight (SCRO) committees are the localised bodies that put into action the NAS Guidelines. During the stem cell controversies that characterised the USA, the Academy recommended that all research involving the combination of human stem cells with non-human embryos, fetuses, or adult vertebrate animals must be submitted to not only the local Institutional Animal Care and Use Committee (IACUC) for review of animal welfare issues, but also to a Stem Cell Research Oversight (SCRO) Committee for consideration of the consequences of the 'human contributions' to any non-human animal.[14] Thus, SCRO committees need to meet to discuss any experiment where there is a possibility that human cells could contribute in a major organised way to the brain or reproductive capacities particularly.

In late September 2015, the National Institutes of Health (NIH) in the USA declared a moratorium on funding chimeric research where human stem cells are inserted into very early embryos from other animals. However, like other instances where federal research monies were removed from controversial research – e.g. human embryonic stem cell lines – such research continued, but with private monies. The moratorium was met with scepticism and criticism of researchers working in this domain who, in a letter to *Science*, argued that such a moratorium impeded the progress of regenerative medicine.[15] Following a consultation period in 2016, the NIH announced that it would replace the moratorium with a new kind of review for specific types of chimera research, including experiments where human stem cells are mixed with nonhuman vertebrate embryos and for studies that introduce human cells into the brains of mammals – except rodents, which will be exempt from extra review.

As the UK's predominant predicament demonstrated, currently it is difficult to predict how and where human cells will populate in another species – when cells are added at the embryonic or very early fetal stages of life. This predicament was recently characterised as the problem of 'off-target' humanised tissues in non-human animals.[16] Currently, animal embryos with human cells are only allowed to develop for a period of twenty-eight days – four weeks – in the USA. As we explained above, animal embryos fall under a separate legal and regulatory structure from human embryos, which traditionally have been allowed to develop for fourteen days, though this number is subject to increasing debate (see McMillan, Chapter 37 in this volume).[17] In practice, this means that assessments of human contributions to an animal embryo are restricted to counting human cells in an animal embryo. Current published research puts the human contribution to the host animal embryo at 0.01–.1.[18] This is primarily because a chimeric embryo is only allowed to gestate for twenty-eight days.

[14] National Academy of Sciences 'Final Report of The National Academies' Human Embryonic Stem Cell Research Advisory Committee and 2010 Amendments to The National Academies' Guidelines for Human Embryonic Stem Cell Research', (National Academies Press, 2010).

[15] A. Sharma et al., 'Lift NIH Restrictions on Chimera Research', (2015) *Science* 350(6261), 640.

[16] I. Hyun, 'What's Wrong with Human/Nonhuman Chimera Research?' (2016) *PLoS biology*, 14(8).

[17] See: B. Hurlbut, *Experiments in Democracy: Human Embryo Research and the Politics of Bioethics* (Columbia University Press, 2017); G. Cavaliere, 'A 14-day Limit for Bioethics: The Debate over Human Embryo Research', (2017) *BMC Medical Ethics*, 18(1) 38.

[18] T. Rashid et al., 'Revisiting the Flight of Icarus: Making Human Organs from PSCs with Large Animal Chimeras', (2014) *Cell Stem Cell*, 15(4), 406–409.

In 2017, privately funded researchers in the US published findings from the first human–pig embryos. While labs have previously created human–animal chimeras, such as mice transplanted with human cancer cells or immune systems or even brain cells, this new experiment was unique because the researchers placed human stem cells — which can grow to become any of the different types of cells in the human body — into animal embryos at their earliest stages of life. Broadly, the making of these human–pig chimeras included collecting pig zygotes (eggs), that were then fertilised *in vitro* to become blastocysts – a progressive phase in embryonic development. Human induced pluripotent stem cells were then pipetted into the developing pig embryo that had been genetically modified. That embryo was then put into a female pig and left to develop for twenty-eight days. After twenty-eight days the animal was sacrificed, and the entire reproductive tract of the animal removed and studied to see where the human cells developed and grew in the embryo.

This study, and others like it, raised ethical concerns relating to 'off-target' humanised tissue with concern for an animal's central nervous system (brain) and reproductive capacities. In the below excerpt, the study leader explains how these concerns of 'off-target' humanisation can be handled in the development of human-pig chimeric embryos:

> . . . we must pay special attention to three types – nerves, sperm and eggs – because humanizing these tissues in animals could give rise to creatures that no one wants to create . . . We can forestall that problem by deleting the genetic program for neural development from all human iPSCs before we inject them. Then, even if human stem cells managed to migrate to the embryonic niche responsible for growing the brain, they would be unable to develop further. The only neurons that could grow would be 100 percent pig.[19]

Scientists are developing a variety of techniques to ensure 'on target' organ complementation so that a fully human organ can be grown inside an animal, and to avoid any 'off-target' problems that could potentially confer human qualities to the non-human experimental animal.

Possible ethical breaches relating to human research subjects and chimeras have been intensely discussed by scholars; however, until recently, concerns over animal welfare have largely taken a back seat in the regulatory and ethical debates over interspecies chimera. When we turn from the regulation of stem cells to the regulation of the organism – or animal – a new set of concerns open. The overwhelming emphasis on avoiding risky humanisation by measuring and counting the number of human cells in a non-human animal can obfuscate the crucial discussion about how animal welfare staff members might monitor changes in behaviour and attributes of experimental chimeric animals. For example, bioethicist Insoo Hyun[20] has argued that people tend to assume the presence of human cells in an animal's brain might enhance it above its typical species functioning. This 'anthropocentric arrogance' is, he points out, completely unfounded.[21] Why, he asks 'should we assume that the presence of human neural matter in an otherwise nonhuman brain will end up improving the animal's moral and cognitive status?'[22] The much more likely outcome, he suggests, of neurological chimerism is not a cognitive humanisation of the animal but 'rather an increased chance of animal suffering and acute biological dysfunction and disequilibrium, if our experience with transgenic animals can be a guide'.[23]

[19] J. C. I. Belmonte, 'Human Organs from Animal Bodies', (2016) *Scientific American*, 315(5), 32–37, 36.
[20] Hyun, 'What's Wrong with Human/Nonhuman Chimera Research?'
[21] Ibid., 3.
[22] Ibid.
[23] Ibid., 4.

Animal care and use committees are less interested in cell counts and more interested in whether potential 'human contributions' may cause unnecessary pain and distress in an animal. Further, the question of how 'human contributions' might be measured in the behaviour of non-human animals is difficult and requires expert knowledge related to the species in question. If highly integrated chimeras are allowed to develop, the role of animal husbandry staff will be crucial in assessing and monitoring the behaviours and states of experimental animals – thus, animal behaviour and animal welfare knowledge may be a significant emerging component of measuring 'humanisation' in health research regulation.

36.5 SHIFTING REGULATORY BOUNDARIES BETWEEN CELLS, HUMAN SUBJECTS AND EXPERIMENTAL ANIMALS

As a domain of science that is continually reinventing and reconceiving the human body and its potentials, the futures of stem cell science and its regulation is not easy to predict or assess. However, it is in this context of ambiguity and change that we situate our discussion. First, theoretically and conceptually, chimera-based research has given rise to new living entities, from 'animals containing human material' to 'human contributions to other animals' that challenge assumed regulatory boundaries, rights, and protections provided for human subjects in contemporary societies. By tracing out how the categories human and animal are enacted in health research regulation we have shown that interspecies chimera requires a double-move on the part of regulators and researchers: animals must be kept animals, and humans must be kept humans. From this vantage point, we can see that interspecies chimera are not so marginal to health research regulation. The regulatory deliberations they elicit require re-examining the most basic and foundational structures of contemporary biomedical research – both human subjects research regulation, and animal care and welfare.

In health research regulation, animals are often defined in law; however, what constitutes or defines a human subject is generally not written down in law or legislation. What constitutes the human is, almost always, taken for granted or tacit regulatory knowledge. The national snapshots we examined here encompass Euro-American political and cultural contexts where regulatory containers, such as the human research subject, are shown to be potentially variable – or at least, drawn into question. For example, deliberations in the USA over what constitutes a 'human contribution' to another animal brings to light how the human subject is not a universal given, but a legal and regulatory designation that has the potential to be made and remade. Scientists, policy-makers and regulators approach the category human and animal differently across cellular and organismal levels, showing that these categories do not precede health research regulation but are actively co-produced within it.

Second, on the technical front, our review of current scientific practice shows how life scientists increasingly work according to the consensus that life is a continuum where species differences do not travel all the way down to the level of cells and tissues, thus destabilising assumed species differences and raising new questions about cell integration and containment across species. Third, politically, we are witnessing increasing agitation around both human and animal rights, in a context where bioscience is taking a significant role in the public sphere by not only informing debates about what life is, but also what life should be for.[24]

The stem cell techniques we have discussed above were first developed not with human materials but with animal. While dilemmas over the humanisation of other animals may appear

---

[24] S. Jasanoff, *Can Science Make Sense of Life?* (Cambridge, UK: John Wiley & Sons, 2019).

to be new, these technical possibilities only exist because of previous animal research, such as the creation of mice–rat chimeras. For example, rat embryonic stem cells were injected into a mouse blastocyst carrying a mutation that blocked the pancreas development of a mouse, resulting in mice with a pancreas entirely composed of rat cells. These rat–mouse chimeras developed into adult animals with a normal functional pancreas, demonstrating that xenogeneic organ complementation is achievable.[25] Recent media coverage of the first human–pig interspecies chimera can conceal from view these longer and less discussed histories of biological research. To come to grips with the regulatory dilemmas elicited by interspecies chimera then, we must be attentive to biomedical research itself and the many kinds of living organisms used to advance scientific knowledge and to develop therapeutic applications for human health problems.

As we have shown, the USA established new private committees where members must assess whether an experiment might give 'human contributions' to experimental animals. Whereas governance in the UK clearly defines and names new legal and regulatory categories such as 'human admixed embryo' or 'animals containing human material'. In contrast, the phrasing 'human contributions' in the USA is suggestive of more of a spectrum rather than new legal and regulatory containers for boundary-crossing biological objects, such as in the UK. Chimeric organisms embody new articulations about the plasticity of biology and the recognition that assumed species differences do not travel all the way down to the molecular level. Consequently, explicit deliberations for regulation and governing procedures are also pushed and pulled in new directions. This remodelling of boundaries in biological practice and state governance has consequences for humans and animals alike.

36.6 CONCLUSION: REALIGNING HUMAN AND ANIMAL VULNERABILITIES

With the advent of new and sophisticated forms of human and animal integration for the study of disease, drug development and generation of human organs for transplants, keeping the human separate from the animal, in regulation, becomes increasingly difficult. The disruptions posed by interspecies chimeras give rise to growing conundrums as disparate regulatory actors try to accommodate chimeric entities within existing health research regulation structures that enact a clear division between the human/animal opposition.

In Europe and North America, the regulation of therapeutically oriented biomedicine has historically been split into two vast and abstracted categories: human and animal. Numerous legal and regulatory processes work to disentangle human material, bodies and donors from organisms and parts categorised as animal. Regulators and policymakers thus find themselves in a tricky situation needing to sustain the regulatory and legal estrangement between humans and other animals, while facilitating basic and applied research on human health – such as the kind described above – that relies on the incorporation of human and animal material in new biological entities.

Our explorations above suggest that health research regulation will need to be sufficiently reflexive on the limits of boundaries that reify the foundational human/animal division and be flexible enough to allow a re-consideration of classificatory tools and instruments to measure the extent and consequences of prospective interspecies chimera research. If human/animal chimeras provide to be an efficient route to engineering human organs, as opposed to genetically

[25] T. Kobayashi et al., 'Generation of Rat Pancreas in Mouse by Interspecific Blastocyst Injection or Pluripotent Stem Cells', (2010) *Cell*, 142(5), 787–799.

modified pigs or organoids,[26] then the humanisation of experimental animals will likely develop further. An ethical and effective health research regulation system will need to be simultaneously reactive and protective of both human and animal vulnerabilities at stake.

In practice, this implies that regulatory efforts could be directed at fostering and maintaining dynamic collaborative relationships between regulatory actors that often work separately, such as stem cell research oversight, human subjects and animal care and use committees. Establishing efficient communicative pathways across disparate regulatory authorities and institutional bodies will demand a mutual disposition to consider and incorporate divergent and emerging concerns. The collaborative relationships should also be invested in the development of novel regulatory tools capable of addressing the present and coming challenges raised both at the level of the cell and at the level of the organism by interspecies research. This means going beyond the existing instruments to measure 'human contributions' at the cellular level to monitor 'on target' human organ generation as well as 'off-target' proliferation of human tissue in experimental animals. Collaboration between regulatory actors that have traditionally operated separately would also need to integrate the knowledge and expertise from animal behaviour and welfare professionals, such as animal husbandry staff.

A learning health research regulation system that operationalises the multi-level collaborative relationships across regulatory actors complemented by the introduction of animal care experts would be better prepared to engage in the disruptions that interspecies chimera research poses to existing regulatory mechanisms, actors, relations and tools. The direction and increased traction of stem cell biotechnologies clearly signposts that the development and growth of human health applications of interspecies chimera research requires a gradual intensification of entanglements between animals and humans. Health research regulation will thus need to reflect on the ethical and practical consequences for experimental animals' and human research subjects' vulnerabilities and address the shifting boundary between experimental animals and human subjects in biomedicine to make room for the new life forms in the making.

[26] S. Camporesi, 'Crispr Pigs, Pigoons and the Future of Organ Transplantation: An Ethical Investigation of the Creation of Crispr-Engineered Humanised Organs in Pigs', (2018) *Etica & Politica/Ethics & Politics*, 20(3), 35–52. Latest predictions are that a combination between genetically modified pigs and interspecies chimera organogenesis could deliver regenerative medicine solutions for transplantation, see F. Suchy and H. Nakauchi, 'Interspecies Chimeras', (2018) *Current Opinion in Genetics & Development*, 52, 36–41.

37

When Is Human?

Rethinking the Fourteen-Day Rule

Catriona McMillan

37.1 INTRODUCTION

The processual, rapidly changing nature of the early stages of human life has provided recurring challenges for the way in which we legally justify the use of embryos *in vitro* for reproduction and research. When the latter was regulated under the Human Fertilisation and Embryology Act 1990 (as amended) ('the HFE Act'), not only did regulators attempt to navigate what we should or should not do at the margins of human life, but they also tried to navigate the various thresholds that occur in embryonic and research processes. In doing so, the response of law-makers was to provide clear-cut boundaries, the most well-known of these being the fourteen-day rule.[1]

This chapter offers an examination of this rule as a contemporary example of an existing mechanism in health research that is being pushed to its scientific limits. This steadfast legal boundary, faced by a relatively novel challenge,[2] requires reflection on appropriate regulatory responses to embryo research, including the revisitation of ethical concerns, and an *examination* of the acceptability of carrying out research on embryos for longer than fourteen days. The discussion below does not challenge the fourteen-day rule, or research and reproductive practices *in vitro* more generally *per se*, but rather explores the ways in which law could engage with embryonic (and legal) processes through attention to thresholds (as a key facet of these processes). This framing has the *potential* to justify extension, but not without proper public deliberation, and sound scientific and ethical basis. The *deliberation* and *revisitation* – not necessarily the revision – of the law is the key part to this liminal analysis.

To begin, this chapter gives an overview of how the fourteen-day rule came into fruition, before going on to summarise the research, published in early 2017, that has given rise to new discussions about the appropriateness of the rule, twenty-seven years after it first came into force. Thereafter, the rest of the chapter builds on the theme of 'processes' from Part I of this volume, and asks, briefly: what might we gain from thinking beyond boundaries in this context? Moreover, what might doing so add to contemporary ethical, legal and scientific discourse about research on human embryos? I argue that recognising the inherent link between processes and the regulation of the margins of human life, enables us to ask more nuanced questions

---

[1] Human Fertilisation and Embryology Act 1990 (as amended), s3(4).
[2] A. Deglincerti, et al., 'Self-organization of the In Vitro Attached Human Embryo', (2016) *Nature*, 533(7602), 251; M. Shahbazi et al. 'Self-organization of the Human Embryo in the Absence of Maternal Tissues', (2016) *Nature Cell Biology*, 18(6), 700–708.

about what we want for future frameworks, for example, 'when is human?',[3] one that legal discussion often shies away from. Instead I will argue that viewing regulation of embryo research as an instance of both processual regulation and regulating for process has the potential to disrupt existing regulatory paradigms in embryo research, and enable us to think about how we can, or perhaps whether we should, implement lasting frameworks in this field.

37.2 BEHIND THE FOURTEEN-DAY RULE: THE WARNOCK REPORT, AND A 'SPECIAL STATUS'

The fourteen-day time limit on embryo research is of global significance. It is one of the most internationally agreed rules in reproductive science thus far,[4] with countries such as the UK, the USA, Australia, Japan, Canada, the Netherlands and India all upholding the rule in their own frameworks for embryo research[5] The catalysts for the implementation of the rule into many of these public policies are often accredited to two key reports:[6] the US Report on embryo research of the Ethics Advisory Board to the Department of Health, Education and Welfare,[7] and the UK report of the Warnock Committee of Inquiry into Human Fertilisation and Embryology.[8] This chapter will focus on the latter.

In 1984, the Warnock Committee published the Report of the Committee of Inquiry into Human Fertilisation and Embryology, also known as 'the Warnock Report'. This deliberative, interdisciplinary process was a keystone to law-making in this area in the UK. As a direct result of these deliberations, the use and production of embryos *in vitro* is governed by the HFE Act. This Act, which stands fast over thirty years later, brought legal and scientific practice out of uncertainty – due to the lack of a statutory framework for IVF and research pre-1990 – to a new state of being where embryos can be used, legally, for reproductive and research purposes under certain specified circumstances.

The Warnock Report was quite explicit that it was not going to tackle questions of the meaning of human 'life' or of 'personhood'. Instead, it articulated its remit as 'how it is right to treat the human embryo'.[9] The Report examined the arguments for and against the use of human embryos for research. Here, the Committee noted the plethora of views on the embryo's status, evidenced by the submissions received prior to the Report. They discussed each position in turn, before concluding that while the embryo deserves some protection in law, this protection should not be absolute. Notably, the source of this protection is not entirely clear from the Report. It cited the state of law at the time, which afforded some protection to the embryo, but not absolute protection.[10] Nonetheless, one can glean from their recommendations that this protection is sourced – at least in part – by virtue of embryos membership of the human species.

---

[3] For some, embryos are inherently 'human', and this chapter does not intend to support or negate this case.

[4] J. Appleby and A. Bredenoord, 'Should the 14-day Rule for Embryo Research Become the 28-day Rule?', (2018) *EMBO Molecular Medicine*, 10(9), e9437.

[5] I. Hyun et al., 'Embryology Policy: Revisit the 14 day Rule', (2016) *Nature*, 533(7602), 169–171.

[6] S. Chan, 'How and Why to Replace the 14-Day Rule', (2018) *Current Stem Cell Reports*, 4(3), 228–234.

[7] Ethics Advisory Board, 'Education and Welfare. Report and Conclusions: HEW Support of Research Involving Human In Vitro Fertilization and Embryo Transfer', (Department of Health, Education and Welfare, 1979).

[8] Committee of Inquiry into Human Fertilisation and Embryology, 'Report of the Committee of Inquiry into Human Fertilisation and Embryology', (Department of Health and Social Security, 1984), Cmnd 9314, 1984, (hereafter 'Warnock Report').

[9] Ibid., 11.9.

[10] Ibid., 11.16.

It is important to note that the Warnock Report did not explicitly answer the question of 'when does life begin to matter morally?', but rather considered the viewpoints submitted and 'provide[d] the human embryo with a special status without actually defining that moral status'.[11] Thus, in the HFE Act's first iteration,[12] not only did regulators attempt to navigate what we should or should not do at the margins of human life, but also the rapidly changing nature of those margins. The regulatory response to this has been to provide clear-cut boundaries surrounding what researchers can and cannot do, in reference to embryos *in vitro*, the most well known of these being the subject of this chapter, the fourteen-day rule, as contained in s3(4) of the HFE Act. The rule reads as follows:

3. Prohibitions in connection with embryos.
   (1) No person shall bring about the creation of an embryo except in pursuance of a licence.

   . . .

   (3) A licence cannot authorise—
       (a) keeping or using an embryo after the appearance of the primitive streak,
       (b) placing an embryo in any animal
       (c) keeping or using an embryo in any circumstances in which regulations prohibit its keeping or use,
   (4) **For the purposes of subsection (3)(a) above, the primitive streak is to be taken to have appeared in an embryo not later than the end of the period of 14 days beginning with the day on which the process of creating the embryo began, not counting any time during which the embryo is stored.** [emphasis added]

This section of the HFE Act also introduced the subsection that famously embodies the Warnock Report's abovementioned 'compromise position', which affords human embryos some 'respect'. It placed a clear *boundary* to the process of research: it is illegal to carry out research on an embryo beyond fourteen days, or after the primitive streak has formed, whichever occurs sooner. After that, the embryo cannot be used for any other purpose, and must be disposed of. In other words, as discussed further below, if decidedly an embryo created and/or used for research purposes, it may only ever be destroyed at the end of the research process.

Why fourteen days? The rule is based upon the evidence given to the Warnock Committee that it is around this stage that the 'primitive streak' tends to develop. It is also the approximate stage at which the embryonic cells can no longer split and thus produce twins or triplets, etc.[13] It was thus felt that this stage was morally significant, reinforced by the belief that this was the earliest known moment when the central nervous system was likely to have formed. This stage also marks the beginning of gastrulation, the process by which cell differentiation occurs. At the time, it was seen as a way to avoid, with absolute certainty, anyone carrying out research on those in the early stages of human life with any level of sentience or ability to experience pain.[14,15] In this way, as a reflection of the Committee's recommendations, embryos *in vitro* are often described as having a 'special status' in law; not that of one with personhood – attained at birth –

[11] N. Hammond-Browning, 'Ethics, Embryos and Evidence: A Look Back at Warnock', (2015) *Medical Law Review*, 23 (4), 588–619, 605.
[12] Human Fertilisation and Embryology Act 1990.
[13] P. Monahan, 'Human Embryo Research Confronts Ethical "Rule"', (2016) *Science*, 352(6286), 640.
[14] Nuffield Council on Bioethics, 'Human Embryo Culture', (Nuffield Council on Bioethics, 2017).
[15] It is worth noting that in 2017 Hulbert et al. found that there are no sensory systems or functional neural connections in embryos at the twenty-eight-day stage. For more discussion on this see Appleby and Bredenoord 'The 14-day Rule'.

but still *protected* in some sense. This in and of itself may be described as recognising the processual; it is implicit in the Committee's efforts to replicate a somewhat gradualist approach that recognises embryonic development – and any 'significant' markers within it.

While many would agree that a 'special status' in law results from this rule, the word 'status' – or any other of similar meaning – does not appear at all in the HFE Act (as amended) in reference to the embryo. It is clear, however, that the recommendations of the Warnock Report, made in light of its proposal for a 'special status', are reflected in this steadfast piece of legislation to this day, operationalised through provisions such as the fourteen-day rule.

Despite their contentions (see above), the Committee can arguably be understood as *implicitly* having answered the question of '*when* life begins to matter', by allowing research up to a certain stage in development.[16] In other words, they prescribed that 'as the embryo develops, it should receive greater legal protection due to its increasing moral value and potential'.[17] This policy, known as the gradualist approach, is somewhat in line with the Abortion Act 1967, which affords more protection to the fetus as it reaches later stages in development[18] (although in other ways these laws do not align at all). In doing so, while not explicit, the Act captures the processual aspect of embryonic/fetal development.

It seems that the human embryo hovers between several normative legal categories, i.e. 'subject' and 'object'.[19] While it clearly does not have a legally articulated 'status' under the HFE Act, it occupies a legal – and for some people, moral – threshold between all of these aforementioned categories, which we can see by the special status it has been given in law. Thus, while there is no explicit legal status of the embryo, what we have, legally, is still *something*. By virtue of giving the embryo *in vitro* legal recognition, with attached allowances and limits, it arguably has a status of sorts. Bearing in mind that the law adopted most of the Warnock Report's recommendations, its status may indeed be described as 'special', as the Report prescribed. It is 'not nothing',[20] yet not a 'person': it is the quintessential liminal entity, betwixt and between. From what we have seen, its status remains 'special', the meaning of which is unclear except that it is afforded 'respect' of sorts. Beyond that, we can glean little regarding what the extent or nature of this from domestic law is. It does not have an explicit legal status, but, as some argue, it may have one implicitly.[21] This begs the question: what does it mean to have 'legal status'? Is it enough to be protected by law? Recognised by law? Entitled to something through law? These are the types of questions we may want to consider for any amendments, or new frameworks, going forward.

37.3 BEYOND FOURTEEN DAYS?

As we have seen, the fourteen-day rule is the key legal embodiment of the embryo's decidedly 'special status' and the application of a legal and moral boundary at the earliest stages of human life. Yet throughout the incremental amendments to the HFE Act (e.g. the HFE Act 2008), there has been little enthusiasm among policy-makers for revisiting, let alone revising, the rule. For some, the latter did not necessarily matter, as, for twenty-seven years, this limit was 'largely

---

[16] Hammond-Browning 'Ethics, Embryos and Evidence', 604.

[17] Ibid., 605.

[18] See Abortion Act 1967, s1.

[19] See C. McMillan et al., 'Beyond Categorisation: Refining the Relationship between Subjects and Objects in Health Research Regulation', (2021) Law, Innovation and Technology, doi: 10.1080/17579961.2021.1898314.

[20] *St George's Healthcare NHS Trust* v. *S* [1998] All ER 673, [1998] 3 WLR 936, 952.

[21] Hammond-Browning 'Ethics, Embryos and Evidence', 606.

theoretical';[22] up until very recently, no researcher had been able to culture an embryo up to this limit.

In early 2016, for the first time, research published in *Nature*[23] and *Nature Cell Biology*[24] reported the successful culturing of embryos *in vitro* for thirteen days. With the possibility of finding out more about the early stages of human life beyond this two-week stage, calls have been made to revisit the fourteen-day rule.[25] Why? It appears that some valuable scientific knowledge may lie beyond this bright legal line in the sand, within this relatively unknown 'black box of development'. For example, it would enable the study of gastrulation, which begins when the primitive streak forms (around fourteen days).[26]

Yet what might all of this mean for compromise, respect, and the resulting 'special' legal status of the embryo? If this rule were to change, would the embryo still be 'special'? Moreover, do we believe this matters? These questions should be addressed if we revisit the rule; it seems that discussions surrounding the fourteen-day rule are part of a broader issue that needs to be addressed. There is a very strong case for public and legal discourse on the meaning and 'special' moral status of the embryo in UK law. One question that we may want to revisit is: if we value the recommendations of the Warnock Committee ('special', 'respect', etc.), does it still have resonance with us today? For example, one might ask: even if the 'special' status has a justifiable source, how can we 'value' it in practice except by avoiding harm? It is arguable that the 'special respect' apparently afforded in law seems meaningless in practical terms.[27]

It is difficult to enable a 'middle position' between protection and destruction in practice; we either allow embryos to be destroyed, or we do not. For some, the embryo's 'special status' is thus, arguably, purely rhetorical; it does not oblige us to 'act or refrain in any way'.[28] However, compromise is arguably more nuanced than allowing or disallowing destruction of embryos. Time is an essential component of legal boundaries within the 1990 Act (as amended). Either one can research the embryo for less than fourteen days, or one cannot. This means that we cannot research the embryo for any longer period of time, for example thirty days or sixty days. Rhetoric aside, the concept of the 'special status' is still very powerful and has acted as a tool to 'stop us in our tracks' with regards to research on embryos. It is arguably a precautionary position, which reflects that we as a society afford a degree of moral and legal value to embryos, and thus the special status caveat requires us to proceed cautiously, to reflect, to justify fully, to revisit, to revise and to continue to monitor as we progress scientifically. If we did not value the embryo at all, then we would have carte blanche to treat it however we wished. If that were the case, research at 30 or 60 or 180 days would not present a problem. Therefore, the embryo's special status need not be an all-or-nothing brake on research, nor a green light position. It thus means something in that sense, however (admittedly) meaningless. The 'special status', then, is – in a way – not a 'compromise', but what I would term a legal and ethical *comfort blanket*.[29]

This is not to criticise the language used by the Committee, however. The Warnock Committee's emphasis on 'compromise' was made in the name of moral pluralism. In other

---

[22] S. Chan, 'How to Rethink the Fourteen-Day Rule', (2017) *Hastings Center Report*, 47(3), 5–6.

[23] Deglincerti et al., 'Self-organization', 533

[24] Shahbazi et al., 'Self-organization of the Human Embryo', 700.

[25] Hyun et al., 'Embryology Policy', 169.

[26] Chan, 'How and Why', 228.

[27] See M. Ford, 'Nothing and Not Nothing: Law's Ambivalent Response to Transformation and Transgression at the Beginning of Life' in S. Smith, and R. Deazley (eds), *The Legal, Medical and Cultural Regulation of the Body: Transformation and Transgression* (London: Routledge, 2009), pp. 21–46.

[28] Ibid., 43.

[29] C. McMillan, *The Human Embryo in Vitro: Breaking the Legal Stalemate* (Cambridge University Press, 2021).

words, it emerged as the Warnock Committee's way of navigating the uncertainty/ambivalence surrounding how to treat embryos *in vitro*, legally. This is not to say that poles of opinion between which this compromise was set have changed. The rule, a reflection of this 'compromise' was, in many ways, a new boundary and threshold akin to its historical counterparts (such as quickening). Yet if we decide that it is worth considering this boundary and whether we want to change it, how can – or should – we rethink it? If we believe that the process embryonic and scientific development is a relevant factor in determining an appropriate regulatory response, what might further focus on these key points in transition bring to contemporary debates?

The rest of this chapter argues that if we want to think beyond the boundary of the fourteen-day rule, one way of framing discussion is by recognising the inherent link between processes and regulating of the margins of human life. When considering frameworks, the latter enables us to ask questions surrounding not only 'what is human?', but *'when* is human'? Asking *'when?'* – used here as an example – allows us to re-focus on not only embryonic development as process, but questions surrounding we need to place different boundaries within that process.

37.4 REVISITING THE RULE

Throughout the regulation of the early stages of human life, law has changed to reflect the changing boundaries of what is 'certain' and 'uncertain'.[30] Where new uncertainties arise,[31] some old ones will always remain.[32] We have thus moved, in some ways, from one type of uncertainty to another when it comes to embryo regulation, and this is because what we are dealing with is an inherently processual entity, that in and of itself has not changed. In other words, the complex and relatively uncertain nature of embryos, the stage of human life at which development occurs at its fastest pace, continues to cause widespread ambivalence[33] on how it is right to treat it.

When considering whether to alter the rule, multiple thresholds – such as the threshold for humanity – within embryonic and research processes come into consideration.[34] As we have seen, there was a strong nod to the gradualist approach in the thought behind the fourteen-day rule – an approach that recognises that human development is a process. The Report did not set out to answer *'when* is human?', but pointed to an important stage in the process of *becoming human*, when limiting research to fourteen days. As we have seen, the Warnock Committee used ethical deliberation and evidence available at the time to suggest this boundary beyond which research could not pass. A key part of this deliberation, although not referred to in terms of 'thresholds' *per se*, were particular (perceived) biological thresholds, such as the threshold for experiencing pain – which they associated with the start of the primitive streak – and thresholds for being able to cause harm therein. One might say that if the fourteen-day rule is a limit or a boundary, then something that we may want to consider – if we deem it appropriate to revisit this rule – is the presence of *thresholds* therein, and the importance that we want to attribute to those thresholds.[35] While being in a liminal state or space connotes occupying a threshold, a key part

[30] S. Taylor-Alexander et al., 'Beyond Regulatory Compression: Confronting the Liminal Spaces of Health Research Regulation', (2016) *Law, Innovation and Technology*, 8(2) 149–176; McMillan, 'The Human Embryo'.
[31] I.e. Should research and reproductive embryos be treated the same? Should the fourteen-day rule be extended? What can we find out about time between fourteen and twenty-eight days? Etc.
[32] I.e. The question of how we should treat embryos is, of course, never certain because there is no objective answer; in recognition of moral pluralism it is very much a subjective matter.
[33] See Ford, 'Nothing and Not Nothing', 31
[34] This is not to suggest that we could cross boundaries between research and reproduction, however.
[35] Taylor-Alexander et al., 'Beyond Regulatory Compression'.

of the liminal process is moving out of the liminal state, i.e. over or beyond that threshold. Thinking about liminal beings such as embryos in such terms highlights the presence of these boundaries and their potential for impermanence, especially in a legal context. For example, if we decide it is appropriate to *consider* extending the rule, these types of moral boundaries (i.e. harm, or sentience, etc.) may very well come into play again, for example if a 'twenty-eight-day rule' is proposed. Further, talks around extending it have already given rise to discussion surrounding another kind of boundary: would extending the rule be of adequate benefit to science? Some argue that there is much more that we can learn from extending the limit.[36] Yet, what amount of benefit is enough benefit to justify extension? Therein lies the threshold, a threshold of the reasonable prospect of sufficient scientific 'benefit'.

If the crossing of thresholds within biological and research processes have been implicitly important for us thus far, what might we learn from this? With regards to the legal processes that we *already have*, attention to process – and therefore thresholds – highlight the following:[37]

- Once an embryo created *in vitro* passes the threshold of being determinedly a 'research' embryo, it cannot (legally) be led back past the said threshold, and it can only come out this process as something to be disposed of after being utilised.
- In contrast, there are lots of thresholds that embryos are led through for a 'reproductive' path, for example: (non)selection after PGD, implantation, freezing and unfreezing, implantation, gestation etc. – indeed, this includes the possibility of crossing the threshold from 'reproduction' to 'research' if, say, PGD tests suggest non-suitability for reproduction.
- When the progenitors of embryos are making decisions regarding what to do with their surplus embryos, they may cross various thresholds themselves, e.g to donate or not, either for research or to others seeking to reproduce.

Regarding the last point, persons/actors are, of course, an essential part of these processes and this should not be lost in any renewed discussions surrounding the rule. Considerations for the actors around embryos can be different for each threshold, i.e. we may consider different sets of factors depending on which threshold any particular embryo is at. For example, at the third threshold above, many factors come into consideration for donors, including their attitudes towards research and their feelings about and towards their surplus embryos and their future (non)uses.[38] Moreover, each threshold is coupled with clear boundaries, be it the fourteen-day rule for 'research' embryos, or rules around what may/may not be implanted for 'reproductive' embryos.

Thresholds – or indeed boundaries – are not necessarily 'bad' here, per this work's analysis. Indeed, both moral and legal thresholds are of crucial importance. Rather, I suggest that we should be alive to their presence and their place among the broader network – of actors, or silos, etc. – in order to ask questions about the conditions we want in order to cross those thresholds.[39] As I have argued elsewhere: 'Considering the multiplicity, variability, and in many ways, subjectivity of these thresholds might enable us to regulate in a more flexible and context-specific way that allows us to recognise the multiplicity of processes occurring within the framework of the [HFE Act].'[40] Attention to process cannot necessarily say how any revisitation might turn out.

---

[36] E.g. S. Wong, 'The Limits to Growth', (2016) *New Scientist*, 232(3101), 18–19.
[37] See McMillan, 'The Human Embryo'.
[38] See E. Jonlin, 'The Voices of the Embryo Donors', (2015) *Trends in Molecular Medicine*, 21(2), 55–57; S. Parry, '(Re) Constructing Embryos in Stem Cell Research: Exploring the Meaning of Embryos for People Involved in Fertility Treatments', (2006) *Social Science and Medicine*, 62(10), 2349–2359.
[39] See McMillan, 'The Human Embryo'.
[40] Ibid.

It is important to revisit the intellectual basis for any law, but especially so in a field where technology and science advance so rapidly. If we do not, we cannot ask important questions in light of new information, for example: what or when is the threshold for 'humanity'? Or if indeed we want 'when is human' to factor into how we regulate embryos, as it has done in the past. Discussing questions such as these would be a great disturbance to the policy norm of the past twenty-seven years or so, which have stayed away from these types of questions. But I argue that within disturbance, we can find resolution through proper legal and ethical deliberation, and public dialogue. In other words, it would not be beneficial of us to shy away from disruptions for fear of practice being shut down, as these disturbances present us with a chance to feed experiences and lessons of those involved in – and benefit from – research back into regulatory, research, and – eventually – treatment practice.

37.5 CONCLUSION

While the time limit on embryo research has undoubtedly been a success on many fronts, if it is to remain 'effective and relevant',[41] we must be open to revisiting it, with proper deliberation and public involvement, with openness and transparency.[42] Not only that, but when doing so, we must not shy away from asking difficult questions if law is to adapt to contemporary research.

The latter part of this chapter has argued that a focus on process has the potential to disrupt existing regulatory paradigms in embryo research and enable us to think about how we can, or perhaps whether we should, implement lasting frameworks in this field. The above did not challenge the pros and cons of the fourteen-day rule, or research and reproductive practices *in vitro* more generally *per se*, but rather briefly explored one of the ways which law could engage with embryonic (and legal) processes through attention to thresholds (as a key facet of these processes).

Overall, while the framing offered here has the *potential* to justify an extension of the fourteen-day rule, it cannot be done without proper public deliberation. This deliberation would need to be subject to sound scientific objections, and perhaps most importantly, subject to scrutiny regarding prevalent moral concern over pain and sentience.[43] This analysis challenges us to *deliberate*, and *revisit* – not necessarily the revise – the law surrounding this longstanding rule. Responsive regulation, per the title of this section of the volume, need not respond to every 'shiny new thing' (e.g. advances in research, such as those discussed in this chapter), but be reflexive (and reflective) so that HRR does not become stagnant.

[41] Appleby and Bredenoord, 'The 14-day Rule'.
[42] G. Cavaliere, 'A 14-day Limit for Bioethics: The Debate over Human Embryo Research', (2017) *BMC Medical Ethics*, 18(38).
[43] See McMillan, 'The Human Embryo'.

38

A Perfect Storm

Non-evidence-Based Medicine in the Fertility Clinic

Emily Jackson

38.1 INTRODUCTION

In vitro fertilisation (IVF) did not start with the birth of Louise Brown on 25 July 1978. Nine years earlier, Robert Edwards and others had reported the first *in vitro* fertilisation of human eggs,[1] and before Joy Brown's treatment worked, 282 other women had undergone 457 unsuccessful IVF cycles.[2] None of these cycles was part of a randomised controlled trial (RCT), however. After decades of clinical use, it is now widely accepted that IVF is a safe and effective fertility treatment, but it is worth noting that there have been studies that have suggested that the live birth rate among couples who use IVF after a year of failing to conceive naturally is not, in fact, any higher than the live birth rate among those who simply carry on having unprotected sexual intercourse for another year.[3]

Reproductive medicine is not limited to the relatively simple practice of fertilising an egg *in vitro*, and then transferring one or two embryos to the woman's uterus. Rather, there are now multiple additional interventions that are intended to improve the success rates of IVF. Culturing embryos to the blastocyst stage before transfer, for example, appears to have increased success rates because by the five-day stage, it is easier to tell whether the embryo is developing normally.[4]

Whenever a new practice or technique is introduced in the fertility clinic, in an ideal world, it would have been preceded by a sufficiently statistically powered RCT that demonstrated its safety and efficacy. In practice, large-scale RCTs are the exception rather than the norm in reproductive medicine.[5] There have been some large trials, and meta-analyses of smaller trials,

---

[1] R. G. Edwards et al., 'Early Stages of Fertilization In Vitro of Human Oocytes Matured In Vitro', (1969) *Nature*, 221, 632–635.

[2] S. Franklin, 'Louise Brown: My Life as the World's First Test-Tube Baby by Louise Brown and Martin Powell, Bristol Books (2015)', (2016) *Reproductive Biomedicine and Society Online*, 3, 142–144.

[3] H. K. Snick et al., 'The Spontaneous Pregnancy Prognosis in Untreated Subfertile Couples: The Walcheren Primary Care Study', (1997) *Human Reproduction*, 12(7), 1582–1588; E. R. te Velde et al., 'Variation in Couple Fecundity and Time to Pregnancy: an Essential Concept in Human Reproduction', (2000) *Lancet*, 355(9219), 1928–1929.

[4] E. G. Papanikolaou et al., 'Live Birth Rate Is Significantly Higher after Blastocyst Transfer than after Cleavage-Stage Embryo Transfer When at Least Four Embryos Are Available on Day 3 of Embryo Culture. A Randomized Prospective Study', (2005) *Human Reproduction*, 20(11), 3198–3203.

[5] K. Stocking et al., 'Are Interventions in Reproductive Medicine Assessed for Plausible and Clinically Relevant Effects? A Systematic Review of Power and Precision in Trials and Meta-Analyses', (2019) *Human Reproduction*, 34(4), 659–665.

but it would not be unreasonable to describe treatment for infertility as one of the least evidence-based branches of medicine.[6]

In addition to an absence of evidence, another important feature of reproductive medicine is patients' willingness to 'try anything'. Inadequate NHS funding means that most fertility treatment is provided in the private sector, with patients paying 'out of pocket' for every aspect of their IVF cycle, from the initial consultation to scans, drugs and an ever-increasing list of 'add-on' services, such as assisted hatching; preimplantation genetic screening; endometrial scratch; time-lapse imaging; embryo glue and reproductive immunology. The combination of a poor evidence base, commercialisation and patients' enthusiasm for anything that might improve their chance of success, results in a 'perfect storm' in which dubious and sometimes positively harmful treatments are routinely both under-researched and oversold.

Added to this, although clinics must have a licence from the Human Fertilisation and Embryology Authority (HFEA) before they can offer IVF, the HFEA does not have the power to license, or refuse to license the use of add-on treatments. Its powers are limited to ensuring that, before a patient receives treatment in a licensed centre, patients are provided with 'such relevant information as is proper', and that 'the individual under whose supervision the activities authorised by a licence are carried on' (referred to as the Person Responsible), ensures that 'suitable practices' are used in the clinic.[7] In this chapter, I will argue that, although giving patients information about the inadequacy of the evidence-base behind add-on treatments is important and necessary, this should not be regarded as a mechanism through which their inappropriate use can be controlled. Instead, it may be necessary for the HFEA to categorise non-evidence-based and potentially harmful treatments as 'unsuitable' practices, which should not be provided at all, rather than as treatments that simply need to be accompanied by a health warning.

38.2 A PERFECT STORM?

In order to be appropriately statistically powered, it has been estimated that a trial of a new fertility intervention should recruit at least 2610 women.[8] Trials of this size are exceptional, however, and it is much more common for smaller statistically underpowered trials to be carried out. Nor does meta-analysis of these smaller trials necessarily offer a solution, in part because their outcomes are not always reported consistently, and the meta-analyses themselves may not be sufficiently large to overcome the limitations of the smaller studies.[9]

Fertility patients are often keen to 'try something new', even if it has not been proven to be safe and effective in a large-scale RCT.[10] IVF patients are often in a hurry. Most people take their fertility for granted, and after years of trying to prevent conception, they assume that conception

---

[6] Archie Cochrane famously said that obstetrics deserved 'the wooden spoon' for being the least scientific medical speciality. A. L. Cochrane, '1931–1971: A Critical Review with Particular Reference to the Medical Profession' in G. Teeling-Smith and N. E. J. Wells (eds), *Medicines for the Year 2000* (London: Office of Health Economics, 1979), pp. 2–12.

[7] Human Fertilisation and Embryology Act 1990, sections 13(6), 17(1) and 17(1)(d).

[8] J. Wilkinson et al., 'Reproductive Medicine: Still More ART than Science?', (2019) *British Journal of Obstetrics and Gynaecology*, 126(2), 138–141.

[9] Stocking et al., 'Interventions in Reproductive Medicine'; J. M. N. Duffy et al., 'Core Outcome Sets in Women's and Newborn Health: A Systematic Review', (2017) *British Journal of Obstetrics and Gynaecology*, 124(10), 1481–1489.

[10] J. Rayner et al., 'Australian Women's Use of Complementary and Alternative Medicines to Enhance Fertility: Exploring the Experiences of Women and Practitioners', (2009) *BMC Complementary and Alternative Medicine*, 9 (1), 52.

will happen soon after they stop using contraception. By the time a woman realises that she may need medical assistance in order to conceive, her plan to start a family will already have been delayed for a year or more. At the same time, women's age-related fertility decline means that they are often acutely aware of their need to start treatment as soon as possible.

Although, in theory, fertility treatment is available within the NHS, it is certainly not available to everyone who needs it. The National Institute for Health and Care Excellence's (NICE) 2013 clinical guideline recommended that the NHS should fund three full cycles of IVF (i.e. a fresh cycle followed by further cycles using the frozen embryos) for women under 40 years old, and one full cycle for women aged 40–42, who must additionally not have received IVF treatment before and not have low ovarian reserve.[11] Implementation of this NICE guideline is not mandatory, however, and in 2018 it was reported that only 13 per cent of Clinical Commissioning Groups (CCGs) provide three full cycles of IVF to eligible women; 60 per cent offer one NHS-funded cycle – most of which fund only one fresh cycle – and 4 per cent provide no cycles at all.[12]

The majority of IVF cycles in the UK are self-funded,[13] and although the average cycle costs around £3350,[14] costs of more than £5000 per cycle are not uncommon. As well as simply wanting to have a baby, IVF patients are therefore commonly also under considerable financial pressure to ensure that each IVF cycle has the best possible chance of success. In these circumstances, it is not surprising that patients are keen to do whatever they can to increase the odds that a single cycle of IVF will lead to a pregnancy and birth.

One of the principal obstacles to making single embryo transfer the norm was that, for many patients, the birth of twins was regarded as an ideal outcome.[15] Most patients want to have more than one child, so if one cycle of treatment could create a two-child family, this appeared to be a 'buy one, get one free' bargain. In order to persuade women of the merits of the 'one at a time' approach, it was not enough to tell them about the risks of multiple pregnancy and multiple birth, both for them and their offspring. Many women are prepared to undergo considerable risks in pursuit of a much-wanted family. Instead, the 'one at a time' campaign emphasised the fact that a properly implemented 'elective single embryo transfer' policy did not reduce birth rates, and tried to persuade NHS funders that a full cycle of IVF was not just one embryo transfer, but that it should include the subsequent frozen embryo transfers.[16]

Not only are patients understandably keen to try anything that might improve their chance of success, they are also paying for these extra services out of pocket. As consumers, we are used to paying more to upgrade to a better service, so this additional expense can appear to be a 'sign of quality'.[17] Rather than putting patients off, charging them several hundred pounds for endometrial scratch and assisted hatching may make these additional services appear even more desirable. New techniques often generate extensive media coverage, leading patients actively to seek

---

[11] National Institute for Health and Care Excellence, 'Fertility: Assessment and Treatment for People with Fertility Problems', (NICE, 2013).
[12] See further www.fertilityfairness.co.uk.
[13] Human Fertilisation and Embryology Authority, 'State of the Fertility Sector 2016–7', (HFEA, 2017).
[14] S. Howard, 'The Hidden Costs of Infertility Treatment', (2018) *British Medical Journal*, 361.
[15] G. M. Hartshorne and R. J. Lilford, 'Different Perspectives of Patients and Health Care Professionals on the Potential Benefits and Risks of Blastocyst Culture and Multiple Embryo Transfer', (2002) *Human Reproduction*, 17(4), 1023–1030.
[16] P. Braude, 'One Child at a Time: Reducing Multiple Births through IVF, Report of the Expert Group on Multiple Births after IVF', (Expert Group on Multiple Births after IVF, 2006).
[17] Wilkinson et al., 'Reproductive Medicine'.

out clinics that offer the new treatment.[18] Clinics that offer the non-evidence-based new intervention are therefore able to say that they are simply responding to patient demand.

As well as the appeal of expensive high-tech interventions, patients are also attracted to simple and apparently plausible explanations for IVF failure. If an IVF cycle does not lead to a pregnancy because the embryo fails to attach to the lining of the woman's uterus, it is easy to understand why patients might be persuaded of the benefits of 'embryo glue', in order to increase adhesion rates. Alternative therapists have also flourished in this market: acupuncturists are said to be able to 'remove blocks to conception'; and hypnotherapists treat women 'with a subconscious fear of pregnancy'.[19] In practice, however, the evidence indicates not only that complementary and alternative medicine (CAM) does not work, but that live birth rates are lower for patients who use CAM services.[20]

Perhaps the most egregious example of an apparently simple and plausible explanation for IVF failure being used to market a non-evidence-based and potentially harmful intervention is reproductive immunology. The existence of the unfortunately named 'natural killer cells' in the uterus has helped to persuade patients that these cells might – unless identified by expensive tests and suppressed by expensive medications – 'attack' the embryo and prevent it from implanting. News stories with headlines like 'The Killer Cells That Robbed Me of Four Babies'[21] and 'My Body Tried to Kill My Baby'[22] suggest a very direct link between NK cells and IVF failure. The idea that the embryo is a genetically 'foreign' body that the woman's uterine cells will attack, unless their immune response is suppressed, sounds plausible,[23] and as Datta et al. point out, 'couples seeking a reason for IVF failure find the rationale of immune rejection very appealing'. It has no basis in fact, however.[24]

There is no evidence that natural killer cells have any role in causing miscarriage; rather despite their name, they may simply help to regulate the formation of the placenta. As Moffett and Shreeve explain, regardless of this lack of evidence, 'a large industry has grown up to treat women deemed to have excessively potent uterine "killers"'.[25] In addition to the absence of RCTs establishing that reproductive immunology increases success rates, the medicines used – which include intravenous immunoglobulins, TNF-α inhibitors, granulocyte-colony stimulating factor, lymphocyte immune therapy, leukaemia inhibitory factor, peripheral blood mononuclear cells, intralipids, glucocorticoids, vitamin D supplementation and steroids – may pose a risk of significant harm to women.[26]

The lack of evidence for reproductive immunology, and the existence of significant risks, has been known for some time. In 2005, Rai and others described reproductive immunology as

---

[18] A. K. Datta et al., 'Add-Ons in IVF Programme – Hype or Hope?', (2015) *Facts, Views & Vision in ObGyn*, 7(4), 241–250.

[19] C. N. M. Renckens, 'Alternative Treatments in Reproductive Medicine: Much Ado About Nothing: "The Fact That Millions of People Do Not Master Arithmetic Does Not Prove That Two Times Two Is Anything Else than Four": W. F. Hermans', (2002) *Human Reproduction*, 17(3), 528–533.

[20] J. Boivin and L. Schmidt, 'Use of Complementary and Alternative Medicines Associated with a 30% Lower Ongoing Pregnancy/Live Birth Rate during 12 Months of Fertility Treatment', (2009) *Human Reproduction*, 24(7), 1626–1631.

[21] R. Barber, 'The Killer Cells That Robbed Me of Four Babies', *Daily Mail* (2 January 2011).

[22] J. Fricker, 'My Body Tried to Kill My Baby', *Daily Mail* (2 July 2007).

[23] See, for example, H. Shehata, quoted in BBC News, 'Baby Born to Woman Who Suffered 20 Miscarriages', *BBC News* (17 January 2014): 'We found that some women's natural killer cells are so aggressive they attack the pregnancy, thinking the foetus is a foreign body'.

[24] Datta et al., 'Add-Ons in IVF Programme'.

[25] A. Moffett and N. Shreeve, 'First Do No Harm: Uterine Natural Killer (NK) Cells in Assisted Reproduction', (2015) *Human Reproduction*, 30(7), 1519–1525.

[26] Datta et al., 'Add-Ons in IVF Programme'.

'pseudo-science', pointing out that 'Not only is there no evidence base for these interventions, which are potentially associated with significant morbidity, the rationale for their use may be false'.[27] The HFEA's most recent advice to patients is also clear and unequivocal:

> There is no convincing evidence that a woman's immune system will fail to accept an embryo due to differences in their genetic codes. In fact, scientists now know that during pregnancy the mother's immune system works with the embryo to support its development. Not only will reproductive immunology treatments not improve your chances of getting pregnant, there are risks attached to these treatments, some of which are very serious.[28]

Despite this, patients continue to be persuaded by a simple, albeit false, explanation for IVF failure, and by 'evidence' from fertility clinics that is better described as anecdote. The Zita West fertility clinic blog, for example, contains accounts from satisfied ex-patients with headlines like 'I Was Born to Be a Mum and Couldn't Have Done It Without Reproductive Immunology'.[29] It is not uncommon for clinics' websites to 'speak of "dreams" and "miracles", rather than RCTs[30] Spencer and others analysed 74 fertility centre websites, and found 276 claims of benefit relating to 41 different fertility interventions, but with only 16 published references to support these, of which only five were high level systematic reviews.[31]

From the point of view of a for-profit company selling fertility services, why bother to do expensive large-scale RCTs, when it is possible to sell a new therapy to patients in the absence of such trials? If patients do not care about the lack of evidence, and are happy to rely upon a clinician's anecdotal report that X therapy has had some success in their clinic, the clinic has no incentive to carry out trials, which may indicate that X therapy does not increase live birth rates.

A free market in goods and services relies upon consumers choosing not to buy useless products. If a mobile phone company were to produce a high-tech new phone that does not work, then after an initial flurry of interest in a shiny new product, its failings would become apparent and the market for it would disappear. Because there can be no guarantee that any cycle of IVF will lead to the birth of a baby, and almost every cycle is more likely to fail than it is to work, it is much harder for consumers of fertility services to tell whether an add-on service is worth purchasing. Rather than relying on individual patients 'voting with their feet' in order to crowd out useless interventions, it may be important instead for an expert regulator to choose for them.

38.3 REGULATING ADD-ON SERVICES

There are three mechanisms through which the provision of add-on services in the fertility clinic is regulated. First, if it involves the use of a medicinal product, that product must have a product licence from the European Medicines Agency or the Medicines and Healthcare products Regulatory Agency. The Human Medicines Regulations 2012 specify that, before a new medicine can receive a product licence, the licensing authority must be satisfied that 'the applicant

---

[27] R. Rai et al., 'Natural Killer Cells and Reproductive Failure – Theory, Practice and Prejudice', (2005) *Human Reproduction*, 20(5), 1123–1126.

[28] HFEA, 'Treatment Add-On', (HFEA, 2019).

[29] 'I was Born to Be a Mum – And Couldn't Have Done It without Reproductive Immunology', (Zita West), www.zitawest.com/i-was-born-to-be-a-mum-and-couldnt-have-done-it-without-reproductive-immunology/.

[30] J. Hawkins, 'Selling ART: An Empirical Assessment of Advertising on Fertility Clinics' Websites', (2013) *Indiana Law Journal*, 88(4), 1147–1179.

[31] E. A. Spencer et al., 'Claims for Fertility Interventions: A Systematic Assessment of Statements on UK Fertility Centre Websites', (2016) *BMJ Open*, 6(11).

has established the therapeutic efficacy of the product to which the application relates', and 'the positive therapeutic effects of the product outweigh the risks to the health of patients or of the public associated with the product'.[32] In short, it must be established that the product works for the indication for which the product licence is sought, and that its benefits outweigh its risks.

In practice, however, the use of medicines as add-ons to fertility treatment generally involves their 'off-label' use. Reproductive immunology, for example, may involve the use of steroids, anticoagulants and monoclonal antibodies. Although efficacy and a positive risk–benefit profile may exist for these medicines' licensed use, this is not the same as establishing that they work or are safe for their off-label use in the fertility clinic. There are comparatively few controls over doctors' freedom to prescribe drugs off-label, even though, when there has not been any assessment of the safety or efficacy of a drug's off-label use, it may pose an unknown and unjustifiable risk of harm to patients.

The General Medical Council (GMC) has issued guidance to doctors on the off-label prescription of medicines which states that:

> You should usually prescribe licensed medicines in accordance with the terms of their licence. However, you may prescribe unlicensed medicines where, on the basis of an assessment of the individual patient, you conclude, for medical reasons, that it is *necessary* to do so to meet the specific needs of the patient (my emphasis).[33]

The guidance goes on to set out when prescribing unlicensed medicines could be said to be 'necessary':

> a. There is no suitably licensed medicine that will meet the patient's need ...
> b. Or where a suitably licensed medicine that would meet the patient's need is not available. This may arise where, for example, there is a temporary shortage in supply; or
> c. The prescribing forms part of a properly approved research project.[34]

Doctors must also be satisfied that be 'there is sufficient evidence or experience of using the medicine to demonstrate its safety and efficacy', and patients must be given sufficient information to allow them to make an informed decision.[35] It is possible that a doctor who prescribed medications off-label could have his fitness to practise called into account, although, in practice, it seems likely that clinicians will simply maintain that these medicines 'meet the patient's need', and that they have sufficient experience within their own clinic to 'demonstrate safety and efficacy'.

Second, before receiving treatment services in a licensed centre, section 13(6) of the Human Fertilisation and Embryology Act 1990 specifies that patients must be provided with 'such relevant information as is proper', and that they must give consent in writing.[36] Although add-ons are not licensable treatments, it could be said that the clinician's statutory duty to give patients clear and accurate information extends to the whole course of treatment they receive in the clinic, not just to the treatment for which an HFEA licence is necessary. Indeed, the HFEA's Code of Practice specifies that:

---

[32] Human Medicines Regulations 2012, 58(4)(a) and 58(4)(b).
[33] General Medicine Council, 'Good Practice in Prescribing and Managing Medicines and Devices', (GMC, 2013), para 68.
[34] Ibid., para 69.
[35] Ibid., paras 70(a) and 71.
[36] Human Fertilisation and Embryology Act 1990, Schedule 3, para 1.

> Before treatment is offered, the centre should give the woman seeking treatment and her partner, if applicable, information about … fertility treatments available, including any treatment add ons which may be offered and the evidence supporting their use; any information should explain that treatment add ons refers to the technologies and treatments listed on the treatment add ons page of the HFEA website.[37]

The Code of Practice also requires centres to give patients 'a personalised costed treatment plan', which should 'detail the main elements of the treatment proposed – including investigations and tests – the cost of that treatment and any possible changes to the plan, including their cost implications'.[38] Before offering patient an add-on treatment, clinics should therefore be open and honest with patients about the risks, benefits and costs of the intervention.

In practice, however, patients will not necessarily be put off by underpowered trial data, especially when more optimistic anecdotal accounts of success are readily available online. In order to try to counter the circulation of misinformation about treatment add-ons, the HFEA has recently instituted a 'traffic light' system that is intended to provide clear and unambiguous advice to patients. At the time of writing, no add-on is green. Most are either amber (that is, 'there is a small or conflicting body of evidence, which means further research is still required and the technique cannot be recommended for routine use'), or red (that is, 'there is no evidence to show that it is effective and safe'). The HFEA further recommends that patients who want more detailed information 'may want to contact a clinic to discuss this further with a specialist'.[39]

It is, however, unsatisfactory to rely upon informed patient choice as a mechanism to control the over-selling of unproven add-on treatments. The fertility industry has 'a pronounced predilection for over-diagnosis, over-use and over-treatment', and the widespread adoption of a 'right to try' philosophy in practice translates into clinics profiting from the sale of unproven treatments.[40] For example, the HFEA gives intrauterine culture – in which newly fertilised eggs are placed in a device inside the woman's womb – an amber rating, and informs prospective patients:

> There's currently no evidence to show that intrauterine culture improves birth rates and is safe. This is something you may wish to consider if you are offered intrauterine culture at an additional cost.

It could instead be argued that the fact that a treatment is expensive and is not known to be either safe or effective is not merely something that patient should 'consider' when deciding whether to purchase it, but rather is a reason not to make that treatment available outside of a clinical trial.

Third, while the HFEA does not license add-on services, Persons Responsible are under a duty to ensure that only 'suitable practices are used in the course of the activities'.[41] If the HFEA were to decide that those add-on services that it ranks as red are not suitable practices, then clinicians should not use them in the clinic. It has not (yet) done this.

[37] HFEA, '9th Code of Practice', (HFEA, 2019), para 4.5.
[38] Ibid., para 4(9).
[39] HFEA, 'Treatment Add-Ons'.
[40] Wilkinson et al., 'Reproductive Medicine'.
[41] Human Fertilisation and Embryology Act 1990, section 17(1)(d).

38.4 WON'T PATIENTS GO ELSEWHERE?

Given patients' interest in add-on services, many of the 70 per cent of UK clinics that offer at least one of these treatments claim to be responding to patient choice.[42] Reputable clinicians maintain that if they cease to offer add-on services, patients are likely to go instead to clinics that do provide these treatments, either within the UK – where a clinic does not need a licence from the HFEA if it is only providing add-on services – or overseas.[43] If patients are going to pay for these treatments elsewhere anyway, then, so the argument goes, it is better to provide them in safe, hygienic, regulated clinics, rather than abandoning patients to the wild west of unregulated fertility services.

The easiest way to see why this argument should be dismissed is to imagine that it is being made about a different sort of non-evidence based treatment, such as stem cell therapies for the treatment of spinal injury. Although stem cell therapies hold very great promise for the treatment of a wide range of conditions, most are still at the experimental stage. That does not stop unregulated clinics overseas from marketing stem cell therapies for the treatment of a wide range of conditions, and as a miraculous cure for ageing.[44]

If a UK doctor was to justify injecting stem cells into a patient's spinal column, on the grounds that, if he did not do so, the patient would be likely to travel to China for unproven stem cell treatment, it could be predicted that the GMC might be likely to investigate his fitness to practise. The argument that, if he did not offer unproven and unsafe treatment in the UK, patients might choose to undergo the same unsafe treatment in a foreign clinic, would be likely to be given short shrift.

38.5 CONCLUSION

It is important to remember that add-on treatments are not simply a waste of patients' money, though they are often that as well. Many add-on treatments are also risky. Despite this, patients are enthusiastic purchasers of additional services for which there is little or no good evidence. In such circumstances, where the lack of robust clinical trial data does not appear to dent patients' willingness to buy add-on treatments, there is little 'bottom-up' incentive to carry out large-scale RCTs.

The HFEA's information for patients is clear and authoritative, but it is not the only information that patients will see before deciding whether to pay for additional treatment services. Patients embarking upon fertility treatment also seek out information from other patients and from a wide variety of online sources. It is increasingly common for ill-informed 'discourses of hope' about unproven treatments to circulate in blogs and in Facebook groups, coexisting and competing with evidence-based information from scientists and regulators.[45] Fertility patients often report doing their own 'research' before embarking on treatment, and this generally means

[42] A. J. Rutherford, 'Should the HFEA Be Regulating the Add-On treatments for IVF/ICSI in the UK? FOR: Regulation of the Fertility Add-On Treatments for IVF', (2017) *British Journal of Obstetrics & Gynaecology*, 124(12), 1848.
[43] W. L. Ledger, 'The HFEA Should Be Regulating Add-On Treatments for IVF/ICSI', (2017) *British Journal of Obstetrics & Gynaecology*, 124(12), 1850–1850.
[44] D. Archard, 'Ethics of Regenerative Medicine and Innovative Treatments', (Nuffield Council of Bioethics, 13 October 2017), www.nuffieldbioethics.org/blog/ethics-regenerative-medicine-innovative-stem-cell-treatment.
[45] A. Petersen et al., 'Stem Cell Miracles or Russian Roulette?: Patients' Use of Digital Media to Campaign for Access to Clinically Unproven Treatments', (2016) *Health, Risk and Society*, 17(7–8), 592–604.

gathering material online, from sources where the quality and accuracy of information may be distinctly variable.[46]

In this perfect storm, it is unreasonable to expect patients to be able to protect themselves from exploitation through the application of the principle of *caveat emptor*.[47] On the contrary, what is needed instead is a clear message from the regulator that the routine selling of unproven treatments should not just prompt patients to ask additional questions, but that these treatments should not be sold in the first place. Of course, it is important that reproductive medicine does not stand still, and that new interventions to improve the chance of success are developed. But these should first be tried in the clinic as part of an adequately powered clinical trial. Trial participants must be properly informed that the treatment is still at the experimental stage, and they should not be charged to participate. The GMC also has a role to play in investigating the fitness to practise of doctors who routinely sell, for profit, treatments that are known to be risky and ineffective. As Moffett and Sheeve put it: 'it is surely no longer acceptable for licensed medical practitioners to continue to administer and profit from potentially unsafe and unproven treatments, based on belief and not scientific rationale'.[48]

[46] E. Jackson et al., 'Learning from Cross-Border Reproduction', (2017) *Medical Law Review*, 25(1), 23–46.
[47] Ledger, 'HFEA Should Be Regulating Add-On Treatments '.
[48] Moffett and Shreeve, 'First Do No Harm'.

39

Medical Devices Regulation

New Concepts and Perspectives Needed

Shawn H. E. Harmon

39.1 INTRODUCTION

This section of the Handbook explores how technological innovations and/or social changes create disturbances within regulatory approaches. This chapter considers how innovations *represent* disturbances with which regulatory frameworks must cope, focusing on innovations that can be characterised as 'enhancing'. Human enhancement can no longer be dismissed as something with which serious regulatory frameworks need not engage. Enhancing pursuits increasingly occupy the very centre of human experience and 'being'; one can observe widespread student use of cognitively enhancing stimulants, the increasing prevalence of implanted technologies, and great swathes of people absently navigating the physical while engrossed in the digital.

Given the diversity of activities and technologies implicated, the rise and mores of the 'maker movement',[1] and the capacity of traditional – commercial – health research entities to locate innovation activities to jurisdictions with desirable regulation, it is impossible to point to a single regulatory framework implicated by enhancement research and innovation. Candidates include those governing human tissue use and pharmaceuticals, but could also include those governing intellectual property, data use, or consumer product liability. The medical devices framework, one might think, should offer a good example of a regime that engages directly and usefully with the concepts implicated by enhancement and the socio-technical changes wrought by enhancing technologies. As such, this chapter focuses on the recently reformed European medical devices regime.

After identifying some enhancements that are available and highlighting what they mean for the person, the chapter introduces two concepts that are deeply implicated by enhancing technologies: 'identity' and 'integrity'. If regulation fails to engage with them, it will remain blind to matters that are profoundly important to those people who are using or relying on these technologies. Their observance in EU Regulation 2017/745 on Medical Devices (MDR),[2] and EU Regulation 2017/746 on In Vitro Diagnostic Medical Devices (IVDR),[3] is examined. It is

---

[1] C. Howard et al., 'The Maker Movement: A New Avenue for Competition in the EU', (2014) *European View*, 13(2), 333–340; M. Tan et al., 'The Influence of the Maker Movement on Engineering and Technology Education', (2016) *World Transactions on Engineering and Technology Education*, 14(1), 89–94.

[2] Regulation (EU) 2017/745, 5 April 2017, on medical devices, amending Directive 2001/83/EC, Regulation (EC) No. 178/2002 and Regulation (EC) No. 1223/2009 and repealing Council Directives 90/385/EEC and 93/42/EEC, OJ L 117, 5.5.2017.

[3] Regulation (EU) 2017/746, 5 April 2017, on *in vitro* diagnostic medical devices and repealing Directive 98/79/EC and Commission Decision 2010/227/EU, OJ L 117, 5.5.2017.

concluded that they are, unfortunately, too narrowly framed and too innovator driven, and are therefore largely indifferent to these concepts.

39.2 ENHANCING INNOVATIONS

Since the first use of walking canes, false teeth and spectacles, we have been 'enhancing' ourselves for both medical and social purposes, but the so-called technological 'revolutions' of late modernity – which have relied on and facilitated innovations in computing, biosciences, materials sciences and more – have prompted changes in the nature and prevalence of the enhancements that we adopt. We now redesign and extend our physical scaffolding, we alter its physiological functioning, we extend the will, and we push the potential capacities of the mind and body by linking the biological with the technological or by embedding the latter into the former.

In the 1960s, Foucault anticipated the erasure of the human being.[4] We might now understand this erasure to be the rise of the enhanced human, which includes the techno-human hybrid (cyborg).[5] This 'posthuman' thinks of the body as the original prosthesis, so extending or replacing it with other prostheses becomes a continuation of a process that began pre-natally.[6] Even if one does not subscribe to the posthuman perspective, 'enhancing' technologies are commonly applied,[7] and are becoming more complex and more intrusive, nestling *within* the body, and performing not only for us, but also on and within us.[8] Examples include a wide range of smart physiological sensors, cochlear implants, implanted cardiac defibrillators, deep brain stimulators, complex prostheses like retinal and myoelectric prosthetics, mind stimulating/ expanding interventions like nootropic drugs, neuro-prosthetics, and consciousness-insinuating constructs like digital avatars, which allow us to build and explore wholly new cyber-environments.

These technologies have many labels, but they all become a part of the person through processes of bodily 'incorporation', 'extension' or 'integration'.[9] Depending on the technology, they allow the individual to generate, store, access and transmit data about the physiological self, or the physical or digital realms they occupy/access, making the individual an integral element of the 'internet of things'.[10] The resultant 'enhanced human' not only has new material characteristics, but also new sentient and sapient capacities (i.e. to experience sensation or to reason and cultivate insight). In all cases, the results are new forms of co-dependent

---

[4] M. Foucault, *The Order of Things: An Archaeology of the Human Sciences* (London: Routledge, 1966).

[5] See D. Haraway, *A Cyborg Manifesto: Science, Technology and Social Feminism in the Late Twentieth Century* (London: Routledge, 1991).

[6] N. Hayles, *How We Become Posthuman: Virtual Bodies in Cybernetics, Literature and Informatics* (University of Chicago Press, 1999); S. Wilson, 'The Composition of Posthuman Bodies', (2017) *International Journal of Performance Arts & Digital Media*, 13(2), 137–152.

[7] D. Serlin, *Replaceable You: Engineering the Body in Postwar America* (University of Chicago Press, 2004).

[8] S. Harmon et al., 'New Risks Inadequately Managed: The Case of Smart Implants and Medical Device Regulation', (2015) *Law, Innovation & Technology*, 7(2) 231–252; G. Haddow et al., 'Implantable Smart Technologies: Defining the 'Sting' in Data and Device,' (2016) *Health Care Analysis*, 24(3), 210–227.

[9] M. Donnarumma, 'Beyond the Cyborg: Performance, Attunement and Autonomous Computation', (2017) *International Journal of Performance Arts & Digital Media*, 13(2), 105–119; A. Brown et al., 'Body Extension and the Law: Medical Devices, Intellectual Property, Prosthetics and Marginalisation (Again)', (2018) *Law, Innovation & Technology*, 10(2), 161–184; M. Quigley and S. Ayihongbe, 'Everyday Cyborgs: On Integrated Persons and Integrated Goods', (2018) *Medical Law Review*, 26(2), 276–308.

[10] The billions of objects linked in networks and exchanging information now includes us, all melting into the fabric of our personal, social, and commercial environments: S. Gutwirth, 'Beyond Identity?', (2008) *Identity in the Information Society*, 1(1), 123–133.

human-technology embodiment. Even more radical high-conscious beings can be envisioned. Examples include genetically designed humans, synthetically constructed biological beings, and artificially intelligent constructs with consciousness and self-awareness.[11] The possibility of more radical high-conscious beings raises questions about status that are beyond the scope of this chapter.[12]

39.3 CORE CONCEPTS IMPLICATED BY THE NEW HUMAN ASSEMBLAGE

This increasingly complex and commonplace integration of bodies and technologies has given rise to theories of posthumanism and new materialism to which the law remains largely ignorant.[13] For example, there is a growing understanding of the person as an 'assemblage', a variably integrated collection of physiological, technological and virtual elements that are in fluid relation to one another, with some elements becoming prominent in some contexts and others in other contexts, with no one element being definitive of the 'person'.[14] The person has become protean, with personhood-defining/shaping characteristics that are always shifting, often at the instigation of enhancing technologies. This conditional state – or variable assemblage – with its integration and embodiment of the technological, makes concepts such as autonomy, privacy, integrity, and identity more socially and legally significant than ever before. For reasons of space, I consider just integrity and identity.

Integrity often refers to wholeness or completeness, which has both physical and emotional elements, both of them health-influencing. Having physical integrity is often equated with conformity to the 'normal' body. The normativity of this concept has resulted in prosthetic users being viewed as lacking physical integrity.[15] However, there is a growing body of literature suggesting that physical integrity need not impose compliance with the 'normal' body.[16] Tied to the state of physical integrity – however we might define it – is the imperative to preserve physical integrity (i.e. to respect the individual and avoid impinging on bodily boundaries), and this implicates emotional/mental integrity. One study uncovered twelve conceptions of integrity, concluding that integrity is supported or undermined by one's view of oneself, by others with whom one interacts, and by relationships.[17] Ultimately, integrity is a state of physical and emotional/existential wellness, both of which are influenced by internalities and externalities, including one's relationship with oneself and others. Critical elements of integrity – feelings of

---

[11] The first practical technology for genetically designed humans – CRISPR Cas-9 – is being refined and applied: S. Harmon, 'Gene-Edited Babies: A Cause for Concern', (2019, *Impact Ethics*), www.impactethics.ca/2019/03/08/genome-edited-babies-a-cause-for-concern. Synthetic beings would be the result of designed biological systems relying on existing and new DNA sequences and assembled to support natural evolution: J. Boeke et al., 'The Genome Project—Write', (2016) *Science*, 353(6295), 126–127. Multiple fields are working on artificial human-type cognitive function, which involves perception, processing, planning, retention, reasoning, and subjectivity: V. Müller (ed.), *Fundamental Issues of Artificial Intelligence* (Cham, Switzerland: Springer, 2016).

[12] D. Lawrence and M. Brazier, 'Legally Human? "Novel Beings" and English Law', (2018) *Medical Law Review*, 26(2), 309–327.

[13] R. Braidotti, *The Posthuman* (Polity Press, 2013); R. Dolphijn and I. van der Tuin, *New Materialism: Interviews and Cartographies* (Open Humanities Press, 2012).

[14] G. Deleuze and F. Guattari, *A Thousand Plateaus* (London: Continuum, 1987); M. DeLanda, *Assemblage Theory* (Edinburgh University Press, 2016).

[15] T. Tamari, 'Body Image and Prosthetic Aesthetics: Disability, Technology and Paralympic Culture', (2017) *Body & Society*, 23(2), 25–56.

[16] S. Harmon et al., 'Moving Toward a New Aesthetic' in S. Whately et al. (eds), *Dance, Disability and Law: Invisible Difference* (Bristol: Intellect, 2018) pp. 177–194.

[17] I. Widäng and B. Fridlund, 'Self-Respect, Dignity and Confidence: Conceptions of Integrity among Male Patients', (2003) *Journal of Advanced Nursing*, 42(1), 47–56.

wholeness, of being 'onself', or of physical security,[18] notions of optimal functioning, interactions with others,[19] and so on – are agitated when technologies are introduced into the body, and there is scope for the law to modulate this agitation, and encourage wellness.

Identity has been described as a mix of *ipse* and *idem*[20] (see also Postan, Chapter 23 of this volume). *Ipse* refers to 'self-identity', the sense of self of the human person, which is reflexive and influenced by internalities such as values and self-perceptions. It is the point from which the individual sees the world and herself; there is nothing behind or above it, it is just there at the source of one's will and energy, and it is persistent, continuous through time and space but by no means stable.[21] *Idem* refers to 'sameness identity', or the objectification of the individual that stems from categorisation. One might hold several *idem* identities depending on the social, cultural, religious or administrative groups to which one belongs (i.e. the range of public statuses that may be assigned at birth or throughout life, or imposed by others). It expresses the belonging of one to a category, facilitating social integration. Ultimately, identity is both internal and fluid, and external and equally fluid, but also potentially static.[22] It can be constructed, chosen or imposed. It can be fragmented and aggregated, and it can be commodified. Both *ipse* and *idem* elements will be shaped by enhancing technologies, both mechanical and biological, which have been described as 'undoing the conventional limits of selfhood and identity'.[23] Empirical research has found that both elements of identity in prosthetic users, for example, are deeply entangled with their devices.[24]

Of course, neither integrity nor identity are unknown to the law. Criminal law seeks to protect our physical integrity, and it punishes incursions against it. Human rights law erects rights to private and family life, which encompass moral and physical integrity and the preservation of liberty.[25] Health law erects rights to physical and mental integrity through mechanisms such as consent, best interests and least restrictive means.[26] Law is also a key external shaper of identity, creating groups based on factors such as developmental status (i.e. rights of fetuses to legal

---

[18] G. Haddow et al., 'Cyborgs in the Everyday: Masculinity and Biosensing Prostate Cancer', (2015) *Science as Culture*, 24(4), 484–506.

[19] How others perceive us is linked to how they look at us. Staring is the complex phenomenon of observation and internalisation with many facets: R. Garland-Thomson, *Staring: How We Look* (Oxford University Press, 1996). It is often defined as an oppressive act of disciplinary looking that subordinates the subject: L. Mulvey, 'Visual Pleasure and Narrative Cinema', (1975) *Screen*, 16(3), 6–18; F. Michel, *Foucault Live: Interviews, 1961–1984* (Semiotext(e), 1996); A. Clark, 'Exploring Women's Embodied Experiences of 'The Gaze' in a Mix-Gendered UK Gym', (2017) *Societies*, 8(1), 2.

[20] M. Hildebrandt, 'Profiling and the Identity of the European Citizen' in. M Hildebrandt and S. Gutwirth (eds), *Profiling the European Citizen: Cross-Disciplinary Perspectives* (Berlin: Springer, 2008), pp. 303–326.

[21] Gutwirth, 'Beyond Identity?'

[22] S. Lasch and J. Friedman (eds), *Modernity and Identity* (Oxford: Blackwell, 1992); D. Polkinghorne, 'Explorations of Narrative Identity', (1996) *Psychological Inquiry*, 7(4), 363–367; A. Blasi and K. Glodis, 'The Development of Identity: A Critical Analysis from the Perspective of the Self as Subject', (1995) *Developmental Review*, 15(4), 404–433; L. Huddy, 'From Social to Political Identity: A Critical Examination of Social Identity Theory', (2001) *Political Psychology*, 22(1), 127–156.

[23] M. Shildrick, 'Individuality, Identity and Supplementarity in Transcorporeal Embodiment' in K. Cahill et al. (eds), *Finite but Unbounded: New Approaches in Philosophical Anthropology* (Berlin: de Gruyter, 2017), pp. 153–172, p. 154.

[24] S. Popat et al., 'Bodily Extensions and Performance', (2017) *International Journal of Performance Arts & Digital Media*, 13(2), 101–104.

[25] *Husayn v Poland* (2015) 60 EHRR 16 (ECHR). See also *Dickson v UK* (2008) 46 EHRR 41 (Grand Chamber).

[26] The 'Mental Capacity Act 2005' stipulates that third-party decision-makers must make decisions that are only in the subject person's best interest as understood from the perspective of that person. Where a decision interferes with the person's physical integrity, the option that represents the least restrictive means must be adopted.

standing and protection),[27] sexual orientation (i.e. right to marriage or work benefits)[28] and gender (i.e. right to gender identity recognition).[29] It also defines 'civil identity', a common condition for access to basic services.[30] And notions of identity have been judicially noticed in relation to new technologies: in *Rose* v *Secretary of State for Health*,[31] which concerned disclosure of information about artificial insemination, the court found that information about biological identity went to the heart of identity and the make-up of the person, and that identity included details of origins and opportunity to understand them, physical and social identity, and also psychological integrity.

Unfortunately, these two increasingly important concepts have not been well-handled by the law. They are subject to very different interpretations depending on one's view of human rights as negative or positive.[32] Severe limits have been placed on the law being used to enable or impose those conditions that *facilitate* individuals living lives of meaning and becoming who they are (or wish to be). Narrow views as to what counts as a life of worth have resulted in limitations being placed on what individuals can do to become who they wish to be, with decision-makers often blind to the choices actually available (e.g. consider discourses around a 'good death' and medical assistance in dying). Thus, at present, neither integrity nor identity are consistently articulated or enabled by law. This could be the result of their multifaceted nature, or of the negative approach adopted in protecting them,[33] or of the indirectness of the law's interest in them.[34] The question of their treatment in health research regulation remains, and it is to this that we now turn.

39.4 CORE CONCEPTS AND MEDICAL DEVICE REGULATION

The market authorisation framework for medical devices is an example of health research regulation that shapes the nature, application and integrative characteristics of many enhancing technologies. Thus, it is profoundly linked to practices aimed at expanding and diversifying the human assemblage, and so it might be expected to appreciate, define and/or facilitate the concepts identified above as being critical to the person. In Europe, the development and market authorisation of medical devices is governed by the previously noted MDR and IVDR, both of which came into force in May 2017, but which will not be fully implemented until May 2020 and May 2022 respectively.[35]

---

[27] *Vo* v *France* (2005) 40 EHRR 12 (ECHR).

[28] International Covenant on Civil and Political Rights (1966), Art. 23(2); International Covenant on Economic, Social and Cultural Rights (1966), Arts. 6(1) and 7.

[29] European Convention on Human Rights and Fundamental Rights (1951), Art. 8 (right to private life); *Goodwin* v *United Kingdom* (28957/95) [2002] IRLR 664.

[30] E. Mordini and C. Ottolini, 'Body Identification, Biometrics and Medicine: Ethical and Social Considerations', (2007) *Annali dell'Istituto Superiore di Sanità*, 43(1), 51–60.

[31] [2002] EWHC 1593 (Admin).

[32] J. Marshall, *Personal Freedom through Human Rights Law?* (Leiden: Martinus Nijhoff, 2009).

[33] The existing right to privacy is extremely limited, and predominantly 'negative', not allowing the construction of positive claims related to identity: P. De Hert, *A Right to Identity to Face the Internet of Things* (Strasbourg: Council of Europe Publishing, 2007), www.cris.vub.be/files/43628821/pdh07_Unesco_identity_internet_of_things.pdf.

[34] S. Harmon et al., 'Struggling to be Fit: Identity, Integrity, and the Law', (2017) *Script-ed*, 14(2), 326–344.

[35] European Commission, 'Medical Devices: Regulatory Framework', (European Commission), www.ec.europa.eu/growth/sectors/medical-devices/regulatory-framework_en; CAMD Implementation Taskforce, 'Medical Devices Regulation/In-vitro Diagnostics Regulation (MDR/IVDR) Roadmap', (2018). During the transition, devices can be placed on the market under the new or old regime. It is unclear what impact these Regulations will have post-Brexit, but the UK, which implemented the old regime through the Medical Devices Regulations 2002, will have to comply with EU standards if it wishes to continue to trade within the EU. The Medicines and Healthcare products Regulatory

As will be clear from other chapters, the framing of regulatory frameworks is critical. Framing signals the regime's subject and objective; it shapes how its instruments articulate problems, craft solutions and measure success. It has been observed that the identification, definition and control of 'objects' is a common aim of regulatory instruments; specific objects are chosen because they represent an opportunity for commerce, a hazard to human health, or a boon – or danger – to social architecture.[36] Certain fields focus on certain objects, with the result that silos of regulation emerge, each defined by its existence-justifying object, which might be data, devices, drugs, tissue and embryos, etc., and the activity in relation to that object around which we wish to create boundaries (i.e., production, storage, use).

The MDR and IVDR are shaped by EU imperatives to strengthen the common market and promote innovation and economic growth, and are thus framed as commercial instruments.[37] Their subject is *objects* (e.g. medical devices), not people, not health outcomes and not well-being. MDR Article 1 articulates this frame and subject, stating that it lays down rules concerning placing or making available on the market, or putting into service, medical devices for human use in the EU.[38] MDR Article 2(1) defines medical device as any instrument, apparatus, appliance, software, implant, reagent, material or other article intended to be used, alone or in combination, for human beings for a range of specified medical purposes (e.g. diagnosis, prevention, monitoring, prediction, prognosis, treatment, alleviation of disease, injury or disability, investigation, replacement or modification of the anatomy, providing information derived from the human body) that does not achieve its principal intended action by pharmacological, immunological, or metabolic means.

The MDR and IVDR construct their objects simultaneously as 'risk objects', 'innovation objects' and 'market objects', highlighting one status or another depending on the context and the authorisation stage reached. All three constructions can be seen in MDR Recital 2 (which is mirrored by IVDR Recitals 1 and 2):

> This Regulation aims to ensure the smooth functioning of the internal market as regards medical devices, taking as a base a high level of protection of health for patients and users, and taking into account the ... enterprises that are active in this sector. At the same time, this Regulation sets high standards of quality and safety for medical devices in order to meet common safety concerns as regards such products. Both objectives are being pursued simultaneously and are inseparably linked ...

They then classify devices on a risk basis, and robust evidence and post-market surveillance is imposed to protect users from malfunction. For example, MDR Recital 59 acknowledges the insufficiency of the old regime, stating that it is necessary to introduce specific classification rules sensitive to the level of invasiveness and potential toxicity of devices that are composed of substances that are absorbed by, or locally dispersed in, the human body; where the device

Agency has highlighted its desire to retain a close working partnership with the EU: MHRA, 'Medical Devices: EU Regulations for MDR and IVDR', www.gov.uk/guidance/medical-devices-eu-regulations-for-mdr-and-ivdr; Medical Devices (Amendment etc.) (EU Exit) Regulations 2019, not yet approved.

[36] G. Laurie, 'Liminality and the Limits of Law in Health Research Regulation', (2017) *Medical Law Review*, 25(1), 47–72; C. McMillan et al., 'Beyond Categorisation: Refining the Relationship Between Subjects and Objects in Health Research Regulation', (2021) Law, Innovation and Technology, doi: 10.1080/17579961.2021.1898314.

[37] MDR Recital 2 cites the Treaty of Union as a foundation for its remit to harmonise the rules for market-access and free-movement of goods, and for setting high standards of device quality and safety.

[38] IVDR Article 1 parallels this language for *in vitro* diagnostic medical devices, which are defined as any medical device that is a reagent, reagent product, calibrator, control material, kit, instrument, apparatus, piece of equipment, software or system, whether used alone or in combination, intended to be used *in vitro* for the examination of specimens, including blood and tissue donations, derived from the human body, for a number of purposes.

performs its action, where it is introduced or applied, and whether a systemic absorption is involved are all factors going to risk that must be assessed. MDR Recital 63 states that safety and performance requirements must be complied with, and that, for class III and implantable devices, clinical investigations are expected.[39] These directions are operationalised in MDR Chapters V (Classification and Conformity Assessments),[40] and VI (Clinical Evaluation and Clinical Investigations).[41] IVDR Recitals 55 and 61, and Chapters V and VI are substantively similar.

The above framing imposes a substantial fetter on what these instruments are intended to do, or are capable of doing. It serves to largely erase the person and personal experience from their perspective and remit. The recipient of a device is constructed as little more than a consumer who must be protected from the harm of a malfunctioning device. An example of the impoverished position of the person is the Regulations' treatment of risk. They rely on a narrow understanding of risk, framing it as commercial object safety at various stages of development and roll-out. Other types of risks and harms are marginalised or ignored.[42] Thus, there is no acknowledgement that their objects – medical devices – will not always be – and will really only briefly be – 'market objects'. Many devices will become 'physiological objects' that are profoundly personal to, and intimate with, the recipient. Indeed, many will cease to be 'objects' altogether, becoming instead components of the human assemblage, undermining or facilitating integrity, and exerting pressures on identity. As such, the nature of the risks they pose changes relatively quickly, and more so over time.

Had broader human well-being or flourishing been foregrounded, then greater attention to public interest beyond device safety might have been expected. Had legislators given any consideration to the consequences of these technologies once integrated with the person and becoming a part of that human assemblage, then further conditions for approval might have been expected. Developers might have been asked to present social evidence about the actual need for the device, or the potential social acceptance of the device, or how the device is expected to interact with other major – or common – health or social technologies, systems, or practices. In short, the patient, or the non-patient user, may have featured in the market access assessment.

The one exception to the Regulations' ignorance of social experience is that relating to post-market surveillance. MDR Recital 74 requires manufacturers to play an active role in the post-market phase by systematically gathering information on experiences with their devices via a comprehensive post-market surveillance system. It is operationalised in Chapter VII.[43] However, while these provisions are useful, they fail to acknowledge the now embodied condition of the regulatory object, and the new personal, social, ethical and cultural significance that it holds. In

[39] The insufficient nature of the Regulations' transparency of clinical evidence to front-line actors has been noted: A. Fraser et al., 'The Need for Transparency of Clinical Evidence for Medical Devices in Europe', (2018) *Lancet*, 392 (10146), 521–530.

[40] MDR Arts. 51–60. Art. 51 creates the classes I, IIa, IIb and III, which are informed by the device's intended purposes and inherent risks.

[41] MDR Arts. 61–82. Art. 61 states that clinical data shall inform safety and performance requirements under normal conditions of intended use, the evaluation of undesirable side-effects and the risk/benefit ratio.

[42] This narrowing has been recognized in the broader health technologies context: M. Flear, 'Regulating New Technologies: EU Internal Market Law, Risk and Socio-Technical Order' in M. Cremona (ed.), *New Technologies and EU Law* (Oxford University Press, 2016), pp. 74–122.

[43] MDR Arts. 83–100. Art. 83 states that manufacturers shall plan, establish, document, implement, maintain and update a post-market surveillance system for each device proportionate to the risk class and appropriate for the device type. IVDR Recital 75 and Chapter VII are substantively similar.

other words, they evince an extremely 'bounded' perspective of their objects. Such has been criticised:

> The attention of law and regulation on 'bounded objects' . . . should be questioned on at least two counts: first, for the fallacy of attempting to 'fix' such regulatory objects, and to divorce them from their source and the potential impact on identity for the subjects themselves; and, second, for the failure to see such objects as also experiencing liminality.[44]

This is pertinent to situations where technologies are integrated into the body, situations which exemplify van Gennep's pattern of experience: separation from existing order; liminality; re-integration into a new world.[45] The features of this new world are that the regulatory object (device) becomes embodied and incorporated in multiple ways – physical, functional, psychological and phenomenological.[46] Both the object (device) and subject (host) are transformed as a result of this incorporation such that the typical subject–object dichotomy entrenched in the law is not appropriate;[47] the Regulations' object-characterisations are no longer apropos and their indifference to the subject is potentially unjust given the 'new world' that now exists.

This cursory assessment suggests that the Regulations are insufficient and misdirected from the perspective of ensuring that the full public interest is met through the regulated activity. As previously observed, new and emerging technologies can be conceptually, normatively and practically disruptive.[48] Technologies applied to humans for purposes of integration – treatment or enhancement – are disruptive on all three fronts, particularly once they enter society. Conceptually, they disrupt existing definitions and understandings of the regulatory objects, which are transformed once they form part of the human assemblage. Normatively, they disrupt existing regulatory concepts like risk, which are exposed as being too narrow in light of how these objects might interact with and harm individuals. Practically, they disrupt existing medical practice – blurring the lines between treatment and enhancement – and regulatory practices – troubling the oft-relied-on human/non-human and subject/object dichotomies.

This assessment also suggests that the historical boundaries between, or categories of, 'devices' and 'medicines', are increasingly untenable because of the types of devices being designed (e.g. implanted mechanical devices and mixed material devices that interact with the physiological, sometimes through the release of medicines). This area of human health research therefore highlights both fault-lines within instruments and empty spaces between them. It might be that the devices and medicines regimes need to be brought together, with a realignment of the regulatory objects and a better understanding of where these objects are destined to operate.

39.5 CONCLUSIONS

As the enhanced human becomes more ubiquitous, and the radical posthuman comes into being, narrow or negative views of integrity and identity become ever more attenuated from the technologically shaped lived experience. Moreover, the greater the human/technology integration, the greater the engagement of integrity and identity. Insufficient attention to these concepts in regulatory frames, norms and decisions raises the likelihood that such will undermine rather

---

[44] Laurie, 'Liminality and the Limits of Law', 68. Also, McMillan et al., 'Beyond Categorisation'.

[45] A. van Gennep, *The Rites of Passage* (University of Chicago Press, 1960).

[46] Quigley and Ayihongbe, 'Everyday Cyborgs', 305.

[47] D. Dickenson, *Property in the Body: Feminist Perspectives*, 2nd Edition (Cambridge University Press, 2017).

[48] R. Brownsword et al. (eds), 'Introduction', *Oxford Handbook of the Law and Regulation of Technology* (Oxford University Press, 2017), pp. 3–38.

than support or protect human well-being. Only with clear recognition will the self-creation –
the being and becoming – that they underwrite be facilitated through the positive shaping of
social conditions.[49]

The MDR and IVDR are directly implicated in encouraging, assessing and rolling out
integrative technologies destined for social and clinical uses, but they do not match the
technical innovation they manage with sufficient regulatory recognition of the integrity and
identity that is engaged. Despite their recent reform, they do not evince a greater regulatory
understanding of the *common* natures and consequences of tissue, organ and technological
artifacts, and they therefore do not represent a significantly improved – more holistic and less
silo-reliant – regulatory framework.

Had they adopted a broader perspective and value base, they would have taken notice of
people as subjects, and crafted a framework that contributed to the development of innovations
that are not only safe, but also supportive of – or at least not corrosive to – what people value,
including integrity and identity. At base, they would have benefited from:

- a clearer and broader value base;
- an emphasis on decisional principles rather than narrow (technical) objects on which
  rules are imposed; and
- greater notice of what the devices become once they are through the market-
  access pipeline.

Ultimately, medical device regulation is an example of health research regulation that operates
in an area where innovation has created disturbances, and those disturbances have not been
resolved. Though some have been acknowledged – leading to the new regime – the real
disturbances have hardly been appreciated.

---

[49] De Hert, note 33, argues that there ought to be a clear right to identity because people cannot function without it; it is
like living, breathing, or being free to feel and think, all of which are minimal requirements for social justice in a
rights-conscious society. Such recognition of identity paves the way for identity to be recognised as a right protected by
law. He says that 'states should undertake to respect the right of each person to preserve and develop his or her ipse and
idem identity without unlawful interference' (1). For more on identity as an emerging legal concept: L. Downey,
Emerging Legal Concepts at the Nexus of Law, Technology and Society: A Case Study in Identity, unpublished PhD
thesis, University of Edinburgh (2017).

Afterword

What Could a Learning Health Research Regulation System Look Like?

Graeme Laurie

1 INTRODUCTION

This final chapter of the *Cambridge Handbook of Health Research Regulation* revisits the question posed in the Introduction to the volume: *What could a Learning Health Research Regulation System look like?* The discussion is set against the background of debates about the nature of an effective learning healthcare system,[1] building on the frequently expressed view that any distinction between systems of healthcare and health research should be collapsed or at the very least minimised as far as possible. The analysis draws on many of the contributions in this volume about how health research regulation can be improved, and makes an argument that a framework can be developed around a Learning Health Research Regulation System (LHRRS). Central to this argument is the view that successful implementation of an LHRRS requires full integration of insights from bioethics, law, social sciences and the humanities to complement and support the effective delivery of health and social value from advances in biomedicine, as well as full engagement with those who regulate, are regulated, and are affected by regulation.

2 LESSONS FROM LEARNING HEALTHCARE SYSTEMS AND REGULATORY SCIENCE

The US Institute of Medicine is widely credited for making seminal contributions to debates about the nature of learning healthcare systems, primarily through a series of expert workshops and reports examining the possible contours of such systems. A central feature of the normative frameworks proposed relies on the collapsing – or at least a blurring – of any distinction between objectives in the delivery of healthcare and the objectives of realising value from human health research. The normative ideal has been articulated as follows:

> . . . a system in which advancing science and clinical research would be natural, seamless, and a real-time byproduct of each individual's care experience; highlighted the need for a clinical data trust that fully, accurately, and seamlessly captures health experience and improves society's knowledge resource; recognized the dynamic nature of clinical evidence; noted that standards

[1] As we go to press, we are heartened to read a blog by Natalie Banner, 'A New Approach to Decisions about Data', in which she advocates for the idea of 'learning governance' and with which we broadly agree. N. Banner, 'A New Approach to Decisions about Data' (*Understanding Patient Data*, 2020), www.understandingpatientdata.org.uk/news/new-approach-decisions-about-data.

should be tailored to the data sources and circumstances of the individual to whom they are applied; and articulated the need to develop a supporting research infrastructure.[2]

It is the challenge of developing and delivering a 'supporting research infrastructure' that is the core concern of all contributions to this volume. We have stated at the outset that our approach is determinedly normative in tackling what we believe to be the central features of any ecosystem of health research regulation. The structure and content of the sections of this volume reflect our collective belief that the design and delivery of any effective and justifiable system of human health research must place the *human* at the centre of its endeavours. Also, when seeking to design systems from the bottom-up, so to speak, we contend that this human-centred approach to systems must go beyond patient-centredness and exercises in citizen engagement. In no way is this to suggest that these objectives are unimportant; rather, it is to recognise that these endeavours are only part of the picture and that a commitment to delivering a whole system approach must integrate both these and other elements into any system design.

There is, of course, the fundamental question of where does one begin when attempting system design? Each discipline and field of enquiry will have its own answer. As an illustration, we can consider a further workshop held in 2011 under the auspices of the Institute of Medicine and other bodies; this was a Roundtable on value and science-driven healthcare that sought to 'apply systems engineering principles in the design of a learning healthcare system, one that embeds real-time learning for continuous improvement in the quality, safety, and efficiency of care, while generating new knowledge and evidence about what works best'.[3] Once again, these are manifestly essential elements of any well-designed system, but it is striking that this report makes virtually no mention of the ethical issues at stake. To the extent that ethics are mentioned, this is presented as part of the problem of current fragmentation of systems,[4] rather than as any part of a systems solution: '[e]ach discipline has its own statement of its ethics, and this statement is nowhere unified with another. There is no common, shared description of the ethical center of healthcare that applies to everybody, from a physician to a radiology technician to a manager'.[5]

Furthermore, while the Roundtable was styled as being about value- and science-driven healthcare, it is crucial to ask what is meant by 'value' in this context of systems design. Indeed, the Roundtable participants did call for greater enquiry into the terms, but as it was characterised in various presentations and discussions, the term was used variously to refer to:

- *Value* to consumers;
- *Value* from 'substantially expanded use of clinical data';[6]
- *Value* in accounting for costs in outcomes and innovation
- *Value* in 'health returned for dollars invested';[7]
- *Value* as something to be measured for inclusion in decision-making processes.[8]

[2] Institute of Medicine, 'Patients Charting the Course: Citizen Engagement and the Learning Health System' (*Institute of Medicine*, 2011), 240.
[3] National Academy of Engineering, 'Engineering a Learning Healthcare System: A Look at the Future' (*National Academy of Engineering*, 2011).
[4] For an analysis of ethics as a (problematic?) negotiated regulatory tool in the neurosciences, see Pickersgill, Chapter 31, this volume.
[5] National Academy of Engineering, 'Engineering a Learning Healthcare System', 5.
[6] Ibid., 4.
[7] Ibid., 21.
[8] Ibid., 22.

As extensively demonstrated by the chapters in this volume, there is a crucial distinction between 'value' seen in these terms and the 'values' that underpin any structure or system designed to deliver individual and social benefit through improved health and well-being. This distinction is, accordingly, the focus of the next section of this chapter.

Before this, a further important distinction between healthcare systems and health research systems must be highlighted. As an earlier Institute of Medicine report noted, patient-centred care is of paramount importance in identifying and respecting the preferences, needs and values of patients receiving healthcare.[9] This position has rightly been endorsed in subsequent learning systems reports.[10] However, for LHRRS, and from a values perspectives, there is arguably a wider range of interests and values at stake in conducting health research and delivering benefits to society.[11] This is the principal reason why this volume begins with an account of key concepts in play in human health research (see Section IA), because this provides a solid platform on which to conduct multidisciplinary, multisector discussions about what is important, what is at risk, and what accommodations should be made to take into account the range of interests that are engaged in health research. This is a further reason why, in Section IIA, our contributors engage critically and at length with the private and public dimensions of health research regulation.

From this, two important top-level lessons arise from this volume:

- There is considerable value in taking a multi- and inter-disciplinary approach to systems design that places bioethics, social sciences and humanities at the centre of discussions because these disciplinary perspectives are crucial to ensuring that the *human* remains at the focus of human health research; indeed, an aspiration to trans-disciplinary contributions would not be remiss here.
- There is a need for further and fuller enquiry into ways in which the values underpinning healthcare and health research do, and do not, align, and how these can be mobilised to improve regulatory design.

On this last point, we can look to recent initiatives in Europe and the UK that have as their focus 'regulatory science' and we can ask further how the contributions in this volume can add to these debates.

A July 2020 report from the UK advocated for innovation in 'regulatory science' as it relates to healthcare in order to complement the nation's industrial strategy, to enable accelerated routes to market; to increase benefits to public health; to assure greater levels of patient safety; to influence international practice; and to promote investment in the UK (Executive Summary).[12] 'Regulatory science' is defined therein as '[t]he application of the biological, medical and sociological sciences to enhance the development and regulation of medicines and devices in order to meet the appropriate standards of quality, safety and efficacy'.[13] The authors prefer this definition among others as a good starting point for further deliberation and action, both for its breadth and inclusiveness as to what should be considered to be in play. The report offers a very full account of the present regulatory landscape in the UK and offers a strong set of recommendations for

[9] Institute of Medicine, 'Crossing the Quality Chasm: A New Health System for the 21st Century' (*Institute of Medicine*, 2001).
[10] National Academy of Engineering, 'Engineering a Learning Healthcare System', 5.
[11] For an account of ethical considerations in a learning healthcare system, see R. R. Faden et al., 'An Ethics Framework for a Learning Health Care System: A Departure from Traditional Research Ethics and Clinical Ethics' (2013) *Hastings Center Report*, 43(s1), S16–S27.
[12] M. Calvert et al., 'Advancing Regulatory Science and Innovation in Healthcare' (*Birmingham Health Partners*, 2020).
[13] Calvert et al., 'Advancing Regulatory Science,' 6, citing S. Faulkner, 'The Development of Regulatory Science in the UK: A Scoping Study' (*CASMI*, 2018).

improvement in four areas: (i) strategic leadership and coordinated support, (ii) enabling innovation, (iii) implementation and evaluation, and (iv) workforce development. However, a striking
omission from the report is any direct and explicit mention of how 'sociological sciences', let alone
bioethical inquiry, can contribute to the delivery of these objectives.

Similarly, in March 2020, the European Medicines Agency (EMA) published its strategy,
'Regulatory Science to 2025'. The stated aim is 'to build a more adaptive regulatory system that
will encourage innovation in human and veterinary medicine'. For the EMA, regulatory science
refers to

> the range of scientific disciplines that are applied to the quality, safety and efficacy assessment of
> medicinal products and that inform regulatory decision-making throughout the lifecycle of a
> medicine. It encompasses basic and applied biomedical and social sciences and contributes to
> the development of regulatory standards and tools.[14]

As with the UK report, however, there is no more than a cursory mention of the concrete ways in
which social sciences and bioethics contribute to these objectives.[15]

This returns us to the key questions that frame this Afterword: where is the human in human
health research? Also, what would a whole-system approach look like when we begin with the
human values at stake and design systems accordingly?

3 FROM VALUE TO VALUES

Currently, there is extensive discussion and funding of data-driven innovation, and undoubtedly,
there is considerable value in the raw, aggregate, and Big Data themselves. However, given that in
biomedicine the data in question predominantly come from citizens in the guise of their personal
data emanating from a growing number of areas of their private lives, it is the contention of this
Afterword, and indeed the tenor of this entire volume, that it is ethics and values that must drive
the regulation that accompanies the data science, and not a science paradigm. As the conclusion
to the Introduction of this volume makes clear, public trust is vital to the success of the biomedical
endeavour, and any system of regulation of biomedical research must prove itself to be trustworthy. As further demonstrated by various contributions to this volume,[16] a failure to address
underlying public values and concerns in health research and wider uses of citizens' data can
result in a net failure to secure social licence and doom the initiatives themselves.[17] This has been
re-enforced most recently in February 2020 by an independent report commissioned by
Understanding Patient Data and National Health Service (NHS) England that found, among
other things, that NHS data sharing should be undertaken by partnerships that are transparent and
accountable, and that are governed by a set of shared principles (principles being a main way in
which values are captured and translated into starting points for further deliberation and action).[18]

---

[14] European Medicines Agency, 'EMA Regulatory Science to 2025: Strategic Reflection' (*EMA*, 2020), 5.
[15] For a plea to recognise the value of upstream input from the social sciences and humanities, see M. Pickersgill et al.,
'Biomedicine, Self and Society: An Agenda for Collaboration and Engagement' (2019) *Wellcome Open Research*, 4(9),
https://doi.org/10.12688/wellcomeopenres.15043.1.
[16] Kerasidou, Chapter 8, Aitken and Cunningham-Burley, Chapter 11, Chuong and O'Doherty, Chapter 12, and Burgess,
Chapter 25, this volume.
[17] P. Carter et al., 'The Social Licence for Research: Why *care.data* Ran into Trouble' (2015) *Journal of Medical Ethics*,
40(5), 404–409.
[18] H. Hopkins et al., 'Foundations of Fairness: Views on Uses of NHS Patients' Data and NHS Operational Data'
(*Understanding Patient Data*, 2020), www.understandingpatientdata.org.uk/what-do-people-think-about-third-parties-
using-nhs-data#download-the-research.

Thus, we posit that any learning system for human health research must be *values-driven*. To reiterate, this explains and justifies the contributions in Section IA of the volume that seek to identify and examine the key values and core concepts that are at stake. Normatively, it would not be helpful or appropriate for this chapter to attempt to suggest or prescribe any particular configuration of values to deliver a justifiable learning system. This depends on myriad social, cultural, economic, institutional and ethical factors within a given country or jurisdiction seeking to implement an effective system for itself. Rather, we suggest that *values engagement* is required amongst all stakeholders implicated in, and affected by, such a system in its given context, and this is the work done by Section IB of the volume in identifying key actors, including publics, and demonstrating through examples how regulatory tools and concepts have been used to date to regulate human health research. Many of these remain valid and appropriate after years of experiences, albeit that the analysis herein also reveals limitations and caveats to existing approaches, for example with consent[19] and proportionality,[20] while also demonstrating the means by which institutions can show trustworthiness[21] and/or conduct meaningful engagement with publics and other stakeholders.[22]

But to understand what it means for a system to be truly effective in self-reflection and learning, we can borrow once again from discussion in the learning healthcare context. As Foley and Fairmichael have pointed out: 'Learning Healthcare Systems can take many forms, but each follows a similar cycle of assembling, analysing and interpreting data, followed by feeding it back into practice and creating a change'.[23] The same is true for a HRRLS. Thus, a learning system is one that consists not only of processes designed to deliver particular outcomes, but also one that has feedback loops[24] and processes of capturing evidence of what has worked *less* well.[25] Self-evidently from the above discussion, 'data' in this context will include data and information about *values failure*[26] or incidents or points in the regulatory processes where sight has been lost of the original values that underpin the entire enterprise.

From the perspective of regulatory theory and practice, this issue relates to the ever-present issue of *sequencing*: when, and at what point in a series of processes should certain actions or instruments be engaged to promote key regulatory objectives?[27] In regulatory theory, sequencing is often concerned with escalation of regulatory intervention, that is, invoking a particular regulatory response when (and only when) other regulatory responses fail. However, this need not be the case. Early and sequential feedback loops in the design and delivery of a system can help to prevent wider systemic failure at a later point in time. This is especially the case if ethical sensitivities to core values remain logically prior to techno-scientific considerations of risk management and are

[19] Kaye and Prictor, Chapter 10 this volume.
[20] Schaefer, Chapter 3 this volume.
[21] Kerasidou, Chapter 8 this volume.
[22] Aitken and Cunningham-Burley, Chapter 11, Chuong and O'Doherty, Chapter 12, and Burgess, Chapter 25 this volume.
[23] T. Foley and F. Fairmichael, 'The Potential of Learning Healthcare Systems' (*The Learning Healthcare Project*, 2015), 4.
[24] Further on feedback loops in the health research context, see S. Taylor-Alexander et al., 'Beyond Regulatory Compression: Confronting the Liminal Spaces of Health Research Regulation' (2016) *Law, Innovation and Technology*, 8(2), 149–176.
[25] For a discussion of a learning system in the context of AI and medical devices, see Ho, Chapter 28, this volume.
[26] For a richer conceptualisation of regulatory failure than mere technological risk and safety concerns, see Flear, Chapter 16, this volume.
[27] See, generally, P. Drahos (ed), *Regulatory Theory: Foundations and Applications* (ANU Press, 2017), and more particularly R. Baldwin et al., *Understanding Regulation: Theory, Strategy, and Practice* (Oxford University Press, 2011), p. 158.

part of risk-benefit analysis.[28] Indeed, as pointed out by Swierstra and Rip, human agency can make a difference at an early stage of development/innovation, when issues and directions are still unclear, but much less so in later stages when 'alignments have sedimented'.[29]

Key among the ethical objectives of any health research system is the need to deliver social value (or at least that prospective research has a reasonable chance of doing so).[30] Some of us have argued elsewhere that there is at present an unmet need to appraise social value iteratively throughout the entire research lifecycle,[31] and this builds on existing arguments to see social value as a dynamic concept. The implications of this for a LHRRS are that the research ecosystem would extend from the research design stage through publication and dissemination of research results, to data storage and sharing of findings and new data for future research. This means that social value is not merely something promissory and illusive that is dangled before a research ethics committee as it pores over a research protocol,[32] but that it is potentially generated and transformed multiple times and by a range of actors throughout the entire process of research: from idea to impact. Seen in this way, social value itself becomes a potential metric of success (or failure) of a learning health research system, and opens the possibility that value might emerge at times and in spaces previously unforeseen. Indeed, Section IIB of this volume is replete with examples of the importance of time within good governance and regulation, whether this be about timely research interventions in the face of emergencies,[33] the appropriateness and timing of effective oversight of clinical innovation,[34] or the challenge of 'evidence' when attempting to regulate traditional and non-conventional medicines.[35]

As a final crucial point about how an ethical 'system' might be constructed with legitimacy and with a view to justice for all, we cannot overlook what Kipnis has called *infrastructural vulnerability*:

> At the structural level, essential political, legal, regulative, institutional, and economic resources may be missing, leaving the subject open to heightened risk. The question for the researcher is, 'Does the political, organizational, economic, and social context of the research setting possess the integrity and resources needed to manage the study?'
>   … [c]learly the possibility of infrastructural vulnerability calls for attention to the contexts within which the research will be done.[36]

Questions of the meanings and implication of vulnerability are addressed early in this volume as a crucial framing for the entire volume.[37] It is also clear that this concern is not one for researchers alone. This brings us to the important question of who is implicated in the design and delivery of a LHRSS ecosystem?

---

[28] On such a systemic exercise for risk-benefit see, Coleman, Chapter 13, this volume. On the blinkered view of regulation that reduces assessments only to techno-scientific assessments of risk-benefit, see Haas and Cloatre, Chapter 30.

[29] T. Swierstra and A. Rip, 'Nano-ethics as NEST-ethics: Patterns of Moral Argumentation About New and Emerging Science and Technology' (2007) *Nanoethics*, 1, 3–20, 8.

[30] See further, van Delden and van der Graaf, Chapter 4, this volume.

[31] A. Ganguli-Mitra et al., 'Reconfiguring Social Value in Health Research Through the Lens of Liminality' (2017) *Bioethics*, 31(2), 87–96.

[32] For a more dynamic account of research ethics review, see Dove, Chapter 18, this volume.

[33] Ganguli Mitra and Hunt, Chapter 32, this volume.

[34] Lipworth et al., Chapter 29, this volume.

[35] Haas and Cloatre, Chapter 30, this volume.

[36] K. Kipnis, 'Vulnerability in Research Subjects: A Bioethical Taxonomy' (2001) *Commissioned Paper*, www.aapcho .org/wp/wp-content/uploads/2012/02/Kipnis-VulnerabilityinResearchSubjects.pdf, 9.

[37] See Rogers, Chapter 1, and Brassington, Chapter 9, this volume.

4 WHO IS IMPLICATED IN THIS ECOSYSTEM, AND WITH WHICH CONSEQUENCES?

In 2019, Wellcome published its Blueprint for Dynamic Oversight of emerging science and technologies.[38] This is aimed determinedly at the UK government, and it is founded on four principles with which few could take exception.

Dynamic oversight can be delivered by reforms underpinned by the following principles:

Inclusive: Public groups need to be involved from an early stage to improve the quality of oversight while making it more relevant and trustworthy. The Government should support regulators to involve public groups from an early stage and to maintain engagement as innovation and its oversight is developed.

Anticipatory: Identifying risks and opportunities early makes it easier to develop a suitable approach to oversight. Emerging technology often develops quickly and oversight must develop with it. UK regulators must be equipped by government to anticipate and monitor emerging science and technologies to develop and iterate an appropriate, proportionate approach.

Innovative. Testing experimental oversight approaches provides government and regulators with evidence of real-world impacts to make oversight better. Achieving this needs good collaboration between regulators, industry, academia and public groups. The UK is beginning to support innovative approaches, but the Government needs to create new incentives for the testing of new oversight approaches.

Proportionate. Oversight should foster the potential benefits of emerging science and technologies at the same time as protecting against harms, by being proportionate to predicted risk. The UK should keep up its strong track record in delivering proportionate oversight. These changes will only be delivered effectively if there is clear leadership and accountability for oversight. This requires the Government to be flexible and decisive in responding to regulatory gaps.

Wellcome, A Blueprint for Dynamic Oversight, (2019)

We can contrast this top-down framework with a bottom-up study conducted by the members of the Liminal Spaces team as part of the project funding this volume. The team undertook a Delphi policy[39] study to generate empirical data and a cross-cutting analysis of health research regulation as experienced by stakeholders in the research environment in the United Kingdom. In short, the project found that:

[t]he evidence supports the normative claim that health research regulation should continue to move away from strict, prescriptive rules-based approaches, and towards flexible principle-based regimes[40] that allow researchers, regulators and publics to coproduce regulatory systems serving core principles.[41]

As a concrete illustration of why this is important, we can consider the last criterion listed as part of the Wellcome Dynamic Oversight framing: proportionality. The Delphi study revealed novel insights about how proportionality as a regulatory tool is seen and operationalised in practice.

[38] J. Clift, 'A Blueprint for Dynamic Oversight: How the UK Can Take a Global Lead in Emerging Science and Technologies' (*Wellcome*, 2019).

[39] On further policy perspectives, see Meslin, Chapter 22, this volume.

[40] On Rules, Principles, and Best Practices, see Sethi, Chapter 17, this volume.

[41] I. Fletcher et al., 'Co-production and Managing Uncertainty in Health Research Regulation: A Delphi Study' (2020) *Health Care Analysis*, 28, 99–120, 99.

In contrast to the up-front risk management framing offered above, the Delphi findings suggest that proportionality is often treated as an ethical assessment of the values and risks at stake at multiple junctures in the research trajectory. That is, while it can be easy to reduce proportionality to a techno-bureaucratic risk/benefit assessment, this is to miss the point that the search for proportionality is a moral assessment of whether, when, and how to proceed in the face of uncertainty. Furthermore, the realisation that a role for proportionality can arise at multiple junctures in the research ecosystem, including into the phase about data access[42] and potential feedback of results to research participants, highlights that the range of actors involved in these processes are diverse and often unconnected. For example, Delphi participants frequently stated that reporting of adverse events was a downstream disproportionate activity:

> the definitions of adverse events result in vast numbers of daily events being classed as reportable with result that trials gets bogged down in documenting the utter unrelated trivia that are common in patients with some disorders and unrelated to the drug to the neglect of collecting complete and high quality baseline and outcome data on which the reliability of the results depend (25, researcher).[43]

However, this should be contrasted with the possible identity interests of patients and citizens, which can be impacted by (non)access to biomedical information about them, as argued elsewhere in this volume.[44] The ethics of what is in play are by no means clear-cut. The implication, then, is that a regulatory tool such as proportionality might have far wider reach and significance than has been previously thought; as part of a LHHRS this not only has consequences for a wider range of actors, but the ethical dimensions and sensitivities that surround their (in)action must be duly accounted for.

Regarding possible means to navigate growing complexity within a research regulation ecosystem, some further valuable ideas emerged from the Delphi study. For example, one participant supported the notion of 'regulators etc. becoming helpers and guiding processes to make approval more feasible. Whilst having a proportionate outlook' (27, clinician). Other survey respondents called for 'networked governance' whereby, among other things, 'regulatory agencies in health (broadly understood) would need to engage more with academics and charities, and to look to utilise a broader range of expertise in designing and implementing governance strategies and mechanisms' (5, researcher).[45]

Manifestly, all of this suggests that a robustly designed LHRRS is a complex beast. In the final part of this section, we offer regulatory stewardship as a means of better navigating this complexity for researchers, sponsors, funders and publics, and of closing feedback loops for all stakeholders.

Regulatory stewardship[46] has no unitary meaning, but our previous research has demonstrated that examples from the literature nevertheless point to a commonality of views that cast stewardship as being about 'guiding others with prudence and care across one or more endeavours – without which there is risk of impairment or harm – and with a view to collective betterment'.[47] More work needs to be done on whether and when this role is already undertaken within research ecosystems by certain key actors who may not see themselves as performing such

[42] On Access Governance, see Shabani, Thorogood, Murtagh, Chapter 19, this volume.
[43] Fletcher, 'Co-production and Managing Uncertainty', 109.
[44] See Postan, Chapter 23, this volume.
[45] Quotes in Fletcher et al., 'Co-production and Managing Uncertainty', 109.
[46] See G. Laurie et al., 'Charting Regulatory Stewardship in Health Research: Making the Invisible Visible' (2018) *Cambridge Quarterly of Healthcare Ethics*, 27(2), 333–347; E. S. Dove, *Regulatory Stewardship of Health Research: Navigating Participant Protection and Research Promotion* (Cheltenham: Edward Elgar Publishing, 2020).
[47] Laurie et al., 'Charting Regulatory Stewardship', 338.

a task nor receiving credit for it. One of the Liminal Spaces team has argued that ethics review bodies take on this role to a certain extent – empirical evidence from NHS Research Ethics Committees (RECs) in the UK suggests a far more supportive and less combative relationship with researchers than is anecdotally reported.[48] However, by definition, ethics review bodies can only operate largely at the beginning of the research lifecycle – who is there to assess whether social value was ever actually realised, let alone maximised to the range of potential beneficiaries, including the redressing of social injustices relating to health and even health/wealth generation?

Further empirical research has shown that productive regulation is often only 'instantiated' through practice;[49] that is, it is generated as a by-product of genuine cooperation of regulators and a range of other actors, including researchers, attempting to give effect to regulatory rules or statutory diktats. We suggest, therefore, that there might be a role for regulatory stewardship as part of an LHRRS as a means of giving effect to the multiple dimensions that must interact to give such a system of operating in a genuinely responsive, self-reflexive, and institutionally[50] auto-didactic way.

5 WHAT COULD A LEARNING HEALTH RESEARCH REGULATION SYSTEM LOOK LIKE?

In light of the above, we suggest that the following key features are examples of what we might expect to find in an LHRRS system:

- A system that is **values-driven**, wherein the foundational values of the system reflect those of the range of stakeholders involved;
- A demonstrable commitment to **inclusivity and meaningful participation** in regulatory design, assessment and reform, particularly from patients and publics;
- Robust mechanisms for **evidence gathering** for assessment and review of the workings regulatory processes and relevant laws;
- **Systems-level interconnectivity** to learn lessons across regulatory silos, perhaps supported by a robust system of **regulatory stewardship**;
- **Clear lines of responsibility and accountability** of actors across the entire trajectory of the research enterprise;
- Coordinated efforts to ensure **ethical and regulatory reflexivity**, that is, processes of self-reference of examination and action, requiring institutions and actors to look back at their own regulatory practices, successes and failures;
- Existence of, and where appropriate closing of, **regulatory feedback loops** to deliver authentic learning back to the system and to its users;
- **Appropriate incentives** for actors to contribute to the whole-system approach, whether this be through recognition or reward or by other means, and eschewing a compliance culture that drives a fear of sanction supplanting it with a system that seeks out and celebrates best practice, while not eschewing errors and lessons from failure.[51]

[48] Dove, 'Regulatory Stewardship'.
[49] N. Stephens et al., 'Documenting the Doable and Doing the Documented: Bridging Strategies at the UK Stem Cell Bank' (2011) *Social Studies of Science*, 41(6), 791–813.
[50] On institutional perspectives on regulation, see McMahon, Chapter 21, this volume.
[51] Evidence of the need for such incentives is presented in A. Sorbie et al., 'Examining the Power of the Social Imaginary through Competing Narratives of Data Ownership in Health Research' (2021), Journal of Law and the BioSciences, https://doi.org/10.1093/jlb/lsaa068.

- **Transparency and demonstrated trustworthiness** in the integrity of the regulatory system as a whole;
- **Regulatory responsiveness** to unanticipated events (particularly those that are high risk both as to probability and as to magnitude of impact). The COVID-19 pandemic is one such example – the clamour for a vaccines puts existing systems of regulation and protection under considerable strain, not least for the truncated timeframe for results that is now expected. Values failure in the system itself is something to be avoided at all cost when such events beset our regulatory systems.

6 CONCLUSION

As indicated at the outset of this volume, the golden thread that runs through the contributions is the challenge of examining the possible contours of a Learning Health Research Regulation System. This, admittedly ambitious, task cannot be done justice in a single Afterword, and all chapters in this volume must be read alone for their individual merit. Notwithstanding, an attempt has been made here to draw elements together that re-enforce – and at times challenge – other work in the field that is concerned with how systems learn and to suggest possible ways forward for human health research. And, even if the ambition of a fully-integrated learning system is too vaulting, we suggest nonetheless that adopting a Whole System Approach to health research regulation can promote more joined-up, reflective and responsive systems of regulation. By Whole System Approach we mean that regulatory attention should be paid to capturing and sharing evidence across the entire breadth and complexity of health research, not just of what works well and what does not, but principally of identifying where, when, and how human values are engaged across the entire research lifespan. This approach, we contend, holds the strongest prospect of delivering on the twin ambitions of protecting research participants as robustly as possible while promoting the social value from human health research as widely we possible.

Index